Studies in Computational Intelligence

Volume 1096

Series Editor

Janusz Kacprzyk, Polish Academy of Sciences, Warsaw, Poland

The series "Studies in Computational Intelligence" (SCI) publishes new developments and advances in the various areas of computational intelligence—quickly and with a high quality. The intent is to cover the theory, applications, and design methods of computational intelligence, as embedded in the fields of engineering, computer science, physics and life sciences, as well as the methodologies behind them. The series contains monographs, lecture notes and edited volumes in computational intelligence spanning the areas of neural networks, connectionist systems, genetic algorithms, evolutionary computation, artificial intelligence, cellular automata, self-organizing systems, soft computing, fuzzy systems, and hybrid intelligent systems. Of particular value to both the contributors and the readership are the short publication timeframe and the world-wide distribution, which enable both wide and rapid dissemination of research output.

Indexed by SCOPUS, DBLP, WTI Frankfurt eG, zbMATH, SCImago.

All books published in the series are submitted for consideration in Web of Science.

Oscar Castillo · Patricia Melin
Editors

Hybrid Intelligent Systems Based on Extensions of Fuzzy Logic, Neural Networks and Metaheuristics

Springer

Editors
Oscar Castillo
Division of Graduate Studies and Research
Tijuana Institute of Technology
Tijuana, Baja California, Mexico

Patricia Melin
Division of Graduate Studies and Research
Tijuana Institute of Technology
Tijuana, Baja California, Mexico

ISSN 1860-949X ISSN 1860-9503 (electronic)
Studies in Computational Intelligence
ISBN 978-3-031-28998-9 ISBN 978-3-031-28999-6 (eBook)
https://doi.org/10.1007/978-3-031-28999-6

This Springer imprint is published by the registered company Springer Nature Switzerland AG
The registered company address is: Gewerbestrasse 11, 6330 Cham, Switzerland

Preface

We describe in this book recent theoretical developments on fuzzy logic, neural networks and metaheuristic algorithms, as well as their hybrid combinations, and their application in areas such as intelligent control and robotics, pattern recognition, medical diagnosis, time series prediction and optimization of complex problems. There are papers with the main topics of type-1 and type-2 fuzzy logic, which basically consist of a group of papers that propose new concepts and algorithms based on type-1 and type-2 fuzzy logic and their applications. There are also papers that present theory and practice of metaheuristics in diverse areas of application. There are interesting papers on different applications of fuzzy logic, neural networks and hybrid intelligent systems in medical problems. In addition, we can find papers describing applications of fuzzy logic, neural networks and metaheuristics in robotics problems. Another group of papers present the theory and practice of neural networks in diverse application areas, including convolutional and deep learning neural networks. There are also a group of papers that present the theory and practice of optimization and evolutionary algorithms in different areas of application. Finally, we can find a group of papers describing applications of fuzzy logic, neural networks and metaheuristics in pattern recognition problems. The papers of the book are organized into five parts that group together papers with similar theoretical topics or applications.

In conclusion, the edited book comprises papers on diverse aspects of fuzzy logic, neural networks and nature-inspired optimization metaheuristics for designing optimal hybrid intelligent systems and their application in areas such as intelligent control and robotics, pattern recognition, decision-making, time series prediction and optimization of complex problems. There are theoretical aspects as well as application papers. We expect that the book will serve as a reference for researchers and graduate students working in the computational intelligence area.

Tijuana, Mexico
November 2022

Patricia Melin
Oscar Castillo

About This Book

In this book, recent theoretical developments on fuzzy logic, neural networks and optimization algorithms, as well as their hybrid combinations, are presented. In addition, the above-mentioned methods are considered in application areas such as intelligent control and robotics, pattern recognition, medical diagnosis, decision-making, time series prediction and optimization of complex problems. Nowadays, the main topics of the book are highly relevant, as most current intelligent systems and hardware implementations in use utilize some form of intelligent feature to enhance their performance. In addition, theoretically speaking, new and advanced models and algorithms of type-2 fuzzy logic are presented, which will be of great interest to researchers in these areas. Also, novel nature-inspired optimization algorithms and innovative neural models are put forward in the manuscript, that are very popular subjects, at this moment. There are contributions on theoretical aspects as well as applications, which make the book very appealing to a wide audience, ranging from researchers, to professors and graduate students.

Contents

Neural Networks

A Decision-Making Approach Based on Multiple Neural Networks for Clustering and Prediction of Time Series

Martha Ramirez and Patricia Melin

1 Introduction

The analysis of time series it is important in the decision-making process, a main reason is because organizations generate too much information daily, making the process of collecting, organizing, and analyzing data a complex task [1–3]. Therefore, it contributes to the use of time series, which are represented by a sequence of data recorded chronologically over time [4]. The importance of analyzing time series derives from the fact that relevant information can be obtained from events that have occurred, such as location, population, date, among other attributes. Once this information is available, data can be grouped, by similar attributes or future values of the time series can be predicted. It should be noted that the prediction of time series, is a constant challenge for many researchers [5–7].

There are computational intelligence models to perform clustering [8, 9] used to find hidden patterns or similar values in a dataset [10–12]. As well, there are multiple computational models designed to perform the prediction of time series [13–15]. Sometimes, it is necessary to combine two or more of these techniques to solve complex problems in the real world [16–18].

In the case of the Artificial Neural Networks (ANNs), by simulate the brain's behavior, these computational models [19–21] can learn directly from the data and be trained to pattern recognition or classification tasks [22, 23].

The learning algorithm of supervised neural networks uses input–output training data to model the dynamic system [24–26], on contrary, in the case of the unsupervised neural networks only the input data is given, so they are used to make representative clusters of data sets [27–29].

M. Ramirez · P. Melin (✉)
Tijuana Institute of Technology, Tijuana, Mexico
e-mail: pmelin@tectijuana.mx

O. Castillo and P. Melin (eds.), *Hybrid Intelligent Systems Based on Extensions of Fuzzy Logic, Neural Networks and Metaheuristics*, Studies in Computational Intelligence 1096,
https://doi.org/10.1007/978-3-031-28999-6_1

Multiple models and applications using ANNs have been designed and their efficiency and accuracy to solve diverse problems have been shown [30–32].

The main contribution of this paper is to propose a computational method to perform clustering and prediction of time series by using unsupervised neural networks to generate data clusters, and secondly, by using supervised neural networks to predict future values, with which a support alternative for decision-making is presented. This approach differs from most existing methods [33–35], by combining supervised and unsupervised neural networks since most computational models in the literature focus in the use of supervised training algorithms for prediction of time series.

This paper is organized as follows. In Sect. 2, the case study is described. The methodology used is explained in Sect. 3. The experimental results are shown in Sect. 4. Finally, in Sect. 5, conclusions are presented.

2 Case Study

The Organisation for Economic Co-operation and Development (OECD) is an international organization that works to build better policies for better lives, their goal is to shape policies that foster prosperity, equality, opportunity, and well-being for all. Currently, there are 38 member countries in the organization. Next, the list of members with the name and code of each country are presented (Table 1).

According to the World Health Organization (WHO) air pollution has become recognized as the single biggest environmental threat to human health based on its notable contribution to disease burden [36]. The OECD countries are high-income

Table 1 List of OECD member countries

Country name	Country code	Country name	Country code	Country name	Country code	Country name	Country code
Australia	AUS	Finland	FIN	Korea, Rep	KOR	Slovak Republic	SVK
Austria	AUT	France	FRA	Latvia	LVA	Slovenia	SVN
Belgium	BEL	Germany	DEU	Lithuania	LTU	Spain	ESP
Canada	CAN	Greece	GRC	Luxembourg	LUX	Sweden	SWE
Chile	CHL	Hungary	HUN	Mexico	MEX	Switzerland	CHE
Colombia	COL	Iceland	ISL	Netherlands	NLD	Turkey	TUR
Costa Rica	CRI	Ireland	IRL	New Zealand	NZL	United Kingdom	GBR
Czech Republic	CZE	Israel	ISR	Norway	NOR	United States	USA
Denmark	DNK	Italy	ITA	Poland	POL		
Estonia	EST	Japan	JPN	Portugal	PRT		

economies that remarkably contribute to the global economy, they are responsible for carbon dioxide emissions and their transportation energy consumption is expected to increase annually by 1.4% between 2012 and 2040 [37].

We are working with a data set from the World Bank site, selecting the population and carbon dioxide emissions time series of the 38 member countries of the Organisation for Economic Co-operation and Development (OECD).

Total population time series is based on the de facto definition of population, which counts all residents regardless of legal status or citizenship (Fig. 1). It consists of six attributes for 29 instances from 1990 to 2018 (Table 2) [38].

Carbon dioxide emissions time series are those stemming from the burning of fossil fuels and the manufacture of cement. They include carbon dioxide produced

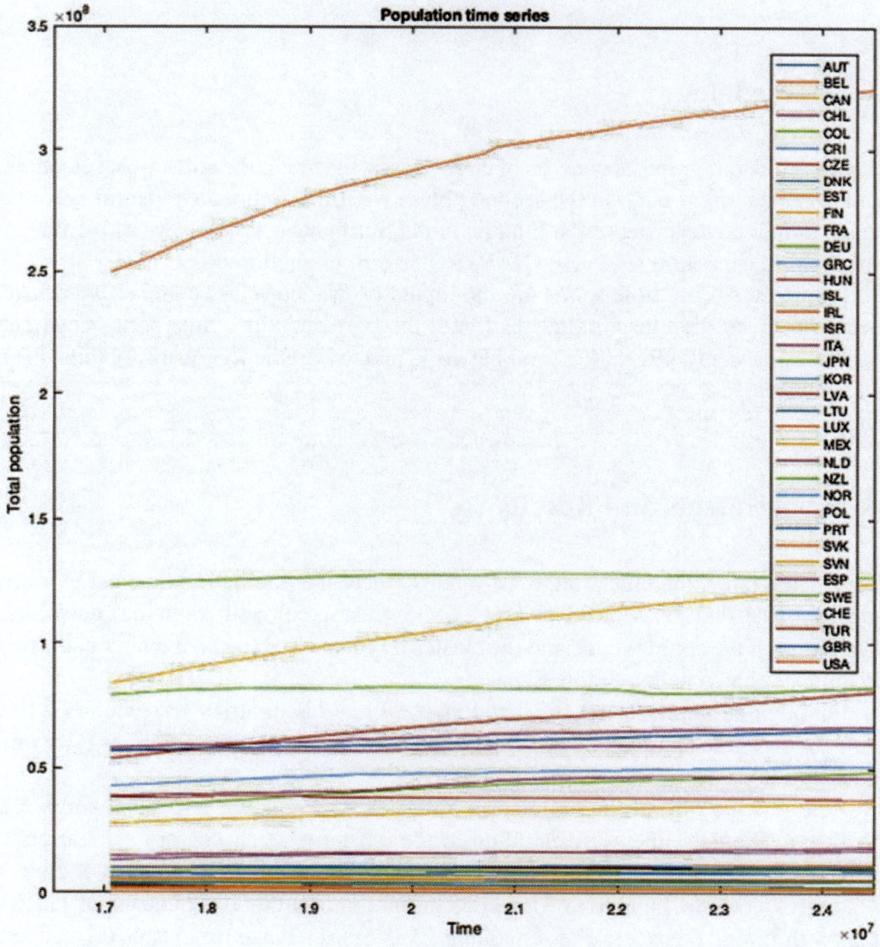

Fig. 1 Population time series

Table 2 Attributes of the annual total population time series

No.	Attributes
1	Country name
2	Country code
3	Indicator name
4	Indicator code
5	Year
6	Total population

during consumption of solid, liquid, and gas fuels and gas flaring (Fig. 2). It consists of six attributes for 29 instances from 1990 to 2018 (Table 3) [39].

3 Methodology

For this work our proposal consists of three phases, the first is the collection, selection, and organization of data. In the second phase, we use a competitive neural network to perform the clustering tasks. Finally, in the third phase we use several Nonlinear Autoregressive neural networks (NAR) to perform prediction tasks.

In this model the time series are the inputs of the four-class competitive neural network. Once the clusters are identified, the corresponding time series segment becomes the input of a NAR neural network, and the prediction results obtained for each cluster (Fig. 3).

4 Experiments and Results

We carried out come clustering experiments where the population time series were grouped into four classes (C1, C2, C3, C4) by using competitive neural networks. The list with the country code and the cluster ID for each of the 38 member countries belongs is shown below (Table 4).

For population time series the first cluster C1 has 25 countries, the second C2 has 10 countries, the third C3 has 2 countries, and finally, the cluster C4 has only one country (Table 5).

Below is the plot of the population time series (Fig. 4), in which we can see a marker in the data line corresponding to the cluster of each country. The asterisk marker (*) of cluster C4 corresponds to the USA. The plus sign marker (+) of cluster C3 corresponds to Japan and Mexico. The plot shows the 10 countries of cluster C2, with the marker (×). The remaining 25 countries make up cluster C1 and are represented by the middle dash marker (−).

As well, for the carbon dioxide emissions time series similar clusters (C1, C2, C3, C4) were formed by using competitive neural networks. For each experiment,

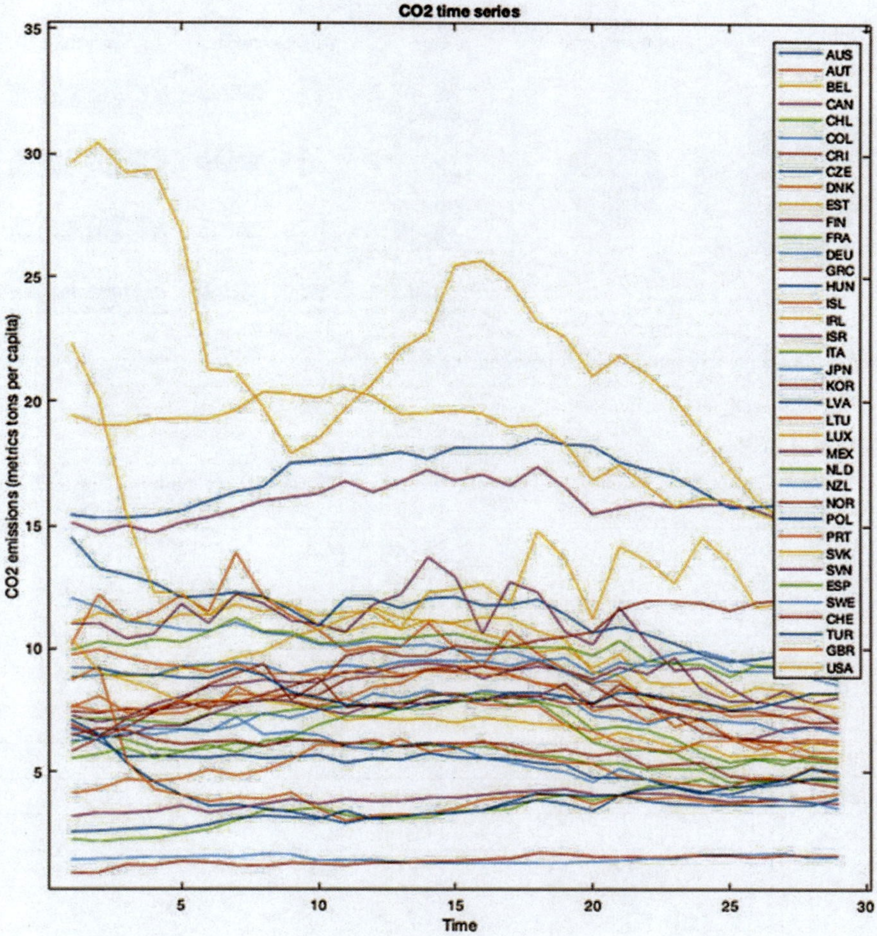

Fig. 2 Carbon dioxide emissions time series

Table 3 Attributes of the carbon dioxide emissions time series

No.	Attributes
1	Country name
2	Country code
3	Indicator name
4	Indicator code
5	Year
6	Carbon dioxide emissions (tons per capita)

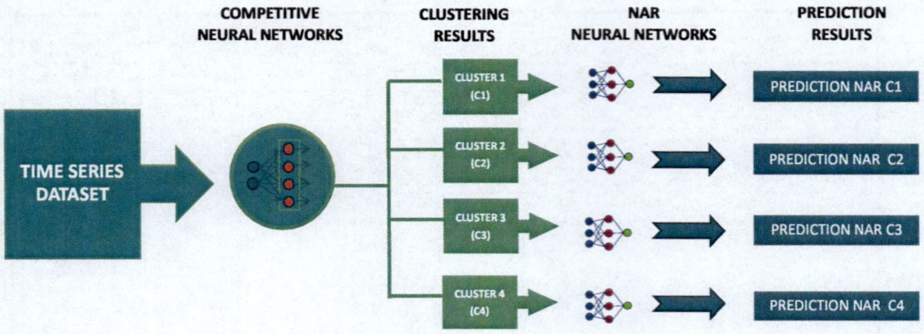

Fig. 3 Conceptual design of the proposed method

Table 4 List of countries per clusters of population time series

Country code	Cluster ID	Country code	Cluster ID	Country code	Cluster ID	Country code	Cluster ID
AUS	C1	FIN	C1	KOR	C2	SVK	C1
AUT	C1	FRA	C2	LVA	C1	SVN	C1
BEL	C1	DEU	C2	LTU	C1	ESP	C2
CAN	C2	GRC	C1	LUX	C1	SWE	C1
CHL	C1	HUN	C1	MEX	C3	CHE	C1
COL	C2	ISL	C1	NLD	C1	TUR	C2
CRI	C1	IRL	C1	NZL	C1	GBR	C2
CZE	C1	ISR	C1	NOR	C1	USA	C4
DNK	C1	ITA	C2	POL	C2		
EST	C1	JPN	C3	PRT	C1		

Table 5 Total countries per clusters of population time series

Cluster Id	Marker	Total countries
C1	–	25
C2	×	10
C3	+	2
C4	*	1

30 executions were performed. The list with the country code and the cluster ID to which each of the 38 member countries belongs is shown below (Table 6).

For carbon dioxide emissions time series, the first cluster C1 has 5 countries, the second C2 has 12 countries, the third C3 has 9 countries and finally the cluster C4 has 12 countries (Table 7).

Same as in the results presented above, the plot corresponding to the carbon dioxide emissions time series is presented (Fig. 5), in which we can see a marker in the data line corresponding to the cluster assigned to each country.

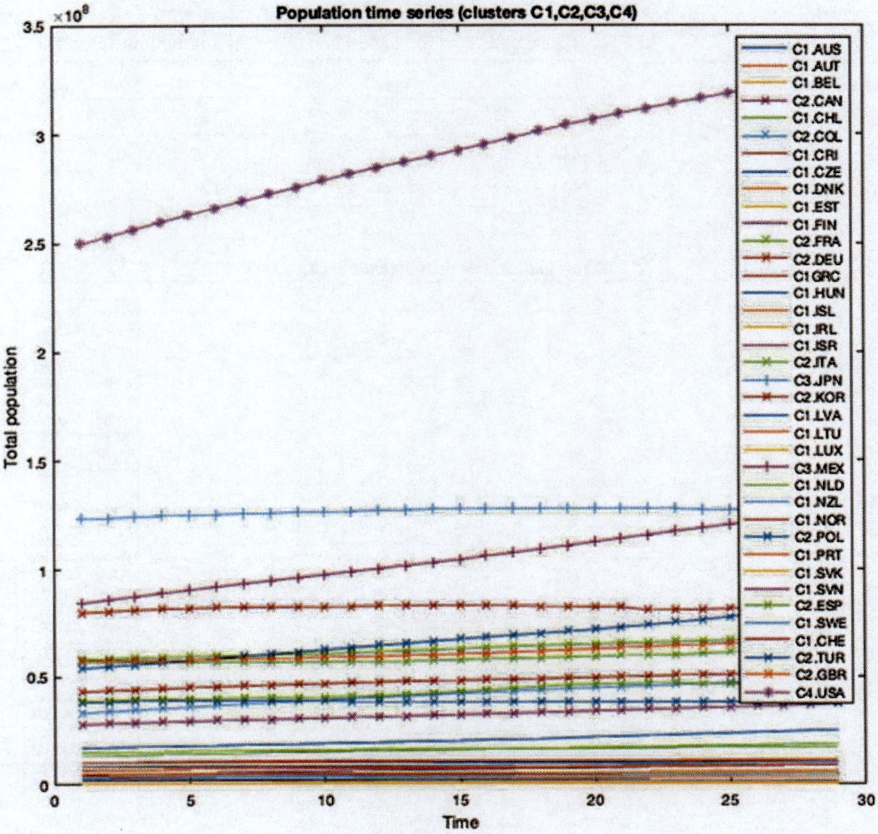

Fig. 4 Population time series (clusters C1, C2, C3, C4)

Table 6 List of countries per clusters of carbon dioxide emissions time series

Country code	Cluster ID	Country code	Cluster ID	Country code	Cluster ID	Country code	Cluster ID
AUS	C1	FIN	C3	KOR	C3	SVK	C2
AUT	C2	FRA	C4	LVA	C4	SVN	C2
BEL	C3	DEU	C3	LTU	C4	ESP	C2
CAN	C1	GRC	C2	LUX	C1	SWE	C4
CHL	C4	HUN	C4	MEX	C4	CHE	C4
COL	C4	ISL	C2	NLD	C3	TUR	C4
CRI	C4	IRL	C3	NZL	C2	GBR	C2
CZE	C3	ISR	C2	NOR	C2	USA	C1
DNK	C3	ITA	C2	POL	C2		
EST	C1	JPN	C3	PRT	C4		

Table 7 Total countries per clusters of carbon dioxide emissions time series

Cluster Id	Marker	Total countries
C1	–	5
C2	×	12
C3	+	9
C4	*	12

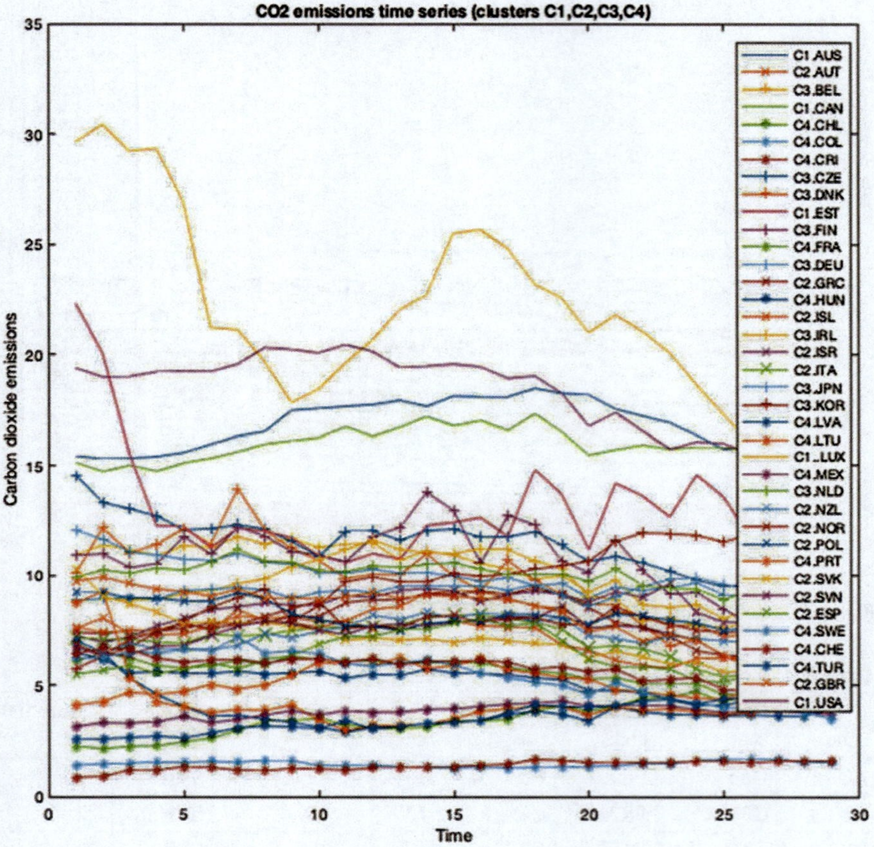

Fig. 5 Carbon dioxide emissions time series (clusters C1, C2, C3, C4)

We used NAR neural network with 10 neurons in the hidden layer to perform the prediction tasks. We considered 70% data for training and 30% for testing. For each experiment, 30 executions were performed. The % Root Mean Square Error (RMSE) was used to measure the performance of each neural network.

First, we present the prediction results by using the complete dataset of the population (Table 8) and carbon dioxide emissions time series (Table 9).

Table 8 Prediction of the population time series

Variable	Average %RMSE	Best %RMSE	Worst %RMSE
Population	0.00699930	0.00488868	0.00884851

Table 9 Prediction of the carbon dioxide emissions time series

Variable	Average %RMSE	Best %RMSE	Worst %RMSE
CO2	0.00044016	0.00031031	0.00079133

Table 10 Prediction of the population time series by clusters

Variable	Average %RMSE	Best %RMSE	Worst %RMSE
Population C1	0.00131000	0.00036386	0.00250869
Population C2	0.00097472	0.00046697	0.00225286
Population C3	0.00042760	0.00005231	0.00379156
Population C4	0.00083594	0.00003595	0.00333030

Table 11 Prediction of the carbon dioxide emissions time series by clusters

Variable	Average %RMSE	Best %RMSE	Worst %RMSE
CO2 C1	0.00087009	0.00033204	0.00179916
CO2 C2	0.00053669	0.00031860	0.00089270
CO2 C3	0.00035440	0.00014393	0.00066555
CO2 C4	0.00133659	0.00074783	0.00203253

Secondly, we used the results of the competitive neural networks to identify the data from each cluster. Then, we use the data corresponding to each cluster to train a set of NAR neural networks for prediction tasks. The prediction results for each cluster of the population time series (Table 10) and carbon dioxide emissions time series (Table 11) are shown.

The results show that it is possible to use competitive neural networks to find clusters based on the population and carbon dioxide emissions recorded in a period for each of the OECD member countries.

Also, for prediction tasks of the population and carbon dioxide time series several Nonlinear Autoregressive Neural Network (NAR) are used, with the advantage that the prediction made would be specific to a particular area or place in comparison with the prediction made using all the values of the time series.

The results show that by combining both methods it is possible to use the unsupervised model to find patterns or similarities in the information and later by using the supervised method to focus on the prediction of the time series segments.

5 Conclusions

We have presented in this work a model for the clustering and prediction of population and carbon dioxide emissions time series of Member countries of the Organization for Economic Co-operation and Development (OECD) using supervised and unsupervised neural networks.

The main contribution of this paper is the used of both unsupervised and supervised neural networks to perform clustering and prediction of time series. As an unsupervised method, we use a competitive neural network to do clustering tasks, and as a supervised method, a set of Nonlinear Autoregressive Neural Networks (NAR) is used to do prediction tasks.

The simulation results have shown the countries with similar populations and carbon dioxide emissions values. Predicting by data clusters with similar attributes allows us to obtain specific results for a particular place or area.

In future work, we consider conducting tests with more variables such as particulate matter and urban population, among others, and the use of Self-organizing maps (SOM) neural networks with other fuzzy logic techniques to improve prediction.

References

1. Tsai, Y., Zeng, Y., & Chang, Y. (2015). Air pollution forecasting using RNN with LSTM. In *2018 IEEE 16th International Conference on Dependable, Autonomic and Secure Computing, 16th International Conference on Pervasive Intelligence and Computing, 4th International Conference on Big Data Intelligence and Computing and Cyber Science and Technology Congress (DASC/PiCom/DataCom/CyberSciTech), 2018* (pp. 1074–1079). https://doi.org/10.1109/DASC/PiCom/DataCom/CyberSciTec.2018.00178.
2. Melin, P., Mancilla, A., Lopez, M., & Mendoza, O. (2007) A hybrid modular neural network architecture with fuzzy Sugeno integration for time series forecasting. *Applied Soft Computing Journal, 7*(4), 1217–1226. https://doi.org/10.1016/j.asoc.2006.01.009. ISSN 1568-4946.
3. Melin, P., & Castillo, O. (2007). An intelligent hybrid approach for industrial quality control combining neural networks, fuzzy logic and fractal theory. *Information Sciences, 177*(7), 1543–1557. https://doi.org/10.1016/j.ins.2006.07.022
4. Prakhar, K., et al. (2022). Effective stock price prediction using time series forecasting. In *2022 6th International Conference on Trends in Electronics and Informatics (ICOEI)* (pp. 1636–1640). https://doi.org/10.1109/ICOEI53556.2022.9776830.
5. Sfetsos, A., & Siriopoulos, C. (2004). Combinatorial time series forecasting based on clustering algorithms and neural networks. *Neural Computing and Applications, 13*, 56–64. https://doi.org/10.1007/s00521-003-0391-y
6. Li, Y., Bao, T., Gong, J., Shu, X., & Zhang, K. (2020). The prediction of dam displacement time series using STL, extra-trees, and stacked LSTM neural network. *IEEE Access, 8*, 94440–94452. https://doi.org/10.1109/ACCESS.2020.2995592
7. Castillo, O., & Melin, P. (2020). Forecasting of COVID-19 time series for countries in the world based on a hybrid approach combining the fractal dimension and fuzzy logic, Chaos, Solitons & Fractals (Vol. 140). https://doi.org/10.1016/j.chaos.2020.110242.
8. Ding, X., Hao, K., Cai, X., Tang, S., Chen, L., & Zhang, H. (2020). A novel similarity measurement and clustering framework for time series based on convolution neural networks. *IEEE Access, 8*, 173158–173168. https://doi.org/10.1109/ACCESS.2020.3025048

9. Melin, P., Amezcua, J., Valdez, F., & Castillo, O. (2014). A new neural network model based on the LVQ algorithm for multi-class classification of arrhythmias. *Information Sciences, 279*, 483–497. https://doi.org/10.1016/j.ins.2014.04.003
10. Austin, E., Coull, B., Zanobetti, A., & Koutrakis, P. (2013). A framework to spatially cluster air pollution monitoring sites in US based on the PM2.5 composition. In *Environment International* (Vol. 59, pp. 244–254). https://doi.org/10.1016/j.envint.2013.06.003.
11. Melin, P., Monica, J. C., Sanchez, D., & Castillo, O. (2020). Analysis of spatial spread relationships of coronavirus (COVID-19) pandemic in the world using self organizing maps, Chaos, Solitons & Fractals (Vol. 138). https://doi.org/10.1016/j.chaos.2020.109917.
12. Melin, P., & Castillo, O. (2021). Spatial and temporal spread of the COVID-19 pandemic using self organizing neural networks and a fuzzy fractal approach. *Sustainability, 13*(8295), 1–17. https://doi.org/10.3390/su13158295
13. Melin, P., Monica, J. C., Sanchez, D., & Castillo, O. (2020). A new prediction approach of the COVID-19 virus pandemic behavior with a hybrid ensemble modular nonlinear autoregressive neural network. *Soft Computing*. https://doi.org/10.1007/s00500-020-05452-z
14. Sánchez, D., & Melin, P. (2015). Modular neural networks for time series prediction using type-1 fuzzy logic integration. In P. Melin, O. Castillo, J. Kacprzyk (Eds.), *Design of intelligent systems based on fuzzy logic, neural networks and nature-inspired optimization* (Vol. 601, pp. 141–154). Studies in Computational Intelligence. Cham: Springer. https://doi.org/10.1007/978-3-319-17747-2_11.
15. Castillo, O., & Melin, P. (2002). Hybrid intelligent systems for time series prediction using neural networks, fuzzy logic, and fractal theory. *IEEE Transactions on Neural Networks, 13*(6), 1395–1408. https://doi.org/10.1109/TNN.2002.804316
16. Chacón, H., Kesici, E., & Najafirad, P. (2020). Improving financial time series prediction accuracy using ensemble empirical mode decomposition and recurrent neural networks. *IEEE Access, 8*, 117133–117145. https://doi.org/10.1109/ACCESS.2020.2996981
17. Melin, P., Soto, J., Castillo, O., & Soria, J. (2012). A new approach for time series prediction using ensembles of ANFIS models. *Expert Systems with Applications, 39*(3), 3494–3506. https://doi.org/10.1016/j.eswa.2011.09.040. ISSN 0957-4174.
18. Soto, J., Melin, P., & Castillo, O. (2014). Time series prediction using ensembles of ANFIS models with genetic optimization of interval Type-2 and Type-1 fuzzy integrators. *Hybrid Intelligent Systems, 11*(3), 211–226. https://doi.org/10.3233/HIS-140196.
19. Valdez, F., Melin, P., & Castillo, O. (2014). Modular neural networks architecture optimization with a new nature inspired method using a fuzzy combination of particle swarm optimization and genetic algorithms. *Information Sciences, 270*, 143–153. https://doi.org/10.1016/j.ins.2014.02.091
20. Pulido, M., & Melin, P. (2021). Comparison of genetic algorithm and particle swarm optimization of ensemble neural networks for complex time series prediction. In: P. Melin, O. Castillo, & J. Kacprzyk (Eds.), *Recent advances of hybrid intelligent systems based on soft computing* (Vol. 915, pp. 51–77). Studies in computational intelligence. Cham: Springer. https://doi.org/10.1007/978-3-030-58728-4_3.
21. Melin, P., Sánchez, D., & Castillo, O. (2012). Genetic optimization of modular neural networks with fuzzy response integration for human recognition. *Information Sciences, 197*, 1–19. https://doi.org/10.1016/j.ins.2012.02.027
22. Sotirov, S., Sotirova, E., Melin, P., Castilo, O., & Atanassov, K. (2016). Modular neural network preprocessing procedure with intuitionistic fuzzy intercriteria analysis method. In T. Andreasen, et al. (Eds.), *Flexible query answering systems 2015*. (Vol. 400, pp. 175–186). Advances in intelligent systems and computing. Cham: Springer. https://doi.org/10.1007/978-3-319-26154-6_14
23. Ramirez, E., Melin, P., & Prado-Arechiga, G. (2019). Hybrid model based on neural networks, type-1 and type-2 fuzzy systems for 2-lead cardiac arrhythmia classification. *Expert Systems with Applications, 126*, 295–307. https://doi.org/10.1016/j.eswa.2019.02.035
24. Moghar, A., & Hamiche, M. (2020). Stock market prediction using LSTM recurrent neural network. In *Procedia computer science* (Vol. 170, pp. 1168–1173). https://doi.org/10.1016/j.procs.2020.03.049. ISSN 1877-0509.

25. Barbounis, T. G., & Theocharis, J. B. (2007). Locally recurrent neural networks for wind speed prediction using spatial correlation. *Information Sciences, 177*(24), 5775–5797. https://doi.org/10.1016/j.ins.2007.05.024. ISSN 0020-0255.
26. Wei, D. (2019). Prediction of stock price based on LSTM neural network. In *2019 International Conference on Artificial Intelligence and Advanced Manufacturing (AIAM)*, (pp. 544-547). https://doi.org/10.1109/AIAM48774.2019.00113.
27. Cherif, A., Cardot, H., & Boné, R. (2011). SOM time series clustering and prediction with recurrent neural networks. *Neurocomputing, 74*(11), 1936–1944. https://doi.org/10.1016/j.neucom.2010.11.026
28. Melin, P., & Castillo, O. (2021). Spatial and temporal spread of the COVID-19 pandemic using self organizing neural networks and a fuzzy fractal approach. *Sustainability, 13*, 8295. https://doi.org/10.3390/su13158295
29. Méndez, E., Lugo, O., & Melin, P. (2012). A competitive modular neural network for long-term time series forecasting. In P. Melin, O. Castillo, & J. Kacprzyk (Eds.), Nature-inspired design of hybrid intelligent systems (Vol. 667, pp. 243–254). Studies in computational intelligence. Springer. https://doi.org/10.1007/978-3-319-47054-2_16.
30. Zhang, J., Chen, F., & Shen, Q. (2019). Cluster-based LSTM network for short-term passenger flow forecasting in urban rail transit. *IEEE Access, 7*, 147653–147671. https://doi.org/10.1109/ACCESS.2019.2941987
31. Li, T., Hua, M., & Wu ,X. (2020). A hybrid CNN-LSTM model for forecasting particulate matter (PM2.5). *IEEE Access, 8*, 26933–26940. https://doi.org/10.1109/ACCESS.2020.2971348.
32. Qian, F., & Chen, X. (2019). Stock prediction based on LSTM under different stability. In *2019 IEEE 4th International Conference on Cloud Computing and Big Data Analysis (ICCCBDA)* (pp. 483–486). https://doi.org/10.1109/ICCCBDA.2019.8725709.
33. Sehrawat, P. K., & Vishwakarma, D. K. (2022). Comparative analysis of time series models on COVID-19 predictions. In *2022 International Conference on Sustainable Computing and Data Communication Systems (ICSCDS)* (pp. 710-715). https://doi.org/10.1109/ICSCDS53736.2022.9760992.
34. Gupta, K., et al. (2022). An experimental analysis of state-of-the-art Time Series Prediction models. In *2022 2nd International Conference on Advance Computing and Innovative Technologies in Engineering (ICACITE)* (pp. 44–47). https://doi.org/10.1109/ICACITE53722.2022.9823455.
35. Prakaisak, I., & Wongchaisuwat, P. (2022). Hydrological time series clustering: A case study of telemetry stations in Thailand. *Water, 14*, 2095. https://doi.org/10.3390/w14132095
36. WHO. (2021). WHO global air quality guidelines: particulate matter (PM2.5 and PM10), ozone, nitrogen dioxide, sulfur dioxide and carbon monoxide. Executive summary, Geneva: World Health Organization (pp. 1–16). https://apps.who.int/iris/handle/10665/345334. ISBN 978-92-4-003443-3.
37. Hussain, Z., Khan, M. K., & Shaheen, W. A. (2022). Effect of economic development, income inequality, transportation, and environmental expenditures on transport emissions: Evidence from OECD countries. *Environmental Science and Pollution Research, 29*, 56642–56657. https://doi.org/10.1007/s11356-022-19580-6
38. The World Bank Data: Population, total (2022). Retrieved from https://data.worldbank.org/indicator/SP.POP.TOTL.
39. The World Bank Data: CO2 emissions (kt). (2022). Retrieved from https://data.worldbank.org/indicator/EN.ATM.CO2E.KT.

Approximation of Physicochemical Properties Based on a Message Passing Neural Network Approach

Leonardo Velazquez-Ruiz, Graciela Ramirez-Alonso, Fernando Gaxiola, Javier Camarillo-Cisneros, Daniel Espinobarro, and Alain Manzo-Martinez

1 Introduction

Machine learning techniques have been introduced into different areas of expertise, forming in recent years strong alliances in physics and chemistry with applications in quantum many-body physics, quantum computing, and chemical and material physics with the aim to accelerate the different processes involved in those tasks such as materials discovery, molecule generation, drug discovery, to mention some [1, 10]. Several papers have been published in the last few years whose solutions to quantum systems are approximated by machine learning algorithms, all using a similar approach. First, a dataset is created by using an ab-initio method of elementary particles or molecules to train a machine learning algorithm whose purpose is to find the physics behind these systems and finally use the previously trained model to predict a molecular property [19].

Researchers from the Perimeter Institute, Vector Institute, and Center for Computational Quantum Physics [16] find that the brute-force Quantum State Tomography (QST) methods require a high computational cost. However, using a machine-learning approach, it is possible to solve QST problems with more than a thousand qubits with a high level of accuracy.

More recently, complex and innovative machine learning models and architectures have been used to model quantum interactions under different laboratory conditions [6]. Furthermore, architectures capable of using as a tool deep neural networks in order to find more complex elements of the systems such as the wave function

L. Velazquez-Ruiz · G. Ramirez-Alonso (✉) · F. Gaxiola · D. Espinobarro · A. Manzo-Martinez
Facultad de Ingeniería, Universidad Autónoma de Chihuahua, Chihuahua 31125, México
e-mail: galonso@uach.mx

J. Camarillo-Cisneros
Laboratorio de Física Química Computacional, Facultad de Medicina y Ciencias Biomédicas, Universidad Autónoma de Chihuahua, Chihuahua 31125, México

O. Castillo and P. Melin (eds.), *Hybrid Intelligent Systems Based on Extensions of Fuzzy Logic, Neural Networks and Metaheuristics*, Studies in Computational Intelligence 1096,
https://doi.org/10.1007/978-3-031-28999-6_2

itself [13] and even state-of-the-art models, such as graph neural networks (GNNs) [14] have already been used for the same purposes. A graph is an abstract structure consisting of nodes (or vertices) and edges where the graph's connections show the nodes' relations.

Message passing neural networks (MPNNs) [5] is the base of most GNNs, models that have achieved comparable accuracy to the Density Functional Theory (DFT) methods in predicting molecule properties [11]. Therefore, it has been demonstrated that GNNs could accurately represent molecules' physical information and interactions. Then, Message passing is a framework of a GNN where messages are passed between the nodes of the graph in each layer of the network. Most GNNs used in molecule prediction tasks use recurrent neural networks (RNNs) such as Long short-term memory (LSTM) [7, 8] or Gate Recurrent Units (GRU) [3] to update the hidden states of the MPNNs. Only few works have incorporated novel architectures, such as the attention mechanisms of Transformers models, in the neural architecture [19].

In this paper, we propose a method based on a MPNN approach that uses an attention mechanism to predict molecule properties, a fundamental task in the field of drug discovery. The attention mechanism of the transformer was selected because of the impressive results reported in Natural Language Processing (NLP) tasks compared with LSTMs or GRUs [2]. Our proposal takes advantage of the multi-head self-attention mechanisms to calculate the hidden states of the MPNNs. Also, to improve our proposal's performance, the teacher forcing strategy (TeFo) was incorporated into its training [4]. TeFo has received great attention in NLP tasks because it uses the ground truth information from a previous time step as input instead of the output from the last time step to calculate the hidden states of the model. As far as we know, this is the first time that a MPNN, in combination with multi-head self-attention mechanisms and the TeFo strategy, is used to predict physicochemical properties. The results of our proposal, named MPNNwA-TeFo, demonstrate that it could surpass a baseline MPNN method; the TeFo strategy is vital to achieving these results.

2 Methodology

The proposed method named MPNNwA-TeFo is shown in Fig. 1a. MPNNwA-TeFo receives as inputs the geometrical information of the molecules in a graph representation. The graph object $G = (V, E)$ has a set of V vertex and edges E that represents the relationship between the atoms of the molecule [9]. Then, an atomic layer performs a linear transformation that will be used as input by the convolution layer of the GNN. The Convolution layer, Fig. 1b, is composed of three processes: message passing, update phase, and readout phase. In this work, the message passing phase consists of two linear layers of 64 and 1024 neurons, respectively. Then, in the update phase, the self-attention mechanism is in charge of updating the hidden states at each node representation. The readout phase, also uses a self-attention mechanism to calculate a characteristic tensor for the entire graph that will be processed by a linear layer. Finally, a regression layer will predict a particular property of the molecule. MPNNwA-TeFo was trained with a teacher forcing strategy that improves the prediction results.

Consider an undirected graph $G = (V, E)$ with a set of vertices $V = \{v_1, v_2, \ldots, v_N\}$ (also called nodes) and edges $e_{ij} \in E$ denoting a connection between nodes v_i and v_j.

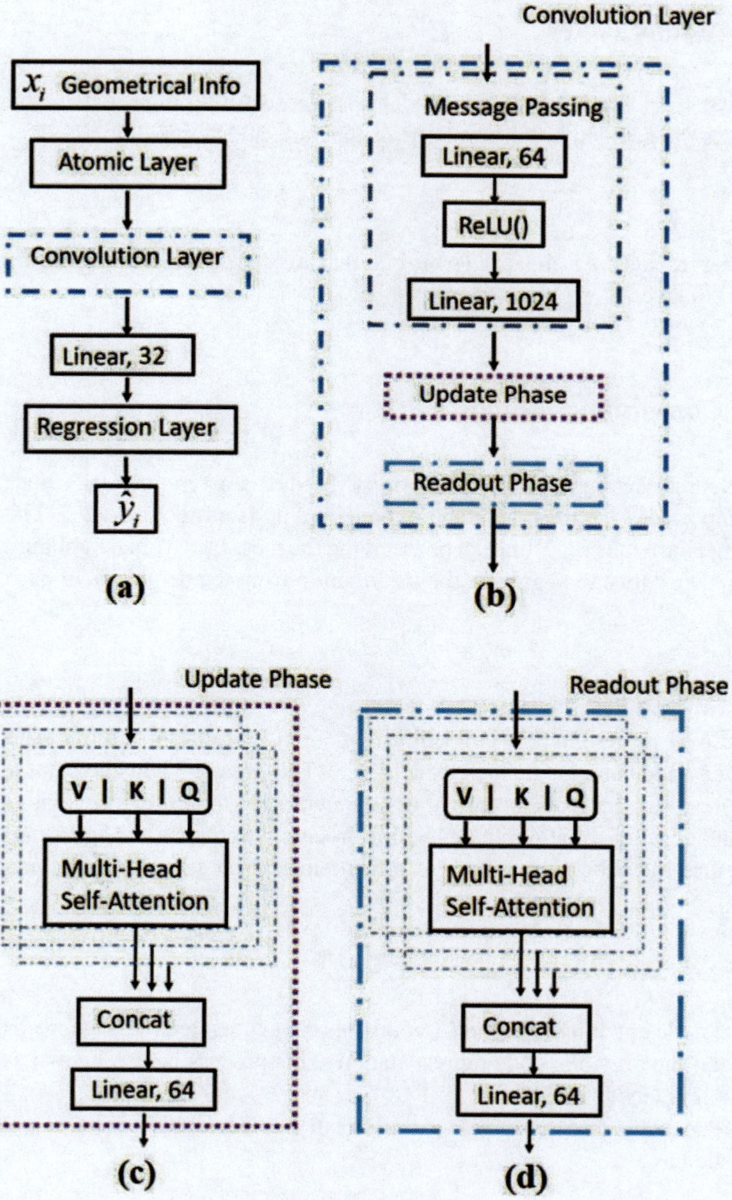

Fig. 1 Block diagram of the proposed method, MPNNwA-TeFo

The attributes of the node v_i is represented by $x_i \in \mathbf{R}$ as an input feature vector. The idea of graph neural networks is to iteratively update the node representation h_i by combining its information with its neighbors [20]. For initialization purposes $h_i{}^0 = x_i$.

2.1 Atomic Layer

The first trainable layer of the model is a dense fully connected module, which analyzes the geometrical information of the molecule defined by

$$h_i^k = W^k h_i^0 + b^k \tag{1}$$

W represents the weighs and b the bias that are shared for all atoms of the molecule in the kth layer.

2.2 Convolutional Layer

The Neural Message Passing Network (MPNN) is a type of GNN that performs two important functions: message passing and update functions [20]. The message passing phase runs for T time steps receiving the messages from neighbouring atoms. The network tries to aggregate the information from the neighbors of each node as:

$$m_i^t = \sum_{i \in N(j)} M_t(h_i^{t-1}, h_j^{t-1}, e_{ij}) \tag{2}$$

where $N(j)$ represents the collection of neighbor atoms of atom j. The message between node i and j is defined by M_t (•,•,•). This message considers the summation of all the messages from the neighbors, depending on their previous node representation and edge information. In GNN, this process is known as the aggregate function. After that, the node representations or hidden states at each node are updated based on

$$h_i^t = U_t\left(h_i^{t-1}, m_i^t\right) \tag{3}$$

The node updating function U_t combines the aggregated message from the neighbors and the previous node representation. This process is also known as combine function in GNN.

The message passing function used in our implementation is based on [5] and it is defined as

$$M_t\left(h_i^t, h_j^t, e_{ij}\right) = A_{e_{ij}} h_j^{t-1} \tag{4}$$

where A_{eij} is the adjacency matrix, one for each edge, that determines how nodes in the graph communicate with each other considering both directions. Typically, the message passing function is implemented via feedforward neural networks. In this work, this information is processed by two fully connected layers of 64 and 1,024 neurons, respectively.

The U_t function is trainable, commonly defined in GNN as

$$U_t = GRU\left(h_i^{t-1}, m_i^t\right) \tag{5}$$

where *GRU* stands for *Gated Recurrent Unit*. In a GNN with a GRU layer, the information is propagated as follows

$$\begin{aligned} z_i^t &= \sigma\left(W^z m_i^t + U^z h_i^{t-1} + b^z\right) \\ r_i^t &= \sigma\left(W^r m_i^t + U^r h_i^{t-1} + b^r\right) \\ \tilde{h}_i^t &= \tanh\left(W^h m_i^t + U^h\left(r_i^t \circ h_i^{t-1}\right) + b^h\right) \\ h_i^t &= \left(1 - z_i^t\right) \circ h_i^{t-1} + z_i^t \circ \tilde{h}_i^t \end{aligned} \tag{6}$$

where $\circ$ stands for the Hadamard product operation, σ is the sigmoid activation function, *tanh* represents the hyperbolic tangent function, W,U and b are parameter weight matrices and a bias vector that will be learned during training. The GRU update function uses information from each node's neighborhood and from previous time-step to update the hidden state. This is the core principle of recurrent neural networks; the output is a function of the current data and previous inputs.

In this work, a self-attention module based on Transformer models [17] replaces the GRU layer. This allows the processing of a sequence as a whole in a parallel way instead of analyzing the input sequence one by one. An instance is linearly transformed in a *value* V_i, *key* K_i, and *query* Q_i vector. Each *query* is matched with each of the different *key* values, and the resulting product (divided by a normalization factor) is passed through the softmax function producing the attention scores or alphas as follows

$$\alpha_0, \alpha_1 = softmax\left(\frac{Q_0 K_0}{\sqrt{d}}, \frac{Q_1 K_1}{\sqrt{d}}, \ldots, \frac{Q_N K_N}{\sqrt{d}}\right) \tag{7}$$

The next step of the algorithm is to perform the product of the alphas with the *values* in order to find a *context vector*.

$$context\ vector = \alpha_0 V_0 + \alpha_1 V_1 \tag{8}$$

Finally, the context vector is fed to a multilayer perceptron (MLP), creating an updated hidden state ${h_i}^k$, this process is shown in Fig. 2. In a self-attention module, the attention mechanism is repeated multiple times in parallel, and each of these is called an *attention head*. The *query*, *key*, and *value* vectors are passed independently to each of these heads. In our proposal, three head attention followed by a fully

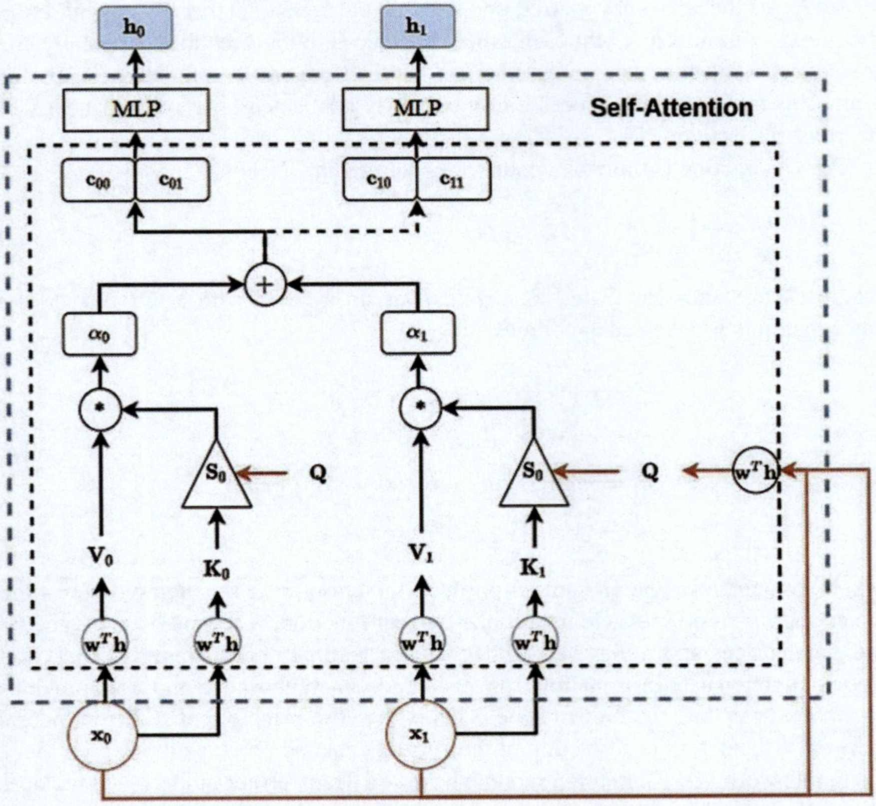

Fig. 2 Self-attention mechanism implemented in our proposal

connected layer of 64 neurons, respectively, are used in the self-attention mechanism. The formal definition of this update process is defined as follows

$$U_t = MultiHead\left(h_i^{t-1}, m_i^t\right) \tag{9}$$

2.3 Readout Phase

The last part of the MPNN is the readout phase which refers to the function of summarizing the node information and calculates a characteristic tensor $\overrightarrow{m}_i$, for the entire graph (molecular tensor) by

$$\vec{m}_i = R\left(\left\{h_i^k | v \in G\right\}\right) \tag{10}$$

In most GNN the *set2set* method [18] is used as a readout function to get graph representations. *Set2set* is based on LSTM, a recurrent neural network designed to model long-distance dependencies [15]. An LSTM updates the hidden state as follows:

$$\begin{aligned}
input_i^t &= \sigma\left(W^{input}x^t + U^{input}h^{t-1} + b^{input}\right)\\
f_i^t &= \sigma\left(W^f x^t + U^f h^{t-1} + b^f\right)\\
o_i^t &= \sigma\left(W^o x^t + U^o h^{t-1} + b^o\right)\\
c_i^t &= f^t \circ c^{t-1} + input^t \circ \tanh\left(W^c x^t + U^c h^{t-1} + b^c\right)\\
h_i^t &= o^t \circ \tanh\left(c^t\right)
\end{aligned} \tag{11}$$

Similar to (6), W, U and b are weight matrices and a bias vector parameter that will be learned during training.

In this work, it is proposed to use a self-attention mechanism that replaces the LSTM layer in the readout function. The self-attention mechanism is implemented with the same parameters as the updating function; 3 self-attention heads with a network architecture of 64 neurons in a feed-forward layer. The tensor $\overrightarrow{m}_i$, is processed by a dense layer of 32 neurons followed by an output of one neuron. The output is in charge of performing the regression task.

2.4 Teacher Forcing

At training, a model usually finds off-target predictions. This problem is exacerbated in recurrent neural network architecture since those results are used as input instances in subsequent steps. A way to accelerate the learning process is to use the target information as input to the decoder to feed the model with the correct answer; this is known as teacher forcing. Obviously, feeding the correct answer to each instance minimizes the loss function in training, but it impoverishes the results in the testing process. In order to reduce this effect, a random process is used in the teacher forcing strategy, which is defined as a probability P of occurrence in the use of the target information.

3 Results

Our proposal was implemented with an Nvidia GPU RTX2060, Intel® Core™ i7-10750H CPU @ 2.60 GHz, and 16 GB of RAM. The QM9 dataset was used to demonstrate the effectiveness of MPNNwA-TeFo. QM9 consists of 134 k stable small organic molecules with 12 target values [12]. The property used as an approximation was the internal energy at 0 K, U_0. Not all of the QM9 dataset was used to train our

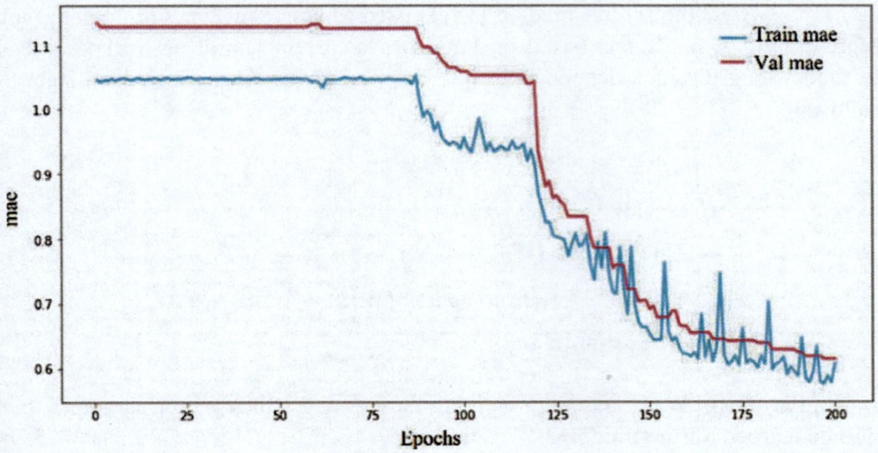

Fig. 3 Plot of the mean absolute error in the training (turquoise) and validation (magenta) sets with 10,000 of molecules considering 200 epochs

proposal because of hardware limitations. Then, on initial experimentation, 10,000, 1,000, and 2,500 instances were used to train, validate, and test our method. The ADAM optimizer was used with β1 = 0.9, β2 = 0.999, epsilon = 1e-07, and the learning rate initialized with 10^{-3} with a decay factor of $f = 0.007$ for each five epochs with a lower limit of 10^{-5}. The batch size used was set to eight.

Figure 3 shows the plot of the mean absolute error (*mae*) in the prediction of the scalar observable U_0 using our proposal. The magenta and cyan lines show the validation *mae* results in validation and training, respectively. It can be seen how the error is maintained with slight variation until it reaches epochs 75–90. After that interval, a decay in error is visualized, which indicates an improvement in the learning process.

In the next experiment, the number of epochs was increased to 400. 30,000, 3,000, and 7,500 molecules were used in the training, validation, and testing set, respectively. In order to find better results without the need to change parameters, a probabilistic teacher forcing system was added where an incidence of 50 percent of ignoring the result found was set at the time of training, and the target was used directly. Figure 4 shows the results obtained with this experiment, and Table 1 presents the corresponding metrics results of all these previous experiments.

MPNN-wATeFo was compared with the MPNN presented in [5] that uses a *GRU* layer in the update phase and LSTM in the readout phase. Table 2 shows the results obtained with this baseline method using the same epochs and instances presented in Table 1. The last column of Table 2 shows the results when a TeFo strategy is included in MPNN. Although the TeFo strategy was not considered in the original MPNN, we include this experiment to demonstrate that this modification could be incorporated into different RNNs architectures.

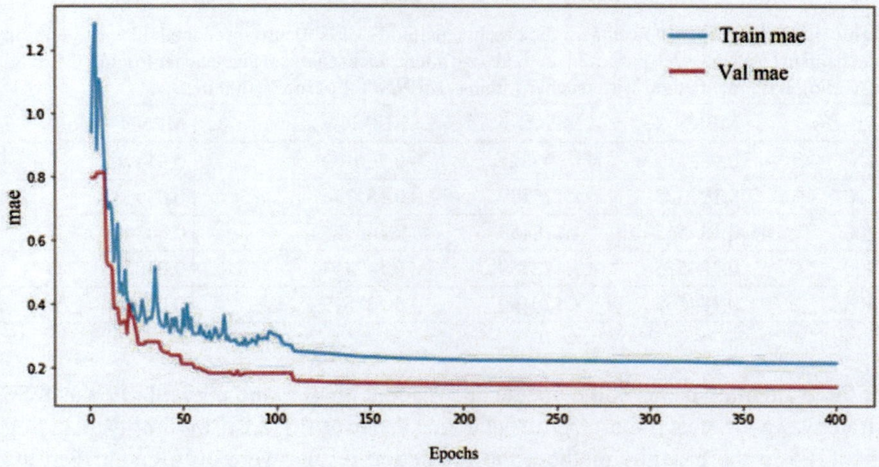

Fig. 4 Graph of the mean absolute error in the training (turquoise) and validation (magenta) sets with 30 K of molecules in the training set + Teacher Forcing with 400 epochs

Table 1 Test results with different epochs and training samples, the last column shows the results when including the teacher forcing strategy

Epochs	MPNN-wA 10,000	MPNN-wA 30,000	MPNN-wATeFo 30,000
1	1.136347	1.197901	0.766109
100	1.059268	1.178459	0.181276
200	0.614349	1.178459	0.1467281
300	–	1.178459	0.142454
400	–	1.170460	0.138679

Table 2 Results obtained with the MPNN baseline, the TeFo strategy was incorporated in this baseline

Epochs	MPNN 10,000	MPNN 30,000	MPNN-TeFo 30,000
1	0.711574	0.9622	0.914762
100	0.147937	0.190375	0.059534
200	0.144883	0.184565	0.043408
300	–	0.184565	0.040288
400	–	0.184565	0.039933

As a comparative analysis, Table 3 presents the summary of results at 400 epochs and 30,000 training instances with the baseline method (MPNN), the proposed method without TeFo only the self- attention mechanisms (MPNN-wA), the proposed method with the TeFo (MPNN-wATeFo), and the baseline method with the incorporation of TeFo (MPNN-TeFo). MPNN-wATeFo surpasses the results of the baseline method MPNN on the prediction of the internal energy at 0 K. The combination

Table 3 Comparison test results for the baseline method (MPNN), proposed method with attention mechanism (MPNN-wA), proposed method with attention mechanism and teacher forcing (MPNN-wATeFo), baseline method with teacher forcing (MPNN-TeFo) for 30,000 instances

Epochs	MPNN	MPNN-wA	MPNN-wATeFo	MPNN-TeFo
1	0.9622	1.197901	0.766109	0.914762
100	0.190375	1.178459	0.181276	0.059534
200	0.184565	1.178459	0.146781	0.043408
300	0.184565	1.178459	0.142454	0.040288
400	0.184565	1.170460	0.138679	0.039933

of three attention heads in the update and readout phases, and particularly, the TeFo strategy, was of great importance to achieving these results. Also, by simply including the TeFo in the baseline method, the prediction results were even better than the original.

4 Conclusions

In this work, we present a new MPNN approach based on attention mechanisms to predict molecular properties, named MPNN-wATeFo. Although GRU and LSTM layers are commonly used in the update and readout phase of MPNNs, in our proposal, these layers were replaced by attention mechanisms based on Transformer methods. Also, the teacher forcing strategy was incorporated in our proposal, where the ground truth information is used as input instead of the output from the previous time step. Comparing the results obtained with MPNN-wATeFo and a baseline method, we confirm that the combination of attention mechanism and TeFo surpass traditional MPNN models. When using attention mechanisms based on transformers, it is common to use 16 or even 254 head attention. In our proposal, were used only three head attention mechanisms. In future work, we will perform a deeper analysis of the appropriate number of head attention needed in this application. Also, an important finding reported in this paper is that we demonstrate that the TeFo strategy could also be incorporated into traditional MPNN architectures. The baseline method could significantly improve its prediction performance in this particular application.

References

1. Carleo, G., Cirac, I., Cranmer, K., Daudet, L., Schuld, M., Tishby, N., VogtMaranto, L., & Zdeborova, L. (2019). Machine learning and the physical sciences. *Reviews of Modern Physics, 91*, 045002. https://doi.org/10.1103/RevModPhys.91.045002
2. Casola, S., Lauriola, I., & Lavelli, A. (2022). Pre-trained transformers: an empirical comparison. *Machine Learning with Applications, 9*, 100334. https://doi.org/10.1016/j.mlwa.2022.100334.

3. Cho, K., van Merrienboer, B., Gulcehre, C., Bahdanau, D., Bougares, F., Schwenk, H., & Bengio, Y. (2014). Learning phrase representations using RNN encoder-decoder for statistical machine translation. In A. Moschitti, B. Pang, & W. Daelemans (Eds.), *Proceedings of the 2014 Conference on Empirical Methods in Natural Language Processing, EMNLP 2014*, October 25–29, 2014, Doha, Qatar, A meeting of SIGDAT, a Special Interest Group of the ACL (pp. 1724–1734). ACL https://doi.org/10.3115/v1/d14-1179.
4. DiPietro, R., & Hager, G. D. (2020). Chapter 21-deep learning: Rnns and lstm. In S. K. Zhou, D. Rueckert, & G. Fichtinger (Eds.), *Handbook of medical image computing and computer assisted intervention* (pp. 503–519). The Elsevier and MICCAI society book series, Academic Press. https://doi.org/10.1016/B978-0-12-816176-0.00026-0.
5. Gilmer, J., Schoenholz, S. S., Riley, P. F., Vinyals, O., & Dahl, G. E. (2017) Neural message passing for quantum chemistry. In *2017, Proceedings of the 34th international conference on machine learning* (Vol. 70). https://doi.org/10.5555/3305381.3305512.
6. Grisoni, F., Moret, M., Lingwood, R., Schneider, G. (2020). Bidirectional molecule generation with recurrent neural networks. *Journal of Chemical Information and Modeling, 60*, 1175–1183.
7. Hochreiter, S., & Schmidhuber, J. (1997). Long short-term memory. *Neural Computation, 9*(8), 1735–1780.
8. Li, Y., Tarlow, D., Brockschmidt, M., & Zemel, R. (2015) Gated graph sequence neural networks. http://arxiv.org/abs/1511.05493. (Comment: Published as a conference paper in ICLR 2016).
9. Liu, Z., & Zhou, J. (2020). Introduction to graph neural networks. *Introduction to Graph Neural Networks*. https://doi.org/10.1007/978-3-031-01587-8
10. Packwood, D., Nguyen, L. T. H., Cesana, P., Zhang, G., Staykov, A., Fukumoto, Y., & Nguyen, D. H. (2022). Machine learning in materials chemistry: An invitation. *Machine Learning with Applications, 8*, 100265. https://doi.org/10.1016/j.mlwa.2022.100265.
11. Qian, C., Xiong, Y., & Chen, X. (2021). Directed graph attention neural network utilizing 3d coordinates for molecular property prediction. *Computational Materials Science, 200*, 110761. https://doi.org/10.1016/j.commatsci.2021.110761.
12. Ramakrishnan, R., Dral, P. O., Rupp, M., Lilienfeld, O. A. V. (2014). Quantum chemistry structures and properties of 134 kilo molecules. *Scientific Data, 1*(1 1), 1–7. https://doi.org/10.1038/sdata.2014.22.
13. Schutt, K. T., Gastegger, M., Tkatchenko, A., M¨uller, K. R., & Maurer, R. J. (2019). Unifying machine learning and quantum chemistry with a deep neural network for molecular wavefunctions. *Nature Communications, 10*(1 10), 1–10. https://doi.org/10.1038/s41467-019-12875-2.
14. Shui, Z., & Karypis, G. (2020). Heterogeneous molecular graph neural networks for predicting molecule properties. In *2020 IEEE International Conference on Data Mining (ICDM)* (pp. 492–500). Los Alamitos, CA, USA: IEEE Computer Society. (Nov 2020). https://doi.org/10.1109/ICDM50108.2020.00058.
15. Taheri, A., Gimpel, K., & Berger-Wolf, T. (2019). Sequence-to-sequence modeling for graph representation learning. *Applied Network Science, 4*, 1–26. https://doi.org/10.1007/S41109-019-0174-8/FIGURES/14.
16. Torlai, G., Mazzola, G., Carrasquilla, J., Troyer, M., Melko, R., & Carleo, G. (2018). Neural network quantum state tomography. *Nature Physics, 14*(5 14), 447–450. https://doi.org/10.1038/s41567-018-0048-5.
17. Vaswani, A., Shazeer, N., Parmar, N., Uszkoreit, J., Jones, L., Gomez, A. N., Kaiser, L. U., & Polosukhin, I. (2017). Attention is all you need. In: I. Guyon, U. V. Luxburg, S. Bengio, H. Wallach, R. Fergus, S. Vishwanathan, & R. Garnett (Eds.), *Advances in neural information processing systems* (Vol. 30). Curran Associates, Inc.
18. Vinyals, O., Bengio, S., & Kudlur, M. (2016). Order matters: Sequence to sequence for sets. In *4th International Conference on Learning Representations, ICLR 2016*, Puerto Rico. 2–4 May 2016

19. Wieder, O., Kohlbacher, S., Kuenemann, M., Garon, A., Ducrot, P., Seidel, T., & Langer, T.: A compact review of molecular property prediction with graph neural networks. *Drug Discovery Today: Technologies, 37*, 1–12. https://doi.org/10.1016/j.ddtec.2020.11.009.
20. Wu, L., Cui, P., Pei, J., & Zhao, L. (2022) *Graph neural networks: foundations, frontiers, and applications.* Singapore: Springer.

Convolutional Neural Networks for Multiclass Classification of Masks

Alexis Campos, Patricia Melin, and Daniela Sánchez

1 Introduction

Because of the new human coronavirus (COVID-19), there have been new respiratory infections [1], and some of the symptoms are dry cough, tiredness, fever, sore throat, etc. The event has paralyzed many activities around the world due to the various effects it has on people.

According to the World Health Organization (as of August 2022), the use of masks serves as a strategy to reduce the spread of the COVID-19 virus, which has infected more than 603 million people worldwide [2]. One of the basic guidelines for the proper placement of the mask is to place it over the nose, mouth and chin [3]. By doing the right thing, people can keep themself and others safe. Failure to comply with these instructions can spread the virus to others.

The use of masks is mandatory in most countries during the COVID-19 pandemic [4], in the month of August 2022 around 27,529,582 new confirmed cases were estimated in the world, so [2] people who are incorrectly using the mask should be identified.

Face masks have many functions, such as preventing airborne virus particles from being transmitted between people and allowing volatile particles to filter into the air. The Centers for Disease Control and Prevention (CDC) recommends [5] wearing surgical masks while exhaling air from your nose and mouth.

A. Campos · P. Melin (✉) · D. Sánchez
Tijuana Institute of Technology, Tijuana, Mexico
e-mail: pmelin@tectijuana.mx

A. Campos
e-mail: alexis.campos@tectijuana.edu.mx

D. Sánchez
e-mail: daniela.sanchez@tectijuana.edu.mx

O. Castillo and P. Melin (eds.), *Hybrid Intelligent Systems Based on Extensions of Fuzzy Logic, Neural Networks and Metaheuristics*, Studies in Computational Intelligence 1096,
https://doi.org/10.1007/978-3-031-28999-6_3

Using deep learning tools through computational methods will allow the system to automate identification. There are several studies that use different deep learning models, such as YOLO [6–8], MobileNet [9, 10], Resnet [11–13] or Inception [14], These deep learning techniques are sometimes preferred by authors because they already have some knowledge of training complex neural networks and maintain good recognition rates.

In this study, a multiclass classification system is proposed to detect the use of masks, with three different categories in the use of face masks, no use of masks and incorrect use of masks. The system is based on convolutional neural networks. A dataset called MaskedFaceNet provided by Cabani was used [15]. Preprocessing is performed on the first 15,000 images in the dataset to identify regions of interest using the Caffe model for face detection and perform image feature extraction, such as resizing and RGB subtraction.

We achieved an accuracy of 99.6865%, improving our accuracy percentage when compared to other authors. Therefore, the initial experimental results of the proposed model are presented.

The rest of this article is organized into the following sections. Section 2 deals with some relevant work on the classification and use of face masks. The proposed method is presented in Sect. 3. In Sect. 4, the results obtained from the experiments and models are shown. Finally, conclusions are drawn with possible future work in Sect. 5.

2 Related Work

Due to the COVID-19 virus, various techniques have been adapted using artificial intelligence to detect and classify people who wear masks. This section describes some of the most relevant studies on the classification of mask use.

A convolutional neural network (CNN) is proposed to identify people with masks who use them correctly, do not have a mask or who are placed incorrectly [16], with the MaskedFaceNet and Flickr-Faces-HQ [17] datasets reaching an accuracy percentage of 98.5%. In [9], the authors present a system to monitor mask protocols, through the Haar Cascades to obtain the region of interest, with the Architecture of MobileNetV2 as a model, reaching 99% accuracy with the Real Facemask dataset and MaskedFaceNet. Also in [11], the authors propose a graphical user interface (GUI) to identify the use of masks by classifying it in the three classes that past authors have proposed, using the ResNet50 architecture with an accuracy of 91%, in the same way in [12] use ResNet50 with 4 different sets of data among them, MaskedFace-Net, MAFA in addition to two Kaggle datasets. The author [7] creates a standard use detection algorithm based on the improved YOLO-v4 for the recognition of mask use, achieving 98.3% accuracy. Other studies [18–20] proposed models of convolutional neural networks for the detection of mask use using machine learning libraries, such as Scikit-Learn, TensorFlow, Keras and OpenCV.

Table 1 Comparison of works related to the classification of masks

Method	Classification	Dataset	Accuracy
MobileNetV2 [9]	Triple	RFMD, MaskedFace-Net	99%
CNN [20]	Binary	MAFA	98.2%
Improved YOLO-v4 [7]	Triple	RFMD, MaskedFace-Net	98.3%
CNN [21]	Triple	Manual, MaskedFace-Net	99.75
ResNet-18 [13]	Triple	MaskedFace-Net, Kaggle	99.05
ResNet50 [12]	Triple	Kaggle [22, 23], MaskedFace-NET, MAFA	94.59%
InceptionV2 [14]	Triple	RMFRD, MAFA, WIDER FACE, MaskedFace-Net	91.1%
CNN [16]	Triple	MaskedFace-Net	98.5%
MobileNet [10]	Triple	RMFRD	90%
ResNet50 [11]	Triple	Kaggle, MaskedFace-Net	91%
Proposed method	Triple	MaskedFace-Net	99.69%

Some differences we found in the published articles were the structure of the model, the dataset and the preprocessing. Table 1 shows a comparison of the recommendations of some authors who conducted similar studies to one presented in this article.

3 Proposed Method

This article presents a model combining deep learning, and machine learning using Python libraries. The proposal encompasses a convolutional neural network that allows the classification and identification of the use of masks in three classes (Mask, No_Mask and Incorrect_Mask).

The workflow of the proposed model is shown in the Fig. 1, where it can be observed that from an input image the identification of the region of interest in this case the faces will be made, once the images are identified the convolutional neural network model will be applied for the detection of the use of masks to be able to classify it within one of the three classes.

3.1 Database

The MaskedFace-Net dataset was used for the Mask and Incorrect_Mask classes, while the dataset known as the Flickr-Faces-HQ Dataset for the No_Mask class,

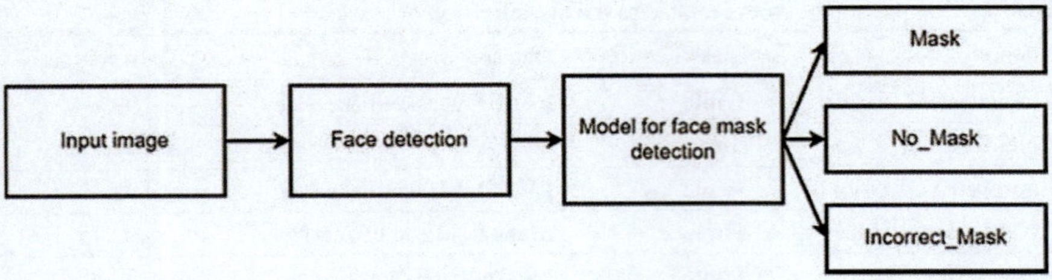

Fig. 1 Basic workflow proposed

Fig. 2 Example of the dataset

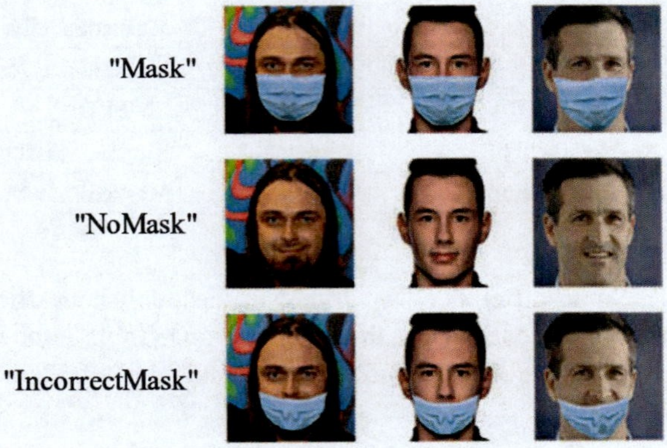

collectively those three classes were used to train and validate the proposed convolutional neural network. In total, 15,000 images were used, distributed in the same quantities for each class, in this case 5000. Figure 2 shows an example of the images contained in these classes. To carry out the training process, the images were divided into different percentages to perform the training, tests and validation, where 70% were assigned for training, while for testing 20% and the rest was used for validation.

The dataset was provided by Cabani [24], and the authors created a dataset called MaskedFace-Net. This dataset contains 137. 016 images with a resolution of 1024 $\times$ 1024, as mentioned by the author, this dataset is based on the Flickr-Faces-HQ (FFHQ) dataset, classified into two groups known as Correctly Masked Face Dataset (CMFD) and Incorrectly Masked Face Dataset (IMFD).

3.2 *Testing Models*

Three different models were proposed to test the effectiveness of each of them with the main database, in this case the MaskedFace-Net images were used exclusively, using the four classes that the dataset provides.

In each model, several experiments were carried out and each of them was executed on 30 occasions, to obtain the mean accuracy and the standard deviation, in addition the best case was detected in each experiment.

3.2.1 Model A-B-C

The MaskedFace-Net dataset is divided into four total classes, where the two main categories are Correctly masked and incorrectly masked, the latter subclassified into 3 classes Undercovered chin, Undercovered nose and Undercovered nose and mouth see Fig. 3. Therefore, from a convolutional neural network it is possible to identify to which category an image belongs.

Three different models were proposed to identify the convolutional neural network architecture that provides the best results.

The first proposed CNN structure named Model A, consists of an in-sized 100 × 100 × 3 input image with 4 convolutional layers, as shown in Fig. 4 In which at the end it is classified into one of the 4 classes of the dataset.

Model B, see Fig. 5, has a larger input image which is 200 × 200 × 3 pixels, and has 4 convolutional layers in the same way, but because it is larger it is computationally heavier in the classification stage.

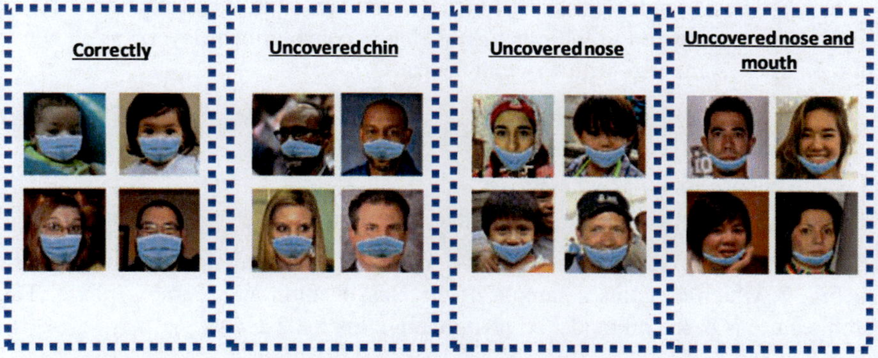

Fig. 3 Example of MaskedFace-Net images

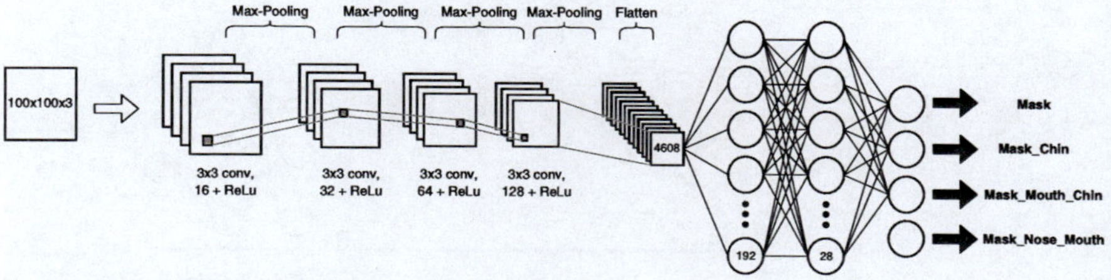

Fig. 4 CNN structure of model A

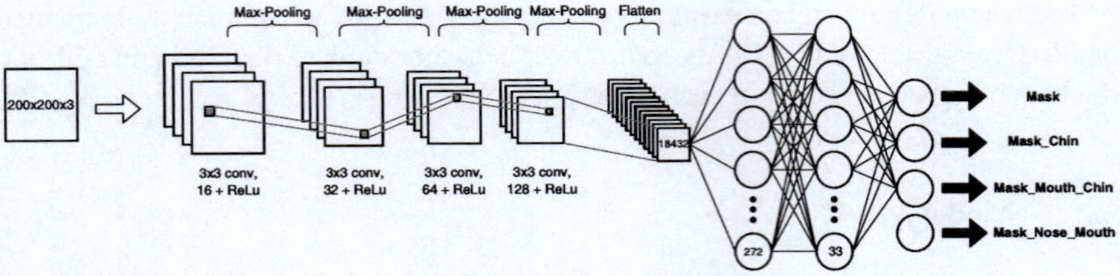

Fig. 5 CNN structure of model B

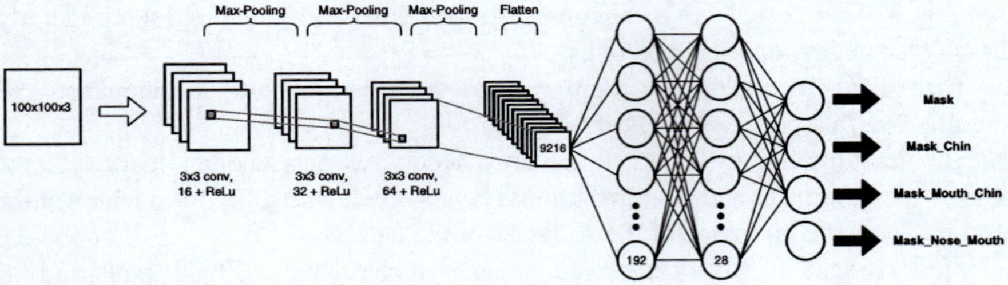

Fig. 6 CNN structure of model C

CNN's latest architecture, called model C uses an input image of 100 × 100 × 3 pixels, but compared to the other 2 this has 3 convolutional layers, as shown in Fig. 6., in the same way with 4 output neurons.

3.3 Architecture Description

The proposed general architecture of the convolutional neural network can be found in Fig. 7, which contains a sample of the classification and learning phase. This architecture is based on model A, mentioned in Sect. 3.2.1.

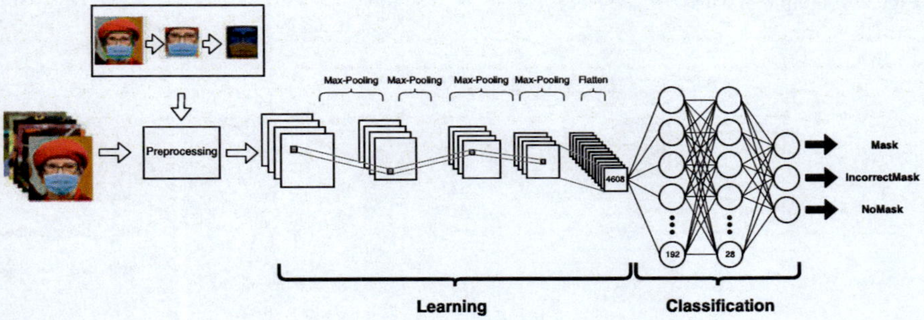

Fig. 7 General architecture of the proposed method

A convolutional neural network has been designed to classify the use of masks and is built as follows: In the first part, you have a data set with which our convolutional neural network will be trained, for each of the images a preprocessing will be applied to highlight the features and identify people with masks, followed by that begins the learning phase in which the number of layers for the neural network and the parameters that will interact in the layers are proposed, followed by the classification stage where from the data obtained in the learning phase it will be identified to which class the input image belongs.

3.4 Models for Classification of Masks

The images that were used for training were classified into three different classes, "No_Mask", "Mask" and "Incorrect_Mask."

An input image of $100 \times 100 \times 3$ pixels is the basis of the model, which is why the images are resized to fit that input. In order to identify the region of interest, for the images of the data set, the Caffe [25] model was applied, which is responsible for identifying the region of the face, as shown in Fig. 8.

As shown in Fig. 9, to help the convolutional neural network, we apply the RGB subtraction technique to the region of interest so that it can counteract slight variations in the image.

4 Results

The results obtained from the A-B-C models were tested through experiments and statistical tests, and the model that proved to have the best results was chosen for the model proposed in this article.

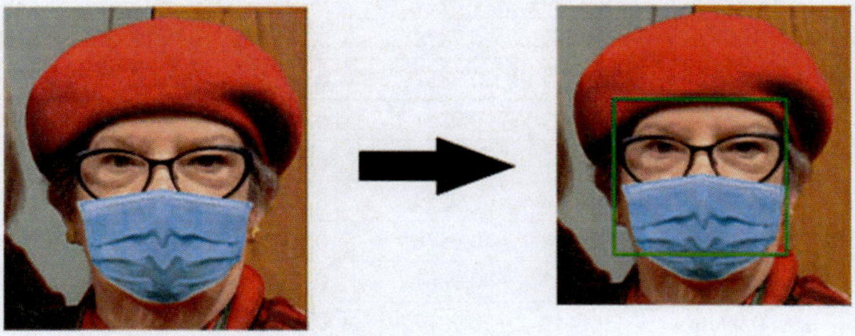

Fig. 8 Face detection sample

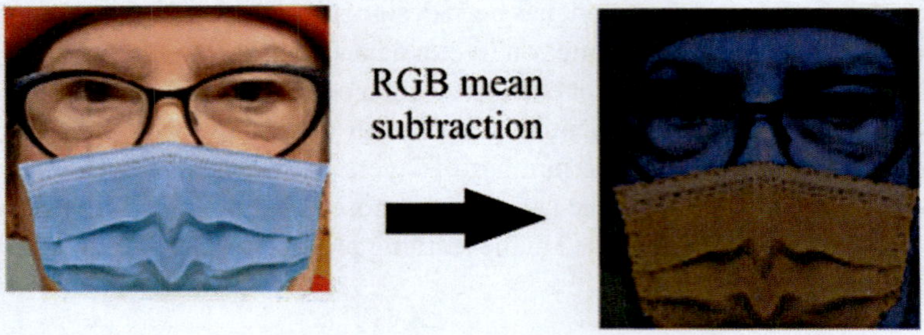

Fig. 9 Preprocessing sample

4.1 Results of Model A-B-C

According to the models proposed, 30 experiments were performed in each of them to identify the one that showed the best results, as shown in Tables 2 and 3, which shows the accuracy and loss of the experiments, calculating the average and standard deviation of each model.

Each model was trained 30 times with different parameters, including the number of times, percentage of training, testing and validation, and batch size, coinciding with the Adam optimizer in this case.

Table 2 Comparing of models (Part #1)

Model	A		B		C	
Training	Accuracy	Loss	Accuracy	Loss	Accuracy	Loss
1	0.9953	0.0363	0.9878	0.0829	0.9888	0.0751
2	0.9890	0.0811	0.9878	0.0995	0.9878	0.1003
3	0.9859	0.1064	0.9878	0.1113	0.9867	0.1276
4	0.9859	0.1318	0.9878	0.1226	0.9857	0.1529
5	0.9906	0.0396	0.9878	0.1342	0.9847	0.1675
6	0.9937	0.0561	0.9878	0.1452	0.9816	0.1183
7	0.9890	0.0895	0.9857	0.0955	0.9806	0.1574
8	0.9875	0.1172	0.9867	0.1159	0.9806	0.1764
9	0.9890	0.1661	0.9867	0.1302	0.9786	0.2343
10	0.9890	0.1826	0.9827	0.1218	0.9745	0.0865
11	0.9718	0.1200	0.9878	0.1079	0.9847	0.1264
12	0.9906	0.1217	0.9867	0.1199	0.9857	0.1639
13	0.9906	0.1443	0.9847	0.1037	0.9857	0.2000
14	0.9906	0.1661	0.9827	0.1311	0.9847	0.2241
15	0.9906	0.0908	0.9827	0.1477	0.9847	0.2374

Table 3 Comparing of models (Part #2)

Model	A		B		C	
Training	Accuracy	Loss	Accuracy	Loss	Accuracy	Loss
16	0.9906	0.0610	0.9878	0.1113	0.9847	0.2469
17	0.9906	0.0740	0.9878	0.1136	0.9847	0.2541
18	0.9906	0.0565	0.9857	0.1342	0.9847	0.2596
19	0.9875	0.0967	0.9878	0.1452	0.9847	0.2644
20	0.9859	0.1180	0.9878	0.1036	0.9847	0.2688
21	0.9859	0.1395	0.9878	0.1237	0.9888	0.1102
22	0.9859	0.1450	0.9878	0.1302	0.9888	0.1456
23	0.9859	0.1489	0.9827	0.1030	0.9878	0.1817
24	0.9859	0.1619	0.9827	0.1203	0.9867	0.2160
25	0.9843	0.1713	0.9827	0.1302	0.9837	0.0950
26	0.9843	0.1827	0.9827	0.1218	0.9857	0.1284
27	0.9843	0.1908	0.9878	0.1070	0.9857	0.1382
28	0.9843	0.1939	0.9878	0.1124	0.9857	0.1698
29	0.9843	0.1964	0.9878	0.1027	0.9847	0.1946
30	0.9843	0.1987	0.9878	0.1164	0.9867	0.2136
	Average	0.9875	Average	0.9862	Average	0.9848
	Standard deviation	0.0042	Standard deviation	0.0021	Standard deviation	0.0031

In Table 4, a summary of the data obtained from the experiments is shown, where the 3 models are compared identifying relevant data, where on average the one that presents the best results is model A.

The best case was obtained in model A; experiment number 1, where the best case was 99.53% with a loss of 3.63%. Outperforming the experiments and their

Table 4 Summary of model comparison

Parameters	A	B	C
Times	20	15	25
Best case	0.9953	0.9878	0.9888
Average	0.9875	0.9862	0.9848
Standard deviation	0.0042	0.0021	0.0031
% Training	70%	72%	72%
% Tests	20%	18%	18%
% Validation	10%	10%	10%
Experiments	30	30	30
Batch	30	50	20
Optimizer	Adam	Adam	Adam

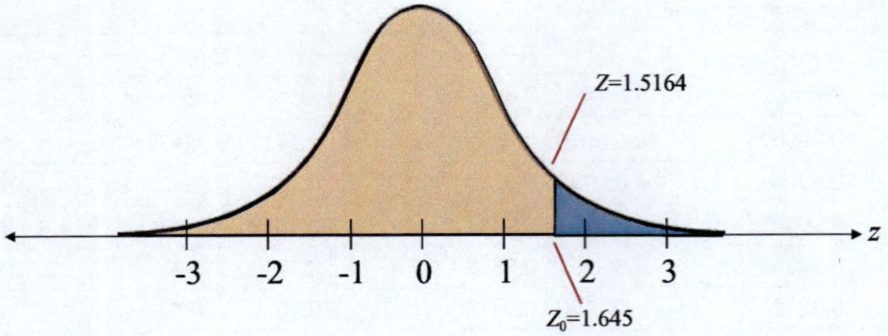

Fig. 10 Statistical test 1 graph

respective models on average, using the MaskedFace-Net database, to prove this, statistical tests were carried out.

4.1.1 Statistical Test 1

Statistical test 1 has the statement that "The mean precision of model A is greater than the mean precision of model B". Therefore, null hypothesis would be that the mean accuracy of model B is greater than or equal to the mean accuracy of model A, to perform the calculations of the statistical test is by means of Eq. 1.

$$z = \frac{\overline{x_1} - \overline{x_2}}{\sqrt{\frac{\sigma_1{}^2}{n_1} + \frac{\sigma_2{}^2}{n_2}}} \tag{1}$$

The mean of the 30 experiments performed for the average precision of model A is 0.9875, while the standard deviation is 0.0042. The mean precision of model B is 0.9862 and the standard deviation is 0.0021.

To reject the null hypothesis, the critical value must be greater than 1.645 because the value of alpha is 0.05. The result of the statistical test is 1.5164 obtained, see Fig. 10, so it fails to reject the null hypothesis. Therefore, there is not enough evidence at 5% significance level to support the claim that the mean accuracy of model A is higher than the mean accuracy of model B.

4.1.2 Statistical Test 2

Hypothesis test 2 has the statement that "The mean accuracy of model A is greater than the mean accuracy of model C". Therefore, the null hypothesis would be that the mean accuracy of model C is greater than or equal to the mean accuracy of model A.

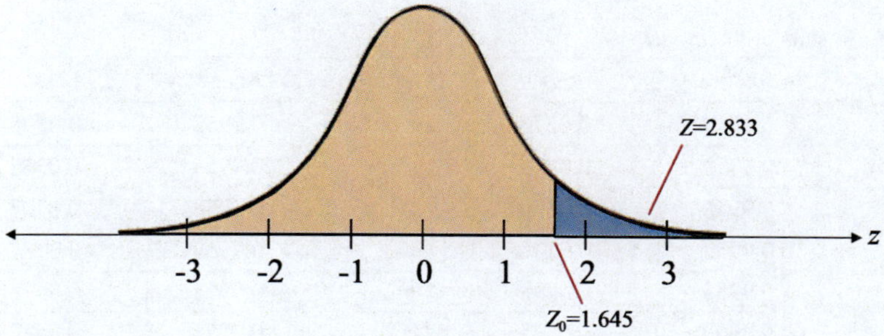

Fig. 11 Statistical test 2 graph

The mean of the 30 experiments performed for the mean precision of model A is 0.9875, while the standard deviation is 0.0042. The mean precision of model C is 0.9848 and the standard deviation is 0.0031.

To reject the null hypothesis, the critical value must be greater than 1.645 because the value of alpha is 0.05. The result of the statistical test is 2.833 obtained, see Fig. 11, so it fails to reject the null hypothesis. Therefore, there is sufficient evidence at a 5% significance level to support the claim that the mean accuracy of model A is greater than the mean accuracy of model C.

According to the data obtained from case study 1, where three different models were evaluated and statistical tests were performed, model A of Fig. 7 is the one with the best results, therefore, this model will be used to perform the training of our proposed method, with the only difference that in this case there will only be three outputs.

4.2 *Results of the Proposed Method*

According to the statistical tests, it was found that model A provided the best results, therefore, this architecture is the one used for the proposed method, experiments and evaluations were also carried out to verify that the results were similar to those obtained in the tested models.

For a fair comparison, the same dataset mentioned in [16] was used, considering the availability characteristics of the images used. This model was trained with the dataset presented in Part 3, using the Python programming language with libraries, such as Tensorflow and Keras. The model was evaluated with 30 training sessions using the proposed model.

The results obtained are shown in Table 5, and it can be seen that the best result was obtained in training exercise 19 with an accuracy of 99.69%.

The accuracy obtained refers to the accuracy of the model with the training data while val_accurracy refers to the accuracy of the validation set. In the same way, which of the, refers to the error made both in the training (loss) and in the validation

Table 5 Results of the model

Training	Accuracy	Loss	Training	Accuracy	Loss
1	0.9969	0.0276	16	0.9958	0.0395
2	0.9958	0.0383	17	0.9958	0.0566
3	0.9958	0.0479	18	0.9958	0.0707
4	0.9958	0.0549	19	0.9969	0.0184
5	0.9958	0.0598	20	0.9958	0.0478
6	0.9969	0.0345	21	0.9958	0.0588
7	0.9969	0.0492	22	0.9958	0.0666
8	0.9969	0.0600	23	0.9948	0.0318
9	0.9916	0.0418	24	0.9948	0.0459
10	0.9958	0.0387	25	0.9948	0.0585
11	0.9958	0.0522	26	0.9958	0.0265
12	0.9958	0.0624	27	0.9958	0.0376
13	0.9969	0.0227	28	0.9958	0.0472
14	0.9969	0.0348	29	0.9958	0.0530
15	0.9969	0.0446	30	0.9948	0.0246

set (val_loss) In general, with each epoch, our neural network improves, Fig. 12 shows how the network converges to achieve the results shown in our best training, in this case training 19.

In Table 6, a confusion matrix was made for evaluating it with the percentage of the testing data, so that, of the 2991 images of that percentage, 2979 were evaluated correctly from which a 99.60% accuracy was obtained.

In this paper, to evaluate the effectiveness of the model, three different parts of the MaskedFace-Net dataset were executed and 15,000 different images were taken for each part of the model evaluation.

From the results in Table 7, we can notice the accuracy and loss values for each part of the assessment. This model yielded the best results with 99.69% used to

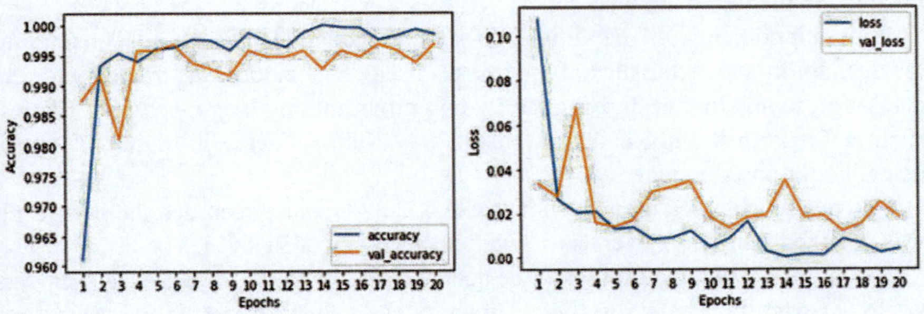

Fig. 12 Evaluating the neural network

Table 6 Confusion matrix of the model

Predicted current	IncorrectMask	Mask	NoMask
IncorrectMask	1007	4	2
Mask	5	1031	1
NoMask	0	0	941

Table 7 Evaluating the model against different parts of the dataset

Part	Average	Loss
1	0.9987	0.0099
2	0.9960	0.0265
3	0.9962	0.0185

evaluate different parts of the dataset and an accuracy achieved of 99.87% when evaluating the first part, la part 2 reaches 99.60% and Part 3 is the final image of the dataset, with accuracy reaching 99.62%.

According to the results obtained by Jones [16], with 98.5% accuracy, a better percentage of accuracy was obtained on average with the proposed neural network model where it was possible to achieve as a best case 99.69% using the MaskedFaceNet database.

5 Conclusions

This article proposes a convolutional neural network to solve classification problems and detect the correct use of masks. Models can be classified into three different layers: Mask, NoMask, and IncorrectMask. In addition, to test the effectiveness of the model, the results were validated by evaluating it in different parts of the MaskedFace-Net dataset and overall, the results show that this model achieves a good rating rate of 99.6865%. it can be applied to real-time applications to slow the spread of the COVID-19 virus. In the future, other databases will be used for classification and a real-time system for monitoring and alarming people wearing masks.

References

1. Pedersen, S. F., & Ho, Y.-C. (2020). SARS-CoV-2: A storm is raging. *Journal of Clinical Investigation, 130*(5), 2202–2205.
2. World Health Organization. (2022). WHO Coronavirus (COVID-19) Dashboard, World Health Organization. Retrieved September 07, 2022, from https://covid19.who.int/.
3. World Health Organization. (2020). Consejos para la población sobre el nuevo coronavirus (2019-nCoV): Cuándo y cómo usar mascarilla. World Health Organization, 1 12 2020.

Retrieved November 03, 2022, from https://www.who.int/es/emergencies/diseases/novel-coronavirus-2019/advice-for-public/when-and-how-to-use-masks.
4. Centres of Disease Control and Prevention. (2021). Erratum. *MMWR. Morbidity and Mortality Weekly Report, 70*(6), 293.
5. CDC. (2021). Healthcare workers. Centers for Disease Control and Prevention, 09 04 2021. Retrieved May 06, 2022, from https://www.cdc.gov/coronavirus/2019-ncov/hcp/respirator-use-faq.html.
6. Singh, S., Ahuja, U., Kumar, M., Kumar, K., & Sachdeva, M. (2021). Face mask detection using YOLOv3 and faster R-CNN models: COVID-19 environment. *Multimedia Tools and Applications, 80*(13), 19753–19768.
7. Yu, J., & Zhang, W. (2021). Face mask wearing detection algorithm based on improved YOLO-v4. *Sensors, 21*(9), 3263.
8. Jiang, X., Gao, T., Zhu, Z., & Zhao, Y. (2021). Real-time face mask detection method based on YOLOv3. *Electronics, 10*(7), 837.
9. Deshmukh, M., Deshmukh, G., Pawar, P., & Deore, P. (2021). Covid-19 mask protocol violation detection using deep learning and computer vision. *International Research Journal of Engineering and Technology (IRJET), 8*(6), 3292–3295.
10. Rudraraju, S., Suryadevara, N., & Negi, A. (2020). Face Mask detection at the fog computing gateway. In *Proceedings of the 2020 Federated Conference on Computer Science and Information Systems.*
11. Bhattarai, B., Raj Pandeya, Y., & Lee, J. (2021). Deep learning-based face mask detection using automated GUI for COVID-19. In *6th International Conference on Machine Learning Technologies.*
12. Pham-Hoang-Nam, A., Le-Thi-Tuong, V., Phung-Khanh, L., & Ly-Tu, N. (2022). Densely populated regions face masks localization and classification using deep learning models. In *Proceedings of the Sixth International Conference on Research in Intelligent and Computing.*
13. Soto-Paredes, C., & Sulla-Torres, J. (2021). Hybrid model of quantum transfer learning to classify face images with a COVID-19 mask. *International Journal of Advanced Computer Science and Applications, 12*(10), 826–836.
14. Wang, B., Zhao, Y., & Chen, P. (2021). Hybrid transfer learning and broad learning system for wearing mask detection in the COVID-19 era. *IEEE Transactions on Instrumentation and Measurement, 70*, 1–12.
15. Cabani, A., Hammoudi, K., Benhabiles, H., & Melkemi, M. (2020). MaskedFace-Net–A dataset of correctly/incorrectly masked face images in the context of COVID-19. *Smart Health, 19*, 1–6.
16. Jones, D., & Christoforou, C. (2021). Mask recognition with computer vision in the age of a pandemic. *The International FLAIRS Conference Proceedings, 34*(1), 1–6.
17. Karras, T., Laine, S., & Aila, T. (2021). A style-based generator architecture for generative adversarial networks. *IEEE Transactions on Pattern Analysis and Machine Intelligence, 43*(12), 4217–4228.
18. Das, A., Wasif Ansari, M., & Basak, R. (2020). Covid-19 face mask detection using TensorFlow, Keras and OpenCV. In *2020 IEEE 17th India Council International Conference (INDICON).*
19. Kaur, G., Sinha, R., Tiwari, P., Yadav, S., Pandey, P., Raj, R., Vashisth, A., & Rakhra, M. (2022). Face mask recognition system using CNN model. *Neuroscience Informatics, 2*(3), 100035.
20. Sethi, S., Kathuria, M., & Mamta, T. (2021). A real-time integrated face mask detector to curtail spread of coronavirus. *Computer Modelling in Engineering & Sciences, 127*(2), 389–409.
21. Aydemir, E., Yalcinkaya, M., Barua, P., Baygin, M., Faust, O., Dogan, S., Chakraborty, S., Tuncer, T., & Acharya, R. (2022). Hybrid deep feature generation for appropriate face mask use detection. *International Journal of Environmental Research and Public Health, 19*(4), 1939.
22. Larxel. (2022). Face mask detection. Kaggle, 22 05 2022. Retrieved March 22, 2022, from https://www.kaggle.com/datasets/andrewmvd/face-mask-detection.
23. Jangra, A. (2020). Face mask detection 12K images dataset. Kaggle, 25 05 2020. Retrieved March 22, 2022, from https://www.kaggle.com/datasets/ashishjangra27/face-mask-12k-images-dataset/metadata.

24. Cabani, A. (2021). MaskedFace-Net. Github, 28 05 2021. Retrieved March 11, 2022, from https://github.com/cabani/MaskedFace-Net.
25. Jia, Y., Shelhamer, E., Donahue, J., Karayev, S., Long, J., Girshick, R., Guadarrama, S., & Darrell, T. (2014). Caffe: Convolutional architecture for fast feature embedding. In *MM 2014-Proceedings of the 2014 ACM Conference on Multimedia.*

Quanvolutional Neural Network Applied to MNIST

Daniel Alejandro Lopez, Oscar Montiel, Miguel Lopez-Montiel, Moisés Sánchez-Adame, and Oscar Castillo

1 Introduction

The Quantum Machine Learning field has several approaches due to the computing capabilities of quantum systems and their advantages [1], having advancements in the machine learning field for near-term quantum computers [2]. Later, more Quantum Neural Network approaches were proposed [3]. However, even though these approaches and implementations are fairly recent, the concepts of quantum neural networks have been present in quantum computing for longer, presumably since 1995, as the evidence of Kak's [4] work suggests. Quantum computing examines the flow and processing of data following the laws of quantum mechanics, specifically, the concept of *superposition*, which refers to the ability of simultaneously being in multiple different states [5]. For this reason, quantum computing presents a promising performance in the resolution of previously unsolvable problems in classical computing, such as integer factorization in Shor's work [6], leading to tackling machine learning problems in new ways [7]. Nowadays, the focus of quantum machine learn-

D. A. Lopez · O. Montiel (✉) · M. Lopez-Montiel · M. Sánchez-Adame
CITEDI, Instituto Politécnico Nacional, Ave. Instituto Politécnico Nacional 1310, 22435 Tijuana, BC, Mexico
e-mail: oross@ipn.mx

D. A. Lopez
e-mail: dlopez@citedi.mx

M. Lopez-Montiel
e-mail: milopez@citedi.mx

M. Sánchez-Adame
e-mail: msanchez@citedi.mx

O. Castillo
Instituto Tecnológico de Tijuana, Calzada del Tecnológico, 22414 Tijuana, BC, Mexico
e-mail: ocastillo@tectijuana.mx

O. Castillo and P. Melin (eds.), *Hybrid Intelligent Systems Based on Extensions of Fuzzy Logic, Neural Networks and Metaheuristics*, Studies in Computational Intelligence 1096,
https://doi.org/10.1007/978-3-031-28999-6_4

ing is shifting towards hybrid models, combining different types of quantum layers with neural networks. In this work, a quanvolutional layer is implemented in the Convolutional Neural Network (CNN) LeNet5 CNN, evaluating its performance in classification and sample processing time on the MNIST dataset, and comparing it to its classical counterpart. Unlike other works, this approach evaluates a portion of the entire MNIST dataset in the multiclass classification task. It also compares model inference and throughput to analyze a possible quantum speedup at adding a quanvolutional layer in contrast to its CNN version.

The structure of this chapter is organized as follows: Sect. 2 presents a selection of relevant works where the main quantum machine learning algorithms are summarized. Section 3 describes a selection of basic concepts of quantum computing. Section 4 introduces the proposed quanvolutional neural network approach. Section 5 describes the performed experiments and obtained results. Section 6 examines the obtained results and compares them to those mentioned in related works. Finally, Sect. 7 presents the conclusions and future work.

2 Related Work

The focus of Quantum Machine Learning depends on the nature of the implementation, as several authors have demonstrated in recent years. In 2020, Henderson [8] compares the performance of a classical CNN, a Quanvolutional Neural Network (QNN), and a CNN with non-linearities for the image classification task on the MNIST dataset. This work established some of the foundations of quantum convolution, showing that the QNN model attains higher accuracy and faster training compared to the classical CNN. The implemented model architectures are almost the same; the main difference lies in the first layer for both the QNN and random model, where the quanvolutional and random non-linear transformation layers are applied, respectively. The entirety of the MNIST dataset is used, and the number of implemented quanvolutional filters is modified, reaching a test set accuracy of 95% or higher. Later in 2021, Amin [9] investigated the Quantum Machine Learning (QML) and Classical Machine Learning (CML) applications in the classification of Computed Tomography (CT) of COVID-19 images. The approach is divided into two phases: The data augmentation phase through a Conditional Generative Adversarial Network (CGAN); and the classification phase for both QML and CML models. The experiments are carried out using the publically available Chinese Hospital Dataset, and the Private Collected Hospital Dataset. The cross-validation technique is also implemented to improve model performance. The obtained results show that a cross-validation threshold of 0.5 together with the QNN model attain a precision, recall, and F1 score of 0.94 for the public dataset and 0.96 for the private one, proving that the QNN model performs better than the CNN. Also, in 2021, Houssein [10] presents a Hybrid Quantum-Classical Convolutional Neural Network (HQ-CNN) model to detect COVID-19 patients; the dataset includes more than five thousand of chest X-ray images, divided into normal, COVID-19, and viral and bacterial pneumonia. The proposed HQ-CNN attains high accuracy (98%) and recall (99%) values in all the binary classification experiments, but in the multiclass dataset achieves lower ones; exhibiting that the HQ-CNN performs well in predicting positive COVID-19 cases but does not perform equally as well in multiclass tasks. In 2022, Zheng [11] presents a Quantum Convolutional Neural Network (QCNN) for real quantum circuits; the quantum blocks are designed for the encoding, model design, and parameter tuning stages. The MNIST dataset is used for the experiments, demonstrating the proposed model's representation, learning, and optimization capacity. Testing results are split in three phases: The "Representation phase", where the capability to prepare an arbitrary quantum state by a shallow quantum circuit while maintaining main features and high accuracy is tested; the results show that increasing the number of layers reduces the average preparation error. The "Learning phase", where the capability to learn features and achieve classification tasks is tested; here, euclidean and non-euclidean data is used, attaining accuracy of 85% and 96.65% for each data type, respectively. Finally, the "Optimization phase" refers to the capability to guarantee trainability with optimization algorithms. The evidence shows that Root Mean Square Propagation (RMSProp), Adaptive Moment Estimation (Adam), and Gradient Descent Optimization (GDO) can achieve high classification accuracy, depending more on the model structure and number of qubits. Lastly, in 2022, Huang [12] proposes a Variational Convolutional Neural Network (VCNN), which allows efficient training and implementation on near-term quantum devices. This algorithm combines multi-scale entanglement renormalization, and it is deployed on the TensorFlow Quantum platform. The used datasets are MNIST and Fashion MNIST, where the VCNN attains an average classification accuracy of 96.41%, higher and in fewer epochs than other QNN algorithms. The experiments compare results for a regular CNN, and several QNN algorithms in binary and multiclass classification tasks. The VCNN algorithm is also tested with different levels of depolarizing noise, showing excellent resilience to noise as it can still classify with high probabilities.

3 Fundamentals

3.1 Convolutional Neural Network

The CNN is characterized by its **convolutional layers** dedicated to extracting the input features and generating a new feature map through convolution kernels. The convolutional layers learn input characteristic representations [13]. Applying an activation function to the convolved results creates a new feature map. Equation 1 shows the feature value at position (i, j) in the kth feature map of lth layer.

$$Z^l_{i,j,k} = (w^l_k)^T X^l_{i,j} + b^l_k \tag{1}$$

where the weight vector is represented by $(w_k^l)^T$, the bias term is b_k^l, and $X_{i,j}^l$ is the input centered at the location (i, j) of the lth layer [13]. **Activation functions** allow the networks to detect nonlinear features. Defining a convolutional feature as $Z_{i,j,k}^l$ and $a(\cdot)$ as the nonlinear activation function. Equation 2 shows the activation value computation.

$$a_{i,j,k}^l = a(Z_{i,j,k}^l) \tag{2}$$

The convolution layers are followed by **pooling layers**, which reduce the size of the feature map by lowering its resolution, achieving shift-invariance. For an output of the feature map $Z_{i,j,k}^l$ with an activation function $a_{i,j,k}^l$, the pooling function $pool(\cdot)$ is computed by:

$$y_{i,j,k}^l = pool(a_{m,n,k}^l), \qquad \forall (m,n) \in R_{i,j} \tag{3}$$

being $R_{i,j}$ a local neighborhood around location (i, j). After the resolution of the feature maps is reduced, the **fully connected layers** join the previous neurons with the current layer producing a one-dimensional vector with all obtained characteristics and followed by the last layers of the networks, the output layers. Minimizing the objective function allows the parameter optimization of a specific task. Let C be defined as the cost function, l as the loss function to optimize for each relation of N, denote all parameters for a CNN as θ, and have N input-output relations (x^n, y^n); $n \in [1, \ldots, N]$; where x^n is the nth input data sample and y^n its corresponding label, and o^n the output of the model [13]; as shown in Eq. 4, the cost function can be computed by:

$$C = \frac{1}{N} \sum_{n=1}^{N} l(\theta; y^n, o^n). \tag{4}$$

3.2 *Quantum Computing*

The quantum bit (**qubit**) is the elemental unit of information in quantum computing. In general, a qubit can be defined using different orthogonal basis states; however, it is common to say that the qubit is defined in the basis state when referring to the z−basis states as the computational basis states. This is a practical convention since there is a direct relationship to the classical "0" and "1" bits. The state of a qubit can be envisioned into a bounding sphere known as Bloch sphere, where the basis states are located at opposite positions, i.e., separated π radians. The ground state $|0\rangle$ is one of the basis states, and it is known as the north pole, whereas the basis state $|1\rangle$ is the south pole.

A qubit $|\psi\rangle$ can be in a superposition state of any two basis states; here, in any combination of the basis state $|0\rangle$ and $|1\rangle$. Equation 5 represents this relationship.

$$|\psi\rangle = \alpha|0\rangle + \beta|1\rangle \tag{5}$$

where α and β are complex numbers and they refer to the probability amplitudes and satisfy the normalization condition $|\alpha|^2 + |\beta|^2 = 1$. The Bloch sphere is defined in the spherical coordinate system (r, θ, ϕ). r refers to the radial distance; θ is the latitude angle measured from North-pole to South-pole, $0 \geq \theta \leq \phi$; and ϕ is the longitude angle measured clockwise around the z-axis [14]. Due to the normalization condition, $x^2 + y^2 + z^2 = 1$, so $r = 1$. Converting spherical coordinates to Cartesian coordinates leads to the Eq. 6.

$$\begin{aligned} x &= r\sin(\theta)\cos(\theta) \\ y &= r\sin(\theta)\sin(\theta) \\ z &= r\cos(\theta) \end{aligned} \tag{6}$$

Multiple qubits produce a **composite quantum system** Ψ which quantum space mathematically can be represented as a tensor product of the quantum state of each qubit [11], as it is depicted on Eq. 7.

$$|\Psi\rangle = |\psi_1\rangle \otimes \cdots \otimes |\psi_n\rangle \tag{7}$$

The quantum state Ψ can also be expressed in base-2 notation; where the complex number c_i is the amplitude of the corresponding i state [14].

$$|\Psi\rangle = c_0|00\ldots0\rangle + c_1|00\ldots1\rangle + \cdots + c_{2^n-1}|11\ldots1\rangle = \sum_{i=0}^{2^n-1} c_i|i\rangle_2 \tag{8}$$

Quantum computing uses **quantum gates** that transform the quantum states of qubits; they can be composed of single-qubit or multiple qubits; in general, they can be represented as U, and they must satisfy the expression $UU^\dagger = I$, where $U^\dagger$ refers to the conjugate transposition of U, and I to the identity matrix. Mathematically, the application of a quantum gate to a quantum state is expressed by Eq. 9.

$$U|\Psi\rangle = U\sum_{i=0}^{2^n-1} \alpha_i|i\rangle = U\sum_{i=0}^{2^n-1} \beta_i|i\rangle \tag{9}$$

The main single-qubit rotation gates based on the Pauli matrices [15] are:

$$R_x(\theta) = \begin{bmatrix} cos\frac{\theta}{2} & -isin\frac{\theta}{2} \\ -isin\frac{\theta}{2} & cos\frac{\theta}{2} \end{bmatrix}$$

$$R_y(\theta) = \begin{bmatrix} cos\frac{\theta}{2} & -sin\frac{\theta}{2} \\ sin\frac{\theta}{2} & cos\frac{\theta}{2} \end{bmatrix}$$

$$R_z(\theta) = \begin{bmatrix} e^{-\frac{i\theta}{2}} & 0 \\ 0 & e^{\frac{i\theta}{2}} \end{bmatrix} \tag{10}$$

where $R_x(\theta)$ refers to a rotation of θ radians around the x-axis, $R_y(\theta)$ to a rotation of θ radians around the y-axis, and $R_z(\theta)$ to a rotation of θ radians around the z-axis. The **quantum measurement** extracts classical information from quantum states; the projection measurement is commonly used. Let M be a Hermitian operator that represents a quantum observable that satisfies the spectral decomposition:

$$M = \sum_m m P_m \tag{11}$$

where P_m projects onto the eigenspace M along the eigenvalue m [11]. The possible measurement outcomes correspond to the observable eigenvalues, and the probability of getting the result m by measuring state $|\psi\rangle$ is given by:

$$p(m) = \langle\psi|P_m|\psi\rangle. \tag{12}$$

4 Methods

This work compares the multiclass image classification performance of a classical CNN model and a QNN model, where a quanvolutional layer is implemented. The followed methodology is specified in Fig. 1, where the main stages of the work can be identified.

4.1 Dataset

The experiments were carried out using the MNIST hand-written digits dataset [16], comprised of 70,000 grayscale images of digits going from 0 to 9 split into ten classes

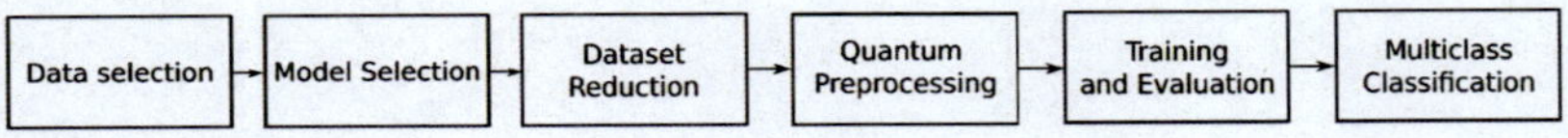

Fig. 1 Proposed methodology

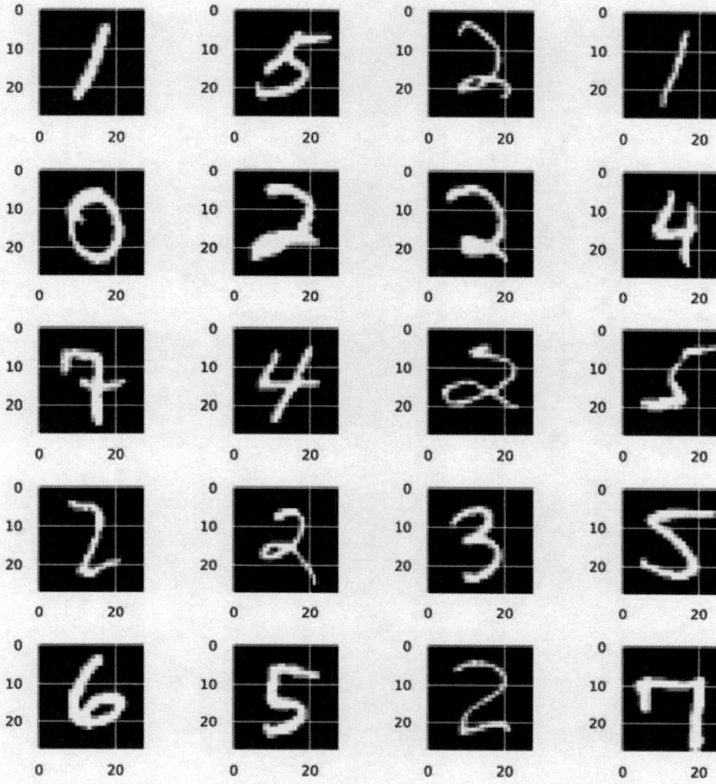

Fig. 2 Samples of MNIST dataset

for each of the numbers. We subdivided this dataset into a smaller portion of 50%, which is further reduced through class balancing, leaving a total of 31,540 images. This was done to counteract the lengthy quanvolutional preprocessing stage in which all the images in the experimental database go through. Figure 2 shows the dataset samples, as well as their 28×28 pixels of resolution.

The proposed class distribution can be seen in Fig. 3, where the number of images per class is balanced to be the same across all classes in order to improve model performance.

4.2 *Model Architecture*

We implemented the LeNet5 for this dataset as it was originally intended to perform this kind of classification [16]. Figure 4 shows the different layers that each sample goes through. The LeNet5 architecture is comprised of three types of layers, convolutional, pooling, and fully connected layers. The convolutional layers are responsible

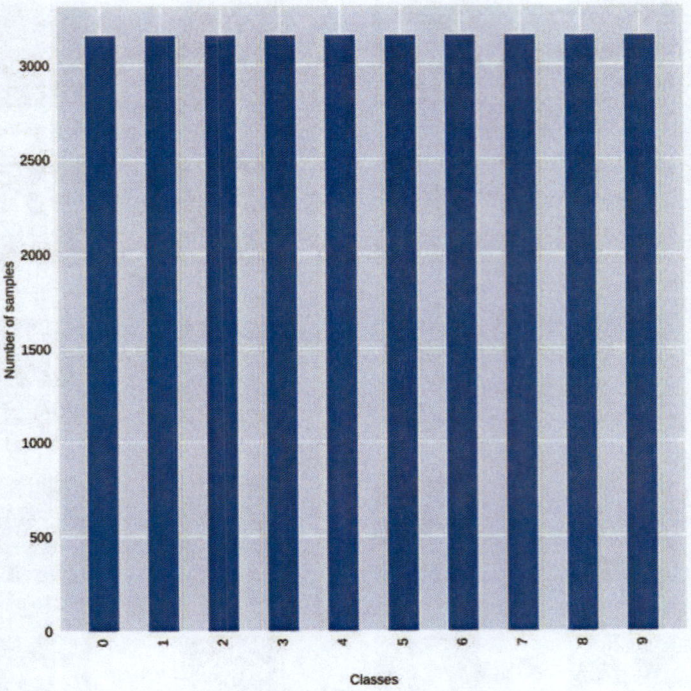

Fig. 3 Class distribution of the reduced MNIST dataset

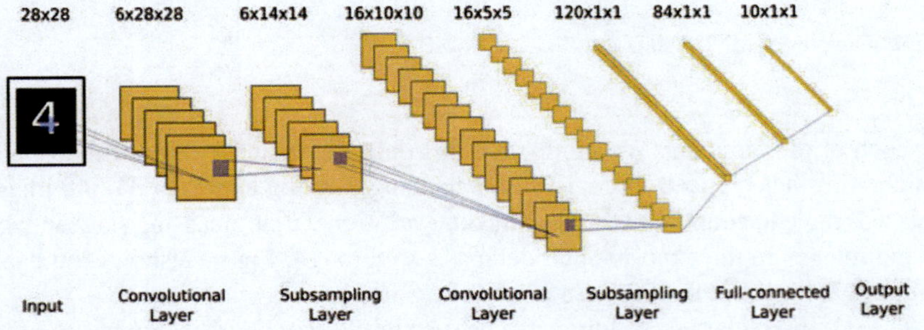

Fig. 4 LeNet5 model architecture [16]

for extracting the main features of the input images and generating new feature maps; the pooling layers reduce these feature maps, leaving only relevant data according to the computation executed. The most frequent pooling operations are average, and max pooling [17]. The fully connected layers join the previous layer with all the neurons, producing a one-dimensional vector that is passed onto the output. The model's output delivers the prediction, and depending on the implemented activation function and the number of output neurons, the problem can be binary or multiclass in the case of classification.

4.3 Quanvolutional Preprocessing

In this work, quantum Machine Learning is implemented by using a quanvolutional layer at the beginning of the LeNet5 model architecture, where the samples are preprocessed before passing them onto subsequent layers of the neural network. This layer is labeled as *quanvolutional* because it behaves similarly to a convolutional layer. Classical convolution receives an input image, and the established filter processes this image to extract its features, processing sequentially small local regions of the entire sample. The results for each region output a single pixel to each of the different channels contained in the image, and the union of all these pixels produces a new image which can be further processed in the following layers. Quantum convolution embeds small pieces of the image into a quantum circuit through parameterized rotations applied to qubits initialized in the ground state, as demonstrated in Fig. 5. After the encoding phase, further quantum computations may be performed on the random quantum circuit, where layers of randomly chosen rotations act on randomly chosen qubits. Following this phase, the quantum system is measured, producing a set of classical expectation values that are mapped to a different channel of a single output pixel [8]. Iterating this procedure over different regions allows the input image to be fully scanned and produce an output object structured as a multi-channel image that goes into the LeNet5 architecture.

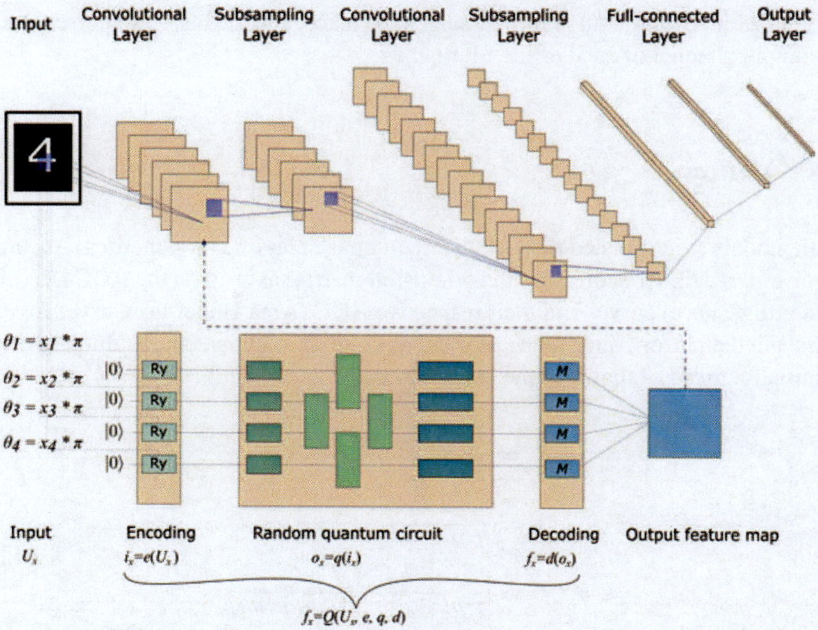

Fig. 5 Quanvolutional layer in LeNet5 model achitecture

As shown in Fig. 5, the proposed quanvolutional filter uses a random quantum circuit q, that takes subsections of the input images defined by U_x. This U_x is a matrix of dimensions (n, n) determined by the defined kernel to use in the quanvolutional preprocessing. The initialization of input U_x is represented by the function e, where the embedding of the input matrix is done through parameterized rotations, establishing the encoded initialization state as $i_x = e(U_x)$. After the encoding phase, the output of the carried out quantum computations are represented by the quantum output state o_x, leading to the equality $o_x = q(i_x) = q(e(U_x))$. The decoding phase measures the output of the random quantum circuit, and it is defined as $f_x = d(o_x) = d(q(e(U_x)))$, where d and f_x represent the decoding function and the measured scalar value respectively. At last, the entire quanvolutional filter can be represented by:

$$f_x = Q(U_x, e, q, d) \tag{13}$$

where Q summarizes the entire quanvolutional filter process comprised of encoding, quantum circuit, and decoding. This preprocessing halves the resolution of the input image, corresponding to a (2, 2) kernel and stride of 2 convolutions in a classical CNN, giving the quanvolutional name to the implemented layer. The lower resolution for the QNN offers less preprocessing time in the quanvolutional layer, and faster processing time in training. A comparison of the regular input images and the preprocessed ones are shown in Fig. 6, in which the input row corresponds to the MNIST samples before the quanvolutional preprocessing takes place, and the last rows correspond to the output channel of each of the four qubits.

4.4 *Metrics*

Both models are evaluated through the main image classification metrics: Accuracy, precision, recall, F1 score, and the confusion matrix, as well as the ROC (Receiver Operating Curve) curves and their respective AUC (Area Under the Curve) for each class and the performance metrics of latency and throughput. The multiclass classification metrics are shown in the set of Eq. 14.

$$\begin{aligned} P &= \frac{TP}{TP+FP} \\ R &= \frac{TP}{TP+FN} \\ ACC &= \frac{TP+TN}{TP+TN+FP+FN} \\ F1 &= 2\left(\frac{P \times R}{P+R}\right) \end{aligned} \tag{14}$$

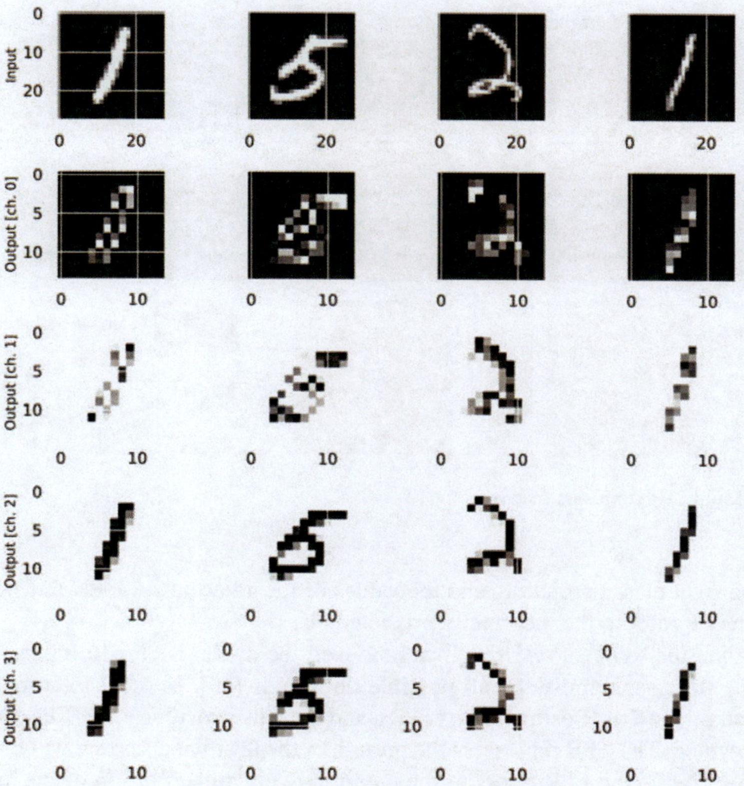

Fig. 6 Comparison of MNIST samples before and after quanvolutional preprocessing

The True Positive (TP) values are the correctly classified digits, and the True Negative (TN) values are the correctly classified digits into the non-desirable class. On the other hand, False Positive (FP) values represent the digits belonging to other classes that were incorrectly predicted as positive; i.e. the desirable class. The False Negative (FN) values are the digits incorrectly classified as belonging to other classes being that they belong to the desirable class. The Precision (P) and Recall (R) metrics shown in Eq. 14, tell the model's accuracy in identifying positive samples, meaning, whether the model is able to correctly classify positive samples of the dataset [18]. In contrast, the Accuracy (ACC) of the model considers all classified samples, measuring the total of correct predictions, as shown in Eq. 14. The F1 score metric combines both the P and R value into a single metric, illustrating the tradeoff of the two via the obtained value.

The confusion matrix is also used to evaluate the total number of correctly classified samples [18]. This metric illustrates all predicted values of the dataset, showing which classes are more difficult to classify for the model, as well as the ones that are easier. Figure 7 shows a multiclass confusion matrix, where the diagonals repre-

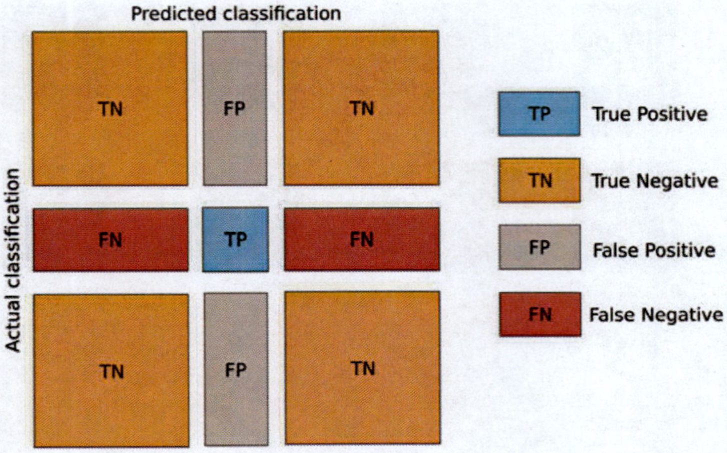

Fig. 7 Multiclass confusion matrix

sent the right class prediction, and the values of the adjacent rows FN and adjacent columns FP refer to the incorrectly predicted classes.

Further, the ROC and AUC tell us how well the model is classifying each class showing their separability by all possible thresholds [19]. In order to compute the ROC curve, the True Positive Rate (TPR), and the False Positive Rate (FPR) must be first obtained. The TPR represents the portion of the positive class that was correctly classified, while the FPR refers to the incorrectly classified portion of the negative class. The Equations in (15) show their computation.

$$
\begin{aligned}
TPR &= \frac{TP}{TP + FN} \\
FPR &= \frac{FP}{TN + FP}
\end{aligned} \tag{15}
$$

Once the FPR and TPR values are computed, the ROC curve may be obtained. This metric plots the TPR and FPR at every possible threshold, showing the capability of the model to separate noise from the positive samples in the dataset. The ROC plot can be separated in two parts, the main diagonal, and the resulting ROC curve from all the plotted threshold points, as it is shown in Fig. 8. Every threshold value establishes the minimum probability that a sample must reach to be categorized in the positive or negative class. For instance, given a threshold value $\sigma = 0.1$, all classified samples with a prediction value below 0.1 are predicted as negative, and all with a prediction value above 0.1 are predicted as positive. Once a threshold value is determined, its corresponding TPR and FPR are computed, and a point of the ROC is plotted. The main diagonal represents the point of the plot in which both the TPR and FPR are equal to 1. Iterating this procedure for all possible thresholds produces

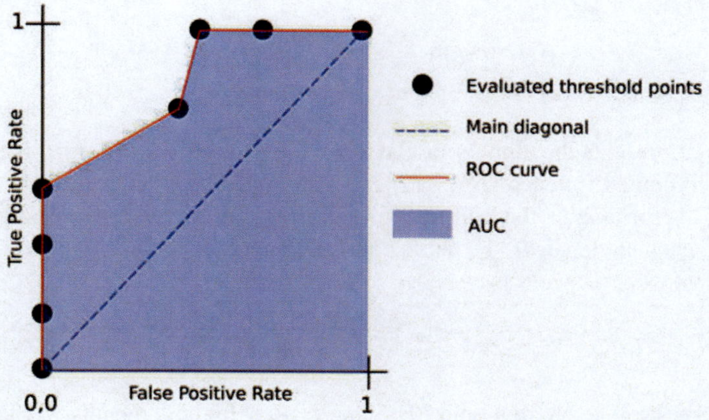

Fig. 8 ROC curve example

the ROC curve. This process is similar to calculating all possible confusion matrices for every threshold value.

Given a threshold $\sigma \in [0, 1]$, the ROC curve is computed by

$$ROC(\sigma) = (FPR(\sigma), TPR(\sigma)) \tag{16}$$

where FPR and TPR are evaluated for every given threshold σ [20]. After the ROC curve is generated, the AUC value for it can be determined by computing the area under said curve, as shown in Fig. 8 by the shadowed area. The AUC equation for a given threshold σ value is shown in Eq. 17

$$AUC(\sigma) = \int_a^b TPR(\sigma)\, d(FPR(\sigma)) \tag{17}$$

where a and b are 0 and 1, and TPR is the function to integrate along its differential $d(FPR)$ on the x-axis. However, the ROC and AUC are not immediately applicable for multiclass tasks, leading to the OvR and OvO regimes. OvR stands for "One vs. Rest", and it is a method to evaluate multiclass models by comparing each class against all other remaining classes simultaneously. One class is taken as positive, while the rest are considered negative. This derives both ROC and AUC into multiclass metrics. Given an OvR score function $f = (f^1, \ldots, f^{N_C})$ allows to form a pair of AUC score for each f^i. The positive class can be represented as the ith class, while the negative instances are represented as the jth classes where $j \neq i$. The number of obtained OvR scores corresponds to the number of classes in the dataset. The overall AUC score is the mean of all N_C pairs of AUC scores [21]. Equation 18 shows the computation of this metric.

$$AUC^{ovr}(f) = \frac{1}{N_C}\sum_{i=1}^{N_C} AUC_{i|\neg i}(f^i) \tag{18}$$

where N_C represents the number of classes in the dataset, $AUC_{i|\neg i}(f^i)$ is the AUC function computed for each OvR pair [21]. Alternatively, OvO stands for "One vs. One", and it formulates the multiclass AUC metric as an average of binary AUC scores for each class pair (i, j), defined in Eq. 19 as:

$$AUC^{ovo}(f) = \frac{\sum_{i=1}^{N_c}\sum_{j\neq i} AUC_{i|j}(f^i)}{N_C(N_C - 1)} \tag{19}$$

In this work, we generate ROC and AUC One vs. Rest for both models. Model performance is evaluated using latency and throughput. Latency, also known as inference time, refers to the time it takes the model to process a single sample [22], which means the time required for the model to receive an input sample and deliver an output response. Let latency be L and the number of processed samples ns, as well as the beginning and ending time of inference as s_{infer} and e_{infer} respectively, Eq. 20 shows *latency* computation is given by:

$$L = \frac{ns}{e_{infer} - s_{infer}} \tag{20}$$

Throughput evaluates processing efficiency for both models. It describes the number of samples a model can process in a given time, that is, the number of inferences the model can deliver [23]. In this work, experiments are carried out for inference time and throughput in the training and testing stages, measuring the time it takes both models to process a single sample, and the number of samples in a specific time. Equation 21 shows throughput calculation represented by THR, (nb) and bsz refer to the number of samples to process and how they are fed to the model, while L refers to the time to process a single sample, obtained with Eq. 20. Time is measured in milliseconds.

$$THR = \frac{nb \times bsz}{L}. \tag{21}$$

5 Experiments and Results

Experiments are carried out in an Ubuntu system 20.04 compatible with the CUDA toolkit 11.2 and CUDNN 8.1, using the TensorFlow-GPU and Keras 2.9.0 framework. The used library for the quantum computing simulations is PennyLane 0.25.1. Hardware specifications are an Intel Core i7-4770 CPU and an NVIDIA GeForce

GTX 1080 Ti GPU. The reduced MNIST dataset obtained after balancing the 50% of the entire dataset is used, leaving the model training phase with 25560 and 2790 samples for the training and validation sets, respectively, and the testing phase with 3190 samples. After the classes in the dataset are balanced, all samples go through quanvolutional preprocessing, preparing the data delivered to the quanvolutional model. In this case, a single quanvolutional layer is implemented together with a quantum circuit of 4 qubits. A set of random parameters is also generated in the interval of $[0, 2\pi)$, which is applied to the random quantum computations used in the quantum circuit. The quanvolutional layer is produced by looping a (2, 2) kernel over the input image with a stride of 2, reducing image resolution to (14, 14). The input image is first encoded by rotating by π the classical input values of the established kernel over the y-axis, embedding them onto the quantum circuit, which then goes through the random layers that apply qubit computations using the random set of parameters. Afterward, the expectation values are measured, producing four classical output values later assigned to each of the four output channels. Model training for LeNet5 comprises two stages: The quantum phase, where the quanvolutional preprocessed set of samples is used to feed the model; and the classical phase, where the classical data is used instead. The model architecture has an input layer whose input size is (14, 14, 4) or (28, 28, 1), depending on whether the model will work with quantum or classical data. The input is followed by a pair of convolutional layers of 6 and 16 filters, respectively, then a kernel of size (5, 5), and stride of 1. Each convolution layer is accompanied by an average pooling layer following it. At last, the flattening layer is followed by three dense layers of 120, 84, and 10 neurons. The used activation function is $tanh$, with the exception of the dense output layer that uses $softmax$ activation function for multiclass classification. In this model we used a first-order gradient-based optimization known as Adam [24]. The hyperparameters are a batch size of 32, 986 steps, and 50 epochs.

Figures 9 and 10 show the obtained accuracy results in the training phase of both models, where the blue and green colored curves represent the model with and without the quantum layer.

As it can be seen, both models attain good accuracy in both the train and validation sets, and although the quanvolutional model begins with slightly lower accuracy, in the end, it follows the classical's model tendency to 1. Furthermore, as it can be seen in Figs. 11 and 12, the loss of both models shows similar behavior.

The evaluation phase is divided into two stages: The classification evaluation stage, where the classification metrics demonstrate the model's performance in the classifying task, and the model performance evaluation stage, where the processing capabilities of both models in terms of time and data are tested. As can be seen in the obtained results in Table 1, the classical CNN model attains slightly better accuracy and loss than the QNN model, meaning the CNN model outperforms the QNN model by a small margin.

Moreover, model predictions are evaluated considering all obtained TP, TN, FP, and FN values from the test set to generate the main classification metrics shown in Table 2.

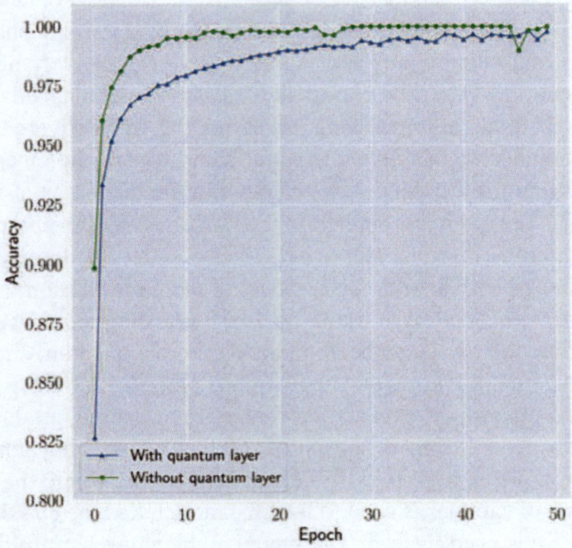

Fig. 9 Training set accuracy of CNN and QNN

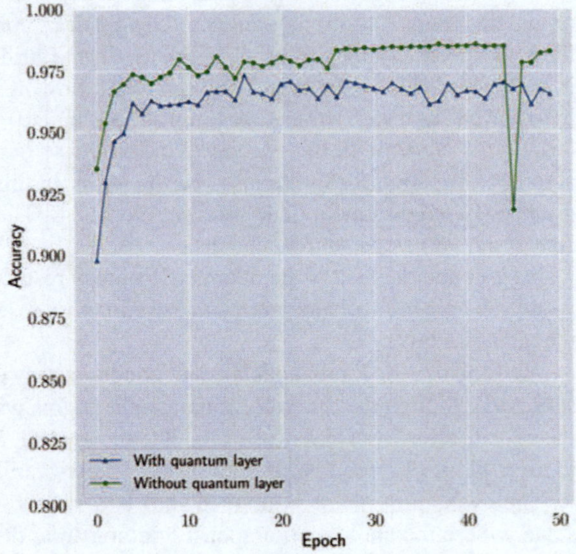

Fig. 10 Validation set accuracy of CNN and QNN

Fig. 11 Training set loss of CNN and QNN

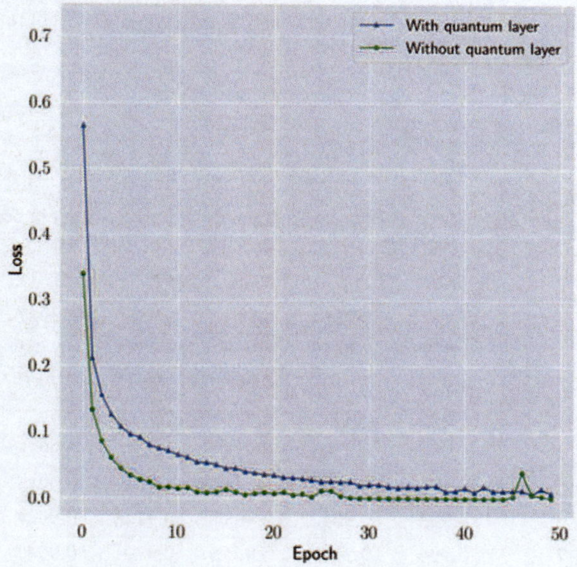

Fig. 12 Validation set loss of CNN and QNN

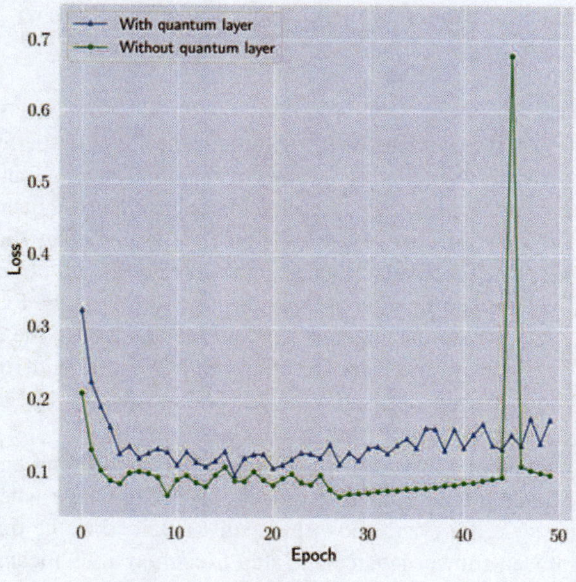

Table 1 Model evaluation results. Best results written in bold

Model	*ACC*	*Loss*
QNN	0.9652	0.1786
CNN	**0.9787**	**0.1085**

Table 2 Model classification metrics. Best results written in bold

Class	CNN				QNN			
	ACC	*P*	*R*	*F1*	*ACC*	*P*	*R*	*F1*
0	**0.9972**	**0.9784**	**0.9937**	**0.986**	0.9947	0.9689	0.9781	0.9735
1	**0.9978**	**0.9815**	0.9969	**0.9891**	0.9975	0.9785	0.9969	0.9876
2	**0.9984**	**0.9906**	**0.9937**	**0.9921**	0.9962	0.9752	0.9875	0.9813
3	**0.9953**	**0.972**	**0.9812**	**0.9766**	0.9922	0.9455	0.9781	0.9615
4	**0.9966**	**0.9904**	**0.9749**	**0.9826**	0.9928	0.9713	0.9561	0.9636
5	**0.9959**	**0.9842**	**0.9749**	**0.9795**	0.9915	0.965	0.9498	0.9573
6	**0.995**	**0.981**	0.9687	**0.9748**	0.9944	0.9631	**0.9812**	0.9721
7	**0.994**	**0.9747**	0.9655	0.9701	0.9947	0.9604	**0.9875**	**0.9738**
8	**0.994**	**0.9777**	**0.9624**	**0.97**	0.9893	0.9495	0.9436	0.9465
9	**0.9931**	0.9569	**0.9749**	**0.9658**	0.9871	**0.976**	0.8934	0.9329

As it is demonstrated in Table 2, the CNN outperforms the QNN in most of the metrics for each class; however, the difference is minimal in most cases. To better appreciate the results presented in Table 2, the confusion matrices of Figs. 13 and 14 show the number of correctly predicted samples by the classical and quanvolutional model respectively, illustrating the more difficult classes to predict.

The classification evaluation stage ends with the ROC curves and their respective AUC, where the attained AUC score can identify the classification performance for every class. The behavior observed in Figs. 15 and 16 is the desired one from both models; even though the results obtained by the QNN are faintly lower than the CNN, they are both proficient in the classifying task.

Model performance is evaluated in the training and testing phase. This demonstrates the difference in processing performance when training and evaluating the models. Plot bars show the results obtained in the training phase, where the QNN presents lower latency and step execution time, meaning that it is faster in processing 1 and 32 samples than the CNN, according to Fig. 17. Consequently, it can also process a higher number of samples in 1 millisecond, outperforming the CNN in each batch sizes scale, as shown in Fig. 18. It is worth mentioning that the carried out experiments were only for a batch size of 32, while the rest are computed scaling the attained results for the smallest batch size.

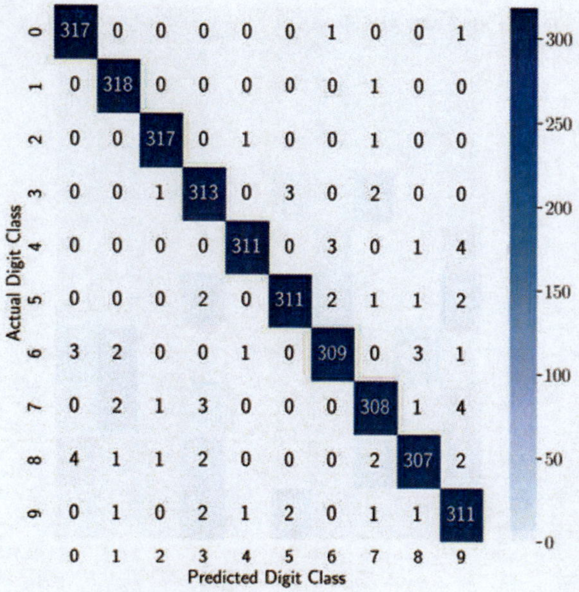

Fig. 13 CNN confusion matrix

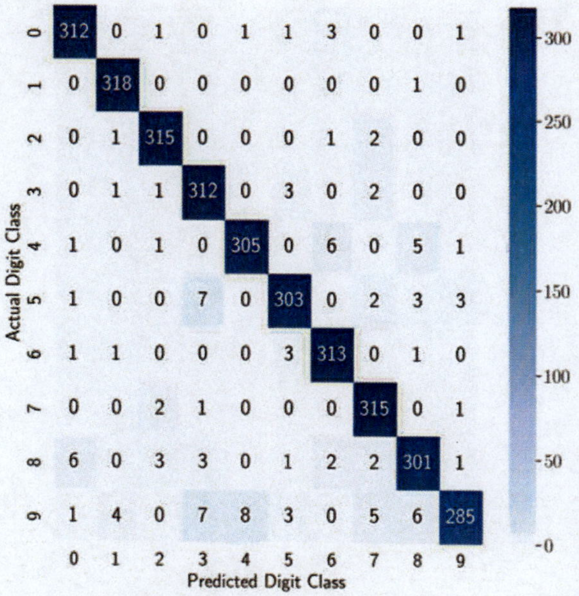

Fig. 14 QNN confusion matrix

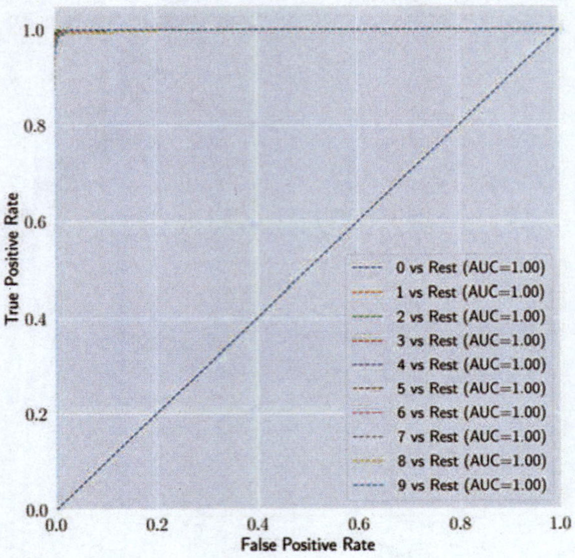

Fig. 15 ROC and AUC for CNN

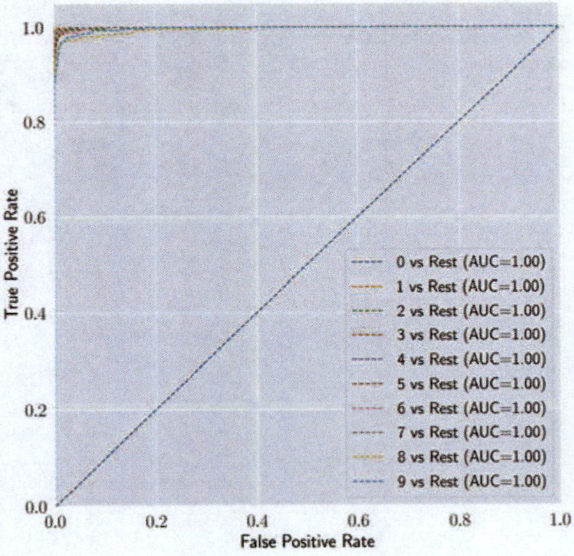

Fig. 16 ROC and AUC for QNN

On the other hand, in the test set, the QNN presents a higher latency and step execution time since quantum operations are performed at the simulation level and not in a quantum computer, leading to a lower throughput value than the CNN, as seen in Figs. 19 and 20.

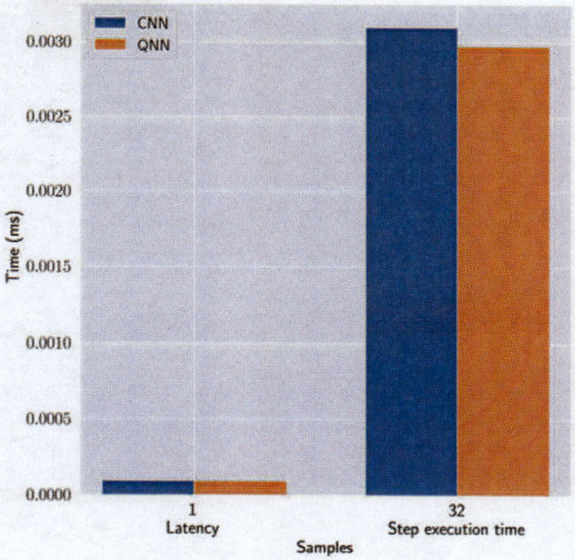

Fig. 17 Model time performance in training set

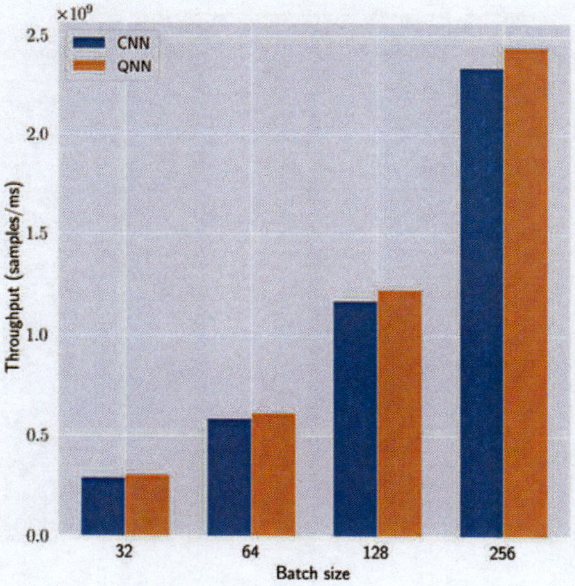

Fig. 18 Model sample performance in training set

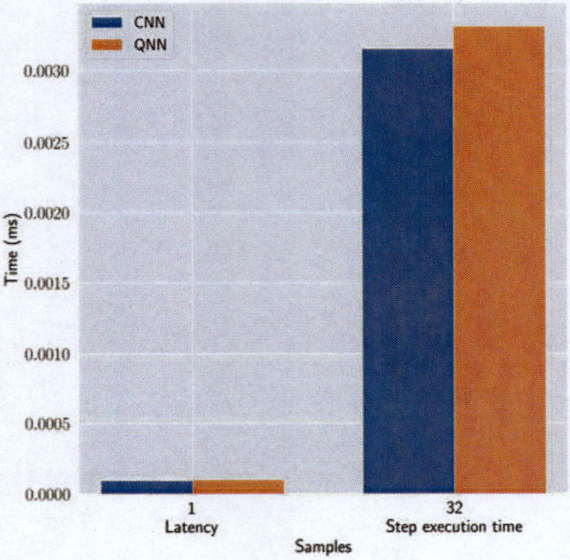

Fig. 19 Model time performance in test set

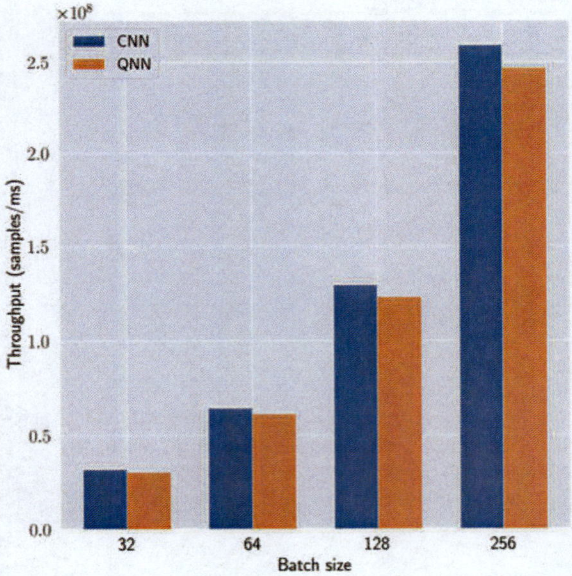

Fig. 20 Model sample performance in test set

6 Discussion

The attained results in the accuracy and loss curves show that the QNN performs almost exactly as the CNN, and even though it is slightly outperformed in most of the metrics of Tables 1 and 2, the fact that it is processing samples with 50% fewer data and performing quantum operations at simulation level illustrates the potential of the quanvolutional model. The confusion matrices, as well as the ROC curves, and AUC scores of Figs. 13, 14, 15, and 16 further demonstrate the nearly identical behavior of both models. It is in the time and sample performance plots where the potential benefits of the QNN lie, as shown in both Figs. 17, and 18, where the QNN shows better processing capabilities in terms of time and number of samples. This outcome shows that the QNN presents faster training time and consequently, should exhibit better sample processing capabilities. However, according to the latency and throughput presented in Figs. 19, and 20, the test set values of the QNN are actually worse than the CNN in both time and sample performance, showing that the training results do not align with the evaluation ones. This can be attributed to a number of factors, but mostly due to the input shape of the QNN including four expectation values measured at the output of the quanvolutional layer and delaying model prediction, as well as executing all quantum computations at simulation level through the PennyLane libraries instead of implementing the proposed QNN on a real quantum computer. Nevertheless, the results demonstrate a favorable tradeoff between input resolution, model accuracy, and performance in comparison to the CNN.

Table 3 shows a results comparison between some of the related works and the proposed methodology. As it can be seen, Henderson's work [8] shows results for only the accuracy metric; however, the evaluation of the model lies in the number of filters used for the quanvolutional layer, proportionally improving the attained results. The focus of the classification task is also important, as it determines the appropriate metrics to evaluate model performance. In the case of Huang's work [12], both binary and multiclass classification is done, evaluating the performance of the VCNN with the corresponding metrics of each task. For the multiclass classification task, only the samples from classes 0, 1, 2, 3, 4, and 6 of the MNIST dataset are included. Compared to the other works, and the proposed one, Zheng's [11] scale is significantly

Table 3 Related work results comparison

Work	Classification task	Model	*ACC*	*P*	*R*	*F1*	*AUC*	*L* (ms)	*THR* (32 samples/ms)
Henderson [8]	Multiclass	QNN	**0.97**	–	–	–	–	–	–
Huang [12]	Multiclass (0, 1, 2, 3, 4, 6)	VCNN	0.9249	–	0.9302	0.9301	–	–	–
Zheng [11]	Binary (3, 6)	QCNN	0.9665	–	–	–	–	–	–
Proposed	Multiclass	QNN	0.9652	0.9655	**0.9652**	**0.9650**	**1.00**	**0.0001**	**2.49×10^8**

smaller due to implementing the QCNN in real quantum circuits, evaluating its performance through different phases. The proposed QNN model attains better results in the multiclass classification task of the MNIST dataset, mainly due to the fact of the LeNet5 architecture, as it is designed for this specific classification problem. Furthermore, it evaluates the time and sample performance of the model, establishing another point of comparison between a classical CNN and a QNN.

7 Conclusion and Future Work

The implemented QNN performs at a similar level to the classical model, even though input image resolution is halved during quanvolutional preprocessing. The obtained OvR ROC curves, and their AUC scores for multiclass classification prove that implementing the correct model in a dataset attains good results, even when working with fewer data. Training results show signals of the QNN performing better than the CNN in time and sample performance. Regardless, test set evaluation proves otherwise, mostly due to the fundamental differences in input layer sizes between models, as well as the implementation of the proposed QNN at simulation level instead of a real quantum computer. Overall, the QNN offers a favorable tradeoff between input resolution, model accuracy, and time and sample performance compared to the CNN. For future work, the CNN's input image resolution could also be reduced, allowing a more objective comparison between models. In addition, performance can be evaluated for the bigger batch size stages, leading to a better latency and throughput comparison. Quanvolutional layers can be increased in future experiments, and they can be carried out on real quantum devices, adding more quantum aspects to the proposed QNN framework.

References

1. Dunjko, V., & Briegel, H. J. (2018). Machine learning & artificial intelligence in the quantum domain: A review of recent progress. *Reports on Progress in Physics, 81*, 074001. https://doi.org/10.1088/1361-6633/AAB406
2. Killoran, N., Bromley, T. R., Arrazola, J. M., Schuld, M., Quesada, N., & Lloyd, S. (2019). Continuous-variable quantum neural networks. *Physical Review Research, 1*(3), 033063. https://doi.org/10.1103/PHYSREVRESEARCH.1.033063/FIGURES/11/MEDIUM
3. Farhi, E., & Neven, H.: Classification with quantum neural networks on near term processors (2018). https://doi.org/10.48550/arxiv.1802.06002
4. Kak, S. C. (1995). Quantum neural computing. *Advances in Imaging and Electron Physics, 94*, 259–313. https://doi.org/10.1016/S1076-5670(08)70147-2
5. Gyongyosi, L., & Imre, S. (2019). A survey on quantum computing technology. *Computer Science Review, 31*, 51–71. https://doi.org/10.1016/J.COSREV.2018.11.002
6. Shor, P. W. (1994). Algorithms for quantum computation: Discrete logarithms and factoring. In *Proceedings - Annual IEEE Symposium on Foundations of Computer Science, FOCS* (pp. 124–134). https://doi.org/10.1109/SFCS.1994.365700

7. Biamonte, J., Wittek, P., Pancotti, N., Rebentrost, P., Wiebe, N., & Lloyd, S. (2017). Quantum machine learning. *Nature, 549*(7671), 195–202. https://doi.org/10.1038/nature23474
8. Henderson, M., Shakya, S., Pradhan, S., & Cook, T. (2020). Quanvolutional neural networks: Powering image recognition with quantum circuits. *Quantum Machine Intelligence, 2*. https://doi.org/10.1007/S42484-020-00012-Y
9. Amin, J., Sharif, M., Gul, N., Kadry, S., & Chakraborty, C. (2021). Quantum machine learning architecture for covid-19 classification based on synthetic data generation using conditional adversarial neural network. *Cognitive Computation*. https://doi.org/10.1007/S12559-021-09926-6
10. Houssein, E. H., Abohashima, Z., Elhoseny, M., & Mohamed, W. M. (2022). Hybrid quantum-classical convolutional neural network model for covid-19 prediction using chest x-ray images. *Journal of Computational Design and Engineering, 9*, 343–363. https://doi.org/10.1093/JCDE/QWAC003
11. Zheng, J., Gao, Q., Lü, J., Ogorzałek, M., Pan, Y., & Lü, Y. (2022). Design of a quantum convolutional neural network on quantum circuits. *Journal of the Franklin Institute*. https://doi.org/10.1016/J.JFRANKLIN.2022.07.033
12. Huang, F., Tan, X., Huang, R., & Xu, Q. (2022). Variational convolutional neural networks classifiers. *Physica A: Statistical Mechanics and its Applications, 605*, 128067. https://doi.org/10.1016/J.PHYSA.2022.128067
13. Lopez-Montiel, M., Orozco-Rosas, U., Sanchez-Adame, M., Picos, K., & Ross, O. H. M. (2021). Evaluation method of deep learning-based embedded systems for traffic sign detection. *IEEE Access, 9*, 101217–101238. https://doi.org/10.1109/ACCESS.2021.3097969
14. Ross, O. H. M. (2020). A review of quantum-inspired metaheuristics: Going from classical computers to real quantum computers. *IEEE Access, 8*, 814–838. https://doi.org/10.1109/ACCESS.2019.2962155
15. Light, G. L. (2021). Pauli matrices immersion. *Materials Science and Engineering: B, 264*, 114910. https://doi.org/10.1016/J.MSEB.2020.114910
16. Yann, L., Léon, B., Yoshua, B., & Patrick, H. (1998). Gradient-based learning applied to document recognition. *Proceedings of the IEEE, 86*, 2278–2323. https://doi.org/10.1109/5.726791
17. Boureau, Y.-L., Ponce, J., & Lecun, Y. (2010). A theoretical analysis of feature pooling in visual recognition. In *27th International Conference on Machine Learning Proceedings* (pp. 111–118). Haifa, Israel: ICML.
18. Grandini, M., Bagli, E., & Visani, G.: Metrics for multi-class classification: An overview (2020). https://doi.org/10.48550/arxiv.2008.05756
19. Davis, J., & Goadrich, M. (2006). The relationship between precision-recall and roc curves **06**, 233–240 (2006). https://doi.org/10.1145/1143844.1143874
20. Villardón, G., Zhao, X., Le, P. B., & Nguyen, Z. T. (2022). Roc curves, loss functions, and distorted probabilities in binary classification. https://doi.org/10.3390/math10091410
21. Yang, Z., Xu, Q., Bao, S., Cao, X., & Huang, Q. (2021). Learning with multiclass auc: Theory and algorithms. *IEEE Transactions on Pattern Analysis and Machine Intelligence*. https://doi.org/10.1109/TPAMI.2021.3101125
22. Liu, J., Liu, J., Du, W., & Li, D.: Performance analysis and characterization of training deep learning models on mobile devices. In *Proceedings of the International Conference on Parallel and Distributed Systems - ICPADS*, 2019-December (pp. 506–515). https://doi.org/10.48550/arxiv.1906.04278
23. Hanhirova, J., Kämäräinen, T., Seppälä, S., Siekkinen, M., Hirvisalo, V., & Ylä-Jääski, A. (2018). Latency and throughput characterization of convolutional neural networks for mobile computer vision. In *Proceedings of the 9th ACM Multimedia Systems Conference, MMSys 2018* (vol. 18, pp. 204–215). https://doi.org/10.48550/arxiv.1803.09492
24. Kingma, D. P., & Ba, J. L. (2014). Adam: A method for stochastic optimization. In *3rd International Conference on Learning Representations, ICLR 2015 - Conference Track Proceedings*. https://doi.org/10.48550/arxiv.1412.6980

Traffic Sign Recognition Using Fuzzy Preprocessing and Deep Neural Networks

Cesar Torres, Claudia I. Gonzalez, and Gabriela E. Martinez

1 Introduction

Deep Neural Networks (DNNs) have proven valuable in many fields due to their feature extraction capabilities; enabling us to execute automation jobs on subjects we thought would be impossible. Computer vision is a branch of Artificial Intelligence (AI) that enables machines to carry out vision-oriented activities, such as object recognition or localization in digital images that were previously only possible by humans.

The application of intelligent vision-based systems with domain-specific knowledge has spiked in recent years, allowing the industry to create systems that benefit us in key aspects of society. Buyukarikan et al. [1] proposed a vision-based system utilizing a DNN that identifies damages in fruits, specifically apples, utilizing raw images for the system, the damage could be caused by external factors, preventing them from being sent to markets, preventing economic loss, and obtaining favorable results in their predictions. Ramesh et al. [2] proposed a model that localizes and classifies breast cancer from mammogram images, the modification of a GoogLeNet to perform a semantic segmentation, obtaining a 99.12% of segmentation accuracy, aiming to help medical professionals to locate, and examine more efficiently breast cancer.

The automotive area has specifically benefited from this by developing state-of-the-art deep learning models [3] to implement autonomous and/or assisted navigation

C. Torres · C. I. Gonzalez (✉) · G. E. Martinez
TECNM/Tijuana Institute of Technology, Tijuana, Mexico
e-mail: cgonzalez@tectijuana.mx

C. Torres
e-mail: m21210008@tectijuana.edu.mx

G. E. Martinez
e-mail: gmartinez@tectijuana.mx

O. Castillo and P. Melin (eds.), *Hybrid Intelligent Systems Based on Extensions of Fuzzy Logic, Neural Networks and Metaheuristics*, Studies in Computational Intelligence 1096,
https://doi.org/10.1007/978-3-031-28999-6_5

systems, which are integrated into the new generation of vehicles to complement or replace the ability to drive.

Current implementations of self-driving technologies consist in the divide and conquer methodology where the task is decomposed in many areas, like localization, classification, mapping, prediction, planning between many others [4]. In the case of detection tasks, innovations like VGG [5] or the ResNet [6] models have proven to be a cornerstone in the development of those systems due to the design with millions of trainable parameters with their considerable number of hidden layers and information that the models are exposed to when training, with databases such as ILSVRC (ImageNet) [7], CIFAR-100 [8], or PASCAL VOC [9] that contain from hundreds to thousands of categories with millions of images to train and test the models.

Previous authors tackled the traffic sign classification with classical computer vision strategies, that include color segmentation or morphologic segmentation [10–12], due to the other approaches include domain transfer learning or the utilization of big convolutional neural networks to solve the task. The work of Bi et al. [13], utilizes a modification in the VGG16 architecture, reducing the number or convolution layers, as well as de introduction of two new layers, to reduce the computational cost of the model without reducing the accuracy. The proposed model achieves an improvement of accuracy when compared to the state-of-the-art model, still utilizing a considerable amount of computation due to the size of the blocks of the VGG16 model.

Khan [12] proposed an optimization to the YOLOv3 framework work with small-sized traffic signs, their proposition includes a pre-processing layer that incudes grid optimization and anchor box optimization to the detection module of YOLO. They reported an improvement of the mean average precision (MAP) of almost 24% with the proposed system.

The utilization of pre-processing techniques for image enhancement is an important step in multiple classification/detection pipelines, the utilization of filters to highlight features is an important tool that can improve the accuracy of a model significantly. The use of edge detection filters is an essential piece in intelligent systems due to the capabilities of reducing the information necessary to analyze an image. The combination of edge detection filters with fuzzy logic is a common practice to reduce uncertainty that could be present in the information. In Versaci et al. [14] proposed an approach to produce fuzzy edge detection by performing a noise reduction utilizing a S type membership function, maximizing entropy of the image. The method was tested in EC maps, thermal IR images and ES images, the results obtained with the proposed methodology deliver a comparable result with popular edge detection filters like Canny with Otsu thresholding. Melin and Gonzalez [15, 16] worked on a generalized type-2 fuzzy edge detection methodology that to improve the result of edge detection methodologies, testing in benchmark images. The results obtained were compared against their classical counterpart, type-1, interval type-2 and generalized type-2, achieving a reduction in the noise present in the image after the edge detection.

For this paper we focused only on traffic sign classification (TSC), a challenging problem due to its nature of the information, where images can be distorted due to the perspective from which they are captured as well as the elements that could be present in the background of the traffic sign. To mitigate the uncertainty present in the information, we proposed the utilization of multiple fuzzy edge detection methodology based on a type-1 fuzzy inference system, fuzzy Prewitt, fuzzy Sobel, and fuzzy morphological gradient edge detectors. We combined the utilization of the preprocessing layer with four different convolutional neural networks to perform the classification task. The filters were applied to multiple publicly available databases.

2 Background

2.1 Fuzzy Logic

Fuzzy logic is a branch of soft computing that aims to create non-linear representations, similar to human reasoning. Fuzzy logic was proposed by Lotfi Zadeh in the 1970's [17] first, with the introduction of fuzzy sets, due to the inconformity of traditional sets that only offer binary values (crisp sets), and then was followed by fuzzy reasoning. Fuzzy sets allow for uncertainty, utilizing linguistic terms, differ from traditional logic by offering a gradual representation in a continuous space, allowing their elements to be part of multiple classes with different levels of membership. The range of these sets is usually defined by experts, depending on the concept being applied.

A fuzzy set can be defined with a Universe X with range of [0,1] with a continuous function $\mu_A : X \to [0, 1]$. The range can be defined with Eq. 1.

$$A = \{(x, \mu_A(x))|x \in X\} \tag{1}$$

Membership functions (MFs) are utilized to create representations of traditional sets into fuzzy sets. Popular membership functions like the one demonstrated in Fig. 1. may include Gaussian MF, Triangular MF, Trapezoidal MF, between many others. In this case we will focus into the Gaussian MF, denoted by Eq. 2.

$$gaussian(x; c, \sigma) = exp - \frac{1}{2}\left(\frac{x - c}{\sigma}\right)^2 \tag{2}$$

Fuzzy control is performed by Fuzzy Inference Systems (FIS), that are typically based in If–Then rules [18] with conditionals and conclusions that are based on the knowledge of experts [19]. Popular fuzzy inference systems are Mandani FIS or Takagi–Sugeno-Kang FIS [20].

Fig. 1 Gaussian membership function example

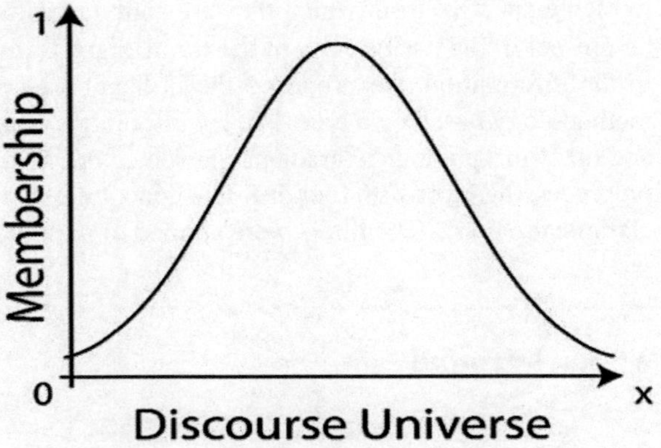

2.2 Data Preprocessing

Data preprocessing is fundamental to improving the performance of intelligent systems. Preprocessing allows us to manipulate the input data by highlighting or hiding features in images [21, 22]. Noise removal, lighting improvement, or edge detection are common techniques applied to the images to standardize the information.

Preprocessing techniques are commonly used by applying a small filter to the image that multiplies its weights to generate the value of the new (processed) image. Edge detection filters are a good example of this; they are high-pass filters that find high-contrast areas, extracting only the fine details, allowing us to reduce the image to its minimum representation, thus allowing us to reduce the computational power needed to process the information with intelligent systems.

Popular edge detection filters are Prewitt, Sobel, Morphological Gradient, among many others [23]. These types of filters are based on the first and second derivatives; each of these filters is calculated by performing a convolution with a special filter to extract the information. To implement gradient-based edge detectors like Prewitt Eq. 4 or Sobel Eq. 5, we need to convolve the respective kernel for gradient X and Y with Grad_x and Grad_y respectively, utilizing the convolution operation defined in Eq. 3.

$$\mathrm{Grad}_x = \sum_{i=1}^{i=3}\sum_{j=1}^{j=3} \mathrm{kernelx}_{I,j} * f_{x+i-2,y+j-2}$$

$$\mathrm{Grad}_y = \sum_{i=1}^{i=3}\sum_{j=1}^{j=3} \mathrm{kernely}_{i,j} * f_{x+i-2,y+j-2} \tag{3}$$

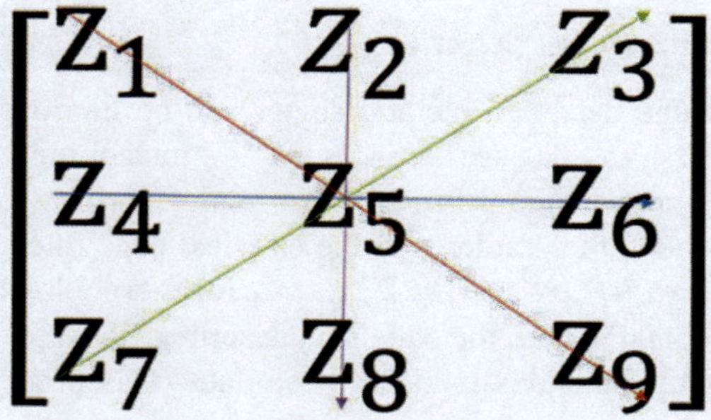

Fig. 2 Morphological gradient directions

$$\text{Prewittx} = \begin{bmatrix} -1 & -1 & -1 \\ 0 & 0 & 0 \\ 1 & 1 & 1 \end{bmatrix}, \text{Prewitty} = \begin{bmatrix} -1 & 0 & 1 \\ -1 & 0 & 1 \\ -1 & 0 & 1 \end{bmatrix} \tag{4}$$

$$\text{Sobelx} = \begin{bmatrix} -1 & -2 & -1 \\ 0 & 0 & 0 \\ 1 & 2 & 1 \end{bmatrix}, Sobely = \begin{bmatrix} -1 & 0 & 1 \\ -2 & 0 & 2 \\ -1 & 0 & 1 \end{bmatrix} \tag{5}$$

Edge detection by morphological gradient differs from classical edge detections like Sobel or Prewitt, where it exists a defined kernel. In this case, we calculate the gradients denoted by G_1, G_2, G_3, G_4 with the directions specified in Fig. 2 ($0^{\circ}, 45^{\circ}, 90^{\circ}, \text{and} 135^{\circ}$), the gradients are calculated with a sliding window applying Eq. 7 with the coordinates defined in Eq. 6, for each gradient, once all gradients are calculated, we perform a sum of all gradients with Eq. 8 to obtain the edges.

$$z_1 = f(x-1, y-1), z_2 = f(x, y-1), z_3 = f(x+1, y-1)$$

$$z_4 = f(x-1, y), z_5 = f(x, y), z_6 = f(x+1, y) \tag{6}$$

$$z_7 = f(x-1, y+1), z_8 = f(x, y+1), z_9 = f(x+1, y+1)$$

$$G_1 = \sqrt{(z_5 - z_2)^2 + (z_5 - z_8)^2}, G_2 = \sqrt{(z_5 - z_4)^2 + (z_5 - z_6)^2}$$

$$G_3 = \sqrt{(z_5 - z_1)^2 + (z_5 - z_9)^2}, G_4 = \sqrt{(z_5 - z_3)^2 + (z_5 - z_7)^2} \tag{7}$$

$$\text{Edges} = G_1 + G_2 + G_3 + G_4 \tag{8}$$

Fuzzy edge detection

Edge detection methodologies can be improved by applying fuzzy logic to them [24]; in this case we included a Mamdani type 1 inference system to handle uncertainty in the information. To implement fuzzy edge detection, first is necessary to obtain the gradients of the classical edge filters x and y for Sobel and Prewitt, and $Grad_1$, $Grad_2$, $Grad_3$, $Grad_4$ for morphological gradient. Once the gradients are obtained with the equations described above, it is necessary to implement membership functions to fuzzify the data. This process can be carried out utilizing any membership function defining ranges according to our information. The next step to calculate the fuzzy edges is to utilize the fuzzy data with a fuzzy controller and the rules defined by an expert; this rule set can vary by case bases, and is defined depending on the necessity of the problem.

2.3 *Convolutional Neural Networks*

Convolutional neural networks (CNNs) are a soft computing technique derivate from the classical neural networks (NNs) based on LeCun's [25] works that simulates the brain responses to visual stimuli captured by the visual cortex. In recent years, the utilization of convolutional neural networks has proven to be effective in many fields, like the automotive industry with the implementation of autopilots and driving assistance systems [26–28] or the medical field, with diagnostic and prevention tools [29–31].

CNNs are structured similar to classical NN, with multiple layers to learn generalizations of the training data [32]. The most basic structure of a CNN shown in Fig. 3, contain the following layers:

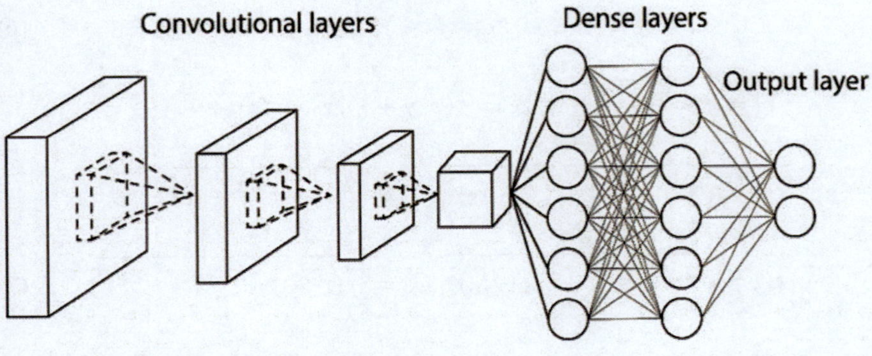

Fig. 3 Example of a basic CNN

- Input layer: this constitutes the input of the network and requires the input to be a fixed size. Typically, it is followed by a convolution layer, but it can be followed by a normalization layer or preprocessing layer before the convolution filters.
- Convolution layer: it contains the random, trainable filters (kernels) applied to the data to perform feature extraction. The filtration process produces a scalar product that generates a new matrix with the extracted content. It can be followed by either more convolution layers to extract more features or a pooling layer to reduce the spatial features.
- Activation: the activation functions are often taken as part of the architecture of a CNN. These functions are thresholds that trigger the neuron when the limit is reached. It allows the network to learn more complex patterns that are passed to the following layers. ReLu and Softmax are popular activation functions utilized in the networks.
- Pooling layer: it is a down sampling operation that uses a sliding filter, similar to the convolution layer, to reduce the tensor size [33]. Pooling layers that are popular include maximum pooling, average pooling, and minimum pooling. In the case of the max pooling, we get the maximum value of the sliding windows, this operation can be replaced to the minimum or the average of the window, depending on the layer selected. The result is stored as the output, creating a new, smaller representation corresponding to the size of the filter.
- Dense layer: contains neurons, like traditional NNs, with weights and biases, the input of this network is the output of the convolution layers, transformed into a single dimension tensor.
- Output layer: the neurons of this layer are responsible of producing the output of the model; it has as many neurons as possible outputs in the model.

3 Methodology

3.1 Proposed System

The proposed methodology consists of the application of different edge detection techniques to traffic sign datasets, both classical and fuzzy, to determine their effects when performing a classification task utilizing a convolutional neural network. The data flow consists in the following steps that can be seen in Fig. 4:

1. Reading the database
2. Applying edge detection techniques (classical or fuzzy edge detection)
3. Perform feature extraction utilizing the convolution layers in the models.
4. Utilizing the dense layers to perform classification of each image.

To test the CNN architectures, we proposed the utilization of multiple images preprocessing techniques. The filters applied to the images were the following:

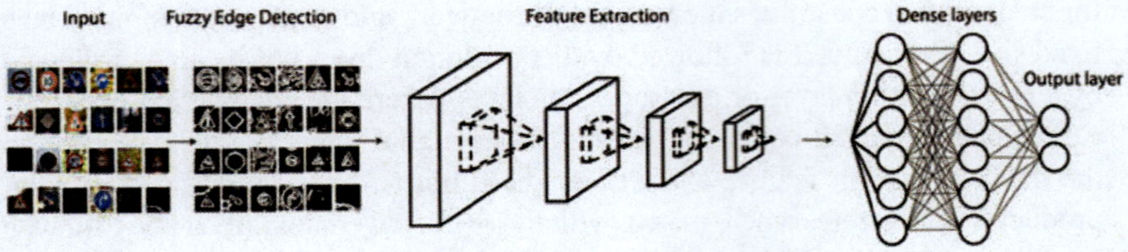

Fig. 4 Proposed approach to perform traffic sign classification

- Color images (only rescaled).
- Grayscale.
- Sobel.
- MG.
- Prewitt.
- Fuzzy MG.
- Fuzzy Sobel.
- Fuzz Sobel Color (Fuzzy Sobel applied to RGB channels).
- Fuzzy Prewitt.

3.2 *Databases*

For this study case, we experimented with three different public databases to test our methodology, GTSRB [34], BelgiumTS [35], and CTSD [36]. Each database contains multiple classes, with images of the signs from their country of origin.

GTSRB

This database contains forty-two different traffic signs from Germany; these were extracted from video frames taken by the Institut Für NeuroInformatik. Table 1 contains the database details, and Fig. 5 illustrates a sample of the images.

BelgiumTS

Similar to the GTSRB, this database contains sixty-two different traffic signs from Belgium. The parameters of this database are shown in Table 2; a sample of the database is illustrated in Fig. 6.

Table 1 GTSRB database description

Parameter	Description
Total images	51,839
Training images	39,209
Test images	12,630
Dimensions	28 px × 28 px
Database format	PPM

Fig. 5 Sample of GTSRB database

Table 2 BelgiumTS database description

Parameter	Description
Total images	7,095
Training images	4,575
Test images	2,520
Dimensions	32 px × 32 px
Database format	PPM

CTSD

It is a small database of Chinese traffic sign with fifty-eight different classes of traffic signs, also segmented from video frames. Figure 7 contains a sample of the database, and Table 3 presents the general parameters of the data.

3.3 Image Preprocessing

We performed edge detection techniques, both classical and fuzzy, to the databases defined in the previous section. To implement the fuzzy edge detection, we first calculated the gradients described in Sect. 2.2.

Fig. 6 Sample of BelgiumTS database

Fig. 7 Sample of CTSD database

Fuzzy Sobel and Fuzzy Prewitt edge detection

In the case of Fuzzy Prewitt and Fuzzy Sobel, we implemented a Mamdani FIS with two inputs, one output and three fuzzy rules. We begin by reading each one of the images from the databases and calculating the gradients (Sobel X, Sobel Y, Prewitt X, and Prewitt Y). The next step is to label the inputs with the linguistic

Table 3 CTSD database description

Parameter	Description
Total images	6,164
Training images	4,170
Test images	1,994
Dimensions	64 px × 64 px
Database format	PPM

variables *Dhorizontal* for X-axis and *Dvertical* for the Y-axis; the inputs stand for a Gaussian MFs which are illustrated in Fig. 8. The *Dhorizontal* input is granulated in three MFs defined as "LowDhorizontal", "MiddleDhorizontal" and "HighDhorizontal". The *Dvertical* input also consists of three MFs labeled "LowDvertical", "MiddleDvertical" and "HighDvertical". The parameters for each axis are calculated with Eqs. (2)–(9), we also obtained the value of σ by using Eq. 13.

$$lowDhorizontal = min(Dhorizontal), lowDvertical = min(Dvertical) \quad (9)$$

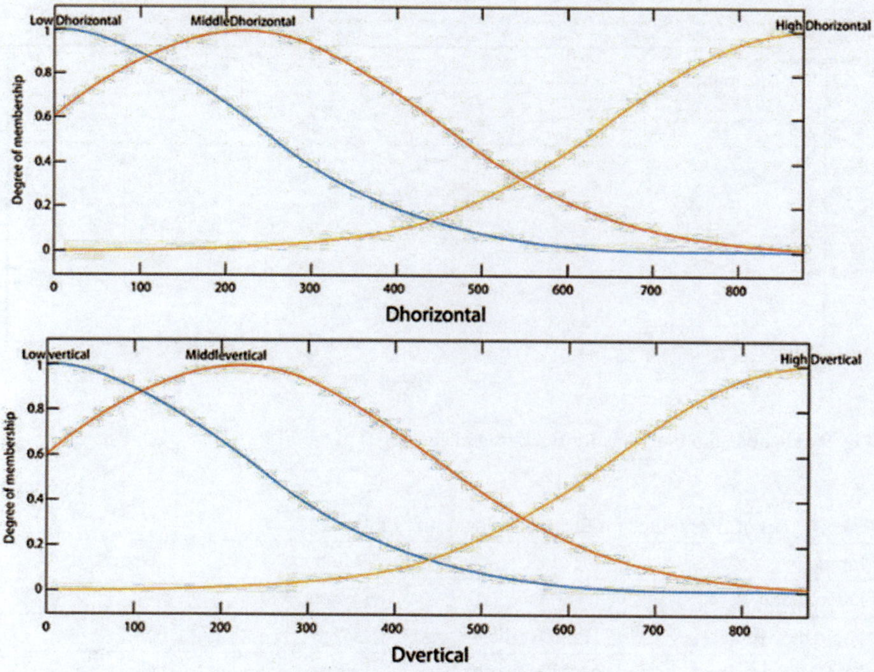

Fig. 8 Membership functions for the input gradients

$$highDhorizontal = max(Dhorizontal), highDvertical = max(Dvertical) \tag{10}$$

$$MiddleDhorizontal = (lowDhorizontal + highDhorizontal)/4 \tag{11}$$

$$middleDvertical = (lowDvertical + highDvertical)/4 \tag{12}$$

$$\sigma Dhorizontal = HighDhorizontal/4, \sigma Dvertical = HighDvertical/4 \tag{13}$$

The FIS produces one output denominated "Edges" with two linguistic variables "Background" and "Edge"; we limited the range between −5, 4.5. cBackground with a center value of = −5 and cEdge = 4.5, Fig. 9 has the visual representation of the output variables. The value of σ for both MF is calculated by Eq. 14. The fuzzy rules were defined by previous experimentation; Table 4 contains the rule set utilized for this study case.

$$\sigma output = abs(cBackground - cEdge)/2 \tag{14}$$

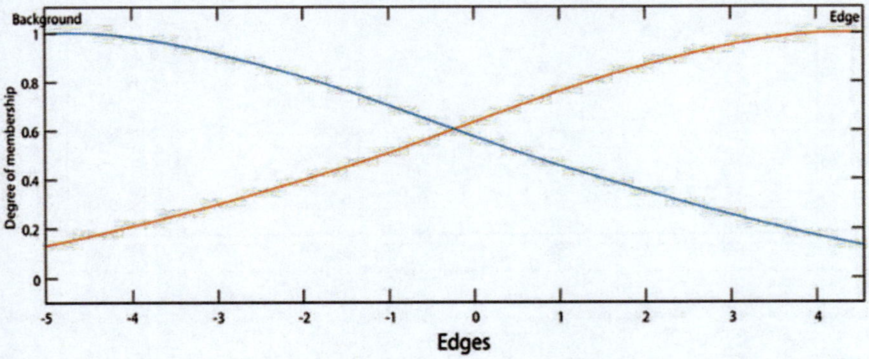

Fig. 9 Membership functions for Background and edges of the FIS

Table 4 Set of fuzzy rules for fuzzy Sobel and fuzzy Prewitt

Inputs		Outputs	Operator
Dhorizontal	Dvertical	Edges	
HighDhorizontal	HighDvertical	Edge	Or
MiddleDhorizontal	MiddleDvertical	Edge	Or
LowDhorizontal	LowDvertical	Background	And

Fuzzy Morphological Gradient edge detection

Like the fuzzy edge detectors defined above, to calculate the fuzzy MG, it is necessary first obtain the four gradients ($Grad_1, Grad_2, Grad_3, Grad_4$), then calculate the values for the linguistic variables Low, Middle, and High for each gradient with Eqs. (15)–(22) and finally obtain the values of σ with Eqs. (23)–(24). The representation of the fuzzy MFs is illustrated in Fig. 10.

$$lowGrad1 = \min(Grad1), lowGrad2 = \min(Grad2) \tag{15}$$

$$lowGrad3 = min(Grad3), lowGrad4 = min(Grad4) \tag{16}$$

$$highGrad1 = \max(Grad1), highGrad2 = \max(Grad2) \tag{17}$$

$$highGrad3 = max(Grad3), highG4 = max(Grad4) \tag{18}$$

$$mediumG1 = (lowGrad1 + highGrad1)/2 \tag{19}$$

$$mediumGrad2 = (lowGrad2 + highGrad2)/2 \tag{20}$$

$$mediumG3 = (lowGrad3 + highGrad3)/2 \tag{21}$$

$$mediumGrad4 = (lowGrad4 + highGrad4)/2 \tag{22}$$

$$\sigma Grad1 = highGrad1/4, \sigma Grad2 = highGrad2/4 \tag{23}$$

$$\sigma Grad3 = highGrad3/4, \sigma Grad4 = highGrad4/4 \tag{24}$$

The MG fuzzy edge detector is a Mamdani FIS, with four inputs (one for each gradient) and one output. Furthermore, the output is divided into two linguistic variables "Background" and "Edge". The outputs are calculated with the same functions as fuzzy Prewitt and fuzzy Sobel with Eq. 14. The fuzzy rules are defined in a similar way, this time taking into consideration the four gradients. Table 5 contains the fuzzy rules utilized in the MG Mamdani FIS.

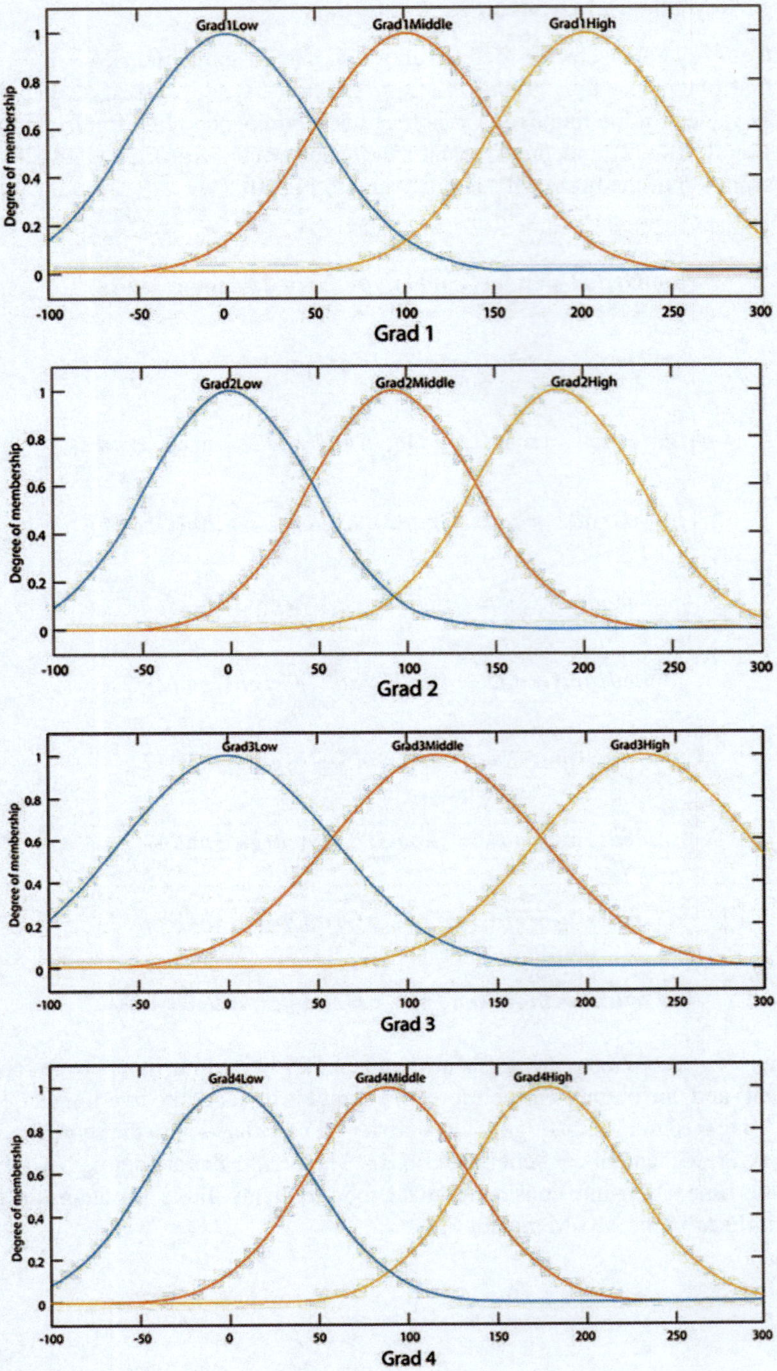

Fig. 10 Membership functions for Morphological Gradient

Table 5 Set of fuzzy rules for fuzzy morphological gradient edge detector

Inputs				Output	Operator
Grad1	Grad2	Grad3	Grad4	Edges	
HighGrad1	HighGrad2	HighGrad3	HighGrad4	Edge	or
MiddleGrad1	MiddleGrad2	MiddleGrad3	MiddleGrad4	Edge	or
LowGrad1	LowGrad2	LowGrad3	LowGrad4	Background	and

3.4 Architecture Definitions

To examine the impact of using fuzzy preprocessing and to observe if the implementation of a preprocessing techniques with the traffic signs results in a meaningful improvement in prediction accuracy with small models, we suggested four CNN models. The architectural definitions of the suggested models are provided in Tables 6, 7, 8 and 9.

Table 6 Architecture CNN-I

Layer/type	Neurons/filters	Filter size
1. Input convolution with ReLu	64	3 × 3
2. Max pooling	n/a	2 × 2
3. Convolution with ReLu	64	3 × 3
4. Max Pooling	n/a	2 × 2
5. Convolution with ReLu	128	3 × 3
6. Max Pooling	n/a	2 × 2
7. Convolution with ReLu	128	3 × 3
8. Max Pooling	n/a	2 × 2
9. Dropout	n/a	0.75
10. Flattening	n/a	n/a
11. Fully Connected	512	n/a
12. Output	N-outputs	n/a

Table 7 Architecture CNN-II

Layer/type	Neurons/filters	Filter size
1. Input Convolution with ReLu	64	3 × 3
2. Max Pooling	n/a	2 × 2
3. Convolution with ReLu	128	3 × 3
4. Max Pooling	n/a	2 × 2
5. Dropout	n/a	0.75
6. Flattening	n/a	n/a
7. Fully Connected	512	n/a
8. Output	N-outputs	n/a

Table 8 Architecture CNN-III

Layer/type	Neurons/filters	Filter size
1. Input convolution with ReLu	64	3 × 3
2. Max pooling	n/a	2 × 2
3. Convolution with ReLu	64	3 × 3
4. Max pooling	n/a	2 × 2
5. Convolution with ReLu	128	3 × 3
6. Max pooling	n/a	2 × 2
7. Convolution with ReLu	128	3 × 3
8. Max pooling	n/a	2 × 2
9. Dropout	n/a	0.45
10. Flattening	n/a	n/a
11. Fully connected	1024	n/a
12. Fully connected	512	n/a
13. Output	N-outputs	n/a

Table 9 Architecture CNN-IV

Layer/type	Neurons/filters	Filter size
1. Input convolution with ReLu	32	3 × 3
2. Max pooling	n/a	2 × 2
3. Convolution with ReLu	64	3 × 3
4. Max pooling	n/a	2 × 2
5. Convolution with ReLu	128	3 × 3
5. Max pooling	n/a	2 × 2
6. Dropout	n/a	0.35
7. Flattening	n/a	n/a
8. Fully connected	256	n/a
9. Output	N-outputs	n/a

The training parameters for the architectures were determined by experimentation. We performed 30 independent experiments with multiple edge detection techniques, both classical and Fuzzy implementations.

- Number of Epochs: 100.
- Loss Function: Sparce-Categorical-Cross-Entropy.
- Training algorithm: Adam with default Keras parameters (Epsilon: $1e^{-7}$, Beta 1: 0.9, Beta 2: 0.999, , Learning rate: 0.0001).
- Batch Size: variable between each dataset (256 for GTSRB and 128 for BelgiumTS and CTSD).
- Number of experiments: 30 experiments.

Table 10 Results architecture CNN-I with GTSRB database

Preprocessing	Minimum	Maximum	Average	Std. Dev.
Color	0.9209	0.9739	0.9650	0.0102
Grayscale	0.9419	0.9738	0.9656	0.0080
Sobel	0.9356	0.9544	0.9460	0.0044
MG	0.6732	0.6948	0.6831	0.0054
Prewitt	0.9458	0.9577	0.9521	0.0038
Fuzzy MG	0.7253	0.7476	0.7337	0.0043
Fuzzy Sobel	0.8707	0.8818	0.8768	0.0029
Fuzz. Sob. Co	0.8114	0.8304	0.8209	0.0046
Fuzzy Prewitt	0.8709	0.8837	0.8773	0.0029

4 Results

The experiments were performed using the Python programming language. The TensorFlow framework and the Keras API are implemented in the generation and training of the models to analyze the impact of various image preprocessing techniques. The experiments were performed independently of each other for each of the previously defined filters. The experiments were carried out on a machine with the following specifications: an Intel core i7-11800H processor, 32 GB of DDR4 3200 MHz RAM, and an Nvidia RTX 3080 Laptop edition GPU with 16 GB of video memory running Ubuntu 20.04 LTS. Tables 10, 11, 12, 13, 14, 15, 16, 17, 18, 19, 20 and 21 show the training results for each architecture; the information is summarized and provides only the minimum, maximum, mean, and standard deviation of the 30 trials.

4.1 Results GTSRB Database

See (Tables 10, 11, 12, 13).

4.2 Results BelgiumTS Database

See (Tables 14, 15, 16, 17).

Table 11 Results architecture CNN-II with GTSRB database

Preprocessing	Minimum	Maximum	Average	Std. Dev.
Color	0.9391	0.9558	0.9495	0.0048
Grayscale	0.9458	0.9614	0.9565	0.0035
Sobel	0.9338	0.9432	0.9393	0.0024
MG	0.6732	0.6948	0.6831	0.0054
Prewitt	0.9353	0.9464	0.9412	0.0029
Fuzzy MG	0.7304	0.7427	0.7356	0.0028
Fuzzy Sobel	0.8584	0.8743	0.8675	0.0036
Fuzz. Sob. Co	0.8021	0.8162	0.8082	0.0031
Fuzzy Prewitt	0.8709	0.8837	0.8773	0.0029

Table 12 Results architecture CNN-III with GTSRB database

Preprocessing	Minimum	Maximum	Average	Std. Dev.
Color	0.9470	0.9723	0.9626	0.0060
Grayscale	0.9457	0.9705	0.9593	0.0065
Sobel	0.9285	0.9488	0.9413	0.0046
MG	0.6732	0.6948	0.6831	0.0054
Prewitt	0.9367	0.9539	0.9469	0.0038
Fuzzy MG	0.7209	0.7426	0.7337	0.0048
Fuzzy Sobel	0.8646	0.8802	0.8723	0.0034
Fuzz. Sob. Co	0.8049	0.8238	0.8180	0.0046
Fuzzy Prewitt	0.8709	0.8837	0.8773	0.0029

Table 13 Results architecture CNN-IV with GTSRB database

Preprocessing	Minimum	Maximum	Average	Std. Dev.
Color	0.9462	0.9689	0.9603	0.0059
Grayscale	0.9495	0.9671	0.9585	0.0046
Sobel	0.9317	0.9482	0.9401	0.0037
MG	0.6732	0.6948	0.6831	0.0054
Prewitt	0.9315	0.9512	0.9424	0.0041
Fuzzy MG	0.7253	0.7476	0.7337	0.0043
Fuzzy Sobel	0.8707	0.8818	0.8768	0.0029
Fuzz. Sob. Co	0.8114	0.8304	0.8209	0.0046
Fuzzy Prewitt	0.8709	0.8837	0.8773	0.0029

Table 14 Results architecture CNN-I with BelgiumTS database

Preprocessing	Minimum	Maximum	Average	Std. Dev.
Color	0.8905	0.9750	0.9574	0.0163
Grayscale	0.9230	0.9746	0.9567	0.0108
Sobel	0.9266	0.9532	0.9391	0.0067
MG	0.7056	0.8929	0.8598	0.0328
Prewitt	0.9202	0.9679	0.9521	0.0107
Fuzzy MG	0.8468	0.8742	0.8592	0.0075
Fuzzy Sobel	0.9135	0.9492	0.9375	0.0077
Fuzz. Sob. Co	0.8730	0.9290	0.9002	0.0109
Fuzzy Prewitt	0.9048	0.9532	0.9339	0.0106

Table 15 Results architecture CNN-II with BelgiumTS database

Preprocessing	Minimum	Maximum	Average	Std. Dev.
Color	0.9032	0.9639	0.9394	0.0128
Grayscale	0.9135	0.9456	0.9319	0.0081
Sobel	0.9000	0.9321	0.9192	0.0078
MG	0.7567	0.8802	0.8487	0.0235
Prewitt	0.9052	0.9619	0.9393	0.0128
Fuzzy MG	0.8377	0.8746	0.8538	0.0068
Fuzzy Sobel	0.9060	0.9369	0.9228	0.0083
Fuzz. Sob. Co	0.8746	0.8988	0.8885	0.0064
Fuzzy Prewitt	0.9091	0.9397	0.9261	0.0068

Table 16 Results architecture CNN-III with BelgiumTS database

Preprocessing	Minimum	Maximum	Average	Std. Dev.
Color	0.9119	0.9702	0.9522	0.0147
Grayscale	0.8873	0.9639	0.9396	0.0186
Sobel	0.8901	0.9472	0.9251	0.0133
MG	0.7754	0.8734	0.8401	0.0241
Prewitt	0.9012	0.9611	0.9391	0.0158
Fuzzy MG	0.7718	0.8563	0.8351	0.0162
Fuzzy Sobel	0.8770	0.9464	0.9198	0.0138
Fuzz. Sob. Co	0.8508	0.9147	0.8877	0.0128
Fuzzy Prewitt	0.8813	0.9429	0.9245	0.0150

Table 17 Results architecture CNN-IV with BelgiumTS database

Preprocessing	Minimum	Maximum	Average	Std. Dev.
Color	0.9159	0.9651	0.9453	0.0135
Grayscale	0.9028	0.9583	0.9341	0.0159
Sobel	0.8587	0.9421	0.9207	0.0180
MG	0.7302	0.8714	0.8344	0.0297
Prewitt	0.8869	0.9488	0.9306	0.0160
Fuzzy MG	0.8103	0.8571	0.8330	0.0111
Fuzzy Sobel	0.8921	0.9302	0.9178	0.0089
Fuzz. Sob. Co	0.8373	0.8956	0.8731	0.0140
Fuzzy Prewitt	0.8956	0.9433	0.9144	0.0121

4.3 Results CTSD Database

We also included a visual representation of the confusion matrix that was created using the best model for each database. The representation is shown in Figs. 11, 12 and 13, where we only included the following preprocessing: color images, fuzzy MG, fuzzy Prewitt, and fuzzy Sobel (Tables 18, 19, 20, 21).

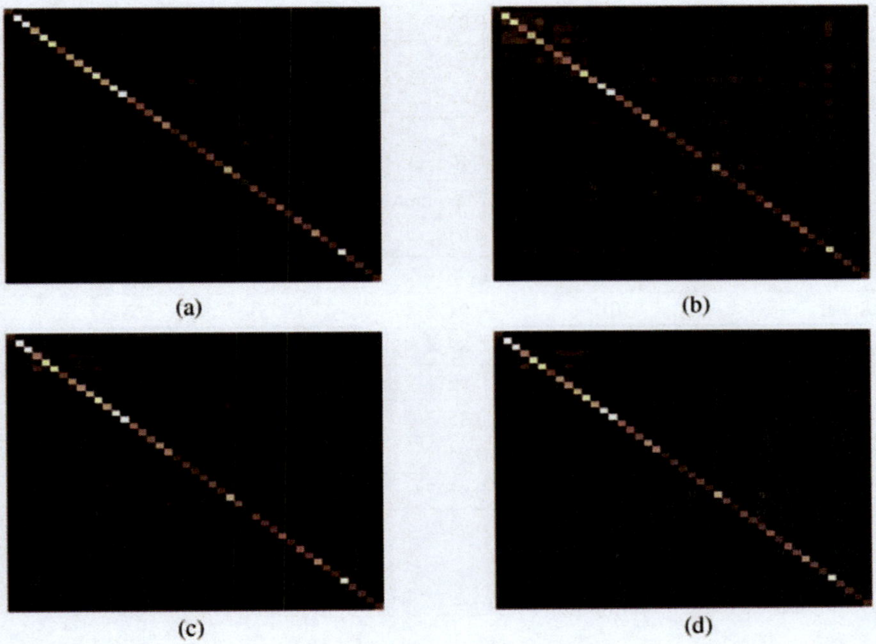

Fig. 11 Confusion matrix GTSRB Database. **a** Color Images, **b** Fuzzy MG, **c** Fuzzy Prewitt, **d** Fuzzy Sobel

(a)

(b)

(c)

(d)

Fig. 12 Confusion matrix BelgiumTS Database. **a** Color Images, **b** Fuzzy MG, **c** Fuzzy Prewitt, **d** Fuzzy Sobel

4.4 Statistical Test

We performed a statistical test to determine the existence of significant evidence between the use of the RGB images and a preprocessing technique. The preprocessing technique evaluated was selected by the highest average obtained. The statistical test is a Z-test with the following parameters.

- Right-tailed test.
- $\alpha = 0.05$ (Level of confidence 95%, rejection zone $z\_c = 1.96$).
- $n = 30$.
- H_0: The utilization of a preprocessing technique (μ_1) offers less or equal precision than utilizing the Color images. (μ_2). $\boldsymbol{H_0}$: $\boldsymbol{\mu_1 \leq \mu_2}$.
- H_a: The utilization of a preprocessing technique (μ_1) offers more precision than the utilization of Color images (μ_2). $\boldsymbol{H_a}$: $\boldsymbol{\mu_1 > \mu_2}$ **(Affirmation)**.

Tables 22, 23, 24 reflect the findings of the statistical tests, for the GTSRB, BelgiumTS and CTSD databases, respectively.

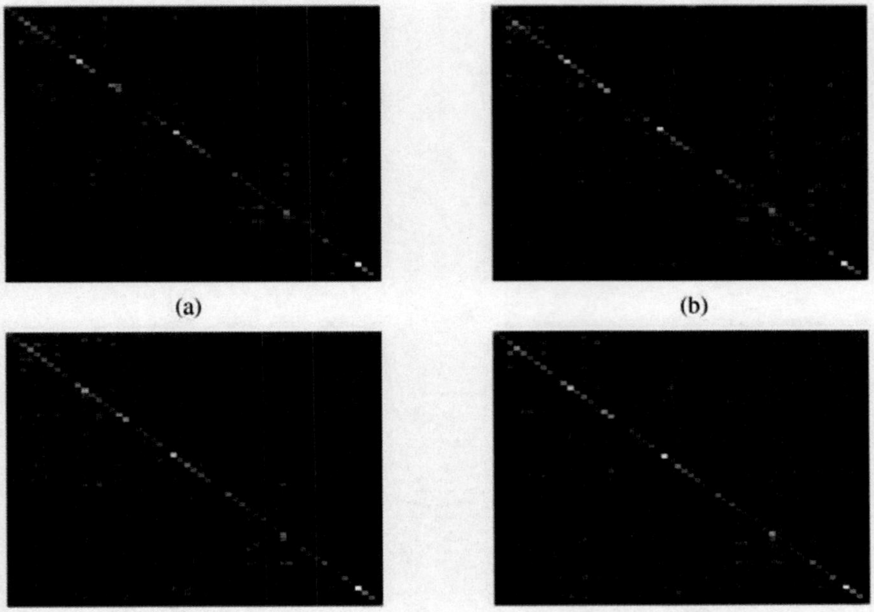

(a) (b) (c) (d)

Fig. 13 Confusion matrix CTSD Database. **a** Color Images, **b** Fuzzy MG, **c** Fuzzy Prewitt, **d** Fuzzy Sobel

Table 18 Results architecture CNN-I with CTSD database

Preprocessing	Minimum	Maximum	Average	Std. Dev.
Color	0.4955	0.7282	0.6684	0.0492
Grayscale	0.6189	0.7232	0.6674	0.0290
Sobel	0.6319	0.7422	0.6863	0.0262
MG	0.5095	0.6219	0.5698	0.0268
Prewitt	0.5878	0.7242	0.6855	0.0310
Fuzzy MG	0.5807	0.6830	0.6364	0.0260
Fuzzy Sobel	0.6620	0.7603	0.7114	0.0263
Fuzz. Sob. Co	0.5838	0.7172	0.6756	0.0293
Fuzzy Prewitt	0.6319	0.7382	0.6985	0.0257

5 Discussion

We were able to determine from the analysis of the data that the adoption of data preparation techniques does not help the GTSRB and BelgiumTS databases. The results obtained with the color and grayscale images yield nearly identical results. Tables 10, 11, 12 and 13, which contain the GTSRB database findings, show that using

Table 19 Results architecture CNN-II with CTSD database

Preprocessing	Minimum	Maximum	Average	Std. Dev.
Color	0.4062	0.5858	0.5318	0.0370
Grayscale	0.4393	0.5707	0.5068	0.0319
Sobel	0.5286	0.6169	0.5599	0.0218
MG	0.3691	0.4885	0.4532	0.0267
Prewitt	0.4955	0.5878	0.5367	0.0229
Fuzzy MG	0.4333	0.5597	0.4947	0.0281
Fuzzy Sobel	0.5045	0.5988	0.5473	0.0213
Fuzz. Sob. Co	0.4694	0.5998	0.5376	0.0233
Fuzzy Prewitt	0.4774	0.5908	0.5333	0.0252

Table 20 Results architecture CNN-III with CTSD database

Preprocessing	Minimum	Maximum	Average	Std. Dev.
Color	0.4885	0.7081	0.6356	0.0565
Grayscale	0.5125	0.7121	0.6345	0.0535
Sobel	0.5246	0.7151	0.6321	0.0483
MG	0.3330	0.5848	0.5256	0.0567
Prewitt	0.5827	0.7031	0.6427	0.0288
Fuzzy MG	0.3480	0.6530	0.5815	0.0551
Fuzzy Sobel	0.5727	0.7252	0.6693	0.0343
Fuzz. Sob. Co	0.5196	0.7202	0.6428	0.0524
Fuzzy Prewitt	0.4885	0.7312	0.6473	0.0581

Table 21 Results architecture CNN-IV with CTSD database

Preprocessing	Minimum	Maximum	Average	Std. Dev.
Color	0.4794	0.6740	0.5645	0.0521
Grayscale	0.4824	0.6530	0.5565	0.0532
Sobel	0.2628	0.6560	0.5701	0.0658
MG	0.2207	0.5807	0.5030	0.0578
Prewitt	0.5206	0.6229	0.5648	0.0250
Fuzzy MG	0.4885	0.5858	0.5222	0.0207
Fuzzy Sobel	0.5446	0.6379	0.5811	0.0216
Fuzz. Sob. Co	0.5035	0.5988	0.5428	0.0200
Fuzzy Prewitt	0.5577	0.6610	0.5928	0.0291

Table 22 Results statistical test for GTSRB database

Model	Preprocessing	$\overline{x_1}$	σ_1	$\overline{x_2}$	σ_2	Z-value
CNN-I	Grayscale versus color	0.9656	0.0080	0.9650	0.0102	0.2535
CNN-II	Grayscale versus color	0.9565	0.0035	0.9495	0.0048	**6.4541**
CNN-III	Grayscale versus color	0.9593	0.0065	0.9626	0.0060	−2.0433
CNN-IV	Grayscale versus color	0.9585	0.0046	0.9603	0.0059	−1.3178

Table 23 Results statistical test for BelgiumTS database

Model	Preprocessing	$\overline{x_1}$	σ_1	$\overline{x_2}$	σ_2	Z-value
CNN-I	Grayscale versus color	0.9567	0.0108	0.9574	0.0163	−0.1961
CNN-II	Grayscale versus color	0.9319	0.0081	0.9394	0.0128	−2.7119
CNN-III	Grayscale versus color	0.9396	0.0186	0.9522	0.0147	– 2.911
CNN-IV	Grayscale versus color	0.9341	0.0159	0.9453	0.0135	−2.9411

Table 24 Results statistical test for CTSD database

Model	Preprocessing	$\overline{x_1}$	σ_1	$\overline{x_2}$	σ_2	Z-Value
CNN-I	Fuzzy Sobel versus color	0.7114	0.0263	0.6684	0.0492	**4.2217**
CNN-II	Fuzzy Sobel versus color	0.5473	0.0213	0.5318	0.0370	**1.9885**
CNN-III	Fuzzy Sobel versus color	0.6693	0.0343	0.6356	0.0565	**2.7926**
CNN-IV	Fuzzy Prewitt versus color	0.59284	0.0291	0.5645	0.0521	**2.6011**

conventional edge detection approaches reduces the accuracy of all architectures by up to 8% with MG and a 2% reduction with Sobel and Prewitt. Similarly, the use of fuzzy edge detectors reduces the accuracy of the classification performance by up to 9% when compared to the unprocessed images. The analysis of the statistical tests in Table 22, where we compared the RGB images against the best preprocessing; in this case, Grayscale images, demonstrates that only in CNN-II we have a significant boost in accuracy when predicting traffic signs with these models.

The results from the experimentation with the BelgiumTS database demonstrated in Tables 14, 15, 16 and 17 share a similar pattern with the GTSRB database, where the utilization of any preprocessing approach reduces the prediction accuracy in all architectures. Traditional edge detectors like Sobel or Prewitt reduce the accuracy between 1–2% depending on the architecture, opposing to MG which affects the performance up to 9% in CNN-III. The utilization of fuzzy edge detection also has a negative impact on this database. The consideration of uncertainty with fuzzy edge detection techniques also did not provide a beneficial contribution to improve the accuracy of the architectures, reducing up to 12% with CNN-III. The statistical test in Table 23 demonstrates that there is no significant evidence to validate the use of a preprocessing technique for this dataset and the proposed models.

Results from the CTSD database have a significant difference when compared to the GTSRB and BelgiumTS databases. In this case, we observe an improvement the prediction accuracy of all four models with many of the edge detection techniques, both traditional and fuzzy edge detection. The most substantial growth is demonstrated by CNN-I in conjunction with the fuzzy Sobel edge detector, where we achieved greater than 3% improvement, with a maximum accuracy reported at 76%. When we examine the statistical test, we observe that across all four architectures there is significant evidence of an improvement in accuracy when a preprocessing approach is used. We achieved better outcomes with the fuzzy Sobel implementation in the CNN-I to CNN-III architectures, and we achieved better results with the fuzzy Prewitt implementation in CNN-IV.

6 Conclusions

Experimentation demonstrated that the CNN-I architecture is the best alternative for traffic sign classification in all three instances. The implementation of preprocessing approaches has been proved to have a harmful effect on the models trained with the GTSRB and BelgiumTS databases. However, the CTSD database demonstrated that combining the models with preprocessing techniques significantly improves the results.

The CTSD database has the particularity that its elements contain a considerable amount of noise (the proportion of background relative to the size of the signal is significant, or the elements are partially covered by external objects such as tree branches). This is when compared against BelgiumTS or GTSRB, which contain the most normalized elements in terms of scale and amount of background with the margin of the image. Therefore, applying external filters to the convolution layer helps to partially remove this external noise, allowing the model to learn important features of the traffic sign, unlike the German or Belgian database, where the filtration part removes data from the sign.

References

1. Buyukarikan, B., & Ulker, E. (2022). Classification of physiological disorders in apples fruit using a hybrid model based on convolutional neural network and machine learning methods. *Neural Computing and Applications, 34*, 16973–16988.
2. Ramesh, S., Sasikala, S., Gomathi, S., Geetha, V., & Anbumani, V. (2022). Segmentation and classification of breast cancer using novel deep learning architecture. *Neural Computing and Applications, 34*, 16533–16545.
3. Bojarski, M., Del Testa, D., Dworakowski, D., Firner, B., Fleep, B., Goyal, P., Jackel, D. L., Momfort, M., Muller, U., Zhang, J., Zhang, X., Zhao, J., & Zieba, K. (2016). End to end learning for self-driving cars. arXiv:1604.07316.
4. Jain, A., Del Pero, L., Grimmett, H., & Ondruska, P. (2021). Autonomy 2.0: Why is self-driving always 5 years away? arXiv:2107.08142v3.

5. Simonyan, K., & Zisserman, A. (2015). Very deep convolutional networks for large-scale image recognition. In *Proceedings of 3rd International Conference on Learning Representations, ICLR 2015*, San Diego, CA, USA.
6. He, K., Zhang, X., Ren, S., & Sun, J. (2016). Deep residual learning for image recognition. In *Proceedings of 2016 IEEE Conference on Computer Vision and Pattern Recognition (CVPR).* IEEE: Las Vegas, NV, USA.
7. Russakovsky, O., Deng, J., Su, H., Krause, J., Satheesh, S., Ma, S., Huang, Z., Karpathy, A., Khosla, A., Bernstein, M., Berg, A. C., & Fei-Fei, L. (2015). ImageNet large scale visual recognition challenge. *International Journal of Computer Vision, 115*, 211–215.
8. Krizhevsky, A. (2012). Learning multiple layers of features from tiny images. University of Toronto.
9. Everingham, M., Van Gool, L., Williams, C. K. I., Winn, J., & Zisserman, A. (2010). The PASCAL visual object classes (VOC) challenge. *International Journal of Computer Vision, 88*, 303–338.
10. Belaroussi, R., Foucher, P., Tarel, J. P., Soheilian, B., Charbonnier, P., & Paparoditis, N.: Road sign detection in images: A case study. In *2010 20th International Conference on Pattern Recognition.* Istanbul, Turkey: IEEE.
11. Larsson, F., & Felsberg, M. (2011) Using Fourier descriptors and spatial models for traffic sign recognition. In *Image analysis.* SCIA 2011, (Vol. 6688, pp. 238–249). Lecture Notes in Computer Science. Heidelberg: Springer
12. Gómez-Moreno, H., Maldonado-Bascón, S., Gil-Jiménez, P., & Lafuente-Arroyo, S. (2010). Goal evaluation of segmentation algorithms for traffic sign recognition. *IEEE Transactions on Intelligent Transportation Systems, 11*(4), 917–930.
13. Bi, Z., Yu, L., Gao, H., Zhuo, P., & Yao, H. (2021). Improved VGG model-based efcient trafc sign recognition for safe driving in 5G scenarios. *International Journal of Machine Learning and Cybernetics, 12*, 3069–3080.
14. Versaci, M., & Morabito, F. C. (2021). Image edge detection: A new approach based on fuzzy entropy and fuzzy divergence. *International Journal of Fuzzy Systems, 23*, 918–936.
15. Melin, P., Gonzalez, C. I., Castro, J. R., Mendoza, O., & Castillo, O. (2014). Edge-detection method for image processing based on generalized type-2 fuzzy logic. *EEE Transactions on Fuzzy Systems, 22*(6), 1515–1525.
16. Gonzalez, C. I., Melin, P., Castro, J. R., Mendoza, O., & Castillo, O. (2016). An improved Sobel edge detection method based on generalized type-2 fuzzy logic. *Soft Computing, 20*, 773–784.
17. Bellman, R. E., & Zadeh, L. A. (1970). Decision-making in a fuzzy environment. *Management Science, 4*(17), B144–B164.
18. Pierrard, R., Poli, J., & Hudelot, C. (2018). Learning fuzzy relations and properties for explainable artificial intelligence. In *2018 IEEE International Conference on Fuzzy Systems (FUZZ-IEEE).* Rio de Janeiro, Brazil: IEEE.
19. Kickert, W. J. M., & Mamdani, E. H. (1978). Analysis of a fuzzy logic controller. *Fuzzy Sets and Systems, 1*(1), 29–44.
20. Takagi, T., & Sugeno, M. (1993). Fuzzy identification of systems and its applications to modeling and control. In *Readings in fuzzy sets for intelligent systems*, (pp. 387–403).
21. Sundararajan, D. (2017). *Digital image processing: A signal processing and algorithmic approach* (1st ed.). Springer.
22. Rangaswamy, C., Raju, G. T., & Seshikala, G.: Preprocessing of lung images with a novel image denoising technique for enhancing the quality and performance. In D. Hemanth, & S. Smys, (Eds.), *Computational vision and bio inspired computing* (Vol. 28, pp. 335–348). Lecture Notes in Computational Vision and Biomechanics.
23. Aborisade, D. O. (2010). Fuzzy logic based digital image edge detection. *Global Journal of Computer Science and Technology, 10*(14), 78–83.
24. Alshennawy, A. A., & Aly, A. A. (2019). *Edge detection in digital images using fuzzy logic technique* (Vol. 51). World Academy of Science, Engineering and Technology.

25. LeCun, Y., Boser, B., Denker, J. S., Henderson, D., Howard, R. E., Hubbard, W., & Jackel, L. D. (1989). Backpropagation applied to handwritten zip code recognition. *Neural Computation, 1*(4), 541–551.
26. Do, T., Duong, M., Dang, Q., & Le, M. (2018). Real-time self-driving car navigation using deep neural network. In *2018 4th International Conference on Green Technology and Sustainable Development (GTSD)* (pp. 7–12). Ho Chi Minh City, Vietnam: IEEE.
27. Barua, B., Gomes, C., Baghe, S., & Sisodia, J. (2019). A self-driving car implementation using computer vision for detection and navigation. In *2019 International Conference on Intelligent Computing and Control Systems (ICCS)* (pp. 271–274). Madurai, India: IEEE.
28. Zablocki, É., Ben-Younes, H., Pérez, P., & Cord. M. (2022) Explainability of deep vision-based autonomous driving systems: Review and challenges. *International Journal of Computer Vision, 130*, 2425–2452.
29. Sarki, R., Ahmed, K., Wang, H., Zhang, Y., Ma, J., & Wang, K. (2021). Image preprocessing in classification and identification of diabetic eye diseases. *Data Science and Engineering, 6*, 455–471.
30. Kumari, K. S, Samal, S., Misra, R., Madiraju, G., Mahabob, M. N., & Shivappa, A. B. (2021). Diagnosing COVID-19 from CT image of lung segmentation & classification with deep learning based on convolutional neural networks. Wireless Personal Communications.
31. Abdou, M. A. (2022). Literature review: Efficient deep neural networks techniques for medical image analysis. *Neural Computing and Applications, 34*, 5791–5812.
32. Manaswi, N. K. (2018). *Deep learning with applications using Python, Chatbots and Face, Object, and Speech Recognition with TensorFlow and Keras* (1st ed.). Apress.
33. Diamantis, D. E., & Iakovidis, D. K. (2021). Fuzzy Pooling. *IEEE Transactions on Fuzzy Systems, 29*(11), 3481–3488.
34. Houben, S., Stallkamp, J., Salmen, J., Schlipsing, M., & Igel, C. (2013). Detection of traffic signs in real-world images: In *The German Traffic Sign Detection Benchmark. International Joint Conference on Neural Networks* (Vol. 1288)
35. Timofte, R., Zimmermann, K., & Van Gool, L. (2009). Multi-view traffic sign detection, recognition, and 3D localization. In" *2009 Workshop on Applications of Computer Vision (WACV)* 1–8, (2009).
36. Zhang, Y., Wang, Z, Qi, Y., Liu, J.,& Yang, J.: (2018). CTSD: A dataset for traffic sign recognition in complex real-world images. In 2018 IEEE Visual Communications and Image Processing (VCIP) (pp. 1–4).

Optimization

Fuzzy Dynamic Adaptation of an Artificial Fish Swarm Algorithm for the Optimization of Benchmark Functions

Leticia Amador-Angulo, Patricia Ochoa, Cinthia Peraza, and Oscar Castillo

1 Introduction

In the last decade, the meta-heuristics algorithms have made a significative impact as techniques for solving complex problems. An algorithm that has achieved excellent results is the AFSA in different type of problems, for mentioned some; Chen et al. in [5] implement a finding rough set reducts with AFSA, Gafoor et al. [7] develop an intelligent approach of score-based AFSA, He et al. [9] study a novel AFSA to solve problems based on the large-scale reliability–redundancy, Krishnaraj et al. [11] propose an AFSA to apply in wireless multimedia sensor network, Lin et al. [12] develop an inverse kinematic analysis of bionic hands based on FSA, Liu et al. [13] study an application of an AFSA in symbolic regression, Liu et al. [14] present a hybridization of Pareto AFSA and genetic algorithm for the urban electric transit network problem, Pourpanah et al. [16] propose a review of the family of AFSA, Yuan et al. [22] present an adaptive simulated annealing and AFSA for the optimization of multi-depot express delivery vehicle routing, and Zhou et al. [27] develop a Chaotic Parallel AFSA applied in the Sensor Networks.

Based on the diversity on the applications or problems that AFSA has been studied, in this paper, two important contribution are the motivation such as; the first is the demonstration on the efficiency that the AFSA presents to solve complex problems,

L. Amador-Angulo (✉) · P. Ochoa · C. Peraza · O. Castillo
Division of Graduate Studies, Tijuana Institute of Technology, Tijuana, México
e-mail: gloria.amador@tectijuana.edu.mx

P. Ochoa
e-mail: martha.ochoa18@tectijuana.edu.mx

C. Peraza
e-mail: cinthia.peraza@tectijuana.edu.mx

O. Castillo
e-mail: ocastillo@tectijuana.mx

O. Castillo and P. Melin (eds.), *Hybrid Intelligent Systems Based on Extensions of Fuzzy Logic, Neural Networks and Metaheuristics*, Studies in Computational Intelligence 1096,
https://doi.org/10.1007/978-3-031-28999-6_6

the second and the main goal is to find the optimal values in S and V parameters applied in the optimization of benchmark set of functions, the proposal is using a T1FLS that allow improved the efficiency on the results.

The organization in the paper is the following: Sect. 2 shows some relevant related works about this research. Section 3 describes the proposal methodology for the fuzzy artificial fish swarm optimization. Section 4 describes the proposal problems in this case the benchmark set of functions used. Section 5 outline the experimental results. Section 6 shows the statistical test and a comparative with the original AFSA. Finally, some conclusions and future works are mentioned in the Sect. 7.

2 Related Works

Improve the algorithms is a technique that in the recent years has demonstrated excellent results for different types of problems. Several methodologies are used for some authors, but the technique that presents good results is the implementation of FLS, for mentioned some; an improved fuzzy adaptive firefly algorithm is proposed by Agrawal et al. [1], a fuzzy set to improve the performance of a bee colony optimization (BCO) is presented by Amador-Angulo et al. [2], a shadowed type-2 FS to improve two algorithms: differential evolution and harmony search is studied by Castillo et al. [3], an improved of the BCO through a generalized type-2 FLS is presented by Castillo et al. [4], an improved for the cuckoo search algorithm implementing a FLS is proposed by Guerrero et al. [8], an interval type-2 FLS to improve a gravitational search algorithm is proposed by Olivas et al. [15], and an improved grey wolf optimizer using FLS developed by Rodríguez et al. [17].

Several authors are interested in improving of the AFSA based on different methodologies for solve complex problems, for mentioned some; in [6] An improved AFSA and its application is presented by Gao et al. [10] an improved AFSA for prediction problems is studied by Hua et al. [18] a parameters analysis of AFSA is presented by Wang et al. [19] a modified AFSA is proposed by Xiao et al. [25] a robot path planning based on improved AFSA is presented by Zhang et al. [26] an improved AFSA is presented by Zhang et al. [28] an improved FSA in the study case of neighborhood rough set reduction is developed by Zou et al. Based in the interest of demonstrated that T1FLS presents excellent results when S and V parameters are found, the proposal methodology is improved the AFSA applied in the optimization of mathematical functions.

3 Artificial Fish Swarm Algorithm

Based on the first idea of the inspiration in the analysis with the simulation of a set of individuals in the recollection of food on the water by Li in 2002, then, Yazdani et al. in 2012 creates of the AFSA [20]. This algorithm is focused in to find of food according to its nature, an important characteristic is to observe the prey individually within its visual distance.

3.1 Original Artificial Swarm Algorithm

The idea in the original AFSA is illustrated in Fig. 1 for each movement that a fish performs in the algorithm. X represents the actual position of an individual (artificial fish (AF)). The total the step for a fish that is executed in the algorithm is represented by the Step (S), the visual distance is represented by Visual (V). Each AF inspects the search space around its vision distance. In this case, the best solution in each execution is changed according to the following condition, the criterion in the realization of a move in a fish is if a new position is better that its current location, the movement is given.

In AFSA n represents a random distribution for each generation in the search space. In the identification of a good solution the search space are the positions with major concentration of fish. AFSA has four important behaviors called: *preying, swarming, following, and random* behavior. The dynamism for the AFSA is based in the Eqs. (1)–(7). Equation (1) indicates the preying behavior; the fish try to move to locations with highest food in the search space. Let $X_i^{(t)}$ be the current position of the ith AF, and X_j be a state of an AF randomly selected within the visual distance of the ith AF as follows:

$$X_j = X_i^{(t)} + V \times rand() \tag{1}$$

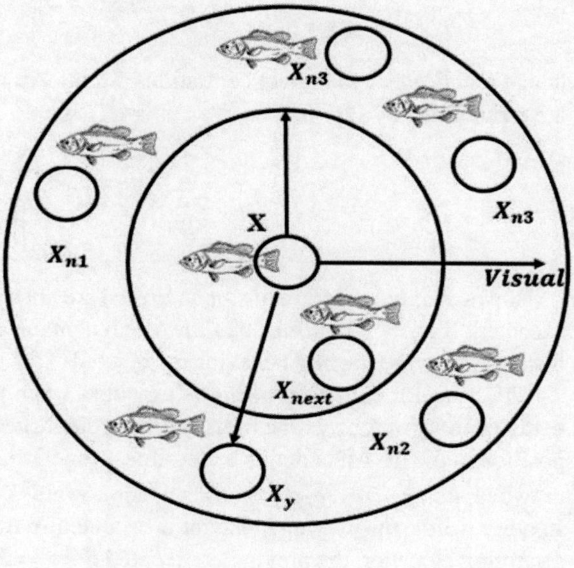

Fig. 1 Graphical representation of the movements for an artificial fish

where the visual distance of the AF is represented by V, $d_{i,j} < V$ and *rand()* is a random vector with 0 and 1 values. If $If Y_i < Y_j$, $X_i^{(t)}$ moves a step toward X_j, the Eq. (2) describes this behavior.

$$X_i^{(t+1)} = X_i^{(t)} + S \times rand().* \frac{X_j - X_i^{(t)}}{\left\| X_j - X_i^{(t)} \right\|} \tag{2}$$

where .∗ represents multiplication for the two vectors. In AFSA if the forward criterion stop is not satisfied, a new state X_j is randomly selected implementing the Eq. (3). If the stop criterion is no found a number determined of time. $X_i^{(t)}$ moves a step randomly, the behaviors is descripted by:

$$X_i^{(t+1)} = X_i^{(t)} + V. * rand() \tag{3}$$

The behavior called *swarming*, which represents in nature a defense mechanism forms its predators. Let *n* be a limit of numbers of AF while $d_{i,j} < V$, and $X_c = (x_{c1}, x_{c2}, \ldots\ldots, x_{cD})$ be the center position, where x_{cd} is the *d*th dimension of X_c, and Eq. (4) indicates this behavior.

$$x_{cd} = \frac{\left(\sum_{j=1}^{n_f} x_{jd}\right)}{n_f} \tag{4}$$

If the criterion of stop in Eq. (5) are satisfied, i.e.;

$$Y_c > Y_i \frac{n_f}{n} > \delta \tag{5}$$

where δ indicates a major concentration of food. As such, $X_i^{(t)}$ moves a step toward are represented by Eq. (6)

$$X_i^{(t+1)} = X_i^{(t)} + S \times rand().* \frac{X_c - X_i^{(t)}}{\left\| X_c - X_i^{(t)} \right\|} \tag{6}$$

However, if the criterion of stop in Eq. (5) are not satisfied, the *preying* behavior is executed. If $n_f = 0$ indicates that *i*th AF no contains a companion within the visual distance, then the *preying* behavior is executed.

The behavior called *following* is executed when a fish in a location contains a better quantity of food, is an observation by other fishes follow. Let $X_i^{(t)}$ be the current position in the *i*th AF contains better fitness function, i.e., $Y_j - Y_i$, and it is not over crowded, i.e., $n_f/n < \delta$, $X_i^{(t)}$ moves a step toward X_j, and the Eq. (2) is executed. In other words, the preying behavior is executed. With a similar methodology to the swarming behavior, the preying is executed if $n_f = 0$.

Fig. 2 Structure of a T1FLS

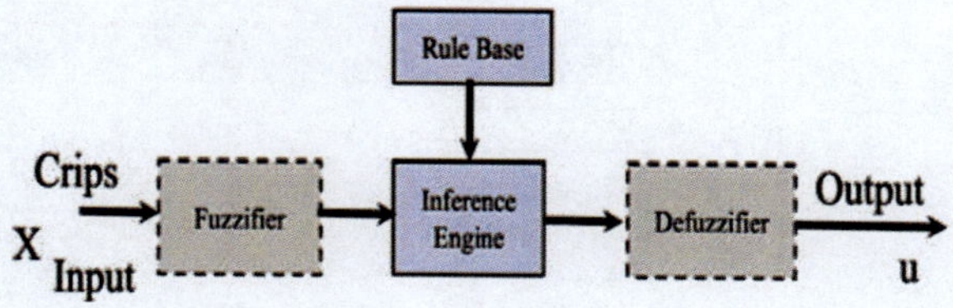

The final behavior called *random,* in nature represents a swarm of fishes moving freely to find food. This behavior allows that each AF to locate its own food. The AF selects a random behavior when the other behaviors are not are satisfied. Equation (7) represents this final behavior:

$$X_i^{(t+1)} = X_i^{(t)} + V \times rand() \tag{7}$$

3.2 Fuzzy Artificial Swarm Algorithm

The proposed idea with the T1FLS was proposed by Zadeh [23] and Zadeh et al. [24]. A T1FLS is defined by the universe X with a MF $\mu_A(x)$ with an interval of [0,1] in the values and is presented by Eq. (8).

$$A = \{(x, \mu_A(x))|x \in X\} \tag{8}$$

where $\mu_A : X \to [0, 1]$.

In this expression $\mu_A(x)$ indicated the membership degree of the element $x \in X$ to the set A. The expression implemented is the following: $A(x) = \mu_A(x)$ for all $x \in X$. Figure 2 illustrates the T1FLS, and the proposed Fuzzy AFSA is illustrates by Fig. 3.

The proposed methodology in the adjustment dynamic for the *S* and *V* parameters of the AFSA is divided in four Fuzzy AFSA with the goal to analyze the impact that S and V parameters reflect in the algorithm. The proposal T1FLS has triangular Membership Functions (MFs). Figures 4, 5, 6 and 7 illustrate the representation for each proposed F-AFSA (a) and the rules (b).

4 Benchmark Sets of Functions

The benchmark sets of functions in this paper is the study case, this section describes a total of six classical benchmark functions with the main objective to validate to AFSA, such as; Ackley, Giewank, Quarticnoise, Rastrigin, Rosenbrock, Spherical;

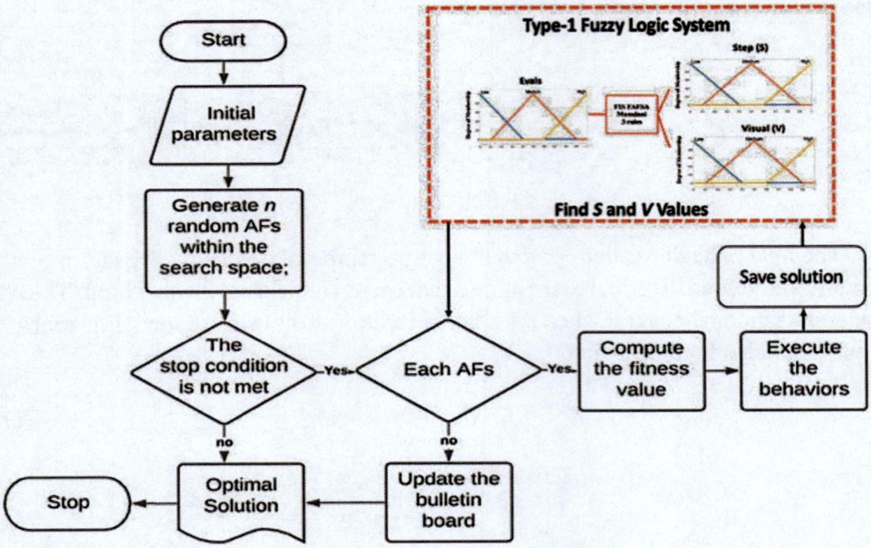

Fig. 3 Flowchart of the procedure in the Fuzzy AFSA

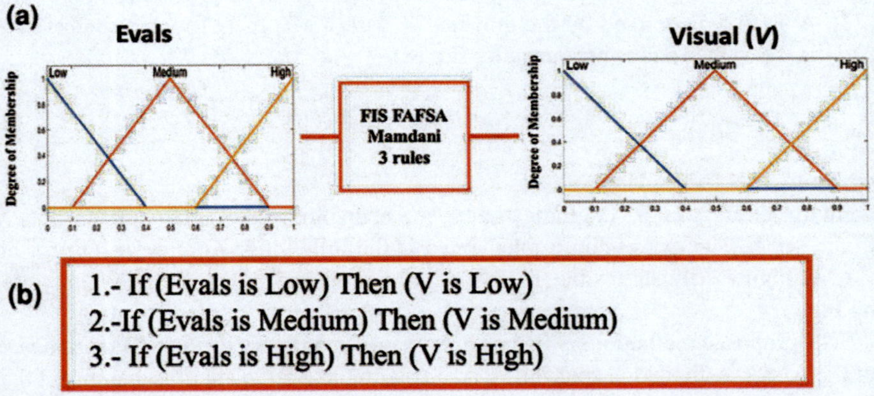

Fig. 4 F-AFSA modifying V parameter incrementally

all functions were evaluated with 10, 30 and 50 dimensions (artificial fish). Figure 8 illustrates the plot, and Table 1 shows the equation for each study case.

5 Experimental Results

For this paper, experimentation was carried out using the T1FLS descripted in the previous section with the 6 benchmark functions.

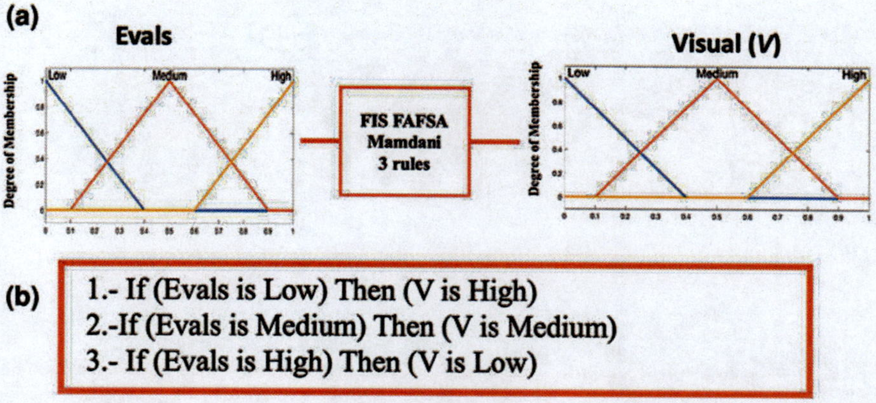

Fig. 5 F-AFSA modifying V in decrement

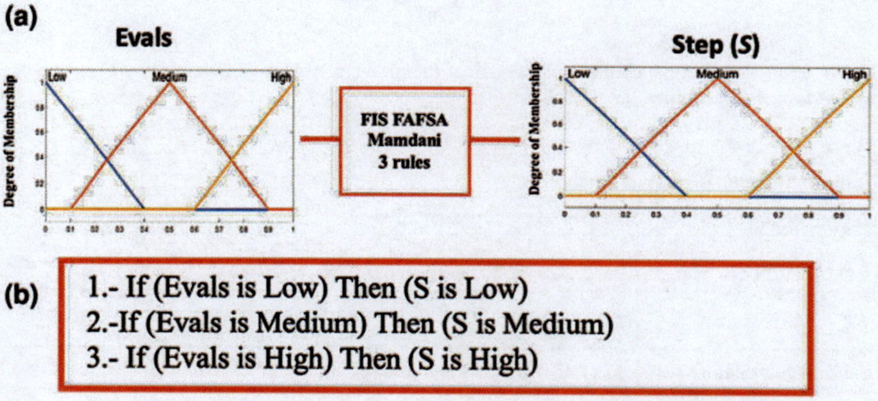

Fig. 6 F-AFSA modifying S incrementally

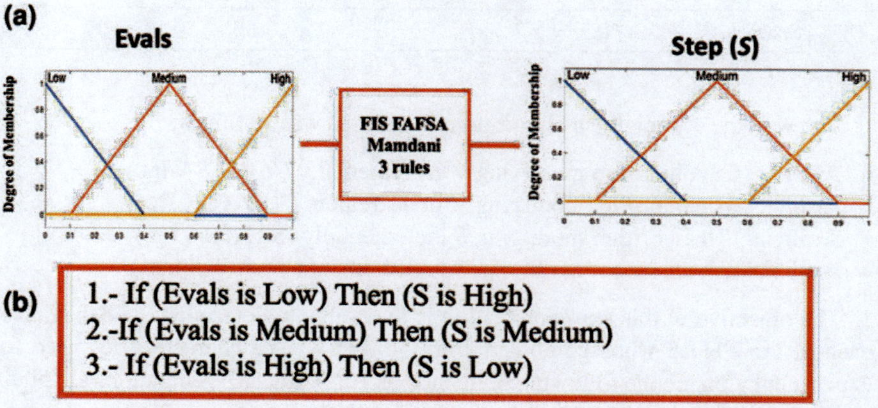

Fig. 7 F-AFSA modifying S in decrement

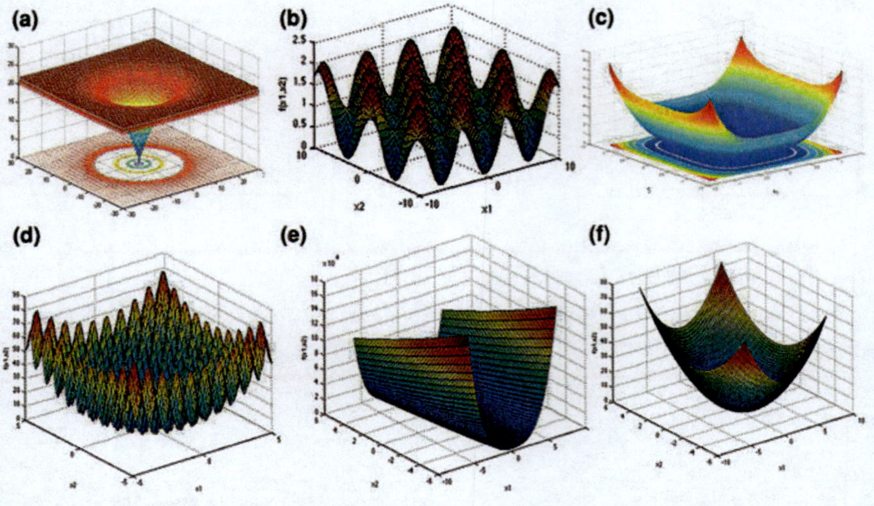

Fig. 8 Plot of the six benchmark mathematical functions; **a** Ackley, **b** Griewank, **c** Quarticnoise, **d** Rastrigin, **e** Rosenbrock, and **f** Sphere

Table 1 Benchmark sets of functions

f_x	Name function	Mathematical representation
F1	Ackley	$f(x) = -20e^{-0.2\sqrt{\frac{1}{n_x}\sum_{j=1}^{n_x} x_j^2}} - \frac{1}{e^{n_x}}\sum_{j=1}^{n_x} cos\left(2\pi x_j\right) + 20 + e$
F2	Griewank	$f(x) = \sum_{i=1}^{d} \frac{x_i^2}{4000} - \prod_{i=1}^{d} cos\left(\frac{x_i}{\sqrt{i}}\right) + 1$
F3	Quarticnoise	$f(x) = \sum_{i=1}^{d} ix_i^4 + rand[0, 1]$
F4	Rastrigin	$f(x) = 10d + \sum_{i=1}^{d}\left[x_i^2 - 10cos(2\pi x_i)\right]$
F5	Rosenbrock	$f(x) = \sum_{i=1}^{d-1}\left[100\left(x_{i+1} - x_i^2\right)^2 + (x_i - 1)^2\right]$
F6	Sphere	$f(x) = \sum_{i=1}^{d} x_i^2$

The way the artificial fish algorithm is combined was as follows:

- Artificial fish algorithm modifying *V* incrementally (V FAFSA Inc)
- Artificial fish algorithm modifying *V* in decrement (V FAFSA Dec)
- Artificial fish algorithm modifying *S* incrementally (S FAFSA Inc)
- Artificial fish algorithm modifying *S* in decrement (V FAFSA Dec).

The objective of this experimentation is to be able to determine the behavior of each of the T1FLS when combined with the AFSA. The characteristics used for experimentation are the following: Dimensions 10, 30 and 50, generations are 5000, and dynamic *S* and *V* depending on the T1FLS used.

The results of the 6 functions are descripted in the Table 2 using a dimension of 10, the results of the 4 T1FLS dynamically varying V and S, as well as the average, the standard deviation, the best result obtained, and the worst result obtained are shown.

From the information shown in the Table 2 we can see in bold the best average between *V FAFSA Inc* and *V FAFSA Dec*, as well as the best average obtained between *S FAFSA Inc* and *V FAFSA Dec*. We can see that the T1FLS called *V FAFSA Dec* obtains better results compared to *V FAFSA Inc*, and in the case of the T1FLS *FAFSA Dec* obtains better results than *V FAFSA Inc*.

The results obtained for dimensions of 30 can be seen in the Table 3, the results are from the 4 T1FLS used to dynamically vary S and V.

The Table 3 shows in bold the best averages for each of the variable S and V, where we can observe that for the T1FLS *S FAFSA Inc* and *V FAFSA Dec* each of

Table 2 Results varying the "*V*" and "*S*" values for 10 dimensions

10 dimensions					
f_x	Performance index	V FAFSA Inc	V FAFSA Dec	S FAFSA Inc	S FAFSA Dec
F1	Average	**1.65E+00**	2.11E+00	2.91E+00	**2.50E+00**
	Standard deviation	1.55E+00	9.11E-01	7.59E-01	1.05E+00
	Best	8.21E-04	8.21E-04	1.16E+00	1.16E+00
	Worst	3.57E+00	3.57E+00	4.30E+00	4.99E+00
F2	Average	3.74E-01	**1.62E-01**	7.68E-02	**3.07E-02**
	Standard deviation	4.41E-01	3.31E-01	2.42E-01	1.65E-01
	Best	8.36E-06	8.36E-06	4.23E-07	6.13E-06
	Worst	1.00E+00	1.00E+00	1.01E+00	9.06E-01
F3	Average	**1.41E+00**	1.87E+00	**1.84E+00**	1.87E+00
	Standard deviation	9.04E-01	2.49E-01	1.83E-01	2.39E-01
	Best	2.49E-01	1.20E+00	1.28E+00	1.18E+00
	Worst	2.33E+00	2.33E+00	2.18E+00	2.30E+00
F4	Average	8.02E+01	**6.98E+01**	1.38E+02	**7.58E+01**
	Standard deviation	6.97E+01	8.13E+01	7.02E+01	8.26E+01
	Best	8.82E-06	8.82E-06	9.16E-04	5.85E-05
	Worst	1.70E+02	1.70E+02	1.84E+02	1.72E+02
F5	Average	2.79E-03	**1.16E-03**	6.76E-03	**1.20E-03**
	Standard deviation	3.79E-03	1.52E-03	1.30E-02	1.85E-03
	Best	6.85E-05	6.85E-05	8.05E-05	9.67E-05
	Worst	8.40E-03	8.40E-03	6.13E-02	9.90E-03
F6	Average	1.36E-03	**9.06E-04**	1.25E-03	**1.20E-03**
	Standard deviation	1.59E-03	5.71E-04	1.09E-03	2.12E-03
	Best	2.48E-04	2.48E-04	2.73E-04	1.22E-06
	Worst	3.70E-03	3.70E-03	5.60E-03	1.21E-02

Table 3 Results varying the "*V*" value for 30 dimensions

30 dimensions					
f_x	Performance index	V FAFSA Inc	V FAFSA Dec	S FAFSA Inc	S FAFSA Dec
F1	Average	**2.24E+00**	2.32E+00	2.61E+00	**2.10E+00**
	Standard deviation	2.00E+00	8.07E-01	8.06E-01	6.41E-01
	Best	8.04E-01	8.04E-01	1.28E+00	6.26E-01
	Worst	5.04E+00	5.04E+00	4.64E+00	3.07E+00
F2	Average	**4.62E-04**	5.38E-04	4.43E-04	**4.17E-04**
	Standard deviation	4.19E-04	3.10E-04	3.33E-04	3.05E-04
	Best	3.37E-06	3.37E-06	1.34E-06	7.80E-07
	Worst	9.97E-04	9.97E-04	9.95E-04	9.71E-04
F3	Average	**7.04E+00**	9.48E+00	**9.32E+00**	9.36E+00
	Standard deviation	4.47E+00	4.34E-01	5.53E-01	4.54E-01
	Best	4.34E-01	8.23E+00	7.32E+00	8.47E+00
	Worst	1.00E+01	1.00E+01	1.01E+01	1.01E+01
F4	Average	7.21E-02	**1.84E-02**	8.57E-02	**3.54E-02**
	Standard deviation	1.05E-01	4.19E-02	2.29E-01	7.34E-02
	Best	6.27E-04	6.27E-04	8.85E-04	7.43E-04
	Worst	2.27E-01	2.27E-01	1.23E+00	3.81E-01
F5	Average	2.20E-02	**9.39E-03**	6.02E-02	**2.79E-03**
	Standard deviation	2.74E-02	1.67E-02	1.09E-01	4.74E-03
	Best	8.23E-05	8.23E-05	7.69E-05	3.31E-05
	Worst	6.18E-02	6.18E-02	4.64E-01	1.92E-02
F6	Average	6.29E-03	**2.32E-03**	**4.88E-03**	7.09E-03
	Standard deviation	8.43E-03	4.04E-03	9.99E-03	2.05E-02
	Best	8.56E-05	8.56E-05	8.10E-05	4.40E-05
	Worst	1.87E-02	1.87E-02	4.50E-02	1.12E-01

them have 3 functions where they are better on average, but nevertheless for T1FLS *S FAFSA Dec* has better results than T1FLS *V FAFSA Inc*.

The summary of the results for dimension of 50 are shown in the Table 4, which offers us the mean, standard deviation, the best result, and the worst result.

For the 50-dimensional experimentation, it is observed that the best results with the variable V are obtained using the FLS *V FAFSA Dec*, while for the variable S the T1FLS that provides us with the best results is *S FAFSA Dec*.

Based on the results of the 4 different FLS used for the variables V and S, a T1FLS with one input (Evaluations or Iterations) and two outputs (V and S) was implemented (FAFSA Complete). This T1FLS is composed in its rules by the T1FLS that best results obtained in the experimentation. Figure 9 illustrate the distribution on the MFs (a) and the rules (b).

Table 4 Results varying the "*S*" value for 50 dimensions

50 dimensions					
f_x	Performance index	V FAFSA Inc	V FAFSA Dec	S FAFSA Inc	S FAFSA Dec
F1	Average	2.82E+00	**2.16E+00**	2.71E+00	**2.38E+00**
	Standard deviation	8.27E-01	4.89E-01	6.65E-01	9.12E-01
	Best	1.54E+00	9.90E-01	1.87E++00	9.90E-01
	Worst	4.81E+00	2.91E+00	4.86E+00	5.33E+00
F2	Average	6.42E-04	**4.64E-04**	6.29E-04	**5.44E-04**
	Standard deviation	3.48E-04	6.69E-04	3.47E-04	3.19E-04
	Best	4.33E-06	−2.60E-03	1.82E-07	4.15E-06
	Worst	9.97E-04	9.98E-04	9.98E-04	9.98E-04
F3	Average	**1.73E+01**	1.76E+01	**1.76E+01**	**1.76E+01**
	Standard deviation	7.74E-01	6.17E-01	6.44E-01	6.77E-01
	Best	1.54E+01	1.58E+01	1.65E+01	1.58E+01
	Worst	1.85E+01	1.86E+01	1.89E+01	1.87E+01
F4	Average	1.83E-01	**8.02E-02**	**6.97E-02**	7.94E-02
	Standard deviation	4.41E-01	1.49E-01	8.25E-02	1.37E-01
	Best	8.00E-04	4.90E-04	9.50E-04	4.90E-04
	Worst	2.24E+00	5.79E-01	3.14E-01	5.79E-01
F5	Average	1.44E-01	**2.45E-02**	1.19E-01	**3.33E-02**
	Standard deviation	2.51E-01	6.40E-02	2.02E-01	7.36E-02
	Best	9.03E-04	2.62E-05	5.98E-06	2.62E-05
	Worst	1.01E+00	2.70E-01	7.46E-01	2.70E-01
F6	Average	9.33E-03	**3.50E-03**	1.57E-02	**4.61E-03**
	Standard deviation	1.52E-02	6.05E-03	3.39E-02	8.18E-03
	Best	3.93E-04	−6.10E-03	1.56E-04	2.71E-04
	Worst	6.87E-02	2.25E-02	1.55E-01	3.77E-02

The Table 5 shows the results with the complete T1FLS with experimentation for 10, 30 and 50 dimensions, information on the average, the standard deviation, the best and worst result of the 6 benchmark functions is presented.

Table 5 shows in bold the best averages of the different dimensions used in the 6 functions, we can see that there is a great improvement when using the FAFSA compared to the 4 T1FLS used where each parameter was dynamically varied (V and S) in an independent way.

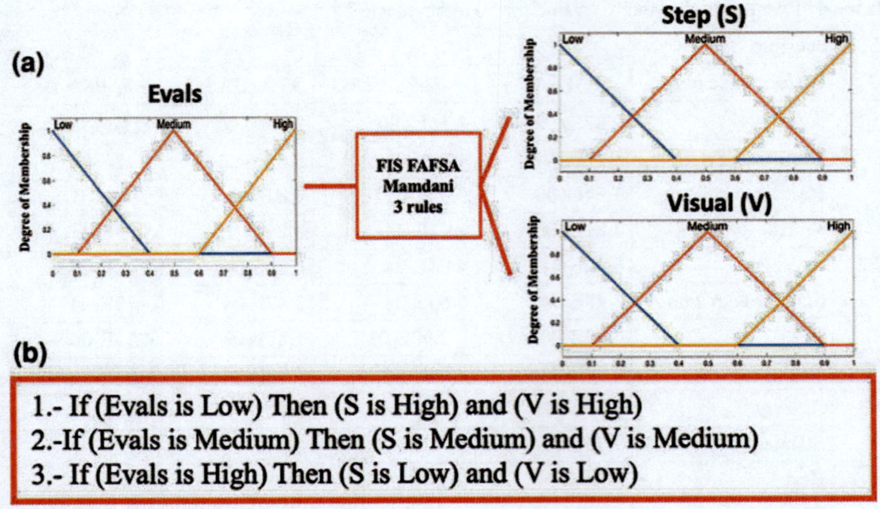

Fig. 9 F-AFSA modifying S decrementally

6 Statistical Test

The objective of a statistical test is to be able to demonstrate that the hypothesis proposed in a work is significant and for this reason we decided to use a statistical test in comparison with an existing work in the literature [21], for the statistical test we used a Z test, the parameters used are summarized in the Table 6.

The statistical test is carried out with an article found in the literature which has the name: Fuzzy Adaptive Artificial Fish Swarm Algorithm (FAF) [21], and offers experimentation with 4 of the 6 functions.

The hypothesis formulated for our statistical test are the following:

Ho: The proposed FAFSA algorithm is greater than or equal to the FAF algorithm.
Ha: The proposed FAFSA algorithm is smaller than the FAF algorithm.

The results of the statistical tests are offered in the Table 7 in which the two algorithms FAFSA and FAF are compared, the Z-value, the mean, the standard deviation, the best and worst error of both algorithms are presented.

The results of the statistical tests confirm that the proposed hypothesis that our proposed method is lower on average than the FAF algorithm, therefore we can affirm that our statistical tests are significant and that they pass the statistical test.

Table 5 Results with the FAFSA complete 10, 30, and 50 dimensions

FAFSA complete 10, 30 and 50 dimensions				
f_x	Performance index	FAFSA Complete D10	FAFSA Complete D30	FAFSA Complete D50
F1	Average	**4.46E-18**	4.46E-13	4.46E-13
	Standard deviation	3.11E-18	3.11E-14	3.11E-14
	Best	6.28E-19	6.28E-15	6.28E-15
	Worst	9.95E-18	9.95E-14	9.95E-14
F2	Average	2.26E-15	3.01E-19	**2.87E-19**
	Standard deviation	1.46E-15	2.72E-19	2.77E-19
	Best	1.00E-20	1.03E-26	4.08E-24
	Worst	7.69E-16	8.7E-24	8.7E-24
F3	Average	**0**	**0**	**0**
	Standard deviation	0	0	0
	Best	0	0	0
	Worst	0	0	0
F4	Average	**0**	**0**	**0**
	Standard deviation	0	0	0
	Best	0	0	0
	Worst	0	0	0
F5	Average	5.42E-09	4.80E-11	**8.32E-14**
	Standard deviation	2.97E-09	8.52E-11	2.54E-13
	Best	1.33E-10	1.49E-17	1.00E-18
	Worst	9.95E-10	3.6E-15	9.76E-18
F6	Average	**0**	**0**	**0**
	Standard deviation	0	0	0
	Best	0	0	0
	Worst	0	0	0

Table 6 Parameters for the Z-Test

Parameter	Value
Level of confidence	95%
Alpha	5%
Ha	$\mu 1 < \mu 2$
Ho	$\mu 1 \geq \mu 2$
Critical Value	−1.645

Table 7 Results of the statistical test FAFSA and FAF

f_x	Performance index	FAFSA Complete	FAF [21]	Z-value
F1	Average	4.46E-13	1.79E-14	−290.854.241
	Standard deviation	3.11E-14	3.37E-15	
	Best	6.28E-15	1.33E-14	
	Worst	9.95E-14		
F2	Average	3.01E-19	7.55E-16	−221.139
	Standard deviation	2.72E-19	1.87E-16	
	Best	1.03E-26	4.44E-16	
	Worst	8.7E-24		
F3	Average	0		
	Standard deviation	0		
	Best	0		
	Worst	0		
F4	Average	0	2.3773	−48.812
	Standard deviation	0	2.6676	
	Best	0	0	
	Worst	0		
F5	Average	4.80E-11		
	Standard deviation	8.52E-11		
	Best	1.49E-17		
	Worst	3.6E-15		
F6	Average	0	8.82E-63	−17.377
	Standard deviation	0	2.78E-62	
	Best	0	1.24E-68	
	Worst	0		

7 Conclusions

The work carried out in this work was thought to carry out experimentation using a T1FLS that dynamically moves each of the variable V and S independently with rules in decrease and increase, this experimentation allowed us to know the behavior of each of the variable V and S. Once the best structure of each variable was selected, a fuzzy system was created that could dynamically modify variable V and S. This system is what we call FAFSA.

The experimentation carried out with the FAFSA fuzzy system shows a great improvement when comparing the results with the 4 different T1FLS: V FAFSA Inc, V FAFSA Dec, S FAFSA Inc and S FAFSA Dec. The statistical test shows that the FAFSA algorithm has an improvement statistically significant compared to the FAF algorithm.

An interesting extension on this work could be to use other study case for example Fuzzy Controllers and analyze the uncertainty with the implementation the Interval Type-2 FLS in the adjustment dynamic of the S and V parameters for the AFSA.

References

1. Agrawal, A., Tripathy, B. K., & Thirunavukarasu, R. (2021). An improved fuzzy adaptive firefly algorithm-based hybrid clustering algorithms. *International Journal of Uncertainty, Fuzziness and Knowledge-Based System, 29*(Suppl-2), 259–278.
2. Amador-Angulo, L., Mendoza, O., Castro, J. R., Rodríguez-Díaz, A., Melin, P., & Castillo, O. (2016). Fuzzy sets in dynamic adaptation of parameters of a bee colony optimization for controlling the trajectory of an autonomous mobile robot. *Sensors, 16*(9), 1458.
3. Castillo, O., Peraza, C., Ochoa, P., Amador-Angulo, L., Melin, P., Park, Y., & Geem, Z. W. (2021). Shadowed type-2 fuzzy systems for dynamic parameter adaptation in harmony search and differential evolution for optimal design of fuzzy controllers. *Mathematics, 9*(19), 2439.
4. Castillo, O., & Amador-Angulo, L. (2018). A generalized type-2 fuzzy logic approach for dynamic parameter adaptation in bee colony optimization applied to fuzzy controller design. *Information Sciences, 460*, 476–496.
5. Chen, Y., Zhu, Q., & Xu, H. (2015). Finding rough set reducts with fish swarm algorithm. *Knowledge-Based Systems, 81*, 22–29.
6. Gao, S., & Wen, Y.: An improved artificial fish swarm algorithm and its application. In *2018 IEEE/ACIS 17th International Conference on Computer and Information Science (ICIS)* (pp. 649–652). IEEE.
7. Gafoor, S. H. A., & Theagarajan, P. (2022). Intelligent approach of score-based artificial fish swarm algorithm (SAFSA) for Parkinson's disease diagnosis. *International Journal of Intelligent Computing and Cybernetics.*
8. Guerrero, M., Castillo, O., & García, M. (2015). Fuzzy dynamic parameters adaptation in the Cuckoo Search Algorithm using fuzzy logic. In *2015 IEEE Congress on Evolutionary Computation (CEC)* (pp. 441–448). IEEE. (May, 2015)
9. He, Q., Hu, X., Ren, H., & Zhang, H. (2015). A novel artificial fish swarm algorithm for solving large-scale reliability–redundancy application problem. *ISA Transactions, 59*, 105–113.
10. Hua, Z., Xiao, Y., & Cao, J. (2021). Misalignment fault prediction of wind turbines based on improved artificial fish swarm algorithm. *Entropy, 23*(6), 692.
11. Krishnaraj, N., Jayasankar, T., Kousik, N. V., & Daniel, A. (2021). 2 Artificial Fish Swarm Optimization Algorithm with Hill Climbing Based Clustering Technique for Throughput Maximization in Wireless Multimedia Sensor Network.
12. Lin, M., Hong, H., Yuan, X., Fan, J., & Ji, Z. (2021) Inverse kinematic analysis of bionic hands based on fish swarm algorithm. In *Journal of Physics: Conference Series* (Vol. 1965, no. 1, p. 012006). IOP Publishing.
13. Liu, Q., Odaka, T., Kuroiwa, J., & Ogura, H. (2013). Application of an artificial fish swarm algorithm in symbolic regression. *IEICE Transactions on Information and Systems, 96*(4), 872–885.
14. Liu, Y., Feng, X., Yang, Y., Ruan, Z., Zhang, L., & Li, K. (2022). Solving urban electric transit network problem by integrating Pareto artificial fish swarm algorithm and genetic algorithm. *Journal of Intelligent Transportation Systems, 26*(3), 253–268.
15. Olivas, F., Valdez, F., Melin, P., Sombra, A., & Castillo, O. (2019). Interval type-2 fuzzy logic for dynamic parameter adaptation in a modified gravitational search algorithm. *Information Sciences, 476*, 159–175.
16. Pourpanah, F., Wang, R., Lim, C. P., & Yazdani, D. (2020). A review of the family of artificial fish swarm algorithms: Recent advances and applications. arXiv:2011.05700.

17. Rodríguez, L., Castillo, O., & Soria, J. (2016). Grey wolf optimizer with dynamic adaptation of parameters using fuzzy logic. In *2016 IEEE Congress on Evolutionary Computation (CEC)* (pp. 3116–3123). IEEE. (July, 2016).
18. Wang, L. G., & Shi, Q. H. (2010). Parameters analysis of artificial fish swarm algorithm. *Computer Engineering, 36*(24), 169–171.
19. Xiao, J., Zheng, X., Wang, X., & Huang, Y. (2006) A modified artificial fish-swarm algorithm. In *2006 6th World Congress on Intelligent Control and Automation* (Vol. 1, pp. 3456–3460). IEEE.
20. Yazdani, D., Akbarzadeh-Totonchi, M. R., Nasiri, B., & Meybodi, M. R. (2012). A new artificial fish swarm algorithm for dynamic optimization problems. In *2012 IEEE Congress on Evolutionary Computation* (pp. 1–8). IEEE.
21. Yazdani, D., Nadjaran Toosi, A., & Meybodi, M. R. (2010). Fuzzy adaptive artificial fish swarm algorithm. In *Australasian Joint Conference on Artificial Intelligence* (pp. 334–343). Berlin, Heidelberg: Springer. (Dec 2010).
22. Yuan, M., Kan, X., Chi, C., Cao, L., Shu, H., & Fan, Y. (2022). An adaptive simulated annealing and artificial fish swarm algorithm for the optimization of multi-depot express delivery vehicle routing. *Intelligent Data Analysis, 26*(1), 239–256.
23. Zadeh, L. A. (1996). Fuzzy sets. In *Fuzzy sets, fuzzy logic, and fuzzy systems: Selected papers by Lotfi A Zadeh* (pp. 394–432).
24. Zadeh, L. A., Klir, G. J., & Yuan, B. (1996). *Fuzzy sets, fuzzy logic, and fuzzy systems: Selected papers* (Vol. 6). World Scientific.
25. Zhang, Y., Guan, G., & Pu, X.: The robot path planning based on improved artificial fish swarm algorithm. *Mathematical Problems in Engineering.*
26. Zhang, C., Zhang, F. M., Li, F., & Wu, H. S. (2014). Improved artificial fish swarm algorithm. In *2014 9th IEEE Conference on Industrial Electronics and Applications* (pp. 748–753). IEEE.
27. Zhou, J., Qi, G., & Liu, C. (2021) A chaotic parallel artificial fish swarm algorithm for water quality monitoring sensor networks 3D coverage optimization. *Journal of Sensors.*
28. Zou, L., Li, H., Jiang, W., & Yang, X. (2019). An improved fish swarm algorithm for neighborhood rough set reduction and its application. *IEEE Access, 7*, 90277–90288.

Particle Swarm Optimization Algorithm with Improved Opposition-Based Learning (IOBL-PSO) to Solve Continuous Problems

Miguel Á. García-Morales, Héctor J. Fraire-Huacuja, José A. Brambila-Hernández, Juan Frausto-Solís, Laura Cruz-Reyes, Claudia G. Gómez-Santillán, and Juan M. Carpio-Valadez

1 Introduction

The use of metaheuristic algorithms to solve problems in different areas of society was introduced at the beginning of this century [1] because metaheuristic algorithms can obtain approximate solutions to the optimal solutions through a fast search, in a reasonable time, and with the flexibility to adapt to different types of problems [2].

The success in using metaheuristic algorithms is because they present a balance between the exploration and exploitation of solution spaces. The exploration process of the metaheuristic algorithms is used to investigate the different solution spaces, and the exploitation process uses the information collected to guide the search toward the objective [1]. For example, Kennedy and Eberhart [3] created the particle swarm Optimization algorithm (PSO); this algorithm mimics the collective behavior of various groups of animals and has used been used to solve different types of problems in multiple areas. However, the PSO has a slow convergence rate and presents specific problems in finding the globally optimal values.

M. Á. García-Morales · H. J. Fraire-Huacuja (✉) · J. A. Brambila-Hernández · J. Frausto-Solís · L. Cruz-Reyes · C. G. Gómez-Santillán
National Technology of Mexico/Madero City Technological Institute, Madero City, Tamaulipas, Mexico
e-mail: automatas2002@yahoo.com.mx

M. Á. García-Morales
e-mail: d21210015@tectijuana.edu.mx

J. M. Carpio-Valadez
National Technology of Mexico/Technological Institute of Leon, León, Guanajuato, Mexico

O. Castillo and P. Melin (eds.), *Hybrid Intelligent Systems Based on Extensions of Fuzzy Logic, Neural Networks and Metaheuristics*, Studies in Computational Intelligence 1096,
https://doi.org/10.1007/978-3-031-28999-6_7

To solve the problem of premature convergence and improve the performance of the PSO, several variants have been proposed in the state-of-the-art that includes different techniques, such as logistic map [4], adaptive strategy [5], and variable velocity strategy [6].

This work proposes a particle swarm optimization algorithm that incorporates the improved opposition-based learning (IOBL) technique; this technique is an improved version of OBL first proposed by Tiz-hoosh [7]. IOBL was proposed by Alomoush et al. [1], introducing randomness to create a new possible solution [1], and it will be used at the end of each iteration of the PSO.

The following section briefly describes the standard PSO, the PSO with variable velocity strategy (VVS-PSO), and the PSO with IOBL (IOBL-PSO) proposed in this research. Subsequently, to verify the feasibility of using the IOBL technique, the evaluation of nine standard benchmark functions will be carried out, and finally, the results obtained from the experimentation and conclusions.

2 Structure of the Standard PSO and the VVS-PSO Variant

In this section, the standard PSO algorithm and its main components are described; later, the variant called VVS-PSO is described.

2.1 *General Structure of the Standard PSO*

Bratton and Kennedy [8] propose a standardized definition of the PSO algorithm initially created by Kennedy and Eberhart [3]. This algorithm also uses the IOBL technique to enable better exploration in the solution space. In steps 1 and 2, an initial population of particles is generated. The best global particle is obtained concerning its objective value. Step 3 calculates the minimum (*velMin*) and maximum (*velMax*) speed with which these particles will move in the solution space. From steps 4–13, each particle is updated concerning its position and velocity. Step 14 calculates the objective function of each particle. From steps 15–22, the best particle is obtained concerning its objective value and is compared with the global solution; if its value is better than the global solution, it is updated. This process continues until the maximum number of iterations (*MaxIt*) is achieved. Finally, the inertial weight (w) is updated, and in step 25, the algorithm returns the best overall solution obtained during the process. This standard definition is described in Algorithm 1.

Algorithm 2.1. Standard particle swarm optimization

Inputs: $MaxIt$: Maximum number of iterations, $nPop$: Population size, w: Inertia weight, w_{min}: Minimum inertia weight, w_{max}: Maximum inertia weight, c_1: Personal learning coefficient, c_2: Global learning coefficient, D: Number of dimensions, $varMin$: The minimum value that a decision variable can take, $varMax$: The maximum value that a decision variable can take.

Outputs: $GlobalBest$: Best solution found.

```
particles = InitializePopulation(nPop)
GlobalBest = getBestGlobal(particles)
velMax = 0.1 * (varMax − varMin), velMin = − velMax
for it = 1 to MaxIt do
  for i = 1 to nPop do
    for j = 1 to D do
      particles_i = w × particles_i.velocity_j + c_1 × random_number [0,1] ×
      (particles_i.bestPosition_j − particles_i.position_j) + c_2 ×
      random_number[0,1] × (GlobalBest.position_j − particles_i.position_j);
      particles_i.velocity_j = Max(particles_i.velocity_j, velMin);
      particles_i.velocity_j = Min(particles_i.velocity_j, velMax);
      particles_i.position_j = particles_i.posicion + particles_i.velocity_j;
      particles_i.position_j = Max(particles_i.position_j, varMin);
      particles_i.position_j = Min(particles_i.position_j, varMax);
    endfor
    ObjectiveFunction(particles_i)
    if particles_i.cost < particles_i.bestCost then
      particles_i.bestPosition = particless_i.position
      particles_i.bestCost = particles_i.cost
      if particles_i.bestCost < GlobalBest.cost then
        GlobalBest = particles_i
      endif
    endif
  endfor
  w = w_min + (k^iter / Max^iter) (w_max − w_min)
endfor
return GlobalBest
```

The algorithm 1 speed is bounded in the range $[velMin, velMax]$. If the speeds obtained are out of range, an adjustment is made so that it is not less than $velMin$ and not greater than $velMax$. On the other hand, when the position of the particle violates the constraints of minimum position $[varMin]$ and maximum position $[varMax]$, an adjustment is also carried out to avoid waste in updating the position. Equations (1) and (2) represent the process of updating the speed and position of the particles:

$$V_i^{k+1} = wV_i^k + c_1r_1\left(L_{Best,i}^k\right) + c_2r_2(G_{Best,i}^k - X_i^k) \tag{1}$$

$$X_i^{k+1} = X_i^k + V_i^k \tag{2}$$

where w corresponds to the inertial weight, r_1 and r_2 are random values in the range [0, 1]. c_1 and $c2$ are user-defined learning factors. $L_{Best,i}^k$ represents the best position of the current particle and $G_{Best,i}^k$ refers to the best global position of the particles. X_i^k refers to the current position of the particle. Usually, the inertial weight w is calculated using Eq. 3.

$$w = w_{min} + \frac{k^{iter}}{Max^{iter}}(w_{max} - w_{min}) \tag{3}$$

where $w_{max} = 0.9$, $w_{min} = 0.4$, k^{iter} represents the current iteration of the algorithm, and Max^{iter} corresponds to the maximum number of iterations of the algorithm.

2.2 Structure of the VVS-PSO Variant

The VVS-PSO variant is presented for the first time in 2021 as a strategy to improve the convergence rate in the original algorithm [6]. VVS-PSO is a recent technique that improves the particle velocity update process; this involves replacing Eq. (1) with Eq. (4).

$$\begin{aligned} V_i^{k+1} &= wV_i^k + c_1r_1\left(L_{Best,i}^k\right) + c_2r_2\left(G_{Best,i}^k - X_i^k\right) \\ &\quad + f(k+1)\left(First_{Best} - \left(X_i^k + f(k+1)R_arand(-1,1)\right)\right) \end{aligned} \tag{4}$$

where $f(k)$ is equal to one in the first iteration and zero in the iteration determined by user through the λ variant, this function is represented in Eq. (5).

$$\begin{cases} f(k) = 1 - \lambda k \, if \, k \leq \frac{1}{\lambda} \\ f(k) = 0 \, if \, k > \frac{1}{\lambda} \end{cases} \tag{5}$$

where $\lambda = [0.1, 0.2, 0.3, 0.4, 0.5, 0.6, 0.7, 0.8, 0.9, 1.0]$.

The value of $First_{Best}$ corresponds to the first k_{Best} solutions (particles) and is calculated using Eq. (6).

$$First_{Best} = \frac{\sum_{i=1}^{k_{Best}} X_i = (x_{i,1}, x_{i,2}, \ldots, x_{i,D})}{k_{Best}} \tag{6}$$

Furthermore, the value of k_{Best} is expressed using Eq. (7).

$$k_{Best} = N - (N - 2)\sqrt{\frac{k}{K^{max}}} \tag{7}$$

where N represents the number of particles, k is the current iteration, and K^{max} is the total number of iterations.

Finally, the variable R_a is the new search radius around X_i^k, expressed using Eq. (8).

$$R_a = 0.1\sqrt{\frac{(X_{max} - X_{min})}{2N}} \tag{8}$$

X_{max} is the upper limit, and X_{min} is the lower limit of the optimization variables. The flowchart of the VVS-PSO algorithm is shown in Fig. 1.

3 Proposed Algorithm

This section describes all the elements that make up the structure of the algorithm proposed in this research, called particle swarm optimization with improved opposition-based learning (IOBL-PSO).

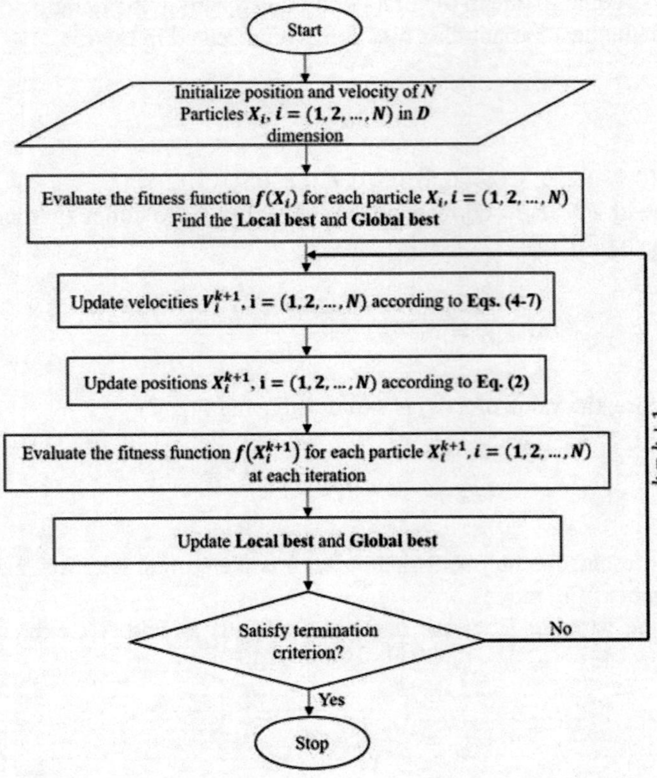

Fig. 1 Flowchart of VVS-PSO algorithm [1]

3.1 *Structure of Improved Opposition-Based Learning (IOBL)*

Alomoush et al. [1] introduce randomness in generating new solutions for the first time to improve the diversification process and present an improved version of opposition-based learning (IOBL). This new technique provides better performance in solving ongoing problems.

Generally, this technique loops through all elements of a solution vector ($particle_d$) and multiplies each element by Random number in the range [0, 1].

Algorithm 2. Improved opposition-based learning method (IOBL)

Inputs: *particles:* population, *D:* number of dimensions.
Outputs: *GlobalBest:* Best solution found.

```
for i = 1 to particles.size do
    r = random_number [0,1]
    While (j < D) do
        X̄_j = particles_i.position[j] * r
    endwhile
    Objectivefunction(X̄)
    if X̄.cost < particles_i.bestCost then
        particles_i = X̄
        if particles_i.cost < GlobalBest.cost then
            GlobalBest = x
        endif
    endif
endfor
return GlobalBest
```

3.2 *Structure of the IOBL Particle Swarm Optimization (IOBL-PSO) Algorithm*

This algorithm also uses the IOBL technique to enable better exploration in the solution space. In steps 1–3, the input parameters are initialized, a random population of particles is generated, and the best global particle is obtained. Step 4 obtains the minimum (*velMin*) and maximum (*velMax*) speed that fit the particles to a given space. From steps 8–13, the position and velocity of are update; it is verified that they fit within a specific range. Step 15 obtains the objective value of the particle. From steps 16–22, compare the best position of the current particle against the global particle; if it has a better objective value, the current particle will replace the global particle. In step 24, the IOBL technique is applied to the entire population and compared against the best global particle; if any of the new IOBL-generated particles have a better objective value, the global best particle will be replaced. Finally, in step 25, the inertial weight (*w*) is updated.

The process continues until the maximum number of iterations (*MaxIt*) is reached.

The algorithm returns the best overall solution obtained during all processes in step 27.

Algorithm 3.2. Particle swarm optimization with IOBL (PSO-IOBL)

Inputs: *MaxIt:* Maximum number of iterations, *nPop:* Population size, *w:* Inertia weight, *wdamp:* Inertia weight damping ratio, *c1:* Personal learning coefficient, *c2:* Global learning coefficient, *D:* Number of dimensions, *varMin:* The minimum value that a decision variable can take, *varMax:* The maximum value that a decision variable can take.

Outputs: *GlobalBest:* Best solution found

```
Initialize parameters MaxI, nPop, w, wdamp, c1, c2, D, varMin, varMax.
particles = GeneratePopulation(HM)
GlobalBest = getBestGlobal(particles)
velMax = 0.1 * (varMax − varMin), velMin = − velMax
for it = 1 to MaxIt do
  for i = 1 to nPop do
    for j = 1 to d do
      particles_i = w × particles_i.velocity_j + c1 × random_number [0,1] ×
      (particles_i.bestPosition_j − particles_i.position_j) + c2 ×
      random_number(0,1) × (GlobalBest.position_j − particles_i.position_j);
      particles_i.velocity_j = Max(particles_i.velocity_j, velMin);
      particles_i.velocity_j = Min(particles_i.velocity_j, velMax);
      particles_i.position_j = particles_i.posicion + particles_i.velocity_j;
      particles_i.position_j = Max(particles_i.position_j, varMin);
      particles_i.position_j = Min(particles_i.position_j, varMax);
    endfor
    Objectivefunction(particles_i)  //calculate the objective function of the particles
    if particles_i.cost < particles_i.bestCost then
      particles_i.bestPosition = particless_i.position
      particles_i.bestCost = particles_i.cost
      if particles_i.bestCost < GlobalBest.cost then
        GlobalBest = particles_i
      endif
    endif
  endfor
  IOBL(particles)  //applies the technique of improving opposition-based learning
  w = w * wdamp
endfor
return GlobalBest
```

4 Computational Experiments

Table 1 shows nine mathematical benchmark functions, their optimal value, and the range of variables used in the experiments. For each algorithm and mathematical function, 30 independent runs were performed.

Table 2 presents the values assigned to the variables and parameters used in the computational experiments for each algorithm analyzed.

Table 1 Mathematical benchmark functions

Function	Lower bound	Upper bound	Global Minimum
Sphere	−100	100	0
Schwefel's 2.22	−10	10	0
Step	−100	100	0
Rosenbrock	−30	30	0
Schwefel's 2.26	−500	500	−12,569.5
Rastrigin	−5.12	5.12	0
Ackleys	−32	32	0
Griewank	−600	600	0
Rotate hyper-ellipsoid	−100	100	0

Table 2 Variables and parameters

variables/parameters	PSO	VVS-PSO	IOBL-PSO
$wdamp$	–	–	0.99
w_{min}	0.4	0.4	0.4
w_{max}	0.9	0.9	0.9
c_1, c_2	2, 2	2, 2	1.5, 2.0
D	500	500	500
$varMin$	Lower bound in Table 4	Lower bound in Table 4	Lower bound in Table 4
$varMax$	Upper bound in Table 4	Upper bound in Table 4	Upper bound in Table 4
$nPop$	30	30	30
$MaxIt$	30	30	30

5 Results

Table 3 shows the results obtained from the computational experiments. The first column identifies each evaluated mathematical function; the second and third columns correspond to the objective values average execution time obtained from the 30 best results for the standard PSO algorithm. On the other hand, the fourth and fifth columns show the information that the third and fourth columns for the proposed reference algorithm (IOBL-PSO). The sixth column corresponds to the *p-value* obtained by applying the non-parametric Wilcoxon test with a significance level of 5%. The shaded cell represents the lowest objective value or shortest execution time, as the case may be. The symbol ↑ represents a significant difference in favor of the reference algorithm.

The results show that the proposed algorithm (IOBL-PSO) is superior to the standard PSO algorithm. Furthermore, quality is superior in all the instances evaluated, and the efficiency is better in three out of nine instances.

Table 4 shows the results obtained from the computational experiments. The first column identifies each evaluated mathematical function; the second and third columns correspond to the objective values average execution time obtained from the 30 best results for the VVS-PSO algorithm. On the other hand, the fourth and fifth columns show the information that the third and fourth columns for the proposed reference algorithm (IOBL-PSO). The sixth column corresponds to the *p-value* obtained by applying the non-parametric Wilcoxon test with a significance level of 5%. The shaded cell represents the lowest objective value or shortest execution time, as the case may be. The symbol ↑ represents a significant difference in favor of the reference algorithm.

The results show that the proposed algorithm (IOBL-PSO) is superior to the VVS-PSO algorithm. Furthermore, quality is superior in all the instances evaluated, and the efficiency is better in six out of nine instances.

Table 3 Results were obtained with the PSO and IOBL-PSO algorithms

Function	PSO		IOBL-PSO		*p-value*
	OV	Time	OV	Time	
Sphere	32,680.518	0.0015	261.055563 ↑	0.00573333	0.00001
Schwefel's 2.22	312.78998	0.0026	3.57836821 ↑	0.00363333	0.00001
Step	31,680.6667	0.00366667	347.366667 ↑	0.00053333	0.00001
Rosenbrock	12,587,324.9	0.00156667	57,012.6648 ↑	0.00426667	0.00001
Schwefel's 2.26	−63,137.0058	0.0069	−12,046.4824 ↑	0.016	0.00001
Rastrigin	2726.61853	0.00206667	135.480273 ↑	0	0.00001
Ackleys	9.88770564	0.002	0.23046396 ↑	0.012	0.00001
Griewank	318.644436	0.00156667	3.50954065 ↑	0	0.00001
Rotate hyper-ellipsoid	8,413,594.45	0.0068	89,038.3048 ↑	0.02336667	0.00001

Table 4 Results were obtained with the VVS-PSO and IOBL-PSO algorithms

Function	VVS-PSO		IOBL-PSO		*p-value*
	OV	Time	OV	Time	
Sphere	51,565.7644	0.00826667	261.055563 ↑	0.00573333	0.00001
Schwefel's 2.22	465.272846	0.00773333	3.57836821 ↑	0.00363333	0.00001
Step	54,335.7	0.0099	347.366667 ↑	0.00053333	0.00001
Rosenbrock	29,222,446.3	0.00516667	57,012.6648 ↑	0.00426667	0.00001
Schwefel's 2.26	−57,387.5437	0.00573333	−12,046.4824 ↑	0.016	0.00001
Rastrigin	3961.65753	0.0079	135.480273 ↑	0	0.00001
Ackleys	11.6386167	0.00783333	0.23046396 ↑	0.012	0.00001
Griewank	544.999098	0.00576667	3.50954065 ↑	0	0.00001
Rotate hyper-ellipsoid	12,824,569.7	0.01093333	89,038.3048 ↑	0.02336667	0.00001

6 Conclusions

Finally, as shown in the results obtained through the experiments carried out, the proposed algorithm (IOBL-PSO) is superior to the standard PSO and VVS-PSO of the state-of-the-art. Moreover, incorporating the IOBL technique at the end of each iteration of the PSO allows for obtaining the best results in evaluating nine benchmark functions. The results show that the proposed algorithm (IOBL-PSO) is superior to the standard PSO algorithm. Furthermore, quality is superior in all the instances evaluated, and the efficiency is better in three out of nine instances. Finally, the results show that the proposed algorithm (IOBL-PSO) is superior to the VVS-PSO algorithm. In quality is superior in all the instances evaluated, and the efficiency is better in six out of nine instances.

References

1. Alomoush, A. A., Alsewari, A. R. A., Zamli, K. Z., Alrosan, A., Alomoush, W., & Alissa, K. (2021). Enhancing three variants of harmony search algorithm for continuous optimization problems. *International Journal of Electrical & Computer Engineering, 11*(3), 2088–8708.
2. Cruz-Reyes, L., Hernández H, P., Melin, P., Fraire H, H. J., & Mar O, J. (2013). Constructive algorithm for a benchmark in ship stowage planning. In *Recent advances on hybrid intelligent systems* (pp. 393–408). Berlin, Heidelberg: Springer.
3. Kennedy, J., & Eberhart, R. (1995). Particle swarm optimization. In *Proceedings of ICNN'95-International Conference on Neural Networks* (Vol. 4, pp. 1942–1948). IEEE.
4. Ajibade, S. M., Chweya, R., Ogunbolu, M. O., & Fadipe, S. R. (2022). Utilizing logistic map to enhance the population diversity of PSO. In *Journal of Physics: Conference Series* (Vol. 2250, No. 1, p. 012016). IOP Publishing.
5. Liu, H., Zhang, X. W., & Tu, L. P. (2020). A modified particle swarm optimization using adaptive strategy. *Expert Systems with Applications, 152*, 113353.

6. Minh, H. L., Khatir, S., Rao, R. V., Abdel Wahab, M., & Cuong-Le, T. (2021). A variable velocity strategy particle swarm optimization algorithm (VVS-PSO) for damage assessment in structures. *Engineering with Computers*, 1–30.
7. Tizhoosh, H. R. (2005). Opposition-based learning: a new scheme for machine intelligence. In *International Conference on Computational Intelligence for Modelling, Control and Automation and International Conference on Intelligent Agents, Web Technologies and Internet Commerce (CIMCA-IAWTIC'06)* (Vol. 1, pp. 695–701). IEEE.
8. Bratton, D., & Kennedy, J. (2007). Defining a standard for particle swarm optimization. In *2007 IEEE Swarm Intelligence Symposium* (pp. 120–127). IEEE.

Study on the Effect of Chaotic Maps in the Formation of New Universes in the Multiverse Optimizer Algorithm

Lucio Amézquita, Oscar Castillo, José Soria, and Prometeo Cortes-Antonio

1 Introduction

Fuzzy Logic has been gaining more popularity since Zadeh [1] proposed it, as a part of computational intelligence techniques [2], in many applications over the past few years, such as scientific research.

One of the multiple applications of Fuzzy Logic [3] is Fuzzy controllers, that can be used in multiple control systems applications, this because Fuzzy Logic resembles the way human beings think to solve problems in many situations, making the resulting systems easier to understand; with the use of uncertainty on these models, we can have more approximate decisions as those made by any person in that place, as the systems use a degree of membership to simulate this.

The metaheuristics are also part of computational intelligence techniques, which can be used to improve other techniques, such as Fuzzy controllers. These techniques can have different inspirations over nature or artificial behaviors [4].

In other works [5–8], we have used the Multi-verse optimizer (MVO) algorithm for benchmark mathematical function tests, because we have been adapting it to perform better in other scenarios, such as fuzzy logic controller design. The MVO [9, 10] algorithm has inspirations in some concepts of cosmology, such as worm holes, black holes and white holes; based on how theoretically they transport matter

L. Amézquita · O. Castillo (✉) · J. Soria · P. Cortes-Antonio
Tijuana Institute of Technology, Tijuana, México
e-mail: ocastillo@tectijuana.mx

L. Amézquita
e-mail: lucio.amezquita19@tectijuana.edu.mx

J. Soria
e-mail: jose.sa@tijuana.tecnm.mx

P. Cortes-Antonio
e-mail: prometeo.cortes@tectijuana.edu.mx

O. Castillo and P. Melin (eds.), *Hybrid Intelligent Systems Based on Extensions of Fuzzy Logic, Neural Networks and Metaheuristics*, Studies in Computational Intelligence 1096,
https://doi.org/10.1007/978-3-031-28999-6_8

in space [11], translating in the algorithm as, how the solutions communicate between them in the metaheuristic process.

The contribution of this work is to analyze the behavior of the MVO algorithm alongside the chaotic maps, and how they can influence in the solutions obtained to obtain the optimal results, by observing the formation of new universes in some specific parameters.

In this paper we organize the sections as follows: Sect. 2 is a review of the state-of-the-art in the MVO algorithm and some adaptations in other works, Sect. 3 Describes the proposed methodology with chaotic maps used for this work, Sect. 4 has some of the main results for our tests and Sect. 5 has our main conclusions in this work.

2 Background and MVO

In computational techniques we have multiple areas, optimization algorithms as one of these techniques focuses on finding the optimal results for the scenario it has been used. The optimization algorithms [12] have numerous inspirations over nature and artificial behaviors, on how they obtain an initial set of solutions, the way they combine them and select, to finally obtain the best solution so far; the Multi-verse optimizer algorithm being one of these algorithms, has its inspirations over worm holes, black holes and white holes, to obtain the best solutions, called here universes, to obtain the fitness of each universe, this translates in the inflation rate for this specific algorithm; in Fig. 1 we can observe a model of the MVO algorithm and his concepts.

The MVO algorithm has been used in other works [5–8], and it has been receiving changes in some of its main parameters, such as Wormhole existence probability (WEP) and Travel distance rate (TDR), where the main objective is to avoid local optima stagnation in the results. In the state-of-the-art, it has been used several improvements used in most metaheuristics, such as levy flights [13], neural networks [14], use of quantum and chaos theory [6]; as for our improvements, we have been using fuzzy logic in some parameters.

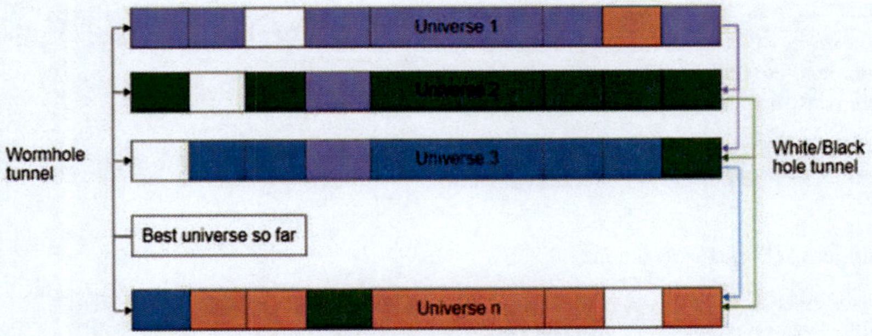

Fig. 1 MVO algorithm concepts interaction

In the research done, these variations of MVO and other algorithms use similar adaptations to improve the overall performance of the algorithm in several situations, having better convergence rate and more consistent results. The adaptations consist on, changing the universes or solutions over the iterations in the algorithm, helping in the formation of new universes in the search of a global optimal solution.

3 FCMVO Proposed Methodology

Our work is focused on adaptation of chaotic maps [15, 16] in the MVO algorithm, this to pay attention to the effects of chaos theory implementing some chaotic maps on the formation of new universes. In other works, we have implemented the MVO algorithm in multiple cases, from benchmark mathematical functions, to fuzzy logic controller optimization [17, 18], resulting in competing results against other optimization algorithms.

As we mentioned in previous chapter, the main adaptations in the algorithm reside over the Wormhole existence probability (WEP) and Travel distance rate (TDR), using a fuzzy inference system instead of an equation. Now, in this new variant of the MVO algorithm, that we are calling Fuzzy-Chaotic Multi-verse Optimizer (FCMVO for short), we are also using fuzzy logic, alongside chaos theory, to have the formation of new universes or solutions for the problems we are testing.

In FCMVO, we use adapt TDR and WEP using fuzzy logic; for this, the original algorithm used the equations presented in (1) and (2). In WEP, a_{min} is the minimum value for WEP, a_{max} is the maximum value for WEP, l is the actual iteration and L is the max iterations; for TDR, l is the actual iteration, L is the max iterations and p is the exploitation precision.

$$WEP = a_{min} + l \times \left(\frac{a_{max} - a_{min}}{L} \right) \tag{1}$$

$$TDR = 1 - \frac{l^{1/p}}{L^{1/p}} \tag{2}$$

For WEP, we modelled a Mamdani [16] fuzzy inference system with one input and one output. The input consists of three triangular membership functions, which represent the lightyears or iterations of the system, granulated in low, medium and high; for the output we have the desired WEP value, granulated in low, medium and high with triangular membership functions. The representation of this fuzzy inference system can be observed in Figs. 2 and 3.

In the case of TDR, we also use a Mamdani fuzzy inference system, which has one input and one output; the input represents the lightyears, that is granulated [16] in three triangular membership functions for low, medium and high, and the output that represents the TDR, is granulated in low, medium and high with triangular

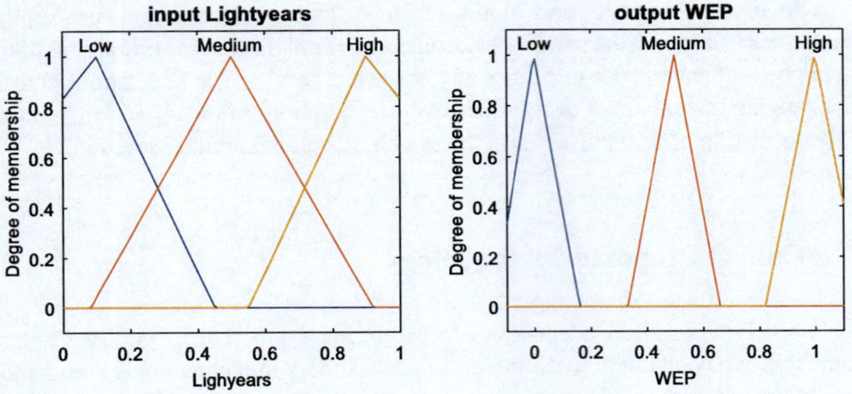

Fig. 2 WEP fuzzy inference system input and output

Fig. 3 WEP surface output

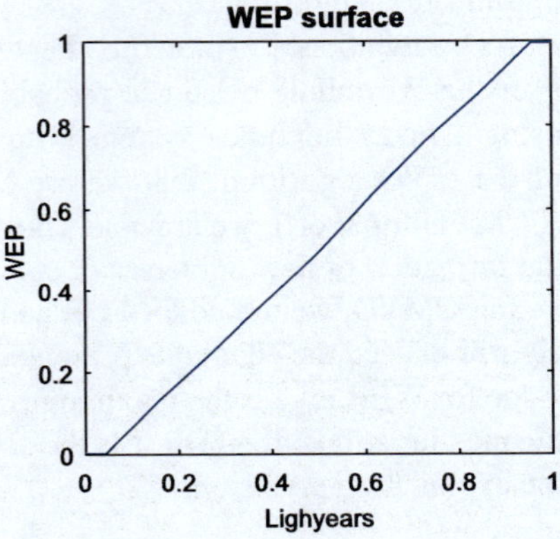

membership functions. This fuzzy inference system can be observed in Figs. 4 and 5.

$$x_i^j = \begin{cases} \begin{cases} xbest_j + TDR \times \left(\left(ub_j - lb_j\right) \times r_4 + lb_j\right) & r_3 < 0.5 \\ xbest_j - TDR \times \left(\left(ub_j - lb_j\right) \times r_4 + lb_j\right) & r_3 \geq 0.5 \end{cases} & r_2 < WEP \\ x_i^j & r_2 \geq WEP \end{cases} \quad (3)$$

The main change in this MVO variation resides in the use of chaos theory, where we use 10 of the most common chaotic maps [15] in the state-of-the-art to observe the effects over the generation of new universes in the MVO algorithm. The chaotic maps used can be observed in Table 1.

We adapt the chaotic maps in (3), that represents a part of the solution of the universe selected. Each parameter represents the following, x_i^j is the jth parameter of the ith solution, $xbest_j$ is the jth of the best universe, lb_j and ub_j are the upper

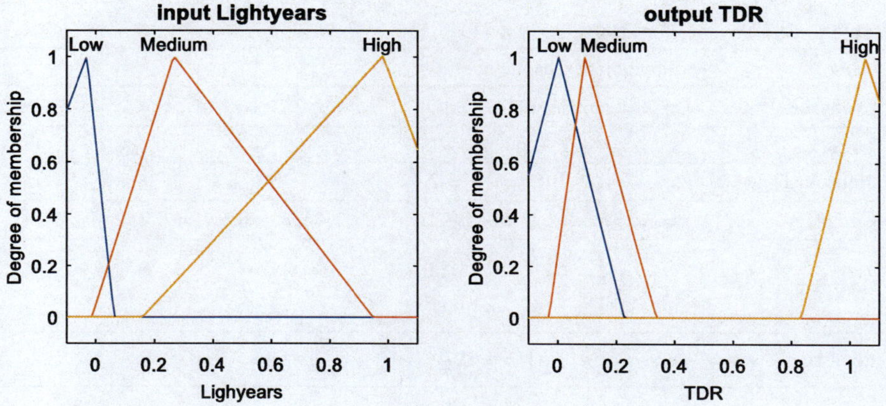

Fig. 4 TDR fuzzy inference system input and output

Fig. 5 TDR surface output

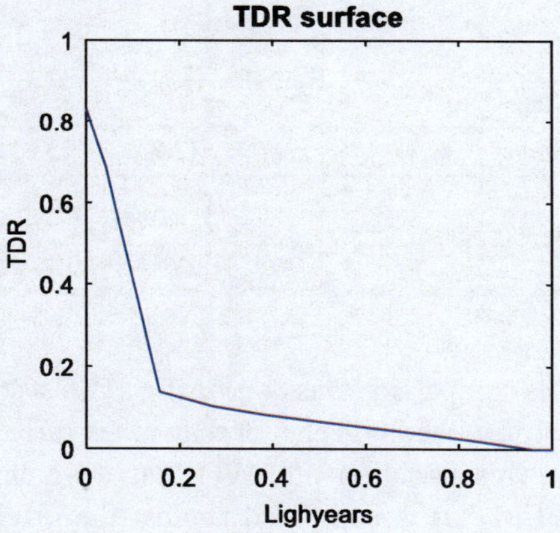

and lower bound of the jth variable, r_2 and r_3 are random numbers between [0, 1], and r_4 is a number generated by the chaotic map between 0 and 1, selecting only one of the 10 maps mentioned for each experiment. In Fig. 6 we can observe a flowchart of the MVO algorithm and where is implemented this equation with chaotic maps.

4 Test and Results

Over this work we made tests with 13 benchmark mathematical functions that we used for this algorithm [9]; we performed 30 tests for each case, with 5, 50 and 100 dimensions, 500 iterations and 50 universes, this to compare with our previous results.

Table 1 Chaotic maps for MVO algorithm [15]

Name	Equation of Chaotic map
Chebyshev	$x_{i+1} = cos\left(icos^{-1}(x_i)\right)$
Logistic	$x_{i+1} = ax_i(1 - x_i), a = 4$
Sinusoidal	$x_{i+1} = ax_i^2 sin(\pi x_i), a = 2.3$
Circle	$x_{i+1} = mod\left(x_i + b - \left(\frac{b}{2\pi}\right)sin(2\pi x_i), 1\right), a = 0.5, b = 0.2$
Gauss/mouse	$x_{i+1} = \begin{cases} 1 & x_i = 0 \\ \frac{1}{mod(x_i,1)} & otherwise \end{cases}$
Iterative	$x_{i+1} = sin\left(\frac{a\pi}{x_i}\right), a = 0.7$
Piecewise	$x_{i+1} = \begin{cases} \frac{x_i}{P} & 0 \leq x_i < P \\ \frac{x_i - P}{0.5 - P} & P \leq x_i < 0.5 \\ \frac{1 - P - x_i}{0.5 - P} & 0.5 \leq x_i < 1 - P \\ \frac{1 - x_i}{P} & 1 - P \leq x_i < 1 \end{cases}, P = 0.4$
Sine	$x_{i+1} = \frac{a}{4} sin(\pi x_i), a = 4$
Singer	$x_{i+1} = \mu\left(7.86x_i - 23.31x_i^2 + 28.75x_i^3 - 13.302875x_i^4\right), \mu = 1.07$
Tent	$x_{i+1} = \begin{cases} \frac{x_i}{0.7} & x_i < 0.7 \\ \frac{10}{3}(1 - x_i) & x_i \geq 0.7 \end{cases}$

The comparison was done using a Z-test statistical test with a 95% of confidence that our new method is better than other variations of the MVO algorithm.

This variation of MVO that we called Fuzzy-Chaotic Multi-verse Optimizer (FCMVO) is compared against the original MVO and other variations, the first test results are presented with a comparison between the Chaotic version (CMVO) in some chaotic maps and its corresponding algorithm, which can be observed in Table 2.

In Table 2, we only presented the two best variants with chaotic maps from the 10 variants proposed for testing, in the case of 50 dimensions. With these results, we observed that there was some improvement over its corresponding algorithm and the chaotic variant CMVO. Our next comparison is between the fuzzy variant FMVO and the original algorithm, that can be observed in Table 3.

From these tests, the fuzzy variant of the Multi-verse Optimizer has improvements over the corresponding algorithm in multiple functions, and from here, we could perform the next tests with the FCMVO variant presented in this work, where we use both fuzzy logic and chaotic maps, this to observe the results improvement. We can observe these results in Table 4.

With our last results, we observed that this new variant of MVO algorithm, FCMVO, has a better performance over its corresponding algorithm, achieving the

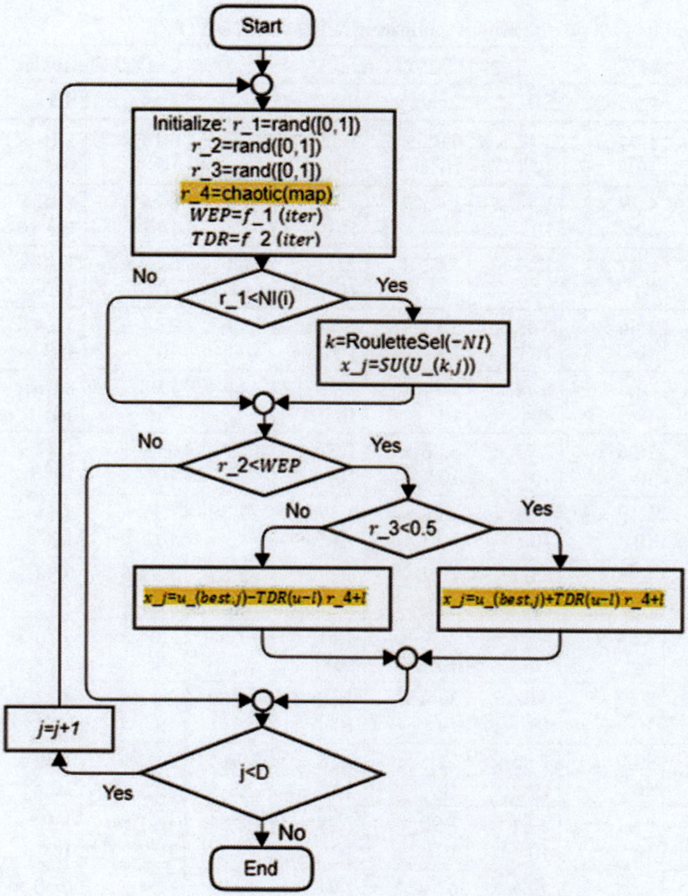

Fig. 6 Flowchart of MVO algorithm with chaotic maps implemented

optimal solution in a more effective way, but this could only be achieved in some chaotic maps, as we presented in our results with the Tent map.

Table 2 Results with 50 dimensions comparing MVO with CMVO

Algorithm	MVO		CMVO circle			CMVO sinusoidal		
Function	Average	SD	Average	SD	z	Average	SD	z
F1	1.04×10^{1}	2.12×10^{0}	6.05×10^{0}	1.32×10^{0}	**9.55**	6.18×10^{0}	1.16×10^{0}	**9.57**
F2	4.29×10^{2}	1.40×10^{3}	1.43×10^{2}	9.27×10^{1}	−1.11	8.24×10^{1}	7.40×10^{1}	−1.35
F3	5.87×10^{3}	1.42×10^{3}	6.62×10^{2}	2.02×10^{3}	1.68	6.50×10^{3}	1.70×10^{3}	1.57
F4	1.66×10^{1}	6.53×10^{0}	3.14×10^{1}	7.54×10^{0}	8.13	2.14×10^{1}	5.14×10^{0}	3.12
F5	6.64×10^{2}	6.96×10^{2}	1.14×10^{3}	9.97×10^{2}	2.14	1.08×10^{3}	8.07×10^{2}	2.12
F6	1.06×10^{1}	2.71×10^{0}	5.86×10^{0}	1.57×10^{0}	**8.23**	6.39×10^{0}	1.37×10^{0}	**7.54**
F7	1.19×10^{-1}	4.03×10^{-2}	1.08×10^{-1}	3.39×10^{-2}	−1.20	1.12×10^{-1}	3.64×10^{-2}	−0.67
F8	1.25×10^{4}	8.00×10^{2}	1.22×10^{4}	8.37×10^{2}	−1.37	1.22×10^{4}	9.58×10^{2}	−1.11
F9	2.54×10^{2}	4.94×10^{1}	2.34×10^{2}	4.61×10^{1}	−1.58	2.04×10^{2}	3.70×10^{1}	**4.42**
F10	3.49×10^{0}	3.08×10^{0}	3.04×10^{0}	4.60×10^{-1}	−0.80	2.43×10^{0}	4.22×10^{-1}	**1.86**
F11	1.09×10^{0}	1.82×10^{-2}	1.05×10^{0}	1.38×10^{-2}	**8.70**	1.05×10^{0}	1.39×10^{-2}	**9.12**
F12	6.57×10^{0}	2.64×10^{0}	7.90×10^{0}	2.78×10^{0}	1.89	4.91×10^{0}	2.50×10^{0}	**2.49**
F13	9.08×10^{0}	1.34×10^{1}	6.56×10^{0}	7.21×10^{0}	−0.91	4.61×10^{0}	6.30×10^{0}	**1.66**

5 Conclusions

For this work, we presented a new variant of the Multi-Verse Optimizer using Chaotic Maps and Fuzzy Logic, this to have better results over benchmark mathematical functions. We focused over a specific parameter called R4, that affects the generation of solutions in the algorithm.

In the results of our tests, we compared the results of previous works in the MVO algorithm and our new results with Chaotic Maps, where we observed some good improvements over the original algorithm and other variations.

Table 3 Results with 50 dimensions comparing MVO with FMVO

Algorithm	MVO		FMVO		
Function	Average	SD	Average	SD	z
F1	1.04×10^{1}	2.12×10^{0}	4.22×10^{0}	1.54×10^{0}	**12.94**
F2	4.29×10^{2}	1.40×10^{3}	1.84×10^{2}	3.50×10^{2}	−0.93
F3	5.87×10^{3}	1.42×10^{3}	4.46×10^{3}	1.32×10^{3}	**3.98**
F4	1.66×10^{1}	6.53×10^{0}	1.69×10^{1}	7.05×10^{0}	0.14
F5	6.64×10^{2}	6.96×10^{2}	5.02×10^{2}	5.32×10^{2}	−1.01
F6	1.06×10^{1}	2.71×10^{0}	4.05×10^{0}	1.38×10^{0}	**11.75**
F7	1.19×10^{-1}	4.03×10^{-2}	7.71×10^{-2}	2.46×10^{-2}	**4.88**
F8	1.25×10^{4}	8.00×10^{2}	1.26×10^{4}	9.34×10^{2}	0.85
F9	2.54×10^{2}	4.94×10^{1}	2.91×10^{2}	4.74×10^{1}	2.99
F10	3.49×10^{0}	3.08×10^{0}	8.15×10^{0}	8.24×10^{0}	2.90
F11	1.09×10^{0}	1.82×10^{-2}	1.00×10^{0}	6.84×10^{-2}	**6.71**
F12	6.57×10^{0}	2.64×10^{0}	4.77×10^{0}	1.64×10^{0}	**3.16**
F13	9.08×10^{0}	1.34×10^{1}	1.95×10^{0}	2.01×10^{0}	**2.88**

We presented only some of the best results in this work, and we observed that the best Chaotic map was the Tent map, giving the best improvements when this was added to the original and Fuzzy variant of MVO for benchmark mathematical functions at any dimensionality studied.

Lastly, in this study we can conclude that when a Chaotic map is used in the R4 parameter for both the MVO and FMVO algorithms, the new version has a better performance over its corresponding algorithm, converging more effectively to the optimal solution. The results of this study have opened a path to more tests with Fuzzy Logic [19, 20] and Chaotic maps, so we can use them over other cases of study, such as Fuzzy controller design [21], and even use it over type-2 Fuzzy Logic [22] or in other application areas [23].

Table 4 Results with 50 dimensions comparing FCMVO with other variants

Algorithm	FCMVO Tent		MVO Tent			FMVO		
Function	Average	SD	Average	SD	z	Average	SD	z
F1	2.06×10^{0}	6.70×10^{-1}	8.93×10^{0}	2.34×10^{0}	**15.47**	4.22×10^{0}	1.54×10^{0}	**7.06**
F2	3.56×10^{2}	9.16×10^{2}	4.13×10^{2}	1.49×10^{3}	−0.18	1.84×10^{2}	3.50×10^{2}	0.96
F3	5.31×10^{3}	1.23×10^{3}	7.04×10^{3}	1.57×10^{3}	**4.75**	4.46×10^{3}	1.32×10^{3}	2.59
F4	3.68×10^{1}	8.56×10^{0}	3.41×10^{1}	8.16×10^{0}	1.26	1.69×10^{1}	7.05×10^{0}	9.84
F5	6.41×10^{2}	6.98×10^{2}	8.09×10^{2}	7.15×10^{2}	−0.92	5.02×10^{2}	5.32×10^{2}	0.87
F6	1.98×10^{0}	6.58×10^{-1}	9.90×10^{0}	1.84×10^{0}	**22.20**	4.05×10^{0}	1.38×10^{0}	**7.46**
F7	9.21×10^{-2}	2.86×10^{-2}	1.04×10^{-1}	3.82×10^{-2}	−1.35	7.71×10^{-2}	2.46×10^{-2}	2.17
F8	1.20×10^{4}	8.69×10^{2}	1.20×10^{4}	8.51×10^{2}	0.08	1.26×10^{4}	9.34×10^{2}	**2.72**
F9	3.08×10^{2}	5.91×10^{1}	2.41×10^{2}	3.86×10^{1}	5.19	2.91×10^{2}	4.74×10^{1}	1.18
F10	6.50×10^{0}	7.08×10^{0}	3.03×10^{0}	3.39×10^{-1}	2.68	8.15×10^{0}	8.24×10^{0}	−0.83
F11	8.49×10^{-1}	6.84×10^{-2}	1.09×10^{0}	1.94×10^{-2}	**18.19**	1.00×10^{2}	6.84×10^{-2}	**8.72**
F12	7.42×10^{0}	2.11×10^{0}	7.83×10^{0}	2.97×10^{0}	−0.62	4.77×10^{0}	1.64×10^{0}	5.42
F13	2.82×10^{0}	2.41×10^{0}	4.66×10^{0}	6.56×10^{0}	−1.44	1.95×10^{0}	2.01×10^{0}	1.52

References

1. Zadeh, L. A. (1968). Fuzzy algorithms. *Information and Control, 12*, 94–102. https://doi.org/10.1016/S0019-9958(68)90211-8
2. Shahid, A. H., & Singh, M. P. (2019). Computational intelligence techniques for medical diagnosis and prognosis: Problems and current developments. *Biocybernetics and Biomedical Engineering, 39*, 638–672. https://doi.org/10.1016/J.BBE.2019.05.010
3. Mittal, K., Jain, A., Vaisla, K. S., Castillo, O., & Kacprzyk, J. (2020). A comprehensive review on type 2 fuzzy logic applications: Past, present and future. *Engineering Applications of Artificial Intelligence, 95*, 103916. https://doi.org/10.1016/J.ENGAPPAI.2020.103916
4. Valdez, F., Castillo, O., & Melin, P. (2021). Bio-inspired algorithms and its applications for optimization in fuzzy clustering. *Algorithms, 14*, 122. https://doi.org/10.3390/A14040122.
5. Amézquita, L., Castillo, O., Soria, J., & Cortes-Antonio, P. (2021). Optimal design of fuzzy controllers using the multiverse optimizer. In *Advances in intelligent systems and computing* (pp. 289–298). Springer Science and Business Media Deutschland GmbH. https://doi.org/10.1007/978-3-030-73050-5_29.

6. Amézquita, L., Castillo, O., & Cortes-Antonio, P. (2022). *Fuzzy-chaotic variant of the multi-verse optimizer algorithm in benchmark function optimization* (Vol. 504, pp. 53–63). Lecture notes networks system, LNNS. https://doi.org/10.1007/978-3-031-09173-5_8/COVER.
7. Amézquita, L., Castillo, O., Soria, J., Cortes-Antonio, P. (2021). A fuzzy variant of the multi-verse optimizer for optimal design of fuzzy controllers. In *Intelligent and fuzzy techniques for emerging conditions and digital transformation* (pp. 537–545). Cham: Springer. https://doi.org/10.1007/978-3-030-85626-7_63.
8. Amézquita, L., Castillo, O., Soria, J., & Cortes-Antonio, P. (2021). Optimization of membership function parameters for fuzzy controllers in cruise control problem using the multi-verse optimizer. In *Studies in computational intelligence* (pp. 15–40). Springer Science and Business Media Deutschland GmbH. https://doi.org/10.1007/978-3-030-68776-2_2.
9. Mirjalili, S., Mirjalili, S. M., & Hatamlou, A. (2016). Multi-verse optimizer: A nature-inspired algorithm for global optimization. *Neural Computing and Applications, 27*, 495–513. https://doi.org/10.1007/s00521-015-1870-7
10. Mirjalili, S., Jangir, P., Mirjalili, S. Z., Saremi, S., & Trivedi, I. N. (2017). Optimization of problems with multiple objectives using the multi-verse optimization algorithm. *Knowledge-Based Syst., 134*, 50–71. https://doi.org/10.1016/j.knosys.2017.07.018
11. Hawking, S. W. (1993). Wormholes in spacetime. In *Euclidean quantum gravity* (pp. 363–369). https://doi.org/10.1142/9789814539395_0024.
12. Hernandez, E., Castillo, O., & Soria, J. (2019). Optimization of fuzzy controllers for autonomous mobile robots using the grey Wolf optimizer. In *IEEE International Conference on Fuzzy Systems* June 2019. https://doi.org/10.1109/FUZZ-IEEE.2019.8858861.
13. Riaz, M. T., Hashmi, W. S., Ahmad, S., Husnain, S., Mujtaba, H., Ali, H., Atiq, S., & Qureshi, M. M.: Design of optimization methodology for economic dispatch of thermal generating units. In *2022 5th international conference on energy conservation and efficiency ICECE 2022-Proceeding*. https://doi.org/10.1109/ICECE54634.2022.9758966.
14. Sánchez, D., Melin, P., & Castillo, O. (2020). Comparison of particle swarm optimization variants with fuzzy dynamic parameter adaptation for modular granular neural networks for human recognition. *Journal of Intelligent & Fuzzy Systems, 38*, 3229–3252. https://doi.org/10.3233/JIFS-191198
15. Saremi, S., Mirjalili, S., & Lewis, A. (2014) Biogeography-based optimisation with chaos. *Neural Computing & Applications, 255*(25), 1077–1097. https://doi.org/10.1007/S00521-014-1597-X.
16. Ewees, A. A., El Aziz, M. A., Hassanien, A. E. (2017). Chaotic multi-verse optimizer-based feature selection. *Neural Computing & Applications, 314*(31), 991–1006. https://doi.org/10.1007/S00521-017-3131-4.
17. Bernal, E., Lagunes, M. L., Castillo, O., Soria, J., & Valdez, F. (2020). Optimization of type-2 fuzzy logic controller design using the GSO and FA algorithms. *The International Journal of Fuzzy Systems, 231*(23), 42–57. https://doi.org/10.1007/S40815-020-00976-W.
18. Ochoa, P., Castillo, O., & Soria, J. (2019) Optimization of fuzzy controller design using a Differential Evolution algorithm with dynamic parameter adaptation based on Type-1 and Interval Type-2 fuzzy systems. *Soft Computing, 241*(24), 193–214. https://doi.org/10.1007/S00500-019-04156-3.
19. Ontiveros, E., Melin, P., & Castillo, O. (2020). Comparative study of interval Type-2 and general Type-2 fuzzy systems in medical diagnosis. *Information Sciences (Ny), 525*, 37–53. https://doi.org/10.1016/J.INS.2020.03.059
20. Olivas, F., Valdez, F., Melin, P., Sombra, A., & Castillo, O. (2019). Interval type-2 fuzzy logic for dynamic parameter adaptation in a modified gravitational search algorithm. *Information Sciences (Ny), 476*, 159–175. https://doi.org/10.1016/J.INS.2018.10.025

21. Valdez, F., Castillo, O., Cortes-Antonio, P., & Melin, P. (2020). A survey of Type-2 fuzzy logic controller design using nature inspired optimization. *Journal of Intelligent & Fuzzy Systems, 39*, 6169–6179. https://doi.org/10.3233/JIFS-189087
22. Castillo, O., Castro, J. R., Melin, P., & Rodriguez-Diaz, A. (2014). Application of interval type-2 fuzzy neural networks in non-linear identification and time series prediction. *Soft Computing, 18*(6), 1213–1224.
23. Castillo, O., & Melin, P. (2003). *Soft computing and fractal theory for intelligent manufacturing*. Springer.

Performance Comparative of Surrogate Models as Fitness Functions for Metaheuristic Algorithms

David Bolaños-Rojas, Jorge A. Soria-Alcaraz, Andrés Espinal, and Marco A. Sotelo-Figueroa

1 Introduction

The current state of the art of surrogate modelling, consists in the creation of "surrogate model ensembles" [1]. Several methods for the creation of surrogate model ensembles exist, such as OTL-PEM [2]. Recently, dynamic ensembles have also been proposed. Such ensembles consider a matrix of global errors together with the local errors. This approach has been proven to be more computationally efficient and better at the generation of solutions than classic surrogate model ensembles [3]. In this work we analyze the performance of several surrogate models trained by well-known continuous functions, our objective is to explore the idea of using surrogate models as objective functions for metaheuristics.

2 PSO

Particle Swarm Optimization is a Bio-inspired metaheuristic in flocks of birds or schools of fish. It was developed by J. Kennedy and R. Eberthart based on a concept called social metaphor. This metaheuristic simulates a society where all individuals contribute with their knowledge to obtain a better solution. There are three factors that influence for change in status or behavior of an individual:

D. Bolaños-Rojas
Licenciatura en Sistemas de información administrativa, Universidad de Guanajuato, Guanajuato, Mexico

J. A. Soria-Alcaraz (✉) · A. Espinal · M. A. Sotelo-Figueroa
Departamento de Estudios Organizacionales, Universidad de Guanajuato, División de ciencias económico-administrativas guanajuato, Mexico
e-mail: jorge.soria@ugto.mx

O. Castillo and P. Melin (eds.), *Hybrid Intelligent Systems Based on Extensions of Fuzzy Logic, Neural Networks and Metaheuristics*, Studies in Computational Intelligence 1096,
https://doi.org/10.1007/978-3-031-28999-6_9

- The knowledge of the environment or adaptation: it is related to the importance given to the experience of the individual
- His Experience or local memory: it is related to the importance given to the best result found by the individual.
- The Experience of their neighbors or Global memory: this is related to how important it is the best result obtained by their neighbors or other individuals.

In this metaheuristic, each individual is considered as a particle, and moves through a multidimensional space that represents the social space or search space depends on the dimension of space which depends on the variables used to represent the problem.

For the update of each particle, we use the velocity vector which tells how fast it will move the particle in each of the dimensions, the method for updating the speed of PSO is given by Eq. (1), and it is updated by the Eq. (2) [4].

$$v_i = wv_i + \varphi_1(x_i - B_{global}) + \varphi_2(x_i - B_{local}) \quad (1)$$

$$x_i = x_i + v_i \quad (2)$$

where

v_i: Velocity of the ith particle.

w: Adjustment factor to environment.

φ_1: Memory Coefficient in the neighborhood.

φ_2: Memory Coefficient.

x_i: Position of the ith particle.

B_{global}: Best solution found so far by all particles.

B_{local}: Best solution found by ith particle.

3 Differential Evolution

Differential Evolution (DE) [5] was developed by R. Storn and K. Price in 1996. It is a vector-based evolutionary algorithm, and it can be considered as a further development to a Genetic Algorithm (GA) [5]. It is a stochastic search algorithm with self-organizing tendency and does not use the information of derivatives. For a d-dimensional problem with d-parameters, a population of n solution are initially generated, so we have xi solution vectors where i = 1, 2, …, n. For each solution xi, at any generation t we use the conventional notation as

$$x_i^t = \left(x_i^{t,} x_2^t x_3^t, \ldots, x_d^t\right) \quad (3)$$

which consists of d-components in the d-dimensional space. This vector can be considered as the chromosomes or genomes. This metaheuristic consists of three main steps: mutations, crossover, and selection.

Mutation is carried out by the mutation scheme. For each vector *xi* at any time or generation *t*, first randomly choose three distinct vector *xp; xq* and *xr* at *t*, and then generate a so-called donor vector by the mutation scheme.

The crossover is controlled by a probability Cr and actual crossover can be carried out in two ways: binomial and exponential. The binomial scheme performs crossover on each of the *d-components* or variables/parameters [5].

4 Surrogate Models

A surrogated model is a numerical method used when an outcome of interest cannot be easily measured or computed so an approximation model is used instead. Surrogate assisted computation was mainly motivated from reducing computational time in optimization of complex problems. There exist a wide variety of surrogate models, from Polynomial response surfaces, Bayesian approaches, gradient-enhanced kriging, radial basis function and random forest. In this initial work we used two of the most well-know surrogated models: Least Squares lineal Model and Polynomial Model.

Least Squares lineal model is an approach to fitting a mathematical or statistical model to data in cases where the idealized value provided for any data point is expressed linearly in terms of the unknown parameters of the model. The output of this approach is a fitting model that can be used to summarize data, predict unobserved values from the system and to understand the mechanisms that may underline the system [6]. In similar way we also used a Second-Order Polynomial Model which estimated polynomial regression coefficients using ordinary least Square estimation.

5 Comparison Metrics Appropriate for Surrogate Models

In their vast study of surrogate models, Alizadeh et al. use a combination of R^2, adjusted R^2, mean absolute error (MAE), maximized mean squared error (MMSE), root mean squared error (RMSE) and cross validation to quantitatively evaluate the performance of surrogate models [6]. They also recommend the use of results graphs to quickly analyze errors, ranges and domains of the studied models.

6 Methodology

This research begun with the implementation of pre-calculated datasets of synthetic problems. The selected problems are the Ackley and Weierstrass functions. The implementation of such functions was adapted to the python language based on the definitions found in the "Virtual Library of Simulation Experiments" [7]. The datasets were generated by sampling the function on their domains at regular intervals, recording the results in csv files with the $x \rightarrow f(x)$ mapping.

After the datasets where generated, the SMT library was used to generate Least Squares linear model and a Polynomial Model using the datasets previously generated.

Next the Pymoo library was used to generate problem classes which used the previously trained surrogate models as fitness functions. The chosen optimization heuristics are Particle Swarm Optimization and Differential Evolution. These optimization algorithms were configured with the following param**eters.**

– Both algorithms where run with the objective of minimizing their fitness functions, with the stop criterion set to number of function evaluations with a parameter of 300 evaluations.

After the algorithms finished their execution, the last 25 individuals from the optimization process, as well as the best individual found, they were evaluated with the original problem function, Ackley and Weierstrass respectively. These evaluations were then recorded in datasets with csv format, containing columns with the individuals' x value, the result of the surrogate model's evaluation s(x) and the result of the original functions evaluation **f(x).**

These datasets were statistically analyzed to determine if the optimization of the surrogate model is an appropriate approach to optimize the original function.

7 Results

In this section discuss the results gathered by the training of Least Squares lineal model and polynomial model under test instances. Once trained, these models were used as fitness functions by PSO and DE. As test instances we used two optimization functions, namely, Ackley and Weierstrass. The Acley function is widely used for testing optimization algorithms. We use its two-dimensional form, which is characterized by a near flat outer region, and a large valley in the center. This function was selected because its many local minima designed to trap Hill climbing algorithm.

As second instance test, we used Weiestrass function, which is a example of a real-valued function that is continuous at any given point but nowhere differentiable. This function is considered as a fractal curve, we selected this function because the high number of non-global optimal points.

Table 1 PSO parameters

Population size	25
w, c_1, c_2	*Adaptive
Max velocity rate	0.20

* Whether w, c1, and c2 are changed dynamically over time. The update uses the spread from the global optimum to determine suitable values [8]

Table 2 DE parameters

Population size	25

We execute experiments in order to get instances of 50, 100 and 200 samples for Ackley and Weiestrass functions. Then we train proposed models to achieve results in Tables 4 for LS and Table 5 for Polynomial Model.

Due quadratic adjustments the polynomial model showed a better fit to Acley and Weierstrass functions. After this process, we used the given LS and polynomial models as fitness functions into PSO and DE using parameters from Tables 1 and 2. We execute 35 experiments to achieve statistical significance. Table 6 shows our best results among several executions of PSO algorithm, column $\boldsymbol{x}$ shows our solution value, *S(x)* shows the evaluation of our solution x under the surrogate model and *F(x)* shows the evaluation of solution x under the actual optimization function. Table 7 shows best results for DE algorithm.

Numerical results show good approximation to real fitness value for Acley and weierstrass functions, also the solutions generated by metaheuristics with surrogated models where close to optimal values for both PSO and DE.

Table 3 Sample of the generated dataset for the Ackley function with the PSO algorithm

x	s(x)	f(x)
0.001001	15.11713	0.004055
0.006853	15.11713	0.029911
0.009939	15.11713	0.045008
0.005746	15.11713	0.024743

Table 4 Adjust error using Least Squares model

Instance	R2	MAE	MER	RMSE
Ackley 50	−99	1114.11	0	193.008
Ackley 100	−199	2206.37	0	270.238
Ackley 200	**−399**	**4390.91**	**2.22E-15**	**380.268**
Weierstrass 50	**−99**	**68**	**0**	**11.780**
Weierstrass 100	**−199**	**134.66**	**−2.27E-15**	**16.4940**
Weierstrass 200	**−399**	**268**	**−2.27E-15**	**23.209**

Table 6 Best results PSO

Experiment	x	S(x)	F(x)
Acley-25	−0.000139	15.11712	0.00055
Acley-50	−0.011142	15.11712	0.05117
Acley-100	0.004824	15.11712	0.02053
Acley-200	−0.062470	15.11716	0.44738
Weierstrass-25	0.0093193	−0.0049545	0.4275948
Weierstrass-50	0.0013246	−0.004955	0.7387444
Weierstrass-100	−0.0062525	−0.0049548	0.6565921
Weierstrass-200	0.0047361	−0.0049549	0.7033211

Table 7 Best results DE

Experiment	X	S(x)	F(x)
Acley-25	−0.000131	15.11766	0.00051
Acley-50	−0.011173	15.11766	0.05121
Acley-100	0.003814	15.11768	0.01833
Acley-200	−0.028670	15.11776	0.24718
Weierstrass-25	0.0003773	−0.008914	0.071529
Weierstrass-50	0.0013567	−0.007755	0.233734
Weierstrass-100	−0.0047767	−0.004128	0.237892
Weierstrass-200	0.0032786	−0.0059134	0.453312

8 Conclusions and Future Work

This work has presented the training and implementation of surrogated models as fitness functions for two metaheuristics PSO and DE. Results shows encouraging results about fitting well-know test fitness functions by surrogate models. Given the simplicity of least square models it performs worse than second-order Polynomial model which achieved least error values when adjusting to both; Acley and Weierstrass functions.

Once the surrogate models were applied as fitness functions to DE and PSO, both models showed promising results in guiding the search toward optimal values. With most of the high-frequency variations of the original functions removed by surrogate models; it was possible for PSO and DE to converge to near-optimal values with less use of computational resources.

For future work we propose to use more complex surrogate models to fit combinatorial fitness functions from scheduling real world problems. Also, we propose to use more specialized metaheuristics for combinatorial optimization.

References

1. Jiang, P., Zhou, Q., & Shao, X. (2020). *Surrogate model-based engineering design and optimization* (1st ed., p. 246) Singapore: Springer.
2. Pang, Y., Wang, Y., Sun, W., & Song, X. (2021). OTL-PEM: An optimization based two-layer pointwise ensemble of surrogate models. *Journal of Mechanical Design, 144*(5). https://doi.org/10.1115/1.4053011
3. Zhou, C., Zhang, H., Chang, Q., Song, X., & Li, C. (2021). An adaptive ensemble of surrogate models based on hybrid measure for reliability analysis. *Structural and Multidisciplinary Optimization, 65*(1), 16. https://doi.org/10.1007/s00158-021-03129-1
4. Kennedy, J., & Eberhart, R. (1995). Particle swarm optimization. In *Proceedings of ICNN'95 - international conference on neural networks, 1995* (Vol. 4, pp. 1942–1948). https://doi.org/10.1109/ICNN.1995.488968
5. Storn, R., & Price, K. (1997). Differential evolution – a simple and efficient heuristic for global optimization over continuous spaces. *Journal of Global Optimization, 11*(4), 341–359. https://doi.org/10.1023/A:1008202821328
6. Alizadeh, R., Allen, J. K., & Mistree, F. (2020). Managing computational complexity using surrogate models: A critical review. *Research in Engineering Design, 31*(3), 275–298. https://doi.org/10.1007/s00163-020-00336-7
7. Surjanovic, S., & Bingham, D. Virtual library of simulation experiments: Ackley function. https://www.sfu.ca/~ssurjano/ackley.html
8. Blank, J., & Deb, K. (2020). PYMOO: Multi-objective optimization in Python. *IEEE Access, 8*, 89497–89509. https://doi.org/10.1109/ACCESS.2020.2990567

A New Continuous Mycorrhiza Optimization Nature-Inspired Algorithm

Hector Carreon-Ortiz, Fevrier Valdez, and Oscar Castillo

1 Introduction

In nature we can find patterns of behavior and organization of living beings that make us think that they perform their work excellently, in any place where there are plants such as forests, we cannot imagine what is happening in the underground, in this inner world there is a communication and exchange of resources through the roots of plants and fungi for the survival of an entire ecosystem, much research has been done on this topic, such as that of Dr. Suzanne Simard, Professor of Forest Ecology at the University of British Columbia [1], in this research this metaheuristic optimization work is inspired by the present metaheuristic work. Suzanne Simard, Professor of Forest Ecology at the University of British Columbia [1], in this research the present metaheuristic optimization work is inspired. Optimization is basically the search for optimal parameters in a system, the parameters are known as design or objective variables. Metaheuristics are general strategies that intelligently combine several techniques to explore the solution space [2]. In design optimization, the design objective could be simply to minimize production cost or to maximize production efficiency. An optimization algorithm is a procedure that is run iteratively by comparing several solutions until an optimal or satisfactory solution is found. There are two types of optimization algorithms widely used today. Deterministic algorithms use specific rules to move from one solution to another. These algorithms are used to adapt sometimes and have been successfully applied for many engineering design problems [3]. Stochastic Algorithms, are random in nature with probabilistic translation rules, therefore, the order of execution or the result of the algorithm may be different for each execution with the same input.

H. Carreon-Ortiz · F. Valdez (✉) · O. Castillo
Tijuana Institute of Technology, Tijuana, Mexico
e-mail: fevrier@tectijuana.mx

O. Castillo and P. Melin (eds.), *Hybrid Intelligent Systems Based on Extensions of Fuzzy Logic, Neural Networks and Metaheuristics*, Studies in Computational Intelligence 1096,
https://doi.org/10.1007/978-3-031-28999-6_10

The algorithm we are proposing in this research is stochastic in nature [4]. To model the subsurface ecosystem, we used the Lotka-Volterra Continuous Equations System, experiments were performed with 5 mathematical functions of the CEC2013, obtaining very good results.

The main contribution of this research is to model the natural process of the understory ecosystem between plant roots and the Mycorrhizal Network by means of the Lotka-Volterra Continuous Equations System such as the Predator–Prey Model (Defense), Cooperative Model (Resource Exchange) and Competitive Model (Colonization).

This article is organized as follows: 1 Introduction, where we briefly present the content of the article; 2. Natural Inspiration, are the bases of the new optimization method that we present; 3. Literature Review, research carried out on the topics that we address here; 4. Continuous Mycorrhiza Optimization Algorithm (CMOA), presentation and development of the new method; 5. Results, exposition and analytical comparison of the experiments carried out; 6. Discussion of Results, analysis and interpretation of the results obtained in the investigation and 7. Conclusions, analytical summary of the experimentation presented in this article.

2 Natural Inspiration

In the subsoil of a forest there is a reality that is present and that the human being cannot see, the roots of the plants and a network of fungi (symbiosis) form an ecosystem (plants united by the network of fungi and the roots of the plants), in which the largest plants (trees) contribute to the fungal network (Mycorrhiza Network—MN) and seedlings (growing plants), Carbon (CO_2) (generated by the photosynthesis of trees) through the MN, the MN in turn provides Water (H_2O), Nitrogen (N), Phosphorus (P), Potassium (K), and other minerals to the entire ecosystem Fig. 1, between the plants united through the MN there is a communication (signaling) to warn the entire ecosystem of any danger or disturbance in the forest such as predators, fires, wind, floods, and man himself Fig. 2, there is also growth in the ecosystem, that is, the incorporation of other plants through the MN (Colonization), this behavior that occurs in the ecosystem inspired us or to model an optimization algorithm using as scenario the Defense of the Plants, the Communication and Exchange of Resources and the Colonization, through the System of Lotka-Volterra Continuous Equations System.

Optimization algorithms inspired by nature have shown flexibility, efficiency and adaptability in solving problems in the real world, for these reasons it is proposed to develop an algorithm inspired by plants and how they adapt to physiological changes, survival and growth through communication and exchange of resources that are transferred through a Mycorrhiza Network. It is well known that today there is no optimization algorithm that can solve all problems (no-free lunch) [5], both in industry, commerce, economy, research, etc., there are daily problems that can be solve with some optimization method, there are many optimization methods in

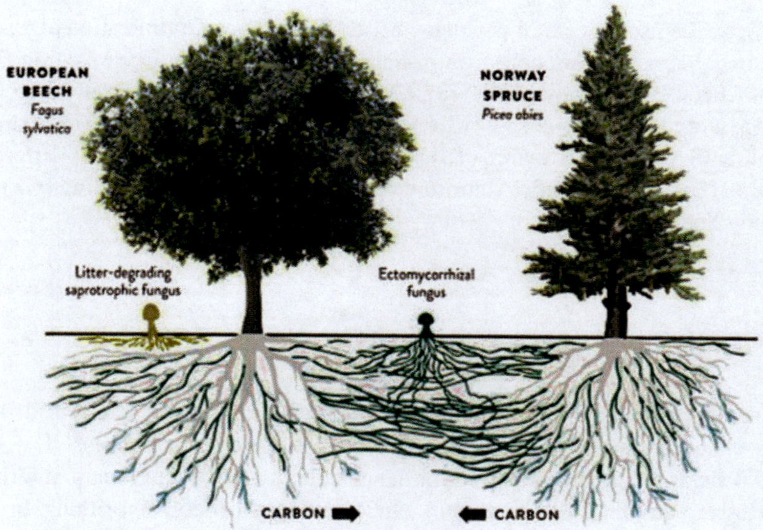

Fig. 1 The exchange of resources between the donor and recipient plants (seedlings) take place through the Mycorrhiza Network (MN) in which CO_2, water, Nitrogen (N), Phosphorus (P), Potassium (K), etc., circulate

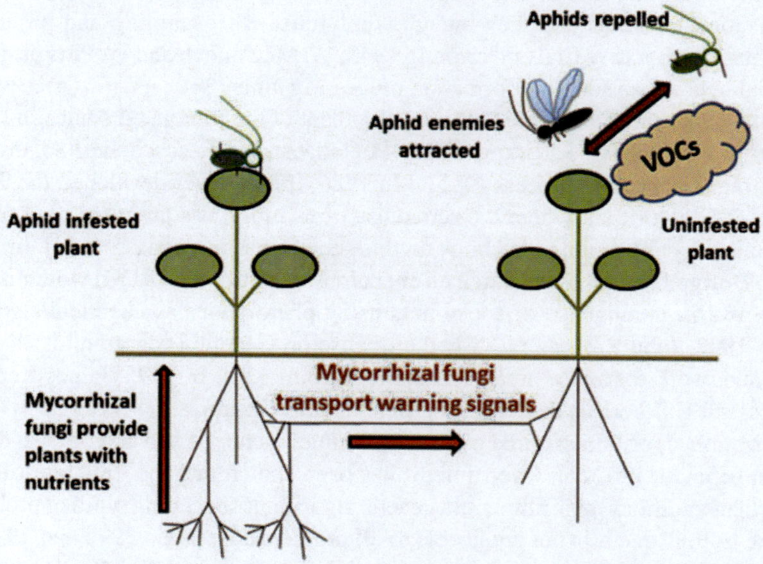

Fig. 2 The signaling of stress that plants suffer from the danger of predators, pests, fire, floods, etc.

which some can solve certain problems better than others. Optimization algorithms with a metaheuristic approach, Swarm-based: Particle Swarm Optimization (PSO) [20], Artificial Bee Colony (ABC) [8], Cuckoo Search (CS) [23], Firefly Algorithm (FA) [15], etc., Trajectory-based: Hill Climbing [29], b-Hill Climbing [37], Simulated Annealing (SA) [9], Tabu Search (TS) [4], etc., Evolutionary Algorithm: Differential Evolution (DE) [38], Genetic Algorithm (GA) [6], Genetic Programming (GP) [28], Harmony Search (HS) [31].

3 Literature Review

Alan Turing was probably the first to use heuristic algorithms during World War II, Turing called his method heuristic search, which was a great success. The 1960s and 1970s was an important period for the development of evolutionary algorithms. John Holland and his collaborators in 1962 developed genetic algorithms. In 1975, Holland published his book outlining his development of Genetic Algorithms (GAs) [6]. Also, in 1975, De Jong finished his research showing the enormous potential and power of genetic algorithms for a wide range of objective functions, be they noisy, multimodal or discontinuous [26]. Artificial Neural Networks, support vector machines and other machine learning techniques can be considered as a heuristic optimization technique, as they intend to minimize their learning and prediction errors through iterative trial and error. In 1943, W. McCulloch and W. Pitts proposed artificial neurons as simple information processing units [27].

In the 1980s and 1990s due to the development of the Simulated Annealing (SA) algorithm developed by S. Kirkpatrick, C. D. Gellat and M. P. Vecchi in 1983, inspired by the metal annealing process [9–11]. In 1986, Fred Glover developed the Taboo Search (TS) algorithm, where the actual use of memory was probably used for the first time, the publication of his book on this subject was not until 1996 [4]. In 1992, Marco Dorigo finished his research on ant colony optimization (ACO) work inspired by the swarm intelligence of social ants using pheromones as chemical signaling [7]. In 1992, John R. Koza published a treatise on Genetic Programming that laid the groundwork for a new area of machine learning [28]. In 1995, James Kennedy and Russell C. Eberhart developed Particle Swarm Optimization (PSO), PSO is an optimization algorithm inspired by the swarm intelligence of fish and birds and even human behavior [20, 21], Over time, it has been shown that the PSO algorithm is better than traditional algorithms and genetic algorithms for certain types of problems and the truth is that it is not applicable to all problems. Between 1996 and 1997, R. Storn and K. Price developed the vector-based Differential Evolution (DE) algorithm, which is more efficient than genetic algorithms in many applications [38]. In 1997, D. H. Wolpert and W. G. Macready published the No-Free Lunch theorems to warn the scientific community that if algorithm A performs better than algorithm B for some optimization functions then B will be better than A for other functions, which means to say that there are no algorithms that are better for all types of optimization problems, the goal of research now is to find the best and most efficient algorithm or algorithms for a given problem [30]. At the beginning of this century Zong Woo Geem et al. in 2001, they developed the Harmony Search (HS) algorithm, with application in water distribution and transport modeling [31]. In 2004, S. Nakrani and C. Tovey proposed the honey bee algorithm and its application to optimize Internet hosting centers [32]. The techniques were inspired by the foraging behavior of real ants in the nature. Artificial Bee Colony algorithm (ABC) was introduced by Karaboga in 2005 [8]. In 2008, Xin-She Yang developed the Firefly Algorithm (FA) at Cambridge University, which was based on the flashing patterns and behavior of fireflies [14–16]. Gravitational Search Algorithm (GSA) introduced by Rashedi et al. in 2009 [12, 13], the algorithm is comprised of collection of searcher agents that interact with each other through the gravity force. In 2009, Xin-She Yang et al. developed the Cuckoo Search (CS) algorithm at Cambridge University in the UK, is based on the obligate brood parasitic behavior of some cuckoo species in combination with the Levy flight behavior of some birds and fruit flies [23, 24]. In 2010, Xin-She Yang developed the Bat Algorithm (BA) for continuous optimization, algorithm inspired by the behavior of bats [33]. In 2012, Xin-She Yang proposed the new Flower Pollination Algorithm (FPA) inspired by the pollination process of flowering plants [17, 18]. Grey Wolf Optimizer (GWO) introduced by Seyedali Mirjalili et al. in 2013 [25], inspired by grey wolves (Canis lupus), mimics the leadership hierarchy and hunting mechanism of grey wolves in nature. In 2015, Mirjalili, Seyedali presented their new Moth-Flame Optimization Algorithm, inspired by the method of navigation of moths in nature called traverse orientation. Moths fly at night keeping a fixed angle with respect to the moon, a very efficient mechanism for traveling long distances in a straight line [19]. In 2016, Seyedali Mirjalili and Andrew Lewis introduced the new The Whale Optimization Algorithm (WOA) that mimics the social behavior of humpback whales [22]. In 2018, Juliano Pierezan and Leandro dos Santos Coelho developed the Coyote Optimization Algorithm, it is a population-based metaheuristic for optimization inspired on the canis latrans species [34]. In 2019, Haiqiang Hao presented his research of An Original Bionic Algorithm, it simulates the existing behavior mechanism in nature [35]. In 2020, Abdolkarim Mohammadi-Balani et al., proposed their new Golden Eagle Optimizer algorithm inspired by the behavior of golden eagles to adjust the speed at different stages of their spiral trajectory for hunting [36]. In 2022, Hector Carreon-Ortiz et al. presented the new Discrete Mycorrhiza Optimization Algorithm inspired by the symbiosis between plant roots and the mycorrhizal network [25], and thus, new metaheuristic algorithms will continue to be developed that can solve some type of problem and can do it better than others for certain and specific cases.

4 Continuous Mycorrhiza Optimization Algorithm (CMOA)

As we have explained before, the CMOA is an optimization algorithm inspired by the symbiotic relationship between plant roots and the mycorrhizal network. In the subsoil of a forest, the largest trees (Mother Trees) that manage to capture sunlight carry out photosynthesis producing carbon dioxide (CO_2) which is essential for the life of plants and fungi, in this symbiosis there is an exchange of resources between plants that provide carbon to fungi and seedlings (small or growing plants), and fungi in turn provide plants with Water (H_2O), Nitrogen (N), Phosphorus (P), Potassium (K), and other minerals. It should be noted that not all large trees are Mother Trees, but those that have more connections with the ecosystem.

Based on the above, we have deduced three basic concepts to model the algorithm using the Lotka-Volterra Continuous Equations System [44–49]:

1. The Predator–Prey model defines the defensive behavior that plants have against any external aggression, such as pests, predators, whether insects or animals Eqs. (1) and (2).
2. The Cooperative model of exchange of resources such as Carbon (CO_2), Water (H_2O), Nitrogen (N), Phosphorus (P), Potassium (K), etc., Fig. 3, Eqs. (3) and (4).
3. The Competitive model (Colonization) with limited resources in an ecosystem, living beings compete to obtain them, Eqs. (5) and (6).

The algorithm starts from two populations (plants and fungi) obtained with random solutions, obtaining the best solution (fitness), with the best solutions found (parameters a and d) we use any of the 3 Lotka-Volterra models through a random decision (diversity d [1–3]), to avoid the stagnation of local minima, these are controlled by the times in which the populations are renewed after each cycle of 30 iterations.

The process of the algorithm you can see in the Figs. 4 and 5.

Fig. 3 Transfer of resources through the NM to all the plants that make up the ecosystem

Fig. 4 Flowchart Continuous Mycorrhiza Optimization Algorithm (CMOA)

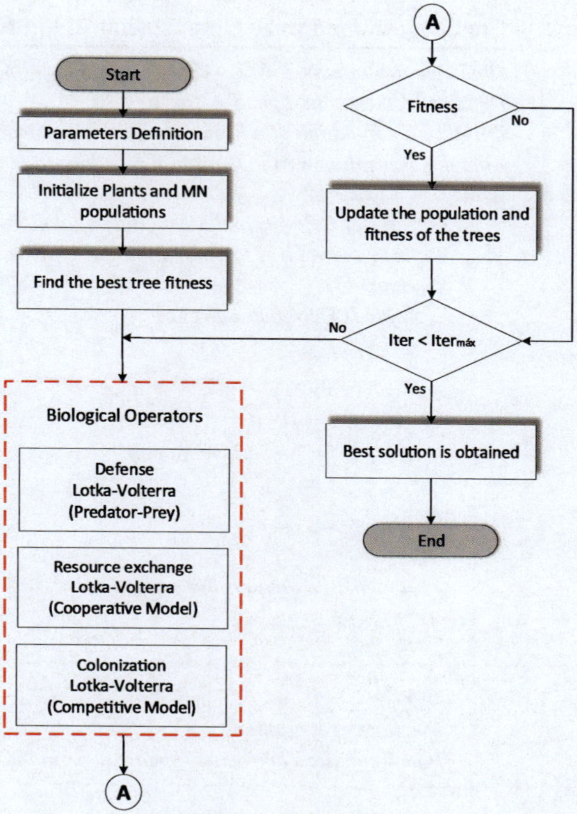

Lotka-Volterra Continuous Equations System:

Defense Model (Predator–Prey) equations are a pair of nonlinear differential equations of first order, which are used to describe the dynamics of biological systems in which two species interact, one as predator and the other as prey where populations change over time Eqs. (1) and (2).

The Cooperative Model of the Lotka-Volterra equations are a pair of nonlinear differential equations of first order, which describe in an ecosystem how the densities of the populations of the interacting species increase considerably Eqs. (3) and (4).

The Competitive Model of the Lotka-Volterra equations (Colonization), are a pair of first-order nonlinear differential equations that model the population dynamics of species that compete for a common resource Eqs. (5) and (6).

Defense (Predator–Prey)

$$\frac{dx}{dt} = x(a - by) \tag{1}$$

Continuous Mycorrhiza Optimization Algorithm (CMOA)

Objective min or max f(x), x = (x_1, x_2, ..., x_d)
Define parameters (a, b, c, d, e, f, x, y)
Initialize a population of n Plants and MN with random solutions
*Find the best solution **fit** in the initial population*
while *(t < maxIter)*
  for *i* = 1:*n (for n Plants and MN in the population)*
    Define rand ∈ [1 3] to apply Biological Operator
    if *(rand = 1)*
      Apply (LV-Predator-Prey)
      $\frac{dx}{dt} = x(a - by)$
      $\frac{dy}{dt} = y(-c + dx)$
    elseif *(rand = 2)*
      Apply (LV-Competitive Model)
      $\frac{dx}{dt} = x(a - bx - cy)$
      $\frac{dy}{dt} = y(d - ex - fy)$
    else
      Apply (LV- Cooperative Model)
      $\frac{dx}{dt} = x(a - bx + cy)$
      $\frac{dy}{dt} = y(d + ex - fy)$
    end if
    Evaluate new solutions
    If new solutions are better, update them in the population
  end for
  *Find the current best **fit** solution*
end while

Fig. 5 CMOA Algorithm Pseudocode

$$\frac{dy}{dt} = y(-c + dx) \tag{2}$$

Cooperative (Resource Exchange)

$$\frac{dx}{dt} = x(a - bx + cy) \tag{3}$$

$$\frac{dy}{dt} = y(d + ex - fy) \tag{4}$$

Competitive (Colonization)

$$\frac{dx}{dt} = x(a - bx - cy) \tag{5}$$

Table 1 Description of the algorithm parameters

Parameter	Description	Values
Population	Population size	20
Populations	Number of populations	2
Dimensions	Dimensions size 5, 10, 15, 20, 30, 50, 100, 500	
Epochs	Number of epochs	30
Iterations	Iterations size 30, 50, 100	
t	Time	
a	Population growth rate x	0.01
b	Influence of population x on itself	0.02
g	Influence of population y on population x	0.06
d	Population growth rate y	0
e	Influence of population x on population y	1.70
h	Influence of population y on itself	0.09
x	Initial population in x	0.0002
y	Initial population in y	0.0006
In the absence of population x = 0, In the absence of population y = 0		
a, b, c, d, e and f—are positive constants		

$$\frac{dy}{dt} = y(d - ex - fy) \tag{6}$$

Table 1 shows the description of all the parameters used by the Continuous Mycorrhiza Optimization Algorithm.

Figure 4 shows the diagram of the Continuous Mycorrhiza Optimization Algorithm.

Figure 5 shows the pseudocode of the Continuous Mycorrhiza Optimization Algorithm.

5 Results

In this section of the work we show the results of the experimentation carried out for this investigation, in Table 2 we can see the 5 mathematical functions of the CEC2013 that we use in the experiments and in Table 1 the parameters used by the CMOA algorithm, in Fig. 6 the graphs and equations of the mathematical functions of the CEC2013 are shown.

Figure 6 shows the graphs and equations of the 5 mathematical functions used in this research.

Experiments were performed with 5 mathematical functions CEC2013 Table 2, and hypothesis tests were also performed by mean difference with other 6 methods

Table 2 Math functions CEC2013

Fun	Name	Range	Nature
F1	Sphere	[−5.12, 5.12]	U
F2	Rosenbrok	[−5, 10]	U
F3	Griewank	[−600, 600]	M
F4	Rastrigin	[−5.12, 5.12]	M
F5	Ackley	[−32.768, 32.768]	M

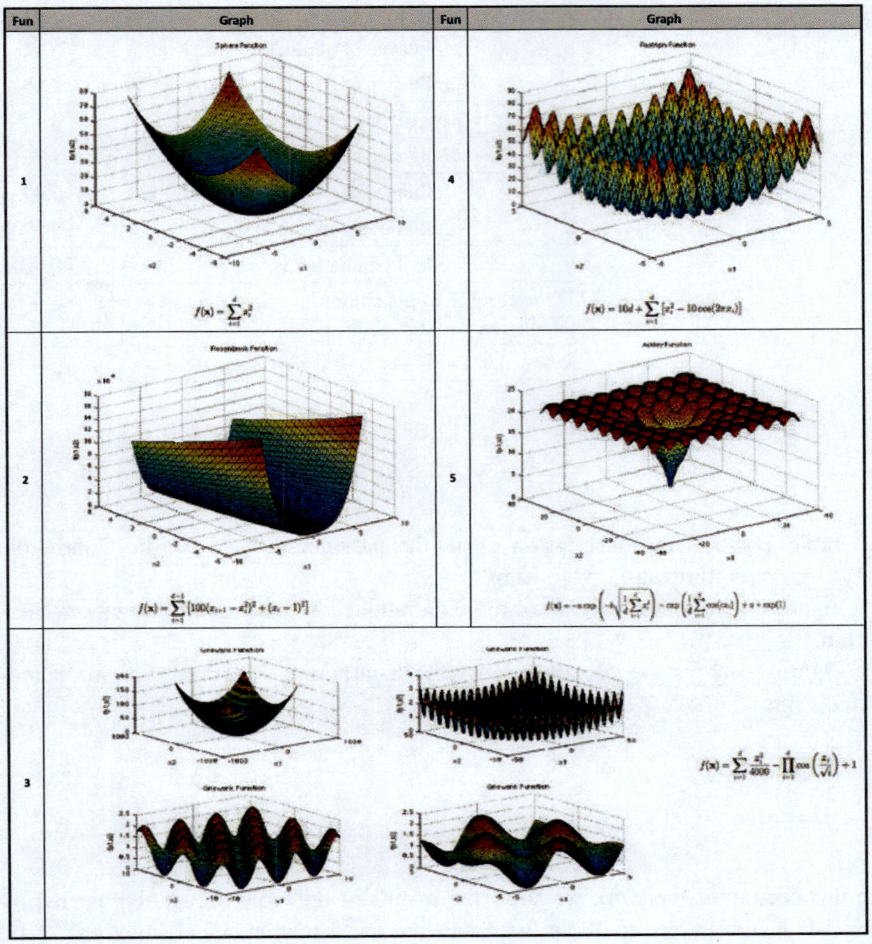

Fig. 6 Graphs of the 5 Mathematical Functions CEC2013

Table 3, Bat Algorithm (BA) [40], Differential Evolution (DE) [42], Cuckoo Search (CS) [41], Genetic Algorithm (GA) [41], Harmony Search (HS) [41] and Particle Swarm Optimization (PSO) [41] for 30 dimensions, in all comparisons there was significant Evidence that our The CMOA method obtained better results, Fig. 7 shows the convergence of the CMOA algorithm for the 5 mathematical functions, observing that the Rosenbrock function is the one that converges faster.

Experiments were performed with 5 mathematical functions CEC2013 Table 2, and hypothesis tests were also performed by mean difference with 6 other methods Table 4, Differential Evolution (DE) [42], Cuckoo Search (CS) [41], Particle Swarm Optimization (PSO) [41], Bat Algorithm (BA) [40], Genetic Algorithm (GA) [41] and Harmony Search (HS) [41] for 50 dimensions, in all comparisons there was significant Evidence that our CMOA method obtained better results, Fig. 8 shows

Table 3 Hypothesis Test for 30 dimensions of the CMOA and other methods

Fun	CMOA		BA			
	Mean	Std Dev	Mean	Std Dev	Z	Evidence
f1	8.11E-09	4.79E-09	2.38E+01	6.85E+00	−19.03	Yes
f2	6.58E-09	3.32E-09	1.52E+01	1.13E+00	−73.68	Yes
f3	1.64E-09	1.39E-09	4.29E+05	3.14E+05	−7.48	Yes
f4	8.13E-09	4.42E-09	4.64E+02	7.87E+01	−32.29	Yes
f5	1.44E-09	1.21E-09	9.09E+01	2.46E+01	−20.24	Yes
Fun	CMOA		DE			
f1	8.11E-09	4.79E-09	1.59E-03	4.28E-04	−20.35	Yes
f2	6.58E-09	3.32E-09	9.78E-03	1.31E-03	−40.89	Yes
f3	1.64E-09	1.39E-09	3.38E+01	2.25E+01	−8.23	Yes
f4	8.13E-09	4.42E-09	3.72E+01	1.17E+01	−17.41	Yes
f5	1.44E-09	1.21E-09	3.18E-03	1.57E-03	−11.09	Yes

CS				GA			
Mean	Std Dev	Z	Evidence	Mean	Std Dev	Z	Evidence
2.13E-04	2.18E-04	−5.35	Yes	4.20E-03	3.88E-03	−5.93	Yes
2.15E+00	1.09E+00	−10.80	Yes	4.71E-01	5.22E-01	−4.94	Yes
2.88E+01	3.29E+01	−4.79	Yes	1.59E+01	2.54E+01	−3.43	Yes
5.34E+01	1.22E+01	−23.97	Yes	1.31E+01	4.13E+00	−17.37	Yes
2.72E-02	5.86E-02	−2.54	Yes	4.43E-04	5.20E-04	−4.67	Yes
HS				PSO			
3.66E-02	5.95E-03	−33.69	Yes	9.28E-05	5.09E-05	−9.99	Yes
2.32E+00	4.68E-01	−27.15	Yes	2.62E-03	5.91E-04	−24.28	Yes
1.02E+02	6.06E+01	−9.22	Yes	5.06E+01	3.83E+01	−7.24	Yes
2.89E+01	4.68E+00	−33.82	Yes	8.76E+01	3.13E+01	−15.33	Yes
2.71E-02	2.34E-02	−6.34	Yes	1.93E-03	3.33E-03	−3.17	Yes

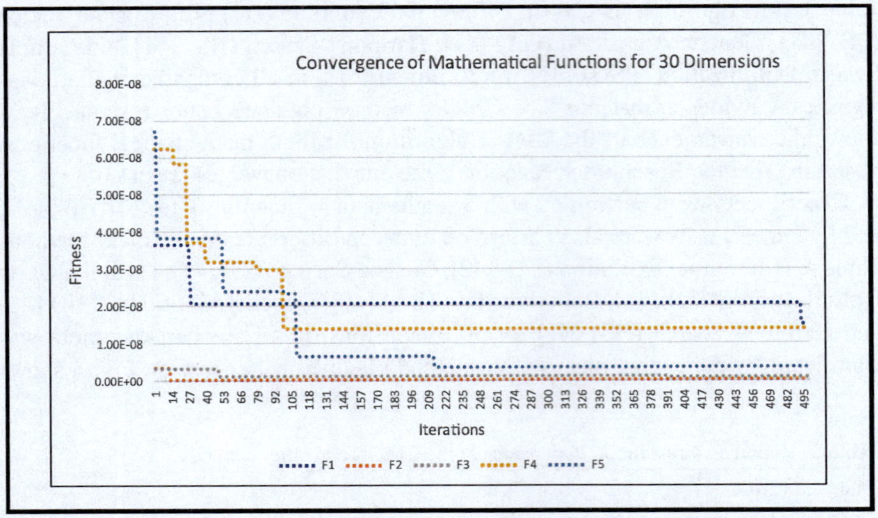

Fig. 7 Convergence of the CMOA algorithm for 30 dimensions and the 5 mathematical functions CEC2013

the convergence of the CMOA algorithm for the 5 mathematical functions, observing that the Griewank function is the one that converges faster.

Experiments were performed with 5 mathematical functions CEC2013 Table 2, and hypothesis tests were also performed by mean difference with 6 other methods Table 5, Flower Pollination Algorithm (FPA) [39], Particle Swarm Optimization (PSO) [43], Genetic Algorithm (GA) [43], Differential Evolution (DE) [43], Whale Swarm Optimization (WSA) [43] and Speciation-based PSO (SPSO) [43] for 100 dimensions, in all comparisons there was significant Evidence that our CMOA method obtained better results, Fig. 9 shows the convergence of the CMOA algorithm for the 5 mathematical functions, observing that the Rosenbrock function is the one that converges faster.

6 Discussion of Results

In the present study it was found that in the 85 hypothesis tests that were performed to compare our method with other 18 different methods had better results, this we can interpret that the CMOA algorithm has a higher degree of convergence than the other methods, of course our method is not better than the other methods because we are sure that they have their utility solving other problems that we have not studied, the best strength of this method is its power of convergence to the global minimum, when performing the hypothesis tests the results were unexpected, Surely if we perform more tests with other methods we will find better methods than ours, by developing this method from our perspective is to contribute to the community a

Table 4 Hypothesis Test for 50 dimensions of the CMOA and other methods

Fun	CMOA		DE			
	Mean	Std Dev	Mean	Std Dev	Z	Evidence
f1	8.56E-09	4.23E-09	7.19E-03	1.79E-03	−22.00	Yes
f2	1.49E-08	5.59E-09	1.86E-02	2.05E-03	−49.70	Yes
f3	8.26E-10	6.32E-10	4.80E+01	7.41E-01	−354.80	Yes
f4	7.36E-09	3.90E-09	7.81E+01	2.27E+01	−18.84	Yes
f5	1.25E-10	9.09E-11	5.91E-03	1.09E-03	−29.70	Yes
Fun	CMOA		CS			
f1	8.56E-09	4.23E-09	9.90E-01	7.89E-01	−6.87	Yes
f2	1.49E-08	5.59E-09	4.84E+00	1.32E+00	−20.08	Yes
f3	8.26E-10	6.32E-10	2.28E+02	1.27E+02	−9.83	Yes
f4	7.36E-09	3.90E-09	1.15E+02	2.07E+01	−30.43	Yes
f5	1.25E-10	9.09E-11	6.21E-01	2.13E-01	−15.97	Yes

PSO				BA			
Mean	Std Dev	Z	Evidence	Mean	Std Dev	Z	Evidence
1.69E-01	6.80E-02	−13.61	Yes	8.90E+00	4.55E+01	−1.07	Yes
1.32E-01	1.09E-01	−6.63	Yes	7.20E-01	1.53E+01	−0.26	Yes
1.71E+02	4.27E+01	−21.93	Yes	3.95E+05	8.78E+05	−2.46	Yes
2.18E+02	5.16E+01	−23.14	Yes	1.06E+02	8.49E+02	−0.68	Yes
1.71E-01	7.64E-02	−12.26	Yes	3.05E+01	1.55E+02	−1.08	Yes
GA				HS			
1.77E+00	6.66E-01	−14.56	Yes	3.54E-01	7.94E-02	−24.42	Yes
1.73E+00	3.15E-01	−30.08	Yes	3.12E+00	3.14E-01	−54.42	Yes
2.96E+01	5.07E+01	−3.20	Yes	2.04E+02	7.36E+01	−15.18	Yes
3.70E+01	7.63E+00	−26.56	Yes	8.26E+01	8.04E+00	−56.27	Yes
5.05E-02	1.61E-02	−17.18	Yes	1.22E+00	7.73E-02	−86.45	Yes

new way to solve optimization problems, it does not pretend to be the best method because there is not one that can solve all problems (No Free Lunch), we are certain that some methods are better than others in some optimization problems, we will continue experimenting to see the scope and limitations that this method has.

7 Conclusions

To verify how competitive the CMOA algorithm is, experiments were carried out with 5 CEC2013 mathematical functions for 30, 50 and 100 dimensions, these results were compared with other methods such as BA, CS, GA, DE, HS and PSO for 30

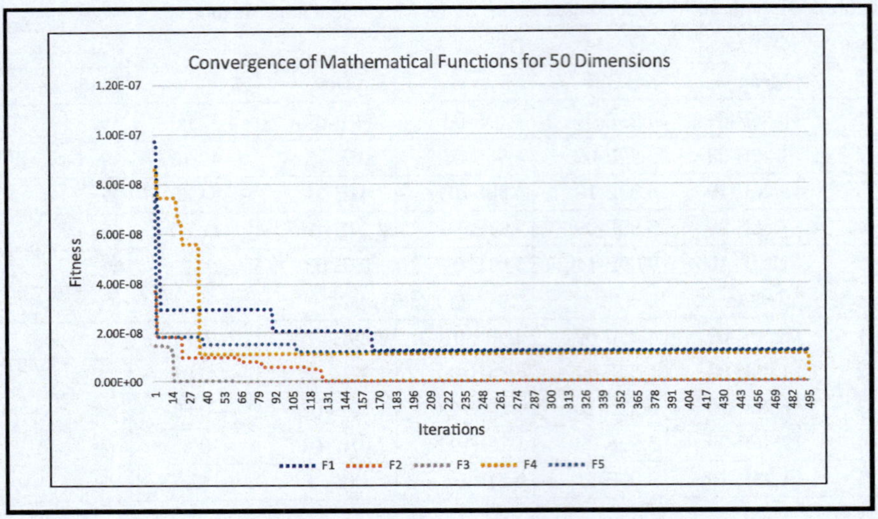

Fig. 8 Convergence of the CMOA algorithm for 50 dimensions and the 5 mathematical functions CEC2013

dimensions, obtaining with 30 hypothesis tests by mean difference, in all cases the CMOA method obtained better results; Comparisons were also made with the DE, CS, PSO, BA, GA and HS methods for 50 dimensions and in the results obtained through 30 hypothesis tests by mean difference and in all cases the CMOA method was better, and finally comparisons with the FPA, PSO, GA, DE, WSA and SPSO methods and in the 25 hypothesis tests by mean difference, the results in all cases favored our CMOA method, we can conclude with the results of the hypothesis tests that were carried out with the CMOA method and the aforementioned methods, that the CMOA method is more competitive than the other methods, surely it is not the best optimization method that exists, but it is a method that can be useful in some optimization problems, we will continue experimenting and publishing the results.

As future work, it is proposed to perform research with the CMOA algorithm to Adaptation of Membership Function Parameters and Optimization of Membership Function Parameters of Type 1 Fuzzy Logic Systems (T1FLS), Interval Type 2 fuzzy Logic Systems (IT2FLS) and Generalized Type 2 Fuzzy Logic Systems (GT2FLS) to solve control problems and also to adapt and optimize LSTM (Long-Short Term Memory) Recurrent Neural Networks (RNN) architectures for Mackey–Glass time series forecasting.

Table 5 Hypothesis Test for 100 dimensions of the CMOA and other methods

Fun	CMOA		FPA			
	Mean	Std Dev	Mean	Std Dev	Z	Evidence
f1	7.00E-09	3.83E-09	4.37E+01	3.21E+01	−7.45	Yes
f2	1.16E-07	2.13E-08	1.59E+00	1.81E-01	−48.06	Yes
f3	1.03E-09	5.81E-10	4.44E+02	8.41E+01	−28.92	Yes
f4	8.40E-09	3.98E-09	3.53E+00	7.74E-01	−24.98	Yes
f5	2.99E-09	4.99E-10	5.10E-02	1.35E-02	−20.65	Yes
Fun	CMOA		PSO			
f1	7.00E-09	3.83E-09	7.21E-06	8.43E-07	−46.80	Yes
f2	1.16E-07	2.13E-08	–	–		
f3	1.03E-09	5.81E-10	5.48E-08	3.26E-08	−9.03	Yes
f4	8.40E-09	3.98E-09	7.17E-06	8.45E-07	−46.42	Yes
f5	2.99E-09	4.99E-10	2.47E-02	3.72E-03	−36.37	Yes

GA				DE			
Mean	Std Dev	Z	Evidence	Mean	Std Dev	Z	Evidence
5.62E-06	2.84E-07	−108.24	Yes	3.99E-05	1.12E-05	−19.51	Yes
–	–			–	–		
1.18E-08	1.00E-09	−50.99	Yes	2.17E-07	7.72E-08	−15.32	Yes
6.04E-06	2.37E-07	−139.37	Yes	4.35E-05	9.03E-06	−26.38	Yes
1.84E-02	7.36E-04	−136.93	Yes	1.12E-01	2.74E-02	−22.39	Yes
WSA				SPSO			
1.00E+00	6.68E-09	−7.11E+08	Yes	4.73E-05	5.86E-06	−44.20	Yes
–	–			–	–		
1.01E-02	8.13E-05	−680.44	Yes	1.54E-05	3.58E-06	−23.56	Yes
1.77E-04	5.07E-05	−19.12	Yes	1.64E-04	1.29E-05	−69.63	Yes
5.00E-01	4.56E-13	−5.48E+09	Yes	2.99E-01	1.32E-02	−124.07	Yes

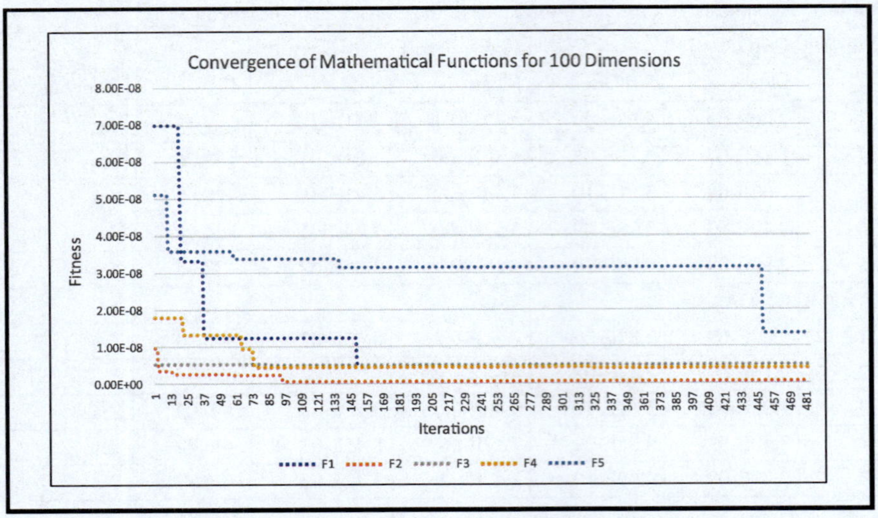

Fig. 9 Convergence of the CMOA algorithm for 100 dimensions and the 5 mathematical functions CEC2013

References

1. Simard, S. W. (2018). *Mycorrhizal networks facilitate tree communication, learning and memory, department of forest and conservation sciences, Faculty of Forestry, University of British Columbia, Vancouver, BC*. Springer International Publishing AG, part of Springer Nature.
2. Osman, I. H., Kelly, J. P. (1996). Meta-heuristics: An overview. In: Osman, I. H., Kelly, J. P. (Eds) *Meta-Heuristics*. Boston, MA: Springer.
3. Yamada, T., & Nakano, R. (2000). Job-shop scheduling by simulated annealing combined with deterministic local search.
4. Glover, F., Mulvey, J. M., Hoyland, K. (1996). Solving dynamic stochastic control problems in finance using tabu search with variable scaling. In: Osman, I. H., Kelly, J. P. (Eds.), *Meta-heuristics*. Boston, MA: Springer.
5. Adam, S. P., Alexandropoulos, S.-A., Pardalos, P., & Vrahatis, M. (2019).No free lunch theorem: A review.
6. Holland, J. H. (1975). *Adaptation in natural and artificial systems: An introductory analysis with applications to biology, control, and artificial intelligence*. U Michigan Press.
7. Dorigo, M., Maniezzo, V., & Colorni, A. (1996). Ant System: optimization by a colony of cooperating agents. *IEEE Transactions on Systems, Man, and Cybernetics-Part B, 26*(1), 29–41.
8. Karaboga, D. (2005). An idea based on honey bee swarm for numerical optimization. Technical report, Technical report-tr06, Erciyes University, Engineering Faculty, Computer Engineering Department.
9. Alrefaei, M. H., & Andradóttir, S. (1999). A simulated annealing algorithm with constant temperature for discrete stochastic optimization. *Management Science, 45*, 748–764.
10. Gelatt, K. S., & C., Vecchi M. (1983). Optimization by simulated annealing, 1983. *Science, 220*, 671–680.
11. Almarashi, M., Deabes, W., Amin, H. H., & Hedar, A. R. (2020). Simulated annealing with exploratory sensing for global optimization. *Algorithms, 13*, 230.
12. Rashedi, E., Nezamabadi-pour, H., & Saryazdi, S. (2009). GSA: A gravitational search algorithm. *Information Sciences, 179*(13), 2232–2248.

13. Sabri, N. M., Puteh, M., & Mahmood, M. R. (2013). A review of gravitational search algorithm. *International Journal of Advances in Soft Computing and its Applications, 5.*
14. Yang, X.-S. (2008). *Nature-inspired metaheuristic algorithms.* Luniver Press.
15. Yang, X.-S. (2009). Firefly algorithms for multimodal optimisation. In O. Watanabe, & T. Zeugmann (Eds.), *Proceedings of the 5th symposium on stochastic algorithms, foundations and applications.* Lecture notes in computer science (Vol. 5792, pp. 169–178).
16. Yang, X. S., & He, X. (2013). Firefly algorithm: Recent advances and applications. *International Journal of Swarm Intelligence, 1.*
17. Yang, X.-S. (2012). Flower pollination algorithm for global optimization. In: Durand-Lose, J., Jonoska, N. (Eds.), *Unconventional computation and natural computation. UCNC 2012.* Lecture notes in computer science (Vol. 7445). Berlin: Springer.
18. Gálvez, J., Cuevas, E., & Avalos, O. (2017). Flower pollination algorithm for multimodal optimization. *International Journal of Computational Intelligence Systems., 10,* 627–646.
19. Mirjalili, S. (2015). Moth-flame optimization algorithm: a novel nature-inspired heuristic paradigm. *Knowledge-Based Systems, 89.*
20. Kennedy, J., & Eberhart, R. (1995). Particle swarm optimization. In *Proceedings of the IEEE international conference on neural networks, Perth, Australia, 27 November–1 December 1995.*
21. Sengupta, S., Basak, S., & Peters II, R. A. (2018). Particle swarm optimization: A survey of historical and recent developments with hybridization perspectives.
22. Mirjalili, S., & Lewis, A. (2016). The Whale optimization algorithm. *Advances in Engineering Software, 95,* 51–67.
23. Yang, X.-S., & Deb, S. (2010). Cuckoo search via levy flights.
24. Yang, X.-S. (2014). Cuckoo search and firefly algorithm: overview and analysis. In: Yang, X. S. (Eds.), *Cuckoo search and firefly algorithm.* Studies in Computational Intelligence (vol. 516). Cham: Springer.
25. Carreon-Ortiz, H., Valdez, F., & Castillo, O. (2022).A new discrete mycorrhiza optimization nature-inspired algorithm. *Axioms, 11.* 391.
26. De Jong, K. (1975). Analysis of the behaviour of a class of genetic adaptive systems, Ph.D., thesis, University of Michigan, Ann Arbor.
27. McCulloch, W. S., & Pitts, W. A. (1943). logical calculus of the ideas immanent in nervous activity. *Bulletin of Mathematical Biophysics, 5,* 115–133.
28. Koza, J. R. (1992). *Genetic programming: One the programming of computers by means of natural selection.* MIT Press.
29. Jacobson, S., & Yücesan, E. (2004). Analyzing the performance of generalized hill climbing algorithms. *Journal of Heuristics, 10,* 387–405.
30. Wolpert, D. H., & Macready, W. G. (1997). No free lunch theorems for optimization. *IEEE Transanction on Evolution Computation, 1,* 67–82.
31. Geem, Z. W., Kim, J. H., & Loganathan, G. V. (2001). A new heuristic optimization algorithm: Harmony search. *Simulation, 76,* 60–68.
32. Nakrani, S., & Tovey, C. (2004). On honey bees and dynamic server allocation in Internet hostubg centers. *Adaptive Behavior, 12,* 223–240.
33. Yang, X.-S. (2010). A new metaheuristic bat-inspired algorithm. In J. R. Gonzalez et al. (Eds.), *Nature inspired cooperative strategies for optimization (NICSO 2010), SCI 284* (pp. 65–74). Springer.
34. Pierezan, J., & Coelho, L. D. S. (2018).. Coyote optimization algorithm: A new metaheuristic for global optimization problems. In *2018 IEEE congress on evolutionary computation (CEC)* (pp. 1–8).
35. Hao, H., & Luo, Y. (2019). An original bionic algorithm: interdependent balance algorithm. In *2019 IEEE 2nd international conference on electronic information and communication technology (ICEICT)* (pp. 584–589).
36. Mohammadi-Balani, A., Nayeri, M. D., Azar, A., & Taghizadeh-Yazdi, M. (2021). Golden eagle optimizer: A nature-inspired metaheuristic algorithm. *Computers and Industrial Engineering, 152,* 107050.

37. Abualigah, L., Hanandeh, E., Khader, A. T., Otair, M. A., & Shandilya, S. K. (2020). An improved b-hill climbing optimization technique for solving the text documents clustering problem. *Current Medical Imaging Reviews, 14*.
38. Storn, R., & Price, K. (1997). Differential evolution - a simple and efficient heuristic for global optimization over continuous spaces. *Journal of Global Optimization, 11*, 341–359.
39. Carreon, H., Valdez, F., & Castillo, O. (2020).*Fuzzy flower pollination algorithm to solve control problems, hybrid intelligent systems in control, pattern recognition and medicine*. Studies in computational intelligence (pp 119–154). Springer Nature Switzerland AG.
40. Sakib, N., Kabir, M. W. U., Subbir, M., & Alam, S. (2014).*A comparative study of flower pollination algorithm and bat algorithm on continuous optimization problems* (Vol. 7– No.9). Department of Computer Science and Engineering Ahsanullah University of Science and Technology Dhaka-1208, Bangladesh, Foundation of Computer Science FCS, New York, USA.
41. Rakhshani, H., & Rahati, A. (2017). Snap-drift cuckoo search: A novel cuckoo search optimization algorithm, Department of Computer Science, Faculty of Mathematics, University of Sistan and Baluchestan, Zahedan 98135–674, Iran, Applied Soft Computing.
42. Tvrdık, J., & Polakova, R. (2013). Competitive differential evolution applied to CEC 2013 problems. In *2013 IEEE congress on evolutionary computation June 20–23*. Cancún, México. (2013).
43. Zeng, B., Gao, L., Li, X. (2017). Whale Swarm for function optimization, Huazhong University of Science and Technology, Wuhan, China. In *Intelligent computing theories and application 13th international conference, ICIC 2017 Liverpool, UK, August 7–10, 2017 proceedings, Part I* (pp. 624 – 639). Springer International Publishing AG.
44. Voroshilova, A., & Wafubwa, J. (2020). Discrete competitive Lotka–Volterra model with controllable phase volume. *Systems, 8*, 17.
45. Lih-Ing, R. (2008). Dynamically consistent discrete Lotka-Volterra competition models derived from nonstandard finite-difference schemes. *Discrete and Continuous Dynamical Systems. Series B. 2*.
46. Farrukh, M., & Mansoor, S. (2012). On discrete Lotka-Volterra type models. *International Journal of Modern Physics Conference Series, 09*, 341–346.
47. Mira-Cristiana, A. (2014). Lotka, Volterra and their model. Didactica. *Mathematica, 32*, 9–17.
48. Mickens, R. E. (2018). A note on exact finite difference schemes for modified Lotka–Volterra differential equations. *Journal of Difference Equations and Applications, 24*(6), 1016–1022.
49. Li, Xu., Jiayi, L., & Guang, Z. (2018). Pattern formation and parameter inversion for a discrete Lotka-Volterra cooperative system. *Chaos, Solitons and Fractals, 110*, 226–231.

Optimal Tuning of an Active Disturbance Rejection Controller Using a Particle Swarm Optimization Algorithm

Olga L. Jiménez Morales, Diego Tristán Rodríguez, Rubén Garrido, and Efrén Mezura-Montes

1 Introduction

Active Disturbance Rejection Control (ADRC) allows compensating internal and external disturbances accurately or approximately in a control system. The main idea is to estimate a term containing the disturbances affecting a plant as well as parameter uncertainties and unmodelled terms by means of a disturbance observer (DOB). The estimate produced by the latter is injected in the plant as part of the control signal with the goal of counteracting the real disturbance. In some cases the ADRC can be shown to be equivalent to classic controllers including the Proportional Integral Derivative Controller (PID) widely used in industry.

There exist works showing the equivalence between a PID controller and a DOB-based controller [1–3]. They conclude that the integral action of the PID controller compensates for constant disturbances and has been considered as an implicit disturbance observer. On the other hand, the DOB-based controller counteracts constant and time-varying disturbances, so it is considered as a robust controller based on online disturbance estimation and compensation.

O. L. Jiménez Morales (✉) · D. Tristán Rodríguez · R. Garrido
Department of Automatic Control, CINVESTAV-IPN, Av. Instituto Politecnico Nacional 2508, Mexico city 07360, Mexico
e-mail: ojimenez@ctrl.cinvestav.mx

D. Tristán Rodríguez
e-mail: dtristan@ctrl.cinvestav.mx

R. Garrido
e-mail: garrido@ctrl.cinvestav.mx

E. Mezura-Montes
Artificial Intelligence Research Institute, University of Veracruz, Veracruz 91223, Mexico
e-mail: emezura@uv.mx

O. Castillo and P. Melin (eds.), *Hybrid Intelligent Systems Based on Extensions of Fuzzy Logic, Neural Networks and Metaheuristics*, Studies in Computational Intelligence 1096,
https://doi.org/10.1007/978-3-031-28999-6_11

For controller design and tuning, in particular the PID controller, there exists a plethora of tuning techniques including the pole assignment technique, the Ziegler and Nichols method, and optimal tuning techniques such as the Linear Quadratic Regulator among others [4–7]. Interestingly enough, in [8, 9], the disturbance observer is used as a tuning technique for fuzzy controllers. However, for the implementation of an ADRC algorithm, obtaining optimum performance is not a trivial task. The above is due to the fact that ADRC requires the simultaneous tuning of the controller gains plus those of the disturbance observer.

Intelligent optimization techniques such as simulated annealing, pattern search, genetic algorithm, and particle swarm optimization (PSO) are currently being used in several areas of automatic control and industrial applications to improve system stability, for parameter identification, and for controller tuning [10–17]. Among the them, the PSO algorithm has been successful due to the simplicity of its implementation, its computational efficiency and its ability to search large spaces of potential solutions.

There exists interesting past literature regarding the parameter tuning of feedback controllers employing intelligent optimization techniques. The PID controllers has been tuned using several algorithms including the Modified Butterfly Optimization algorithm [18], and the PSO algorithm [19, 20]. On the other hand, several metaheuristics including the Genetic Algorithm, Simulated Annealing, and Tabu Search has been used for PID controller tuning applied to bioprocess control [21]. An interesting survey on applying intelligent optimization techniques to PID controller tuning is found in [22] whereas [23] gives an account on how to apply multiobjective optimization to controller tuning.

In the case of ADRC there are several works where the tuning is performed through a metaheuristic [24–26]. The PSO algorithm is employed in [24] for tuning an ADRC algorithm applied to a ship. The controller is based on an extended state observer and the fitness function correspond to the integral time absolute error (ITAE) performance index. The Whale Optimization algorithm is employed in [25] for tuning of an ADRC algorithm applied to a quadrotor, and the integral of the absolute error is used as a fitness function. The tuning of extended state observers is performed in [26] by means of a PSO algorithm, and the fitness function corresponds to the integral of the absolute error. It is interesting to point out that in all the above references the stability conditions associated to the closed-loop system are not considered into the implementation of the optimization algorithms. Moreover, the control effort is not taken into account in the fitness function, and only simulation results are reported.

The objective of this work is to show the tuning of an ADRC algorithm by means of Particle Swarm Optimization techniques. The controller is implemented in real-time and applied on a low-cost educational prototype endowed with a Makeblock servo motor [27]. One of the key features of the proposed tuning methodology, which is a depart from previous controller tuning procedures using intelligent optimization techniques, is to use the conditions derived from a stability analysis of the closed-loop system composed of the servo motor and the ADRC algorithm, to define the feasible set of solutions. Thus, the PSO algorithms would produce particles associated with optimal gains that guarantee closed-loop stability while minimizing a fitness function.

This work is divided as follows. Section 2 describes the Active Disturbance Rejection Controller. Section 3 presents the Particle Swarm Optimization techniques used for ADRC tuning. Section 4 gives details of the experimental platform, exposes the experimental results and presents the performance analysis of the controller. The work ends with some concluding remarks.

2 Active Disturbance Rejection Controller

2.1 Preliminaires

The main idea of the Disturbances Observer (DOB) developed in [28–30] is to use measurements of the input and output of a disturbed plant under control to estimate the disturbance. Then, the disturbance estimate is used to counteract the effects of the real disturbance. According to the diagram shown in Fig. 1 the plant output is given by:

$$Y(s) = P(s)[U(s) + \hat{D}(s)] \tag{1}$$

where s is a complex variable, $G(s)$ is the transfer function which models the plant [4], $Y(s) = \mathcal{L}\{y\}$ and $U(s) = \mathcal{L}\{u\}$ are the Laplace transform of the output y and the input u, $D(s) = \mathcal{L}\{d\}$ and $\hat{D}(s) = \mathcal{L}\{\hat{d}\}$ are respectively the Laplace transform of the disturbance d and its estimate $\hat{d}$, and $\mathcal{L}$ stands for the Laplace operator.

According to Fig. 1, the DOB is composed of the inverse of the nominal model of the plant $P_m(s)$ and a strictly proper stable filter $F(s)$, and the expression for estimating $\hat{D}(s)$ is

$$\hat{D}(s) = [P_m^{-1}(s)Y(s) - U(s)]F(s) \tag{2}$$

Note that the estimation of the disturbance is realized using only input and output measurements.

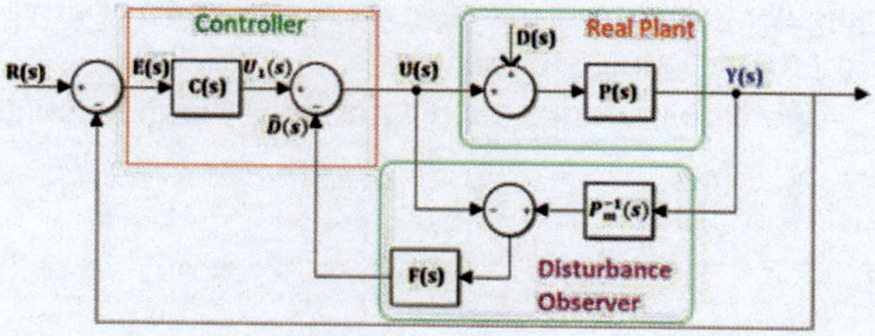

Fig. 1 Diagram of the ADRC algorithm composed of a Disturbance Observer and a controller designed for the nominal model $P_m(s)$ of the plant

2.2 ADRC Applied to a DC Servomotor

Consider the mathematical model of a DC servomotor affected by an unknown disturbance $\bar{d}$ [31]:

$$\ddot{y} = -a\dot{y} + bu + \bar{d} \tag{3}$$

The term $a > 0$ is related to viscous friction and $b >$ is the input gain. It is assumed that a is unknown and the input gain b is known. Moreover, $-a\dot{y}$ and $\bar{d}$ are grouped into a single term $d = \bar{d} - a\dot{y}$, which is assumed to be bounded, i.e. $|d| \leq D$. Then, Eq. (3) simplifies to:

$$\ddot{y} = bu + d \tag{4}$$

The Laplace transform of (4) is given by:

$$s^2Y(s) = bU(s) + D(s) \tag{5}$$

From (5) it follows that the nominal model P_m of the servomechanimsm corresponds to:

$$P_m(s) = \frac{b}{s^2} \tag{6}$$

According to the theory presented previously, and considering [1, 28–30], the filter $F(s)$ is defined as:

$$F(s) = \frac{\beta}{s + \beta} \tag{7}$$

with cutoff frequency $\beta > 0$.

Finally, from (2), (5) and (7), the perturbation estimate is given by:

$$\hat{D}(s) = \frac{\beta s}{s + \beta} sY(s) - \frac{\beta b}{s + \beta} U(s) \tag{8}$$

Note that $sY(s)$ corresponds to the angular velocity of the servomotor. Moreover, for the implementation of the Disturbance Observer, measurements of the angular velocity of the servomotor are required. However, in many practical cases, only angular position measurements are available, so a Luenberguer Observer is designed to provide velocity estimates.

For ease of design we rewrite Eq. (4) in the time domain as:

$$\begin{aligned} \dot{x}_1 &= x_2 \\ \dot{x}_2 &= bu + d \end{aligned} \tag{9}$$

where $x_1 = y$ and $x_2 = \dot{y}$. and the Disturbance Observer (8) in the time domain is given by:

$$\dot{\hat{d}} = -\beta\hat{d} + \beta(\dot{\hat{x}}_2 - bu) \tag{10}$$

Following the ideas in [30] to avoid the use of $\dot{\hat{x}}_2$, define the following auxiliary variable:

$$\omega = \hat{d} - \beta\hat{x}_2 \tag{11}$$

Thus, performing the time derivative of (11) and substituting (10) produces:

$$\begin{aligned} \dot{\omega} &= -\beta\hat{d} - \beta bu \\ \hat{d} &= \omega + \beta\hat{x}_2 \end{aligned} \tag{12}$$

The next control law with state feedback is proposed and allows compensating for the effects of the disturbance d:

$$u = \frac{1}{b}[u_n - \hat{d}] \tag{13}$$

where u_n corresponds to the control signal generated by the controller designed for the nominal plant $P_m(s)$. First, assume that $\hat{d}$ in (13) exactly compensates for the perturbation d in (9), i.e. $\hat{d} - d = 0$. Then, substituting (13) into (9) yields the next undisturbed nominal model of the plant:

$$\begin{aligned} \dot{x}_1 &= x_2 \\ \dot{x}_2 &= u_n \end{aligned} \tag{14}$$

Therefore, the term u_n is given by:

$$\begin{aligned} u_n &= \ddot{r} + \alpha_1\dot{e} + \alpha_2 e \\ &= \ddot{r} + \alpha_1(\dot{r} - x_2) + \alpha_2(r - x_1) \end{aligned} \tag{15}$$

where r, $\dot{r}$ and $\ddot{r}$ are the reference and its derivatives, $\dot{e} = \dot{r} - x_2$, and α_1, α_2 are positive constants. Substituting (15) into (14) yields the dynamics of the closed-loop undisturbed system:

$$\begin{aligned} \dot{x}_1 &= x_2 \\ \dot{x}_2 &= \ddot{r} + \alpha_1\dot{e} + \alpha_2 e \\ &= \ddot{r} + \alpha_1(\dot{r} - x_2) + \alpha_2(r - x_1) \end{aligned} \tag{16}$$

which has the next Hurwitz stable polynomial:

$$P_c(s) = s^2 + \alpha_1 s + \alpha_2 \tag{17}$$

Then, for the implementation of the Disturbance Observer and the ADRC, the following control law is set as:

$$u = \frac{1}{b}[\ddot{r} + \alpha_1(\dot{r} - \hat{x}_2) + \alpha_2(r - \hat{x}_1) - \hat{d}] \tag{18}$$

The estimates $\hat{x}_1$ and $\hat{x}_2$ are produced by the following Luenberger Observer (LO) built for the nominal system (14):

$$\begin{aligned} \dot{\hat{x}}_1 &= \hat{x}_2 + \gamma_1 \epsilon \\ \dot{\hat{x}}_2 &= u_n + \gamma_2 \epsilon \\ \hat{y} &= \hat{x}_1 \\ \epsilon &= y - \hat{y} \end{aligned} \tag{19}$$

where ϵ is the observation error. The gains $\gamma_1 > 0$ and $\gamma_2 > 0$ are the coefficients of the next polynomial associated to the LO, which is Hurwitz stable:

$$P_{do}(s) = s^2 + \gamma_1 s + \gamma_2 \tag{20}$$

Finally, from the stability analysis [30], the following condition must be fulfilled to guarantee closed-loop stability:

$$0 < \beta < \gamma_1 \tag{21}$$

3 Particle Swarm Optimization

The Particle Swarm Optimization (PSO) algorithm is one of the most widely used swarm intelligence methods since it has been able to resolve a variety of complex optimization problems in several different areas [32].

Proposed in 1995 by Kennedy and Eberhart [33], this algorithm was discovered through a simulation of a simplified social model inspired by the behavior of bird flocks. The PSO is composed of a swarm of particles evolving in a set of feasible solutions where it desired to find a minimum or a maximum of a fitness function. Each one of the particles is a potential solution to the problem. The swarm movements in the search space are called flights, and they are produced from the best individual particle positions, from the best global positions corresponding to all the particles in the swarm, and from past position values. The updating of the best positions, individual and global, is done by the evaluation of a fitness function.

This work uses three variations of the PSO algorithm in order to compare the performance of each one.

3.1 PSO with Inertia Weight

The PSO with inertia weight or ω-PSO algorithm [34] is composed of N particles of dimension L. The particles are represented as $z_n(k) = [z_{n,1}, \ldots, z_{n,L}]^T$ where $n \in [1, \ldots, N]$. For each particle, there exists a velocity vector $v_n(k) = [v_{n,1}, \ldots, v_{n,L}]^T$. The algorithm evolves through the following discrete-time dynamic system:

$$\begin{aligned} v_n(k+1) = &\ \omega v_n(k) \\ &+ c_1 \text{rand}()(\text{pBest}(k, n) - z_n(k)) \\ &+ c_2 \text{rand}()(\text{gBest}(k) - z_n(k)) \end{aligned} \tag{22}$$

$$z_n(k+1) = z_n(k) + v_n(k+1) \tag{23}$$

where the initial conditions are $z_{n,l}(0) = \text{rand}()$ and $v_{n,l}(0) = 0$ such that $l \in [1, \ldots, L]$ and rand() is the uniformly distributed random number function. The parameter $\omega \in [0, 1)$ is called the inertia weight and $c_1 \in [0, 1)$, $c_2 \in [0, 1)$ are called the learning factors. The terms $\text{pBest}(k, n) \in \mathbb{R}^L$ and $\text{gBest}(k) \in \mathbb{R}^L$ are defined as follows:

$$\text{pBest}(k, n) = \arg \min_{0 \le s \le k} J(z_n(s)) \tag{24}$$

$$\text{gBest}(k) = \arg \min_{0 \le s \le k, 1 \le j \le n} J(z_j(s)) \tag{25}$$

such that $J(\cdot)$ is a fitness function. Algorithm 1 shows the pseudo code of the ω-PSO algorithm.

Algorithm 1 ω-PSO algorithm

1: Create a random initial swarm $z_n(k)$ $\forall n$
2: Evaluate the fitness function $J(z_n(k))$ $\forall n$
3: Calculate pBest(k, n) $\forall n$
4: Calculate gBest(k)
5: **while** stop condition == false **do**
6:     **for** $n = 1$ to N **do**
7:         Compute the velocity vector $v_n(k+1)$
8:         Perform the flight $z_n(k+1)$
9:         Evaluate the fitness function $J(z_n(k))$
10:        Calculate pBest(k, n)
11:    **end for**
12:    Calculate gBest(k)
13: **end while**

3.2 Fractional PSO

The Fractional PSO (FPSO) was developed from the fractional calculus [35] which uses fractional derivatives to control the convergence rate of the PSO. This work uses the Fractional Velocity PSO, where the fractional derivative is applied to the velocity in the classical PSO considering the first four terms of the fractional differential derivative [36]. This algorithm employs N particles of dimension L, which are represented as $z_n(k) = [z_{n,1}, \ldots, z_{n,L}]^T$ where $n \in [1, \ldots, N]$. For each particle, there exists a velocity vector $v_n(k) = [v_{n,1}, \ldots, v_{n,L}]^T$. The discrete dynamic system that represents the FPSO algorithm is as follows:

$$\begin{aligned} v_n(k+1) =& \alpha v_n(k) + \frac{1}{2}\alpha v_n(k-1) + \frac{1}{6}\alpha(1-\alpha)v_n(k-2) \qquad (26) \\ &+ \frac{1}{24}\alpha(1-\alpha)(2-\alpha)v_n(k-3) \\ &+ c_1\text{rand}()(\text{pBest}(k,i) - z_n(k)) + c_2\text{rand}()(\text{gBest}(k) - z_n(k)) \\ z_n(k+1) =& z_n(k) + v_n(k+1) \qquad (27) \end{aligned}$$

where the initial conditions are $z_{n,l}(0) = \text{rand}()$ and $v_{n,l}(0) = 0$ such that $l \in [1, \ldots, L]$ and rand() is the uniformly distributed random numbers function, pBest() and gBest() are defined by (24) and (25). The terms c_1 and c_2 are computed through the next formulae:

$$c_j = (c_{ji} - c_{jf})\frac{k_{max} - k}{k_{max}} + c_{jf}, \quad j = 1, 2 \qquad (28)$$

such that c_{ji} and c_{jf} are the initial and final value of the learning factors respectively. The variable α is linearly and adaptively regulated as follows:

$$\alpha = 0.9 - \frac{k}{(1 + e^{-E_f(k)})k_{max}} \qquad (29)$$

where

$$E_f(k) = \frac{di_{gb}(k) - di_{\min}(k)}{di_{\max}(k) - di_{\min}(k)} \qquad (30)$$

$$di(z_n(k)) = \frac{1}{N-1}\sum_{j=1, j\neq n}^{N} \|z_n(k) - z_j(k)\| \qquad (31)$$

and

$$di_{gb}(k) = di(\text{gBest}(k)) \qquad (32)$$

$$di_{min}(k) = \min_{1\le s\le n} di(z_s(k)) \qquad (33)$$

$$di_{max}(k) = \max_{1\le s\le n} di(z_s(k)) \qquad (34)$$

3.3 PSO-AWDV

The PSO with an adaptive weighted delay velocity (PSO-AWDV) is proposed by Xiu [37] in order to deal with problems in the optimization as premature convergence and local stagnation. This algorithm is composed of N particles of dimension L. The particles are represented as $z_n(k) = [z_{n,1}, \dots, z_{n,L}]^T$ where $n \in [1, \dots, N]$. For each particle, exists a velocity vector $v_n(k) = [v_{n,1}, \dots, v_{n,L}]^T$. Its discrete-time dynamics are as follows:

$$\begin{aligned} v_n(k+1) =& \omega v_n(k) + (1-\omega) v_n(k-1) \\ &+ c_1 \text{rand}()(\text{pBest}(k,i) - z_n(k)) \\ &+ c_2 \text{rand}()(\text{gBest}(k) - z_n(k)) \end{aligned} \tag{35}$$

$$z_n(k+1) = z_n(k) + v_n(k+1) \tag{36}$$

where the initial conditions are $z_{n,l}(0) = \text{rand}()$ and $v_{n,l}(0) = 0$ such that $l \in [1, \dots, L]$ and rand() is the uniformly distributed random numbers function, pBest() and gBest() are defined by (24) and (25), and c_1 and c_2 decreases linearly as in the FPSO algorithm and ω changes according to the following equation:

$$\omega = 1 - \frac{a}{1 + e^{bE(k)}} \tag{37}$$

$$E(k) = \frac{\max\limits_{1 \le s \le N} J(z_s(k)) - \min\limits_{1 \le s \le N} J(z_s(k))}{\max\limits_{1 \le s \le N} J(z_s(k))} \tag{38}$$

3.4 Fitness Function

Each particle in the PSO algorithms is defined using the terms β in (12) and α_1, α_2 in (18) to minimize a fitness function. To this end, define each particle in the PSO algorithms as $z_n = [\alpha_1, \alpha_2, \beta]^T$. The fitness function is:

$$J(z_n) = \int_0^T \left(w_1 |e_t| + w_2 |e_v| + w_3 |u| + w_4 \left| \frac{du}{dt} \right| \right) dt \tag{39}$$

where $e_t = r - x_1$ and $e_v = \dot{r} - \hat{x}_2$ are the tracking error and the velocity error from the dynamic system (9) in closed-loop with (18) respectively, u is the control signal and the last term is its time derivative. Note that each element of the fitness function has a weight w_i, where $i = 1, \dots, 4$, that gives priority to one of the terms in the integral. Unlike previous works, the control signals and its time derivative are considered in the fitness functions. The control signal is used to avoid excessive energy consumption, and the derivative of the control signal helps reducing the impact of measurement noise on the closed-loop system. If this term is not taken into account, then the DOB term β may take large values that produces high levels of chatter in the control signal thus wreaking havoc the performance of the closed-loop system.

3.5 Set of Feasible Solutions

Since the PSO algorithms does not explicitly take into account the limits in the particles to ensure that the dynamic system is stable in closed-loop, it is important to define a set of feasible solutions. This set is called $\Omega \in \mathbb{R}^D$. In this paper, the Projection Boundary method [38] is implemented in the PSO algorithms to limit the particle positions to the set Ω. In this technique, if an element of the solution $z_n(k+1)$ falls out of Ω, the Projection Boundary method projects it into the boundary of Ω:

$$z_{n,d}(k+1) = \begin{cases} z_{n,d}(k+1) \text{ if } \min(\Omega_d) < z_{n,d}(k+1) < \max(\Omega_d) \\ \min(\Omega_d) \quad \text{if } \min(\Omega_d) > z_{n,d}(k+1) \\ \max(\Omega_d) \quad \text{if } \max(\Omega_d) < z_{n,d}(k+1) \end{cases} \tag{40}$$

where $d = 1, \ldots, D$.

4 Experiments

4.1 Experimental Platform

The experimental platform shown in Fig. 2 used for the evaluation of the algorithm is composed of a personal computer equipped with Matlab/Simulink real-time programming software and the QUARC real-time environment from Quanser consulting. A Quanser Q2 card provides data acquisition. The control signal produced by the acquisition board feeds a low-cost educational prototype composed of a linear power amplifier and a Makeblock DC servomotor. A galvanic isolator is used to protect the computer and the data acquisition board from the power amplifier.

4.1.1 Technical Characteristics of the Makeblock Servomotor

The servomotor presented in Fig. 2 is sold by the Makeblock company, and in the sequel it will be called the Makeblock servomotor. It has a 360 ppr optical encoder that allows the measurement of the angular position of the motor, and a steel output shaft that allows direct coupling of inertial loads, gears, pinions and synchronous belts, among others. Table 1 depicts its technical specifications, availability in the national market and cost in USD.

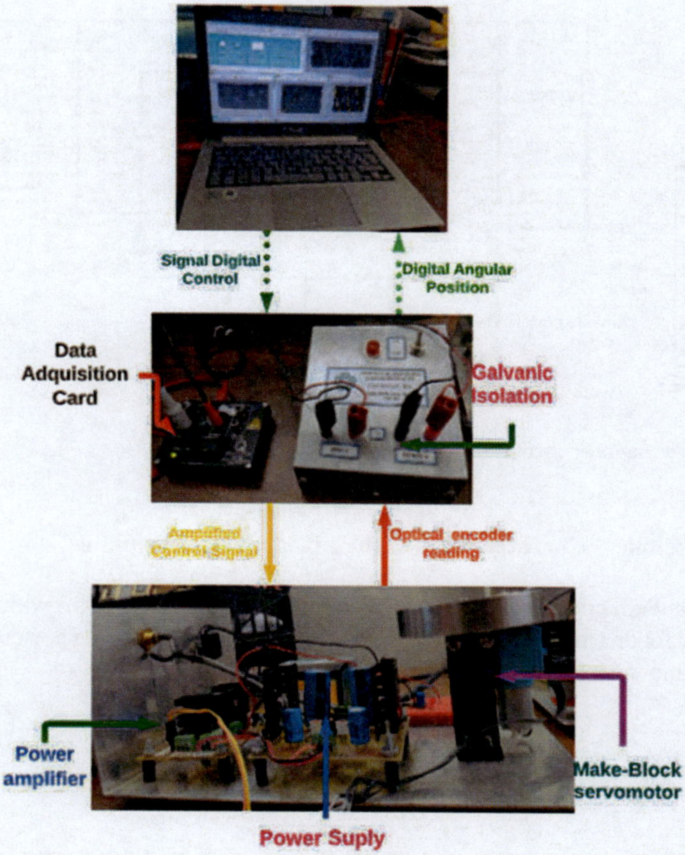

Fig. 2 Experimental platform

Table 1 Technical specifications of the *MakeBlock* servomotor [39]

Technical characteristic	
Rated voltage	7.4 V
No-load current	240 mA
Rotation speed	178 ± 10 RPM
Start torque	5 kg·cm
Feedback	Optical encoder
Rotation	Unlimited
Encoder accuracy	360 ppr
Weight	0.0615 kg kg
Availability	Highly available
Cost	30 USD

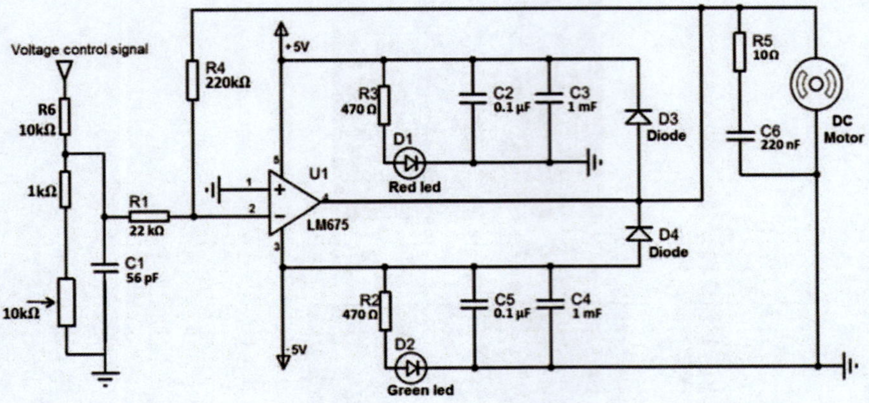

Fig. 3 Power amplifier circuit

4.1.2 Technical Characteristics of the Linear Power Amplifier

A National Semiconductor linear power amplifier model LM675, set with a gain of 10, is used for driving the Makeblock servomotor. The diagram of the power amplifier and its components is shown in Fig. 3.

4.2 *Optimization Procedure*

To perform the optimization by the PSO algorithms to find the parameters α_1, α_2 and β, the following steps are performed:

- Dynamic simulation conditions. In order to perform the dynamic simulation of the DC servomotor employed in the optimization procedure, consider the model (9) under the following conditions. A quantizer with a quantization interval of 1440 is added to the simulation to take into account the output of an optical incremental encoder used to measure the servomotor angular position. The values $a = 19.2519$ and $b = 12.2809$, obtained through a Least Squares algorithm [40], are considered in the simulation. The gains of the Luenberger Observer (19) are set to $\gamma_1 = 160$ and $\gamma_2 = 6400$. The disturbance d is simulated according to the next model:

$$\bar{d} = 0.05 \sin 2t + 0.1 \sin 0.2t + 0.1 \sin 0.5t + 0.1 \tag{41}$$

  This model represents a generic disturbance and it does not necessarily correspond to disturbances found in real servomotors. Nevertheless, it helps tuning de ADRC algorithm.
- Optimization problem definition. Define the feasible set of solutions employing the restrictions on $\alpha_1 > 0, \alpha_2 > 0$ associated to the characteristic polynomial (17),

Table 2 Values of the weights in the fitness function

	Test 1	Test 2	Test 3	Test 4
Value of w_1	100	100	100	100
Value of w_2	10	50	100	10
Value of w_3	0.1	0.1	0.1	0.1
Value of w_4	0.1	0.1	0.1	0

and β, which is a positive constant according to (7), and whose restriction imposed by the stability analysis corresponds to the inequality (21):

$$\Omega = [\alpha_1, \alpha_2, \beta | \alpha_1 \in (0, \infty), \alpha_2 \in (0, \infty), \beta \in (0, \gamma_1)] \tag{42}$$

Therefore, from (39) and (42) the optimization problem is defined as follows:

$$\begin{aligned} &\min J \\ &\text{subject to } \Omega \end{aligned} \tag{43}$$

A dynamic simulation of the servomotor model in closed-loop with the ADRC algorithm is performed with every particle generated by the PSO algorithms. The initial conditions of the DC servomotor model and of the Luenberger Observer are $x(0) = [1, 1]^T$, $\hat{x}(0) = [1, 1]^T$ and the reference and its time derivative in control law (18) are set to $r = 0$ and $\dot{r} = 0$ respectively. Each dynamic simulation lasts 6s. During the simulation the signals $e_t = r - x_1$, $e_v = \dot{r} - \hat{x}_2$, and u are stored in a file. Subsequently, the time derivative $\frac{du}{dt}$ is computed. Then, the above signals allow computing the fitness function (39) and the pBest and gBest terms used in the PSO algorithms.

For the implementation of the PSO algorithms, four tests are performed. Every test uses a different set of weights in the fitness function (39). The values are depicted in Table 2.

The parameters of the PSO algorithms are set using the IRACE package [41]. In the case of the ω-PSO algorithm the parameters are set as $\omega = 0.7$,$c_1 = 0.7$, $c_2 = 0.9$, $N = 23$ and a maximum number of iterations $MI = 180$. In the case of the FPSO algorithm the parameters are $c_{1i} = 0.9$, $c_{1f} = 0.7$, $c_{2i} = 0.9$, $c_{2f} = 0.8$, $N = 18$ and a maximum number of iterations $MI = 180$. The parameters of the PSO-AWDV algorithm are set as $c_{1i} = 0.9$, $c_{1f} = 0.5$, $c_{2i} = 0.9$, $c_{2f} = 0.8$, $N = 16$ and $MI = 180$. Each run of the algorithm stops after (N×MI) evaluations. An statistical test is performed with 30 runs of each algorithm to ensure that it gives congruent results. Using parallel processing with 6 cores of a PC computer, which runs at 4.5 GHz, the 30 runs of every algorithm are executed in 26 min.

For the sake of space, only the time evolution of the particles for the three PSO algorithms corresponding to the Test 1 is reported.

Table 3 Results of Test 1

	ω-PSO	FPSO	PSO-AWDV
Dimension 1 [α_1]	32.62	32.50	32.2454
Dimension 2 [α_2]	307.42	305.69	301.6532
Dimension 3 [β]	71.89	72.0607	72.1876
Minimum evaluation of J	**29.9782**	29.9823	29.9797
Median evaluation of J	29.9888	30.0119	29.9837
Average evaluation of J	29.9985	30.0090	29.9893
Standard deviation evaluation of J	$2.28e^{-2}$	$2.20e^{-2}$	$1.47e^{-2}$

4.2.1 Test 1

Figures 4, 5 and 6 show the time evolution of the particles in each dimension with respect to the iterations in the three different PSO algorithms. In this case the particles in the PSO-AWDV exhibits more oscillations. Table 3 depicts the results for every PSO algorithm where the lowest value of the fitness function is obtained through the ω-PSO algorithm.

4.2.2 Test 2

In the second test, the weight w_2 is increased. This change clearly affects the first dimension of the particles as it is shown in Table 4. Note that the gain α_1 increases compared with the value obtained in the Test 1. Moreover, the PSO-AWDV algorithm gives the lowest value of the fitness function.

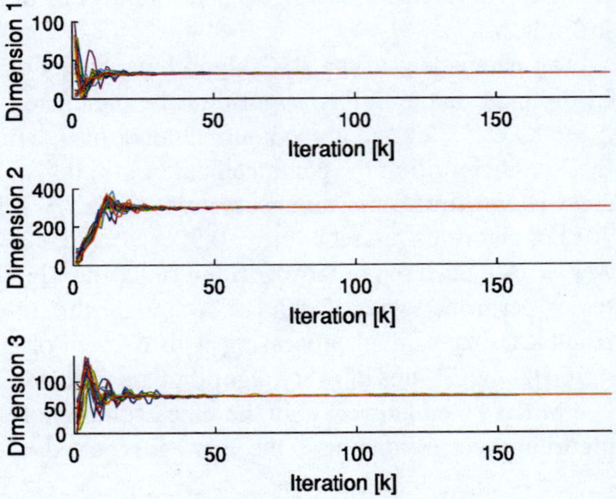

Fig. 4 Evolution of the particles in the ω-PSO algorithm: Test 1

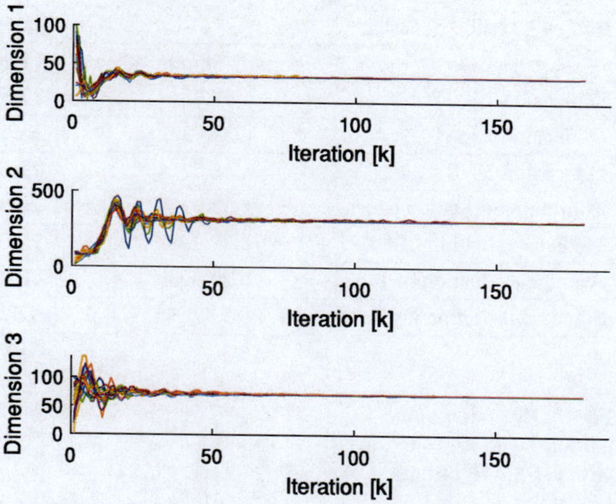

Fig. 5 Evolution of the particles in the FPSO algorithm: Test 1

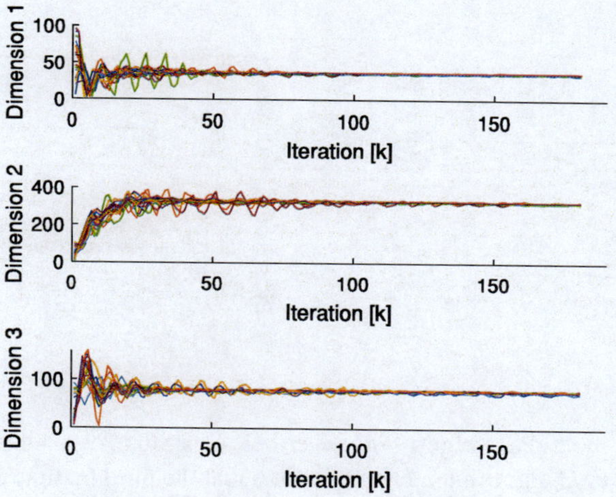

Fig. 6 Evolution of the particles in the PSO-AWDV algorithm: Test 1

4.2.3 Test 3

In the Test 3, the term w_2 is further increased and the corresponding gains are shown in Table 5. In this case, the PSO-AWDV algorithm provided the lowest value of the fitness function.

4.2.4 Test 4

The goal of this test is to show the behavior of the PSO algorithm when the term w_4 of the fitness function (39) is equal to zero. The above means that the fitness function does not take into account the chatter in the control signal. The test is done with

Table 4 Results in Test 2

	ω-PSO	FPSO	PSO-AWDV
Dimension 1 [α_1]	37.26	37.69	36.95
Dimension 2 [α_2]	326.37	327.38	322.14
Dimension 3 [β]	78.52	79.63	79.18
Minimum evaluation of J	69.9758	70.0083	**69.9745**
Median evaluation of J	29.9888	70.0154	69.98045
Average evaluation of J	29.9985	70.0177	69.9860
Standard deviation evaluation of J	$5.42e^{-2}$	$9.52e^{-3}$	$1.44e^{-2}$

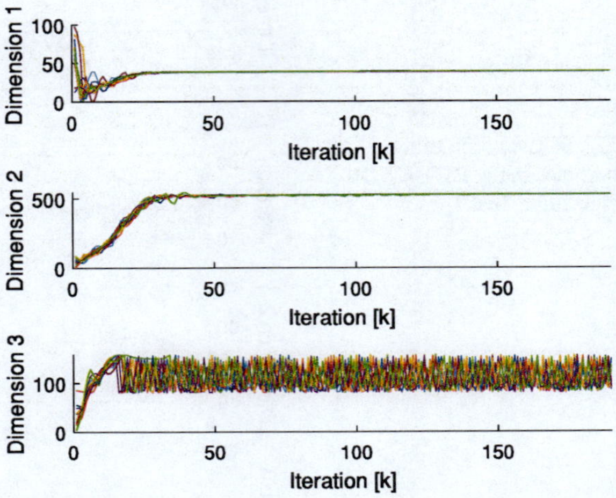

Fig. 7 Evolution of the particles with $w_4 = 0$ proved with ω-PSO algorithm

the ω-PSO algorithm, nevertheless, similar outcomes are obtained with the other PSO algorithms. Figure 7 shows that the third particle never reaches a constant value and holds sustained oscillations. This behaviour shows how important is to take into account the time derivative of the control signal in the fitness function.

Table 5 Results in Test 3

	ω-PSO	FPSO	PSO-AWDV
Dimension 1 [α_1]	38.93	39.98	39.29
Dimension 2 [α_2]	342.87	358.41	346.86
Dimension 3 [β]	92.14	88.12	90.74
Minimum evaluation of J	119.2806	119.3604	**119.2599**
Median evaluation of J	119.3941	119.3978	119.3815
Average evaluation of J	119.4374	119.3963	119.3671
Standard deviation evaluation of J	$1.66e^{-1}$	$1.20e^{-3}$	$4.51e^{-2}$

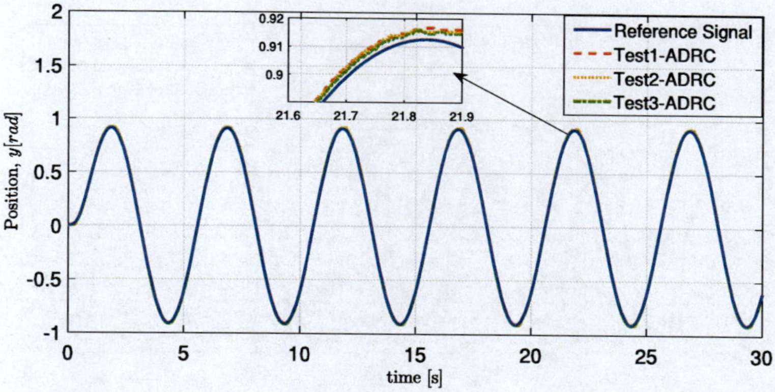

Fig. 8 Reference signal r verses servomotor output signal y using the ADRC corresponding to each of the tests

4.3 Experimental Results

The goal of this section is to show the performance of the ADRC algorithm using the gains obtained through the PSO algorithms. The ADRC algorithm is coded in the MATLAB/SIMULINK programming platform under the QUARC real-time environment from Quanser Consulting with a sampling time of 1 ms and the Euler01 integration method. The Makeblock servomotor used in the experiments drives an inertia disk. A sinusoidal function is used as a reference signal $r = 0.8 \sin 0.2t$, filtered by a first order low-pass filter with a cut-off frequency of 10 rad/s.

Three experiments were performed, in which the controller gains α_1 and α_2 of the control law (18) and the cutoff frequency β of the DOB are tuned using the parameters shown in Table 6. These gains correspond to the PSO algorithms producing the lowest value of the fitness function.

Figure 8 shows the trajectory tracking position results of the DC servomotor. The three set of gains produce similar results. Figure 9 shows the corresponding control signals of the ADRC algorithm. Note that the control signal generated by the gains corresponding to Test 1 produces larger peaks compared with the other signals. The tracking error signals are depicted in Fig. 10. It is worth noting that the tuning produced by the PSO-AWDV algorithm produces the lowest tracking error.

Table 6 Gains used in the implementation of the ADRC algorithm

Gains	Test 1: ω-PSO	Test 2: PSO-AWDV	Test 3: PSO-AWDV
α_1	32.62	36.95	39.29
α_2	307.42	322.14	346.86
β	71.89	79.18	90.74

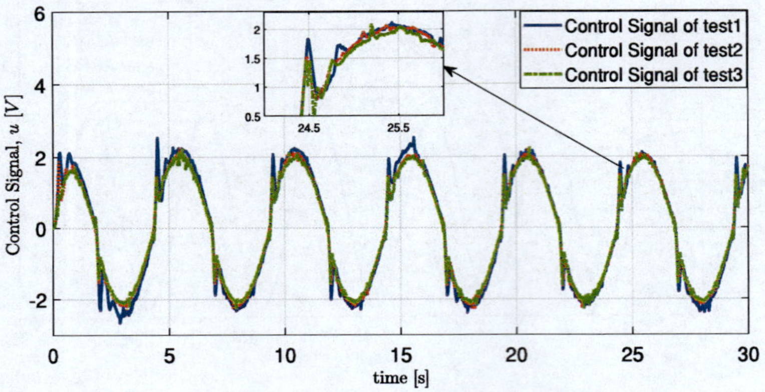

Fig. 9 ADCR control signal u tuning with the ω-PSO and PSO-AWDV algorithms

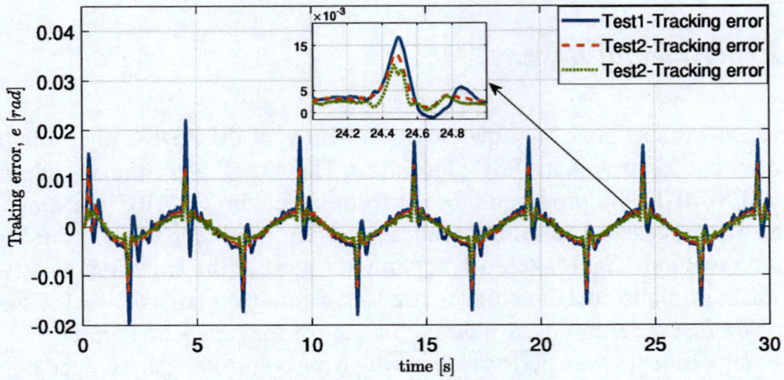

Fig. 10 The tracking error e and corresponding to each one of the tests

Three indices are used to measure the performance of the ADRC, with respect to the error signal, the control signal and its time derivative. These indices are represented mathematically as follows:

$$ISE = \int_{T_1}^{T_2} k[e(t)]^2 dt \tag{44}$$

$$IAC = \int_{T_1}^{T_2} |u(t)| dt \tag{45}$$

$$IACV = \int_{T_1}^{T_2} |\frac{du(t)}{dt}| dt \tag{46}$$

Table 7 ADRC Performance

Indices	Test 1 ω-PSO	Test 2 PSO-AWDV	Test 3 PSO-AWDV
IEC	0.7715	0.5721	**0.3185**
IAC	7.108	6.986	**6.774**
IACV	**38.07**	39.94	47.71

where k represents a scaling factor and (T_1, T_2) defines the time interval during which the performance indices are calculated. The indices are evaluated using $k = 100$ and the interval $T_2 - T_1 = 5$s with $T_1 = 20$s and $T_2 = 25$s.

Table 7 shows the results obtained in the experiments. The smallest Integral Squared Error (ISE) and Integral of the Absolute Value Control (IAC) indices are produced by the gains tuned using the Test 3.

The above is due to the fact that the cutoff frequency $\beta = 90.74$ is the largest of the three values. High values of this term promotes lower tracking errors. On the other hand, the index (IACV), which evaluates the control signal chatter increases for increasing values of β. Note also that a good performance trade-off between signal chatter and tracking error is obtained with the gains corresponding to the Test 2 since the chatter level measured by the IACV index is close to the one obtained with the gains corresponding in the Test 1 and the tracking error quality, measured using the ISE index takes values ISE $= 0.5721$, which is close to the mean of the ISE values obtained using the gains produced by the Tests 1 and 3, i.e. 0.545.

5 Conclusion

The results reported in this work show that the tuning produced by the Particle Swarm Optimization (PSO) algorithm are able to generate optimal gains in an Active Disturbance Rejection Control algorithm. These gains produce good performance in an low-cost educational prototype composed by an low-cost servomotor and a simple linear amplifier. Moreover, the proposed fitness function as well as the use of a realistic servomotor model, which includes the effects of an optical encoder and a disturbances, enhance the optimization procedure. A salient feature of the proposed optimization procedure is that it takes into account the stability conditions of the ADRC algorithm to define the set of feasible solutions. It is also worth remarking that the three PSO algorithms produce similar results, nevertheless, the PSO-AWDV algorithm generated the lowest value of the fitness function in two of three tests.

Acknowledgements The authors would like to thank Gerardo Castro and Jesús Meza for setting up the educational prototype and for installing the MatLab and QUARC software in the computer used in the experiments. The first and second authors thank CONACyT Mexico for its support through the PhD scholarships CVU 860371 and CVU 996816 respectively.

References

1. Luna, L., & Garrido, R. (2018). On the equivalence between P+ DOB and set point weighted PI controllers for velocity control of servodrives under load disturbances. In *2018 XX Congreso Mexicano de Robótica (COMRob)* (pp. 1–6). IEEE.
2. Yamada, K., Komada, S., Ishida, M., & Hori, T. (1997). Analysis and classical control design of servo system using high order disturbance observer. In *Proceedings of the IECON'97 23rd International Conference on Industrial Electronics, Control, and Instrumentation (Cat. No. 97CH36066)* (vol. 1, pp. 4–9). IEEE.
3. Garrido, R., & Luna, J. L. (2018). On the equivalence between PD+ DOB and PID controllers applied to servo drives. *IFAC-PapersOnLine, 51*(4), 95–100.
4. Ogata, K. (2002). *Modern Control Engineering*. Upper Saddle River: Prentice Hall.
5. Meshram, P., & Kanojiya, R. G. (2012). Tuning of pid controller using ziegler-nichols method for speed control of dc motor. In *IEEE-International Conference on Advances in Engineering, Science and Management (ICAESM-2012)* (pp. 117–122). IEEE.
6. Patel, V. V. (2020). Ziegler-nichols tuning method. *Resonance, 25*(10), 1385–1397.
7. Poznyak, A. S. (2009). *Advanced Mathematical Tools for Automatic Control Engineers: Stochastic Techniques*. Oxford: Elsevier.
8. Kim, B. K., Chung, W. K., & Ohba, K. (2009). Design and performance tuning of sliding-mode controller for high-speed and high-accuracy positioning systems in disturbance observer framework. *IEEE Transactions on Industrial Electronics, 56*(10), 3798–3809.
9. Lee, Y. S., Kim, D. S., & Kim, S.-K. (2018). Disturbance observer-based proportional-type position tracking controller for dc motor. *International Journal of Control, Automation and Systems, 16*(5), 2169–2176.
10. Singh, M., Patel, R. N., & Jhapte, R. (2016). Performance comparison of optimized controller tuning techniques for voltage stability. In *2016 IEEE First International Conference on Control, Measurement and Instrumentation (CMI)* (pp. 11–15). https://doi.org/10.1109/CMI.2016.7413701
11. Amine, K. (2019). Multiobjective simulated annealing: Principles and algorithm variants. *Advances in Operations Research 2019*.
12. Mishra, S., Prusty, R. C., & Panda, S. (2020). Design and analysis of 2dof-pid controller for frequency regulation of multi-microgrid using hybrid dragonfly and pattern search algorithm. *Journal of Control, Automation and Electrical Systems, 31*(3), 813–827.
13. Beigi, A. M., & Maroosi, A. (2018). Parameter identification for solar cells and module using a hybrid firefly and pattern search algorithms. *Solar Energy, 171*, 435–446.
14. Ibrahim, M. A., Mahmood, A. K., & Sultan, N. S. (2019). Optimal pid controller of a brushless dc motor using genetic algorithm. *International Journal of Power Electronics and Drive Systems ISSN, 2088*(8694), 8694.
15. Ramesh, H., & Xavier, S. (2022). Optimal tuning of servo motor based linear motion system using optimization algorithm. *Journal of Electrical Engineering & Technology*, 1–16.
16. Nyong-Bassey, B., & Epemu, A. (2022). Systems identification of servomechanism parameters using jellyfish, particle swarm and constraint optimization. *Nigerian Journal of Technology, 41*(3), 569–577.
17. Cortez, R., Garrido, R., & Mezura-Montes, E. (2022). Spectral richness pso algorithm for parameter identification of dynamical systems under non-ideal excitation conditions. *Applied Soft Computing, 128*, 109–490. https://doi.org/10.1016/j.asoc.2022.109490
18. Arulvadivu, J., Manoharan, S., Lal Raja Singh, R., & Giriprasad, S. (2016). Optimal design of proportional integral derivative acceleration controller for higher-order nonlinear time delay system using m-mboa technique. *International Journal of Numerical Modelling: Electronic Networks, Devices and Fields, 3016*.
19. Bouallègue, S., Haggège, J., Ayadi, M., & Benrejeb, M. (2012). Pid-type fuzzy logic controller tuning based on particle swarm optimization. *Engineering Applications of Artificial Intelligence, 25*(3), 484–493.

20. Chang, W.-D. (2022). An improved particle swarm optimization with multiple strategies for pid control system design. *International Journal of Modeling and Optimization, 12*(2)
21. Roeva, O., & Slavov, T. (2012). Pid controller tuning based on metaheuristic algorithms for bioprocess control. *Biotechnology & Biotechnological Equipment, 26*(5), 3267–3277.
22. Joseph, S. B., Dada, E. G., Abidemi, A., Oyewola, D. O., & Khammas, B.M. (2022). Metaheuristic algorithms for pid controller parameters tuning: Review, approaches and open problems. Heliyon, 309–399.
23. Rodríguez-Molina, A., Mezura-Montes, E., Villarreal-Cervantes, M. G., & Aldape-Pérez, M. (2020). Multi-objective meta-heuristic optimization in intelligent control: A survey on the controller tuning problem. *Applied Soft Computing, 93*, 106–342.
24. Hu, J., & Chen, W.: Design of active disturbance rejection controller for dynamic positioning based on improved particle swarm optimization. *Mathematical Problems in Engineering, 2022*.
25. Liu, X., Gao, Q., Ji, Y., Song, Y., & Liu, J. (2022). Active disturbance rejection control of quadrotor uav based on whale optimization algorithm. In *2022 IEEE International Conference on Mechatronics and Automation (ICMA)* (pp. 351–356). IEEE.
26. Zhang, D., Yao, X., & Wu, Q. (2016). Parameter tuning of modified active disturbance rejection control based on the particle swarm optimization algorithm for high-order system. In *2016 IEEE International Conference on Aircraft Utility Systems (AUS)* (pp. 290–294). IEEE.
27. Olga Jimenez, J. M., & Ruben, G. (2020). Estudio comparativo de servomotores de cd orientados a la construccion de prototipos educativos. In *Congreso Internacional de Robotica Y Computacion (CIRC-2020)* (pp. 32–40). IEEE.
28. Ohishi, K., Ohnishi, K., & Miyachi, K. (1988). Adaptive dc servo drive control taking force disturbance suppression into account. *IEEE Transactions on Industry Applications, 24*(1), 171–176.
29. Ohnishi, K., Shibata, M., & Murakami, T. (1996). Motion control for advanced mechatronics. *IEEE/ASME Transactions on Mechatronics, 1*(1), 56–67.
30. Garrido, R., & Luna, L. (2021). Robust ultra-precision motion control of linear ultrasonic motors: A combined adrc-luenberger observer approach. *Control Engineering Practice, 111*, 104–812. https://doi.org/10.1016/j.conengprac.2021.104812
31. Spong, M.W., Hutchinson, S., Vidyasagar, M., et al. (2006). *Robot Modeling and Control* (vol. 3). New York: Wiley.
32. Wang, D., Tan, D., & Liu, L. (2018). Particle swarm optimization algorithm: an overview. *Soft computing, 22*(2), 387–408.
33. Kennedy, J., Eberhart, R. (1995). Particle swarm optimization. In *Proceedings of ICNN'95 - International Conference on Neural Networks* (vol. 4, pp. 1942–1948). https://doi.org/10.1109/ICNN.1995.488968
34. Sidorov, G. (2018). Artificial Intelligence. Alfa-Omega.
35. Pires, E., Tenreiro Machado, J., Moura Oliveira, P., Cunha, J., & Mendes, L. (2010). Particle swarm optimization with fractional-order velocity. *Nonlinear Dynamics, 61*, 295–301. https://doi.org/10.1007/s11071-009-9649-y
36. Song, B., Wang, Z., & Zou, L. (2021). An improved pso algorithm for smooth path planning of mobile robots using continuous high-degree bezier curve. *Applied Soft Computing, 100*, 106–960.
37. Xu, L., Song, B., & Cao, M. (2021). An improved particle swarm optimization algorithm with adaptive weighted delay velocity. *Systems Science & Control Engineering, 9*(1), 188–197. https://doi.org/10.1080/21642583.2021.1891153
38. Juarez-Castillo, E., Acosta-Mesa, H., & Mezura-Montes, E. (2019). Adaptive boundary constraint-handling scheme for constrained optimization. *Soft Computing*, 1–34. https://doi.org/10.1007/s00500-018-3459-4
39. Makeblock. (2019). 180 Optical Encoder Motor. https://store.makeblock.com/180-optical-encoder-motor Consultado: Septiembre 2022. https://store.makeblock.com/180-optical-encoder-motor

40. Ioannou, P. A., & Sun, J. (2012). *Robust Adaptive Control*. Mineola, New York: Courier Corporation.
41. Manuel López Ibáñez, e.a (2020). Leslie Pérez Cáceres: The Irace Package: User Guide. Université Libre de Bruxelles, Brussels, Belgium. https://books.google.com.mx/books?id=pfJHAQAAIAAJ

Fuzzy Logic

Optimization of Fuzzy Controllers Using Distributed Bioinspired Methods with Random Parameters

Alejandra Mancilla, Oscar Castillo, and Mario García-Valdez

1 Introduction

In a recent paper, we presented a distributed algorithm for multi-population metaheuristics [1, 2] using Genetic Algorithms (GAs) [3, 4] and Particle Swarm Optimization (PSO) [5, 6]. We execute this algorithm asynchronously using a queue-based architecture [7, 8]. One of the problems of using a multi-population multi-heuristic approach is that we need to set the algorithms' parameters for each population [9]. Finding these parameters is time-consuming because we need to run several experiments trying different configurations until we find a suitable parametrization. This problem could be especially problematic if we have a multi-population algorithm consisting of several small populations, each with a set of initial parameters we need to set.

To tackle this problem, in the past, we followed a homogeneous approach [10], simply using the same set of parameters for all populations. In this case, we have two types of algorithms: some populations run a GA while others execute a PSO. This situation adds another layer to the problem because each algorithm requires different parameters. In this work, we compare two strategies to set the initial parameters of the algorithms of the multi-populations. The first is a homogeneous approach with fixed values, and the second is a heterogeneous strategy selecting random values from a predefined range of values established for some of the parameters of the algorithms [11, 12]. The main contribution of this papers is to compare two strategies

A. Mancilla (✉) · O. Castillo · M. García-Valdez
Tijuana Institute of Technology/Tecnologico Nacional de Mexico, Tijuana, Mexico
e-mail: alejandra.mancilla@tectijuana.edu.mx

O. Castillo
e-mail: ocastillo@tectijuana.mx

M. García-Valdez
e-mail: mario@tectijuana.edu.mx

O. Castillo and P. Melin (eds.), *Hybrid Intelligent Systems Based on Extensions of Fuzzy Logic, Neural Networks and Metaheuristics*, Studies in Computational Intelligence 1096,
https://doi.org/10.1007/978-3-031-28999-6_12

for distributed parameter specifications, if they have the same performance, this means that for this type of systems, either one can be used. For some applications a random heterogeneous approach will be preferred because the researcher does not need to set the exact parameter for better performance, is only needed to set a suitable range.

We organized this paper as follows. First, in Sect. 2, we describe our proposal and define the experimental setup consisting on optimizing a fuzzy rear-wheel controller using different configurations on a benchmark problem. We show and discuss the results in Sect. 3. Finally, we present our conclusions and future work in Sect. 4.

2 Methodology and Experimental Setup

We continue with our previous work, in which we optimized the parameters for a fuzzy controller used for path tracking by an autonomous mobile robot, this type of optimization has many applications and techniques, and is not limited to type-1 fuzzy systems [13, 14]. This problem is particularly time-consuming because we could follow these three steps to optimize this type of controller: First, we run one or more simulations with the controller; this is time-consuming, then we measure the controller's error in each simulation, calculate an average, and if we are not satisfied with the results, we adjust the parameters and start again. That is why designers of fuzzy controllers often use population-based metaheuristics to adjust these parameters. However, even when using a metaheuristic, fuzzy controller optimization continues to be time-consuming. We briefly explain the multi-population multi-heuristic algorithm used in this paper. The authors more extensively explain the algorithm in [1].

We propose an event-based, distributed algorithm that exchanges data between processes using message queues for asynchronous communication (see Fig. 1). (1) We start the algorithm by executing a single task that pushes a specified n number of populations to the Population Queue. Each population message contains the initial configuration and the type of metaheuristic used by that particular population. The main idea is that each process runs a metaheuristic for a small number of iterations on a population received as a message and then pushes the resulting (evolved) population to the Evolved Population Queue (4). (3) Workers take messages asynchronously to process the data (populations). Each worker inside a container continuously checks for messages in the Population Queue and, after receiving a population, executes a metaheuristic on this population for several iterations. (4) resulting populations from the Evolved Population Queue. (5–6) An essential component of the architecture is the Combinator process, responsible for taking the resulting populations from the Evolved Population Queue and, after reading the message, stopping the algorithm if the number of function evaluations has been reached or if a suitable solution was found.

However, the primary responsibility of this component is to migrate or combine populations arriving from the Evolved Population Queue. The combination method

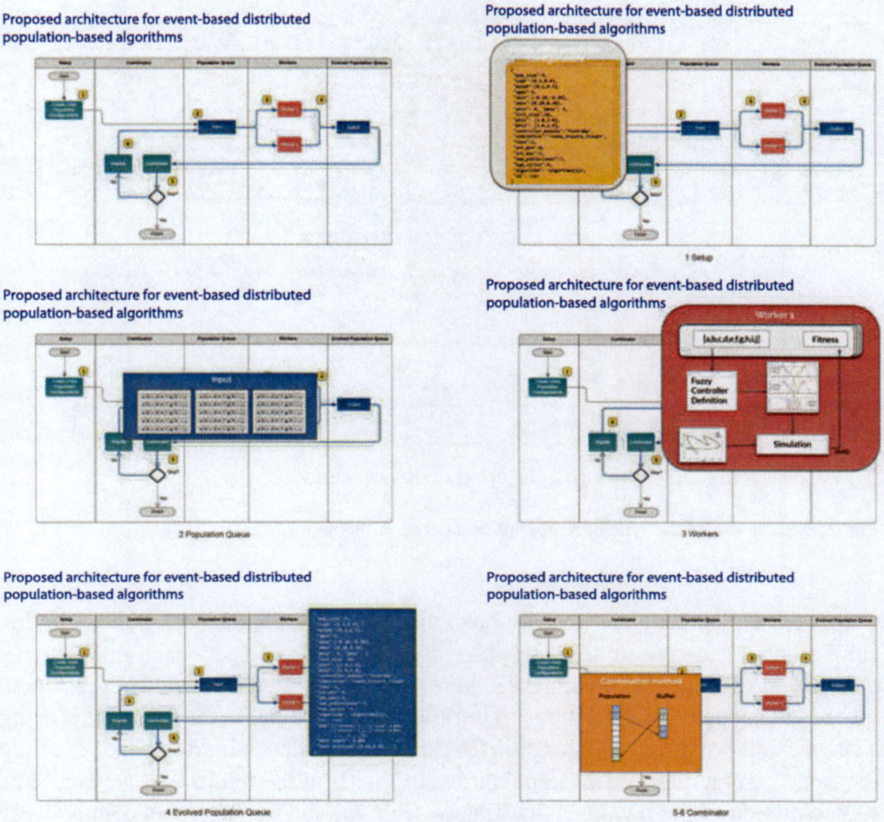

Fig. 1 Proposed architecture for event-based distributed population-based algorithms

we propose in this work is as follows. We take the top two best solutions from each population and insert them in a buffer that always keeps the top-k solutions. Finally, we replace the two worst solutions for the population with the buffer's current first and second-best solutions.

One of the problems of metaheuristics is establishing the initial values of the algorithm's parameters. Usually, bio-inspired metaheuristics have parameters that control how much the algorithm does an exploration or exploitation over the search space [15]. If the algorithm exploits too much, there is a higher risk of premature convergence to a local minimum. On the other hand, if there is too much exploration, the algorithm will search allmost randomly, constantly changing the area of attention. A balance between the two types of search is required to escape local minima while simultaneously doing a local search in a promising area.

When we have several populations, we could use a strategy to have some populations tending to exploration and others to exploitation. The idea is that this strategy keeps a natural balance between the two extremes. The problem remains: how do

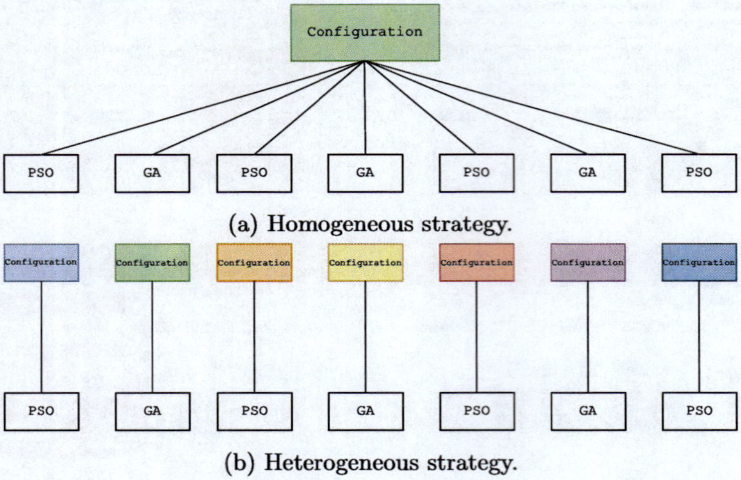

Fig. 2 Illustration of the type of strategies compared in this work

we set the parameters now that we have many more, a set for each population? A simple heterogeneous strategy proposed by [11] randomly sets each population's parameters. Although simple, the random strategy has yielded promising results in other multi-population algorithms . The other strategy found in the literature is using a single, well-balanced configuration for each metaheuristic algorithm and repeating the same configuration in all populations. In Fig. 2 on the top image, we can see a representation of the homogenous strategy (Fig. 2a) and the heterogeneous strategy at the bottom (Fig. 2b).

2.1 Control Problem

The benchmark control problem we use to validate our proposal is the rear-wheel controller described by Paden [16]. In this control problem, we have as input the error e that is the distance between the rear wheel and the desired trajectory. The error is positive if the wheel is to the left of the path and negative if it is to the right. The second input is the angle between the tangent at the nearest point in the trajectory and the bearing vector θ_e. The output is the angular velocity ω. We take the design of the controller from a previous work [2] using five membership functions for each input variable. We have two signs because the error and the heading can be left or right. We are tuning the ten parameters defining each MF. The fixed and variable parameters of the MFs are in Table 1. We proposed a controller with a total of twenty five fuzzy rules.

Table 1 Parameters for the controller's MFs

Variable	Linguistic value	MF	Parameters
θ_e	High negative	μ_{trap}	$[-50, -5, -b, -b+c]$
θ_e	Medium negative	μ_{tria}	$[-d-e, -d, -d+e]$
θ_e	Low	μ_{tria}	$[-a, 0, a]$
θ_e	Medium positive	μ_{tria}	$[d-e, d, d+e]$
θ_e	High positive	μ_{trap}	$[b-c, b, 5, 50]$
$error$	High negative	μ_{trap}	$[-50, -5, -g, -g+h]$
$error$	Medium negative	μ_{tria}	$[-i-j, -i, -i+j]$
$error$	Low	μ_{tria}	$[-f, 0, f]$
$error$	Medium positive	μ_{tria}	$[i-j, i, i+j]$
$error$	High positive	μ_{trap}	$[g-h, g, 5, 50]$
ω	High negative	μ_{trap}	$[-50, -5, -1, -0.5]$
ω	Medium negative	μ_{tria}	$[-1, -0.5, 0]$
ω	Low	μ_{tria}	$[-0.5, 0, 0.5]$
ω	Medium positive	μ_{tria}	$[0, 0.5, 1]$
ω	High positive	μ_{trap}	$[0.5, 1, 5, 50]$

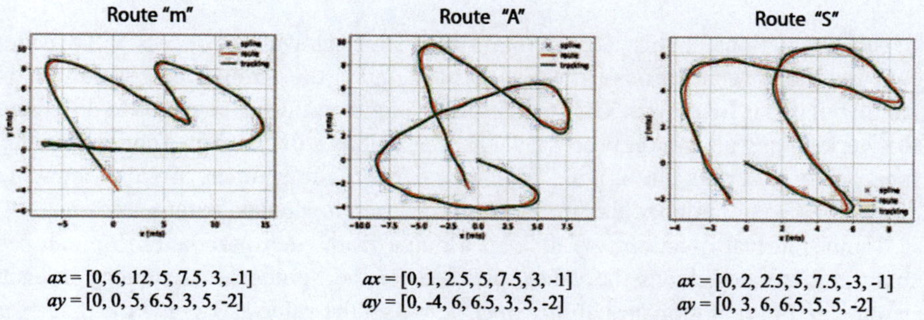

Fig. 3 Paths used for fitness evaluation

The fitness of each candidate solution, in this case, the fuzzy controller generated with the parameterized MFs, is established by running three simulations in these three paths (see Fig. 3).

2.2 Setup

The distributed algorithms' parameters are shown in Table 2. We configured the multi-population algorithm with seven populations of size nine, this population size may seem small, but we need to remember that this is a multi-population algorithm, so the total population size is estimated by multiplying the population size by the number

Table 2 Parameter values for the algorithms compared

Algorithm	Parameter	Fixed value	Random range [min, max]
GA	Selection	Tournament selection (k = 3)	
	Mutation	Gaussian ($\mu = 0.0$ and $\sigma = 0.2$)	
	Mutation probability	0.3	[0.1, 0.5]
	Crossover	One point (Probability = 0.7)	[0.3, 0.9]
PSO	Topology	Fully connected	
	Speed limit	Min = −0.25	Min = [−0.20, −0.30]
		Max = 0.25	Max = [0.20, 0.30]
	Cognitive and social	$C_1 = 2$, $C_2 = 2$	[1.0, 2.0]
Populations	Pop size	9	
	Populations	7	
	Iterations	4	
	Cycles	4	
	#Func. Eval.	1008	

of sub-populations, in this case the total size is 63. Each worker process will execute four iterations (generations) of the algorithm, again the number of generations is small, but this is by design. The small number of generations is required to distribute the work among all worker processes. All populations will complete four cycles; this means they will pass through the combinator module four times. We also show the values we use to initialize the homogenous and heterogeneous parameters.

Using interval notation, we defined a range from each parameter. We can see that the interval includes the value specified for the homogenous strategy in each parameter. For mutation probability in GAs, we set the value at 0.3, and the range for random values is between 0.1 and 0.5. We experimented with widening the range but did not receive better results. We repeated the same setup for the other parameters. We wanted to maintain the speed limit sign for PSO, so we established a range of [−0.20, −0.30] and [0.20, 0.30], respectively.

3 Results

In Table 3, we show the results of the experiments. Results with parameters with random values are compared with the distributed versions with fixed parameters found in our previous work.

We note that the distributed PSO with random parameters implementation yields better results than the distributed homogeneous parameters on average. Here we

Table 3 Results the best RMSE

RMSE						
	Homogeneous parameters			Heterogeneous parameters		
	GA	PSO	PSO-GA	GA	PSO	PSO-GA
AVERAGE	0.01091	0.00645	0.00656	0.01029	0.00632	0.00634
STDDEV	0.00600	0.00148	0.00185	0.00332	0.00165	0.00135
MEDIAN	0.00955	0.00643	0.00625	0.01021	0.00628	0.00610
MIN	0.00384	0.00360	0.00336	0.00399	0.00310	0.00388
MAX	0.03455	0.01000	0.01168	0.01584	0.00891	0.00889

Table 4 Results the time in seconds needed to complete each run

Time (s)						
	Homogeneous parameters			Heterogeneous parameters		
	GA	PSO	PSO-GA	GA	PSO	PSO-GA
AVERAGE	421.6722	415.6491	431.5258	395.7506	415.7994	409.9042
STDDEV	28.5406	23.4744	22.6012	26.5488	20.4787	24.8533
MEDIAN	417.8451	412.6850	432.3661	392.8096	414.6650	408.4821
MIN	364.6546	365.8164	394.6160	340.6548	374.8318	372.2037
MAX	526.7686	467.4715	478.8072	461.4638	461.9615	465.2711

show the average error obtained in 30 runs. Here we show the best RMSE obtained by the best fuzzy controller found in each run.

In Table 4 we show the time needed to complete each run in seconds. When comparing the time required to complete each execution of the algorithms, there is no substantial difference between versions. As expected, the distributed version with random parameters turned out similar to that with fixed parameters. An interesting case is the GA that took less time to complete in the heterogeneous strategy. This shortening in time could be related to the time it takes for a simulation to complete. Because we terminate the simulation early if the robot loses track of the trajectory, the time required to complete the simulations for a population of incapable individuals could take less time. This result, however, does not happen in the PSO case, where the times are similar.

We did the statistical Ztest, comparing the same algorithms but with homogeneous parameters that are those of the column that we see on the left side and the algorithms with heterogeneous parameters that we see in the upper row. We concluded that there is not enough evidence to reject the null hypothesis. The p-value results show that it is not within the acceptable confidence interval. We show in Table 5.

We show a box-plot of the data in Fig. 4, and we can see that there is no significant difference between the median of the same algorithms.

Table 5 Statistical Ztest

p-values, Z-test α=0.05, independent samples, unequal variances, 30 samples

		Heterogeneous parameters		
	$H_a : \mu_{eAi} > \mu_{eAj}$	GA X:0.01029 SD:0.00332	PSO X:0.00632 SD:0.00165	PSO-GA X:0.00634 SD:0.00135
Homogeneous params	GA $X : 0.01091$ $SD : 0.00600$	0.3119		
	PSO $X : 0.00645$ $SD : 0.00148$		0.3819	
	PSO-GA $X : 0.00656$ $SD : 0.00185$			0.3085

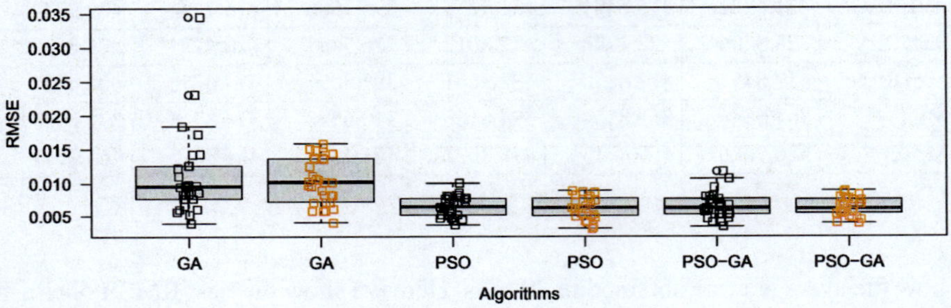

Fig. 4 Box-plot of the data

4 Conclusions and Future Work

In conclusion we presented a design and implementation of a distributed multi-population, multi-algorithm method to optimize the parameters of the MFs for a fuzzy controller.

We run experiments with both heterogeneous and homogeneous configurations of the algorithms.

Preliminary results show that a distributed execution with homogeneous parameters gives similar results to a distributed implementation with random parameters in a similar time. By using random heterogeneous parameters, we do not spend the time needed to find a suitable set of parameters, we only need to specify a range.

For future work, we need to run the experiments in a computer with more cores or on the cloud. As a next step, we could implement other optimization algorithms. Also, there are other parameters that are important like the poplulation size and the number of iterations. Moreover, we can also investigate if a dynamic tuning of parameters yields better results. For instance, the Combinator process could adjust the parameters in each iterations taking into account certain metrics of the population like the diversity or the average fitness.

References

1. Mancilla, A., García-Valdez, M., Castillo, O., & Merelo-Guervós, J. J. (2022). Optimal fuzzy controller design for autonomous robot path tracking using population-based metaheuristics. *Symmetry, 14*(2), 202.
2. Mancilla, A., Castillo, O., & Valdez, M. G. (2022). Evolutionary approach to the optimal design of fuzzy controllers for trajectory tracking. In C. Kahraman, S. Cebi, S. Cevik Onar, B. Oztaysi, A. C. Tolga, & I. U. Sari (Eds.), *Intelligent and fuzzy techniques for emerging conditions and digital transformation* (pp. 461–468). Cham: Springer International Publishing.
3. Back, T. (1996). *Evolutionary algorithms in theory and practice: Evolution strategies, evolutionary programming, genetic algorithms*. Oxford University Press.
4. Holland, J. H., et al. (1992). *Adaptation in natural and artificial systems: An introductory analysis with applications to biology, control, and artificial intelligence*. MIT Press.
5. Kennedy, J.: Swarm intelligence. In *Handbook of nature-inspired and innovative computing* (pp. 187–219). Springer.
6. Clerc, M. (2010). *Particle swarm optimization* (Vol. 93). Wiley.
7. Valdez, M. G., & Guervós, J. J. M. (2021). A container-based cloud-native architecture for the reproducible execution of multi-population optimization algorithms. *Future Generation Computer Systems, 116*, 234–252.
8. Merelo Guervós, J. J., & García-Valdez, J. M.: Introducing an event-based architecture for concurrent and distributed evolutionary algorithms. In *International Conference on Parallel Problem Solving from Nature* (pp. 399–410). Springer.
9. Ma, H., Shigen, S., Mei, Y., Zhile, Y., Minrui, F., & Huiyu, Z.: Multi-population techniques in nature inspired optimization algorithms: A comprehensive survey. *Swarm and Evolutionary Computation*, 365–387.
10. Mancilla, A., Castillo, O., & Valdez, M. G. (2021). *Optimization of fuzzy logic controllers with distributed bio-inspired algorithms* (pp. 1–11). Cham: Springer International Publishing.
11. Gong, Y., & Fukunaga, A. (2011). Distributed island-model genetic algorithms using heterogeneous parameter settings. In *2011 IEEE Congress of Evolutionary Computation (CEC)* (pp. 820–827). IEEE.
12. Hernandez-Aguila, A., Garcia-Valdez, M., Merelo-Guervos, J. J., & Castillo, O. (2017). Randomized parameter settings for a pool-based particle swarm optimization algorithm: A comparison between dynamic adaptation of parameters and randomized parameterization. In *Proceedings of the Genetic and Evolutionary Computation Conference Companion* (pp. 205–206).
13. Cuevas, F., Castillo, O., & Cortés-Antonio, P. (2022). Generalized type-2 fuzzy parameter adaptation in the marine predator algorithm for fuzzy controller parameterization in mobile robots. *Symmetry, 14*(5), 859.
14. Cuevas, F., Castillo, O., & Cortes, P. (2022). Optimal setting of membership functions for interval type-2 fuzzy tracking controllers using a shark smell metaheuristic algorithm. *International Journal of Fuzzy Systems, 24*(2), 799–822.
15. Yang, X. S., Cui, Z., Xiao, R., Gandomi, A. H., & Karamanoglu, M. (2013). *Swarm intelligence and bio-inspired computation: Theory and applications*. Newnes.
16. Paden, B., Čáp, M., Yong, S. Z., Yershov, D., & Frazzoli, E. (2016). A survey of motion planning and control techniques for self-driving urban vehicles. *IEEE Transactions on Intelligent Vehicles, 1*(1), 33–55. Publisher: IEEE.

Application of Compensatory Fuzzy Logic in Diabetes Problem Using Pima-Indians Dataset

José Fernando Padrón-Tristán, Laura Cruz-Reyes, Rafael A. Espin-Andrade, Claudia Guadalupe Gómez Santillán, and Carlos Eric Llorente-Peralta

1 Introduction

Diabetes Mellitus (DM) is a group of metabolic diseases characterized by hyperglycemia resulting from defects in insulin secretion, insulin action, or both. Insulin regulates the amount of glucose in the blood. The chronic hyperglycemia of DM is related to long-term damage, dysfunction, and failure of different organs, including, but not limited to, the eyes, kidneys, nerves, heart, and blood vessels [1].

A glucose test is performed to diagnostic if a person has diabetes; the person taking the test must have fasted for eight hours. The measurement for the glucose found on the blood is in mg/dl, and the classification levels are three [2]: (a) Normal: Below 108 mg/dl; (b) Prediabetes: Between 109 to 125 mg/dl; and (c) Diabetes: Above 126 mg/dl.

DM is considered as one of the leading global health problems because of its high prevalence, -a large number of persons suffer from it-, high economic cost, and the increasing number of premature deaths; a common characteristic is hyperglycemia as a result of defects in insulin secretion [3].

There are mainly three types of DM: type 1, type 2, and gestational diabetes.

J. F. Padrón-Tristán (✉) · C. E. Llorente-Peralta
National Institute of Technology of México and Technological Institute of Tijuana, Tijuana, México
e-mail: jfpt76@gmail.com

L. Cruz-Reyes · C. G. G. Santillán
National Institute of Technology of México and Technological Institute of Ciudad Madero, Ciudad Madero, México
e-mail: lauracruzreyes@itcm.edu.mx

R. A. Espin-Andrade
Autonomous University of Coahuila, Coahuila, México

O. Castillo and P. Melin (eds.), *Hybrid Intelligent Systems Based on Extensions of Fuzzy Logic, Neural Networks and Metaheuristics*, Studies in Computational Intelligence 1096,
https://doi.org/10.1007/978-3-031-28999-6_13

DM type 1 (DMT1) is one of the most common chronic diseases in childhood, although it can be diagnosed at any age [4]. DMT1 occurs when the immunity system destroys pancreatic β cells, which are in charge of produce insulin.

DM type 2 (DMT2) is generally characterized by insulin resistance, where the body does not adequately respond to insulin, causing the blood glucose levels to keep rising, releasing more insulin. This release of insulin can eventually exhaust the pancreas, resulting in the body producing less and less insulin, causing even higher blood sugar levels (hyperglycemia) [5]. Around 90% of diabetes cases are DMT2.Gestational DM (GDM) increases blood sugar; it develops during pregnancy and usually disappears after giving birth. It can occur at any stage of pregnancy but is more common in the second half. It happens when the body cannot produce enough insulin to meet the additional needs in pregnancy. GDM can cause problems for the mother and the baby during and after birth, but the risk of these problems happening can be reduced if it is detected and well treated.

According to the National Institute of Diabetes and Digestive and Kidney Diseases (NIDDKD), the symptoms of diabetes include [6]:

- increased thirst and urination,
- increased hunger,
- fatigue,
- blurred vision,
- numbness or tingling in the feet or hands,
- sores that do not heal and
- unexplained weight loss.

The approach of this work is to complement the knowledge discovery task done in previous works by applying compensatory fuzzy logic (CFL) to the PIMA Indian dataset (PID).

2 Background

Recent works have reported an increment in the quality of the classification over the PID. PID is originally from the National Institute of Diabetes and Digestive and Kidney Diseases. The objective of the dataset is to diagnostically predict whether a patient has diabetes, based on specific diagnostic measurements included in the dataset [7]. PID contains information about PIMA Indian females, located near Phoenix, Arizona, who are under continuous study since 1965 due to the high incidence rate of diabetes [15]. Information about the PID includes integer variables described in Table 1.

The CFL is a methodology derived from Fuzzy Logic (FL) proposed by Lotfi Zadeh in the sixties. CFL is a multivalent logic that allows the modeling of vagueness in the information by normalizing the data in the interval [0, 1], Fig. 1 shows an example modeled with Fuzzy Logic compared to one in classical logic.

Table 1 Description of variables in PID

Variable name	Description	Normal ranges	Ranges in dataset
Pregnancies	Number of times pregnant	Does not apply	[0, 17]
Glucose [101]	Plasma glucose concentration, 2 h in an oral glucose tolerance test	[80, 140]	[44, 199]
Blood-pressure	Diastolic blood pressure (mm/Hg)	Under 80	[24, 122]
Skin thickness	Triceps skin fold thickness (mm)	[12.5, 23.5]	[7, 99]
Serum-Insulin	2 h serum insulin (μU/ml)	Does not apply	[0, 846]
BMI	Body Mass Index (weight in kg/(h^2 in m^2)	[20, 25]	[18.2, 67.1]
Diabetes pedigree function	Synthesis of the history of Diabetes Mellitus in relatives, generic relationship of those relatives to the subject	Does not apply	[0.078, 2.42]
Age	Age of the individual (years)	Does not apply	[21, 81]
Outcome	Class variable, occurrence if a patience has diabetes mellitus or not	[0, 1]	[0, 1]

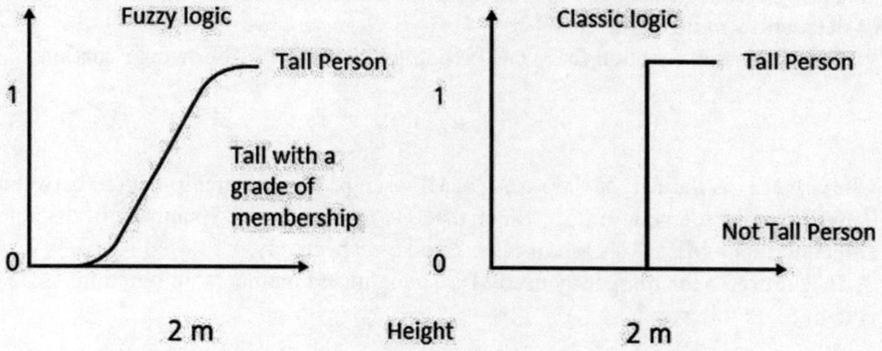

Fig. 1 Comparison fuzzy logic versus classical logic

The FL predicates can have different interpretations depending on its calculated truth value; values close to zero mean that the FL predicate is closer to a false statement whereas a value close to 1 is closer to a true statement, while the value 0.5 represents that the predicate is as false as true.

CFL satisfies the compensation axiom:

$$min(x_1, x_2, \ldots, x_n) \leq c(x_1, x_2, \ldots, x_n) \leq max(x_1, x_2, \ldots, x_n)$$

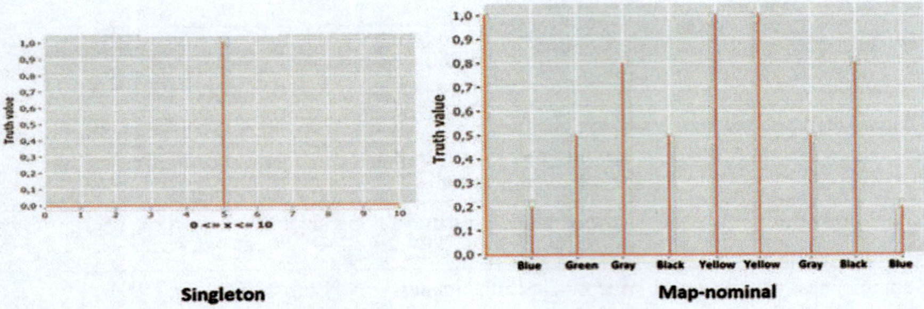

Fig. 2 Examples of discrete MF: singleton and map-nominal functions

This compensation axiom gives the name to the proposed structure of the CFL, in which a variation of a component affects the others [9]. The main characteristics of CFL are flexibility, tolerance with the imprecision of data, the ability to model non-linear problems, and the capacity to express results naturally [10]. The results of the CFL predicates are presented with the semantic structure for the construction of logic predicates, which are easy to understand in natural language, allowing to even model expert knowledge [10]. The construction of predicates uses linguistic states and an operations set (And, Or, Not, Implication, and Equivalence); the linguistic states are constructed through linguistic labels associated with attributes, each of these attributes have an associated membership function, which allow the normalization of data with a multivalent logic [9].

A membership function (MF) can be defined through the following equation:

$$A = \{(x, \mu_A(x))|x \in X\}$$

where $\mu_A(x)$ is the MF of the set A. A MF assigns a membership degree between 0 and 1 to each element of *X*, it can be discrete or continuous. Examples of discrete and continuous MF are shown in Figs. 2 and 3 respectively.

In addition to the frequently used MF, a generalized membership function (GMF) is defined as follows:

$$GMF(x, \beta, \gamma, m) = \frac{sigm(x, \beta, \gamma)^m (1 - sigm(x, \beta, \gamma))^{1-m}}{M}$$

where

$$sigm(x, \beta, \gamma) = \frac{1}{1 + e^{-\alpha(x-\beta)}}$$

$$\alpha = \frac{\ln 0.99 - \ln 0.01}{\beta - \gamma}$$

$$M = m^m (1 - m)^{1-m}$$

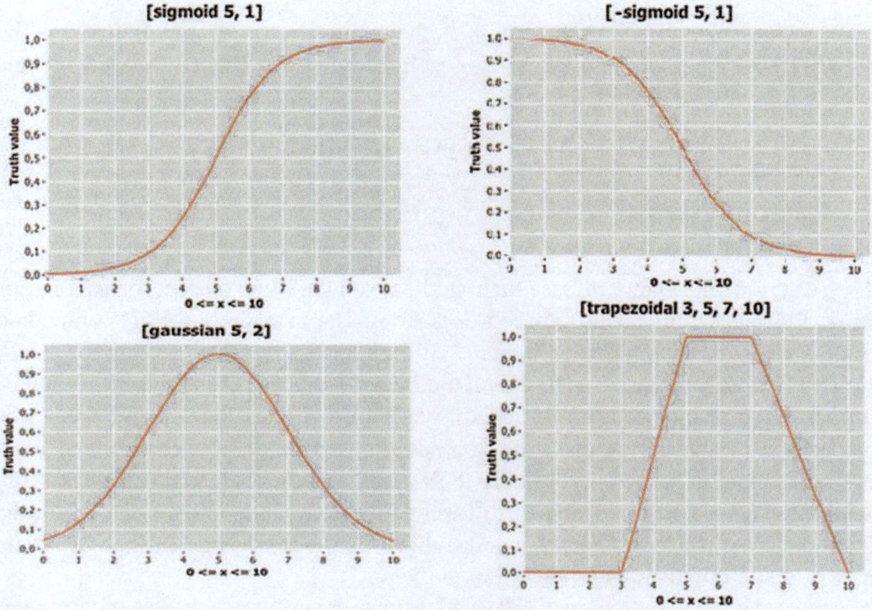

Fig. 3 Examples of continuous MF: sigmoid, negative sigmoid, Gaussian and trapezoid

GMF is a handy function because, depending on its parameters, the shape of the MF is automatically defined and can take the form of different MF, such as a Gaussian, negative sigmoid, or sigmoid, among others [11].

The logic operations are defined according to the selected type of logic, which must fulfill a set of axioms. Some types of logic are the Archimedean, Zadeh's, geometric mean, arithmetic mean. One of the most usual Compensatory Logic is the Geometric Mean Based Compensatory Logic (GMBCL); the conjunction, disjunction, order and negation logic operators associated with this type of logic are defined as follows [12]:

$$C(x_1, x_2, \ldots, x_n) = (x_1, x_2, \ldots, x_n)^{\frac{1}{n}}$$

$$D(x_1, x_2, \ldots, x_n) = 1 - \left[(1 - x_1)(1 - x_2) \ldots (1 - x_n)\right]^{\frac{1}{n}}$$

$$O[x, y] = 0.5[C(x) - C(y)] + 0.5$$

$$N(x_i) = 1 - x_i$$

The universal quantifier for the GMBCL is defined with the following equation:

$\forall_{x \in U} p(x) = c_{x \in U} p(x)$ where $c_{x \in U}$ is the conjunction operator, and *p(x)* is the membership degree of the predicate for each record in the dataset, the result of this operation is known as truth value [9].

3 Related Works

The related works analyzed base their studies with the same dataset focused in this work, PID. The approaches of these works are similar in terms of the type of method used, hence we can compare to them.

The Self-Organized Maps based on the Fuzzy Predicate Clustering (SFPC) [13] proposes the design of a new clustering system that uses fuzzy logic predicates to perform the clustering task, the fuzzy logic used for the construction of logic predicates can be chosen accordingly to the problem, the logic can be: "Max–Min, AMCL or GMCL); it is designed to form the clusters automatically based on the input data The clustering and knowledge discovered needs to be analyzed by a field expert to add linguistic interpretability to them..

Spider-Monkey Rule Miner (SM-RuleMiner) is an approach that incorporates a fitness function to generate a comprehensive rule set that balances accuracy, sensitivity, and specificity [14]. The experimentation is done with a tenfold cross-validation technique (10FCV). For every fold, the SM-RuleMiner use a test dataset and a training dataset generated from the original set, where the data is partitioned in 10 equal parts: 1 partition is used for testing and the other 9 partitions are used for training; the rule set is generated from the training dataset and applied to the test dataset to obtain the accuracy.

Recursive-Rule Extraction (Re-RX) is a model with a high accuracy in its classification, but generates considerably more rules than other algorithms, to deal with this drawback, J48Graft is used in conjunction with Re-RX combined with sampling selection techniques [15].

Table 2 shows a comparison of these works in terms of the algorithms and methods used for classification.

Table 2 Comparison of the method used for the algorithms in the related works

Algorithm	Method
SFPC [13]	CFL
SM-RuleMiner [14]	Optimization-based rule miner
Re-Rx with J48Graft [15]	Sampling Re-Rx with J48Graft
This work	CFL Predicates

4 Proposed Methodology and Algorithms for Classification

In the present work, CFL is implemented for knowledge discovery through the extraction of predicates that are transformed into classification rules. Figure 4 shows the proposed method to obtain the classification model.

The proposed algorithm is based on 10FCV, where the preprocessed dataset is partitioned in nine subsets for training and one subset for testing.

A discovery task is performed as a learning process on each subset of the instance; this task discovers a list of fuzzy predicates that fulfill certain conditions, the conditions are the operator used, implication or equivalence operator; the general structure for the discovery of predicates is as follow:

1. Predicates that implies if a patient is prone to diabetes $P_x - diabetic$ and
2. predicates that equate if a patient is prone to diabetes $P_x \leftrightarrow diabetic$.

The selection of the best predicate per fold is made according to the accuracy of the evaluation of the same learning instances.

In order to increase the accuracy of class separation, this work analyzes several thresholds from a list, instead of using a single threshold value of 0.5; the instances records are evaluated and the truth value calculated for each record is compared with

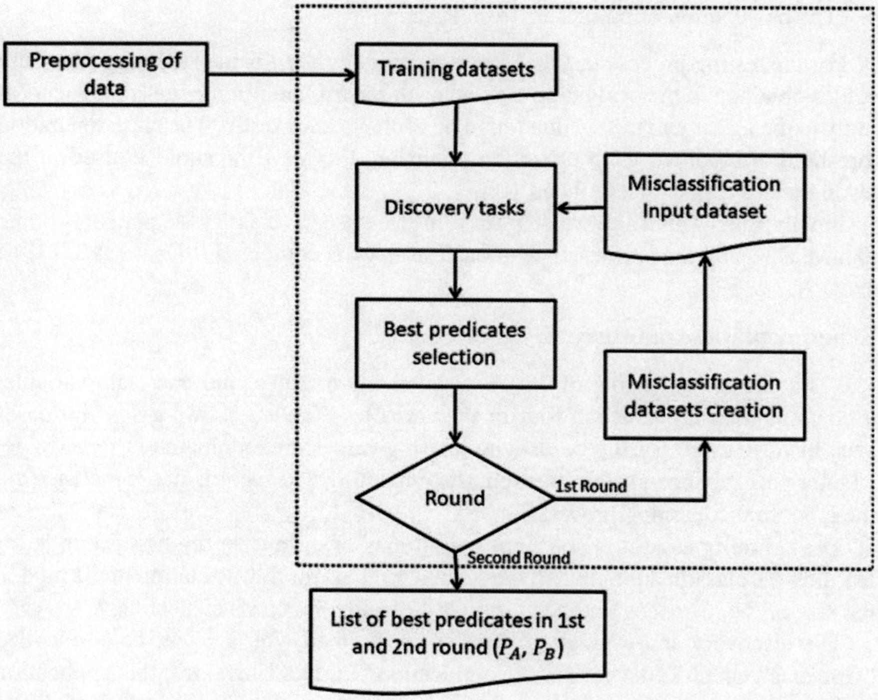

Fig. 4 Method for learning a classification model

each of the thresholds. A predicate is considered as correct for the classification of the record if the truth value of a record is less than the threshold value for the non-diabetic class, or if the truth value of a record is higher than or equals the threshold value for the diabetic class. The average of successful classifications for each predicate and fold creates a list of thresholds performance, from which is selected the one with the best performance.

The best predicate for each fold is the one that gets the higher accuracy using the best threshold obtained; this predicate is added to a list of predicates identified as P_A and is a candidate to be part of the classification model.

The misclassified records create new instances used to perform a second round of the discovery task to obtain a second list of predicates identified as P_B.

The CFL based classification model is applied to unseen datasets during a learning stage; these datasets, along with the lists of predicates P_A and P_B, are the input for the learned model.

The evaluation task of the model uses four types.

The model performs evaluation tasks on the instances using four types of predicates:

1. Predicates from the P_A list.
2. Predicates from the P_B list.
3. The Conjunction of predicates from P_A and P_B.
4. The Disjunction of predicates from P_A and P_B.

For the testing process in Fig. 5, another strategy used is the scaling of the truth values obtained in the evaluation task on each record, the truth value scales accordingly to the maximum truth value found on each instance result. The class separation threshold is modified to improve the precision, this new threshold is used in the testing phase (Figs. 6, 7, 8, 9 and 10).

Finally, the original and scaled truth values are used to classify if a patient is prone to have DM, and the performance of each strategy is compared (Figs. 11, 12, 13, 14 and 15).

Experimentation and Discuss

PID is a dataset consisting of 768 records, eight variables, and one class variable; it contains missing values in four of the variables; Table 3 shows a description of such inconsistencies. All records with missing values are eliminated in order to get a better performance in classification; the remaining 532 records in the instance are the input to the learning process.

The resulting dataset is split into two groups of instances; the first group is for learning the classification model; the second group is for the evaluation of the model, the size of the instances for each group is 479 and 53 records respectively.

The discovery and evaluation tasks are performed with a scientific core called "Universe" which is the core of the application "Eureka Universe", this application can search for unknown predicates using a template, or evaluate a fixed predicate with CFL, calculating the truth value on a given dataset.

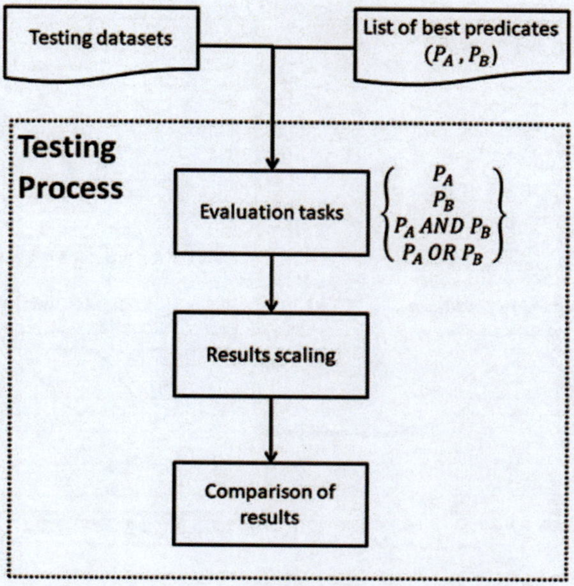

Fig. 5 Method for application a learned classification model

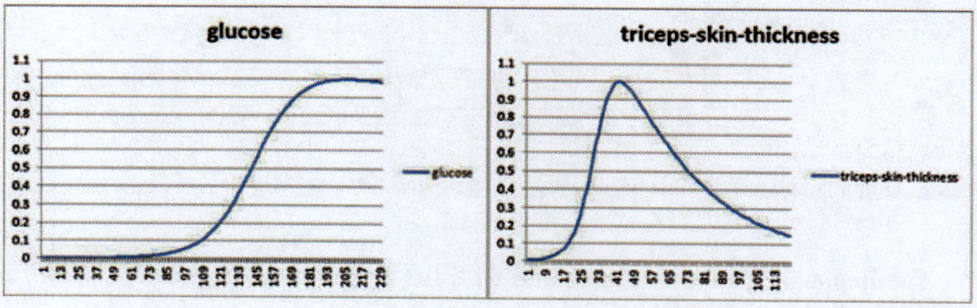

Fig. 6 Graph generated from the GMF using the parameters from Table 16

The configuration for the search algorithm of the discovery task is set as follows:

1. Depth level: 1; specifies the maximum number of operators that can be nested.
2. Minimum truth value: 0.85; The discovered predicates have a truth value greater than or equal to this value;
3. Number of results: 15; the total number of predicates displayed;
4. Type of logic: Geometric mean;
5. Operators; And, Or; The operators used in the search and construction of predicates.
6. Symbolic predicates:

   a. $All\ predicates \longrightarrow diabetic$ (IMP "*" diabetic)
   b. $All\ predicates \longleftrightarrow diabetic$ (EQV "*" diabetic)

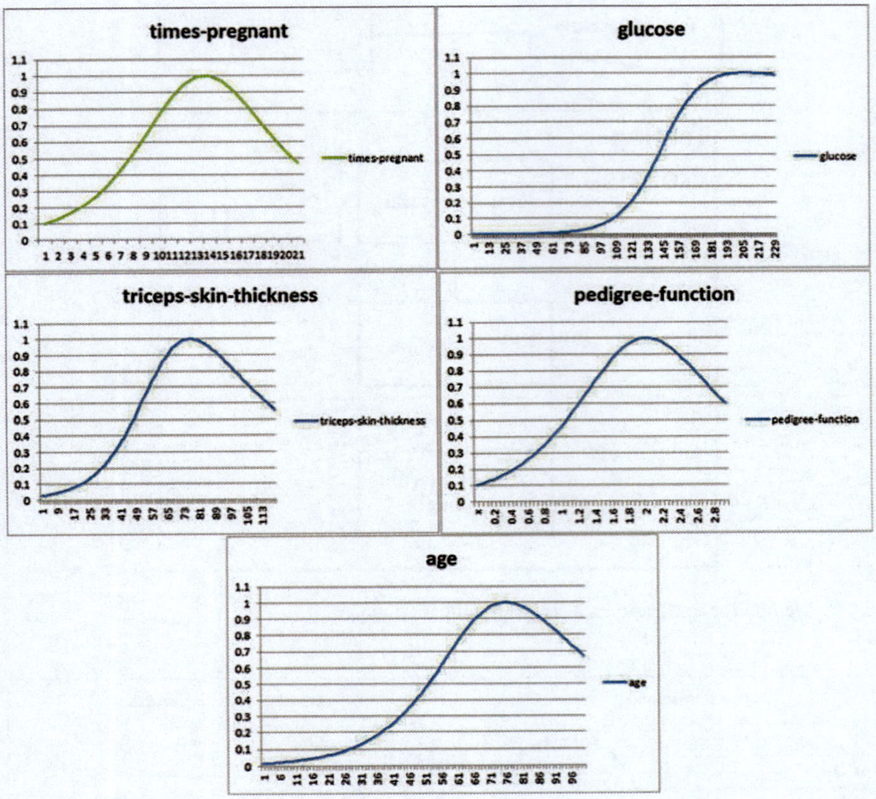

Fig. 7 Graph generated from the GMF using the parameters from Table 17

7. The linguistic states are configured with the GMF, to automatically search for the best parameters for each state on every predicate discovered.

The discovery tasks found in the first round a total of 150 predicates, in each fold discovers 15 predicates for the implication operator and one predicate for the equivalence operator.

The discovered predicates on each fold perform evaluation tasks in their corresponding training instance, the one that was used in the learning process, and the resulting truth values are used to determine to which class the registry belongs and if the patient is prone to diabetes or not.

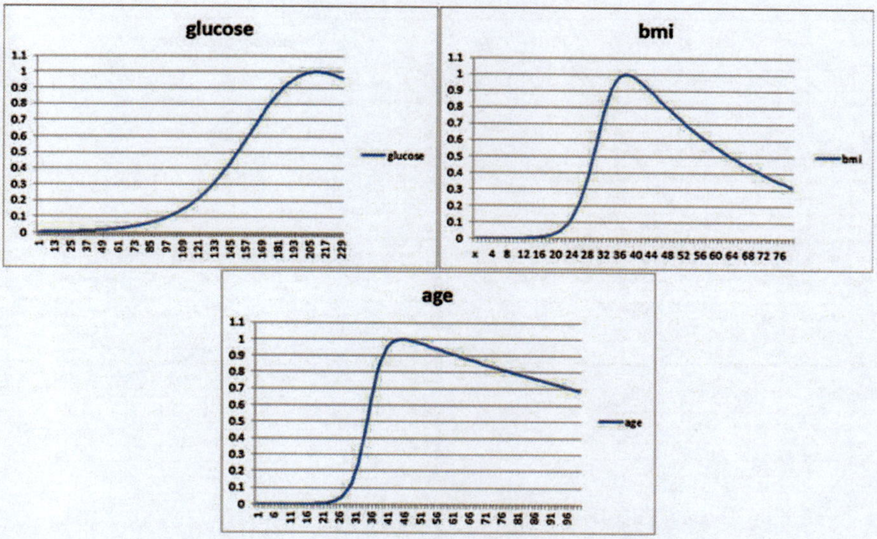

Fig. 8 Graph generated from the GMF using the parameters from Table 18

For the calculus of the truth value from the predicate, is taken only the antecedent of the predicate, for example, for the predicate:

$$(IMP(AND \text{ “pedigree-function\_ A” “serum-insulin\_ A”}) \text{ “}diabetic\text{”})$$

Antecedent taken to perform the evaluation:

$$(AND \text{ “pedigree-function\_ A” “serum-insulin\_ A”})$$

The truth values of each record are compared with the thresholds, and it is determined whether there is a success in the classification, the result matches the class the record belongs to. Table 4 shows a comparison of the accuracy obtained with different thresholds.

The accuracy of the predicates in each fold is examined to select the best one; Table 5 shows the performance table. For equivalence operator, only one predicate per fold was discovered, so it is not necessary to perform this process in the first round. Tables 6 and 7 shows the best predicated discovered for the implication and evaluation operator.

Using the best predicates in every fold, the evaluation task performs its evaluation in their respective training instances; new instances for the second round of training are created when filtering all successful classified records and leaving only the misclassified records; Table 8 is created using only training instances with misclassified records as input data.

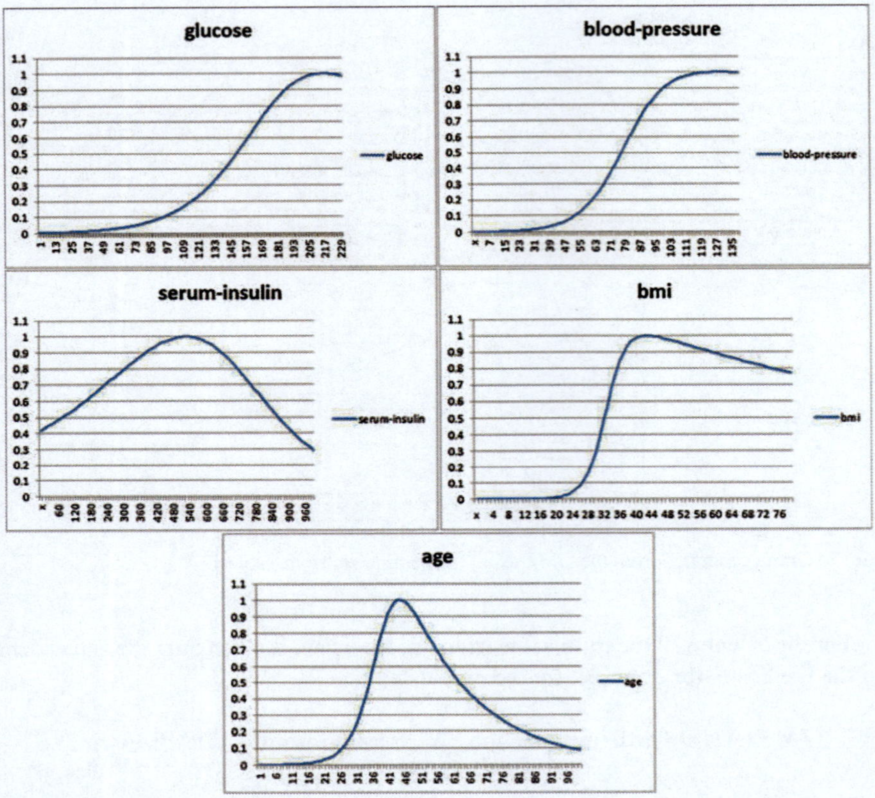

Fig. 9 Graph generated from the GMF using the parameters from Table 19

The newly created instances are used in the second round of discovery tasks; this process is similar the first round of discovery tasks, the difference is that the selection of the best predicates is made using the best threshold found in the first round; Tables 9 and 10 compares their accuracy per fold.

Tables 11 and 12 show the best discovered predicates in the second round of discovery tasks for the implication and equivalence operators, respectively.

Every predicate from P_A and P_B is evaluated in the testing stage, along with the conjunction and disjunction of both lists; the resulting maximum truth-value on each strategy is scaled; Tables 13 and 14 compares the accuracy on each strategy.

Table 15 compares all the explored strategies and rank them from worst to best; there is an improvement in accuracy by changing the threshold from 0.5 to 0.1 and using the equivalence operator instead of the implication operator also improves the accuracy. Finally, combining both strategies enhances the quality of results from 67% to 76.6% (Tables 16, 17, 18, 19, 20, 21, 22, 23, 24 and 25).

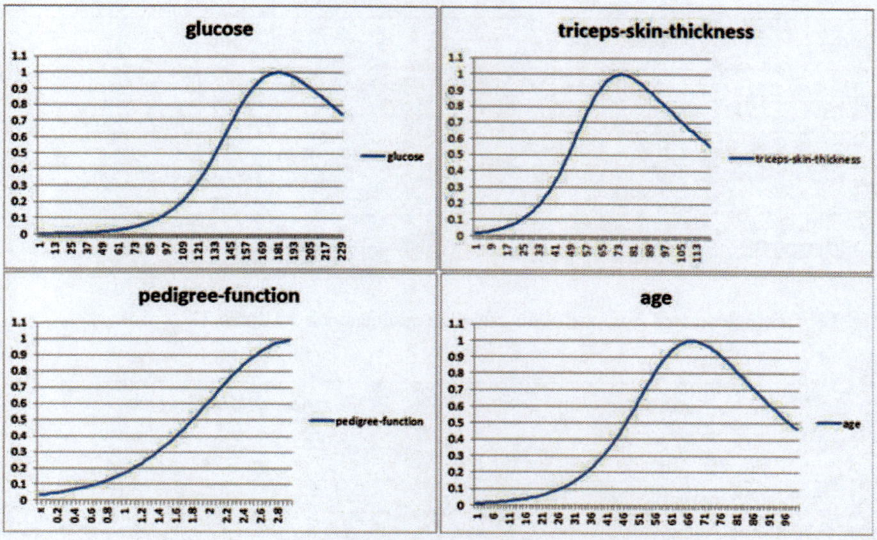

Fig. 10 Graph generated from the GMF using the parameters from Table 20

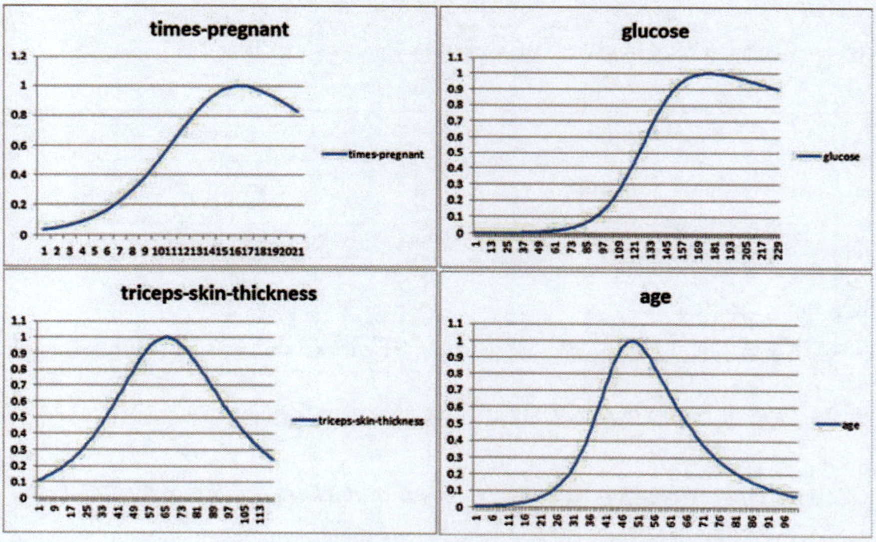

Fig. 11 Graph generated from the GMF using the parameters from Table 21

Following is the list of the evaluated predicated followed by its respective parameters discovered for the GMF, the associated graph for the linguistic states, and its corresponding rule.

1. (AND "triceps-skin-thickness_A" "glucose_A")

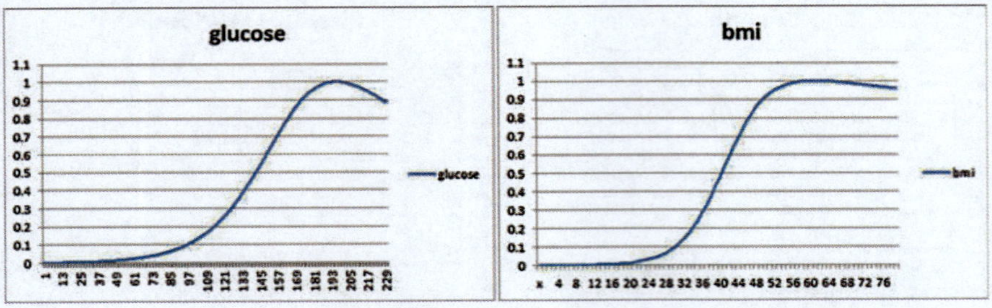

Fig. 12 Graph generated from the GMF using the parameters from Table 22

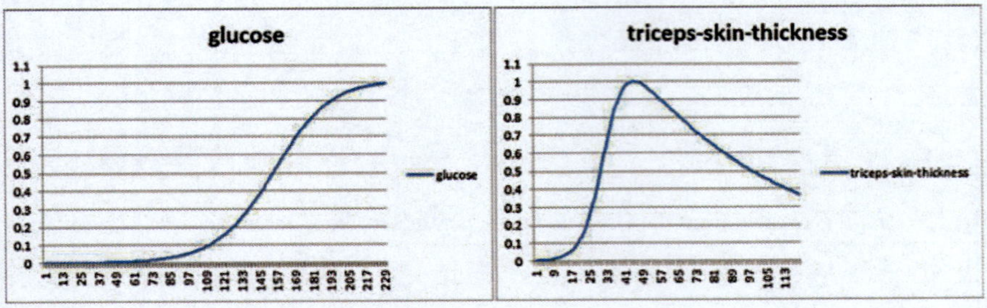

Fig. 13 Graph generated from the GMF using the parameters from Table 23

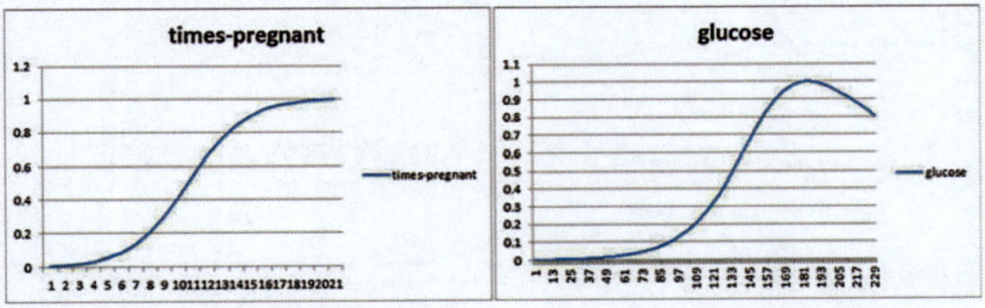

Fig. 14 Graph generated from the GMF using the parameters from Table 24

Rule 1: If *glucose* > 145 AND 33 < *triceps-skin-thickness* < 75 mm then *diabetic.*

2. (OR "triceps-skin-thickness_A" "pedigree-function_A" "glucose_A" "age_A" "times-pregnant_A")

Rule 2: If *times-pregnant* > 8 OR *glucose* > 145 OR *triceps-skin-thickness* > 49 OR *pedigree-function* > 1.1 OR *age* > 51 then *diabetic.*

3. (AND "bmi_A" "glucose_A" "age_A")

Rule 3: If *glucose* > 145 AND *bmi* > 28 AND *age* > 33 then *diabetic.*

4. (AND "serum-insulin_A" "bmi_A" "age_A" "blood-pressure_A" "glucose_A")

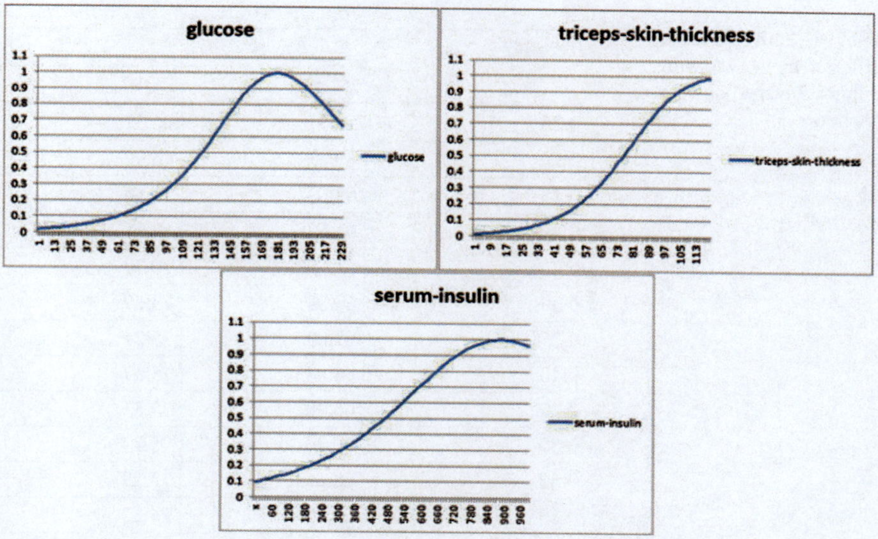

Fig. 15 Graph generated from the GMF using the parameters from Table 25

Table 3 Missing values per variables in PID

Variable	Records with missing values
Glucosa	5
Blood-pressure	2
Triceps-skin-thickness	192
Bmi	2
Blood-pressure and triceps-skin-thickness	28
Blood-pressure, triceps-skin-thickness and bmi	7

Rule 4: If *glucose* > 145 AND *blood-pressure* > 80 AND *serum-insulin* > 60 AND *bmi* > 32 AND *age* > 36 then *diabetic.*

5. (OR "age_A" "glucose_A" "triceps-skin-thickness_A" "pedigree-function_A")

Rule 5: If *glucose* > 133 OR *triceps-skin-thickness* > 49 OR *pedigree-function* > 1.9 OR *age* > 46 then *diabetic.*

6. (OR "times-pregnant_A" "glucose_A" "triceps-skin-thickness_A" "age_A")

Rule 6: If *times-pregnant* > 10 OR *glucose* > 130 OR *triceps-skin-thickness* > 36 OR 37 < *age* < 66 then *diabetic.*

7. (OR "bmi_A" "glucose_A")

Rule 7: If *glucose* > 145 OR *bmi* > 40 then *diabetic.*

Table 4 Performance of different thresholds with implication and equivalence operators

Threshold	% Accuracy	
	IMP	EQV
0.05	59.31	36.31
0.10	69.87	46.92
0.15	69.63	57.56
0.20	68.52	65.98
0.25	67.76	70.79
0.30	67.30	73.69
0.35	67.03	75.21
0.40	66.87	75.67
0.45	66.78	75.81
0.50	66.70	75.79
0.55	66.65	75.23
0.60	66.62	74.52
0.65	66.60	73.79
0.70	66.59	72.77
0.75	66.57	71.60
0.80	66.57	70.71
0.85	66.57	69.40
0.90	66.57	68.19
0.95	66.56	67.42

8. (AND “glucose_A” “triceps-skin-thickness_A”)

Rule 8: If *glucose* > 157 AND *triceps-skin-thickness* > 33 then *diabetic.*

9. (OR “times-pregnant_A” “glucose_A”)

Rule 9: If *times-pregnant* > 10 OR *glucose* > 133 then *diabetic.*

10. (OR “triceps-skin-thickness_A” “glucose_A” “serum-insulin_A”).

Rule 10: If *glucose* > 126 OR *triceps-skin-thickness* > 75 OR *serum-insulin* > 480 then *diabetic.*

Table 5 Performance of predicates per fold with threshold = 0.10 for implication operator

Fold	Predicates							
	1	2	3	4	5	6	7	8
1	68.27	68.27	68.27	68.27	68.27	68.27	68.48	68.27
2	67.01	67.01	67.01	67.01	67.01	67.01	67.01	67.01
3	66.39	66.39	66.39	66.39	66.39	66.39	66.39	66.39
4	65.76	65.76	65.76	65.76	65.76	65.76	65.76	65.76
5	65.34	65.34	65.34	65.34	65.34	65.34	65.34	65.34
6	67.43	67.43	67.43	67.43	67.43	67.43	67.43	67.43
7	66.18	66.18	66.18	66.18	66.18	66.18	66.18	66.18
8	68.06	68.06	68.06	68.06	68.06	68.06	68.06	68.06
9	65.76	65.76	65.76	65.76	65.76	65.55	65.76	65.76
10								

Fold	Predicates						
	9	10	11	12	13	14	15
1	68.48	68.27	68.27	68.27	68.27	68.27	68.27
2	67.01	67.01	67.01	67.01	67.01	67.01	67.01
3	66.39	66.18	66.39	66.39	66.39	66.39	66.39
4	65.76	65.76	65.76	65.76	65.76	65.76	65.76
5	65.34	65.34	65.34	65.34	65.34	65.34	65.34
6	67.43	67.43	67.43	67.43	67.43	67.43	67.43
7	66.18	66.18	66.18	66.18	66.18	66.18	66.18
8	68.06	67.85	68.06	68.06	68.48	68.06	68.06
9	65.76	65.76	65.76	65.76	65.76	65.76	65.76
10	66.81	66.81	66.81	66.81	66.81	66.81	66.81

Table 6 Best discovered predicates per fold with threshold = 0.1 for implication operator

Fold	Number	Predicate
1	7	(AND "serum-insulin_A" "age_A")
2	1	(AND "age_A" "pedigree-function_A" "serum-insulin_A")
3	1	(AND "pedigree-function_A" "times-pregnant_A" "age_A")
4	1	(AND "pedigree-function_A" "age_A")
5	1	(AND "times-pregnant_A" "pedigree-function_A")
6	1	(AND "pedigree-function_A" "serum-insulin_A")
7	1	(AND "age_A" "serum-insulin_A")
8	13	(AND "age_A" "serum-insulin_A")
9	1	(AND "bmi_A" "serum-insulin_A")
10	1	(AND "serum-insulin_A" "age_A")

Table 7 Best discovered predicates per fold with threshold = 0.45 for equivalence operator

Fold	Predicate
1	(AND "triceps-skin-thickness_A" "glucose_A")
2	(OR "triceps-skin-thickness_A" "pedigree-function_A" "glucose_A" "age_A" "times-pregnant_A")
3	(AND "bmi_A" "glucose_A" "age_A")
4	(AND "serum-insulin_A" "bmi_A" "age_A" "blood-pressure_A" "glucose_A")
5	(OR "age_A" "glucose_A" "triceps-skin-thickness_A" "pedigree-function_A")
6	(OR "times-pregnant_A" "glucose_A" "triceps-skin-thickness_A" "age_A")
7	(OR "bmi_A" "glucose_A")
8	(AND "glucose_A" "triceps-skin-thickness_A")
9	(OR "times-pregnant_A" "glucose_A")
10	(OR "triceps-skin-thickness_A" "glucose_A" "serum-insulin_A")

According to the typical ranges of variables shown in Table 1, the found predicates can be interpreted and expressed in natural language through the application of linguistic labels, resulting in the rules listed below:

1. If *glucose* is *high* AND *triceps-skin-thickness* is thick then *diabetic*
2. If *times-pregnant* is *elevated* OR *glucose* is *high* OR *triceps-skin-thickness* is *thick* OR *pedigree-function* is *high* OR *age* is *elder* then *diabetic*
3. If *glucose* is high AND *bmi* is high AND *age* is *adult* then *diabetic*
4. If *glucose* is *high* AND *blood-pressure* is *elevated* AND *serum-insulin not low* AND *bmi is high* AND *age* is *adult* then *diabetic*
5. If *glucose* is *elevated* OR *triceps-skin-thickness* is *thick* OR *pedigree-function* is *high* OR *age* is *older than adult* then *diabetic*
6. If *times-pregnant* is *elevated* OR *glucose* is *elevated* OR *triceps-skin-thickness* is *thick* OR 37 *age* is *adult* then *diabetic*
7. If *glucose* is *high* OR *bmi* is *high* then *diabetic*
8. If *glucose* is *high* AND *triceps-skin-thickness* is *thick* then *diabetic*
9. If *times-pregnant* is *elevated* OR *glucose* is *elevated* then *diabetic*
10. If *glucose* is *elevated* OR *triceps-skin-thickness* is *very thick* OR *serum-insulin* is *high* then *diabetic.*

Table 26 shows a comparison of this work with the related ones in the literature in terms of precision.

This work is interpretable, unlike the other proponents that are not interpretable.

The best work in terms of accuracy obtains IF–THEN-ELSE rules, but the predicates we get are richer linguistic structures, with greater interpretability. The number of predicates needed for good precision is minimal, only one predicate per fold of experimentation.

Table 8 Example of a fragment of a new instance with misclassified records for second round of training

Times-pregnant	Glucose	Blood-pressure	Triceps-skin-thickness	Serum-insulin	BMI	Pedigree-function	Age	Class
6	148	72	35	0	33.6	0.627	50	1
3	78	50	32	88	31	0.248	26	1
0	118	84	47	230	45.8	0.551	31	1
1	115	70	30	96	34.6	0.529	32	1
9	119	80	35	0	29	0.263	29	1
11	143	94	33	146	36.6	0.254	51	1
10	125	70	26	115	31.1	0.205	41	1
3	158	76	36	245	31.6	0.851	28	1
9	102	76	37	0	32.9	0.665	46	1
2	90	68	42	0	38.2	0.503	27	1

Table 9 Accuracy percentage of predicates per fold for implication operator using threshold = 0.1 in second round of discovery tasks

Fold	1	2	3	4	5	6	7	8	9	10	11	12	13	14	15
1	59.1	50.9	56.4	46.4	50.0	60.0	52.7	39.1	50.9	50.0	46.4	50.0	50.0	50.0	50.0
2	50.8	51.7	58.5	52.5	40.7	55.9	54.2	54.2	51.7	54.2	41.5	52.5	44.1	53.4	52.5
3	70.3	77.5	71.7	79.7	78.3	78.3	78.3	74.6	78.3	75.4	78.3	77.5	76.1	78.3	78.3
4	53.3	41.5	58.5	59.3	57.0	57.8	58.5	58.5	58.5	57.8	48.9	57.8	58.5	56.3	57.8
5	72.4	91.3	85.8	82.7	85.0	83.5	86.6	86.6	80.3	86.6	73.2	86.6	85.0	85.0	86.6
6	61.8	61.8	53.7	61.8	59.3	62.6	61.8	61.0	67.5	64.2	62.6	61.0	61.0	61.8	61.0
7	64.0	32.4	62.5	69.1	67.6	66.2	62.5	69.1	69.1	69.1	66.2	69.1	69.1	70.6	69.1
8	50.4	45.2	48.1	48.1	44.4	51.1	49.6	49.6	48.1	48.9	48.9	48.9	48.1	46.7	48.9
9	67.3	62.6	10.9	62.6	63.3	69.4	63.3	62.6	61.2	65.3	63.3	74.1	57.8	63.3	63.3
10	81.2	82.7	82.7	72.9	79.7	79.7	79.7	37.6	79.7	79.7	84.2	79.7	79.7	66.9	72.2

Table 10 Accuracy percentage of predicates per fold for equivalence operator using threshold = 0.45 in second round of discovery tasks

Fold	1	2	3	4	5	6	7	8	9	10	11	12	13	14	15
1	98.3	97.8	97.8	97.4	98.3	97.8	97.8	97.8	97	97.8	98.3	97.8	94.8	97.8	97.8
2	100	99.7	99.4	99.7	98.1	98.1	99.7	99.4	99.7	97.2	98.1	96.9	98.1	98.1	98.8
3	95.4														
4	99.3	100	100	100	99.7	100	98.6	98.3	99.3	98.3	99.3	99.3	99.3	98.6	99.7
5	99.1	99.1	99.1	99.1	99.1	97	97.8	99.1	98.7						
6	99.7	97.5	99.7	99.4	99.7	99.7	100	100	100	99.7	98.7	100	100	100	98.7
7	96.9	96.4													
8	97.7	97.7	97.7	97.7	97.7	97.7	97.2	97.2	97.7	96.7	97.7	97.2	96.7	97.2	96.7
9	94.4														
10	100	99.4	100	100	100	98.4	100	97.4	99.7	98.1	99.7	97.7	98.1	97.1	96.4

5 Conclusions

In this work, the diabetes classification problem was addressed using Pima Indian Dataset by classifying if a patient was prone to diabetes mellitus.

The classification was done using a classification model based on compensatory fuzzy logic complemented with the universe core, and proposing a methodology that included:

(a) The predicate discovery was performed using the equivalence operator.
(b) Accuracy was compared using a list of thresholds.
(c) A second round of discovery tasks was carried out on instances created with incorrectly classified records.

Table 11 Best predicates per fold for implication operator in second round of discovery tasks

Fold	Best predicate	Predicate
1	6	(AND "age_B" "times-pregnant_B")
2	3	(AND "bmi_B" "serum-insulin_B")
3	4	(AND "blood-pressure_B" "serum-insulin_B" "bmi_B")
4	4	(AND "age_B" "blood-pressure_B" "pedigree-function_B" "serum-insulin_B" "bmi_B" "times-pregnant_B")
5	2	(AND "serum-insulin_B" "glucose_B" "pedigree-function_B" "age_B" "bmi_B" "times-pregnant_B")
6	9	(AND "age_B" "glucose_B" "times-pregnant_B" "bmi_B")
7	14	(AND "age_B" "triceps-skin-thickness_B" "blood-pressure_B" "pedigree-function_B" "times-pregnant_B" "bmi_B" "glucose_B" "serum-insulin_B")
8	6	(AND "serum-insulin_B" "triceps-skin-thickness_B" "glucose_B")
9	12	(AND "pedigree-function_B" "age_B" "serum-insulin_B")
10	11	(AND "blood-pressure_B" "triceps-skin-thickness_B" "age_B" "pedigree-function_B" "serum-insulin_B" "times-pregnant_B" "bmi_B" "glucose_B")

Table 12 Best predicates per fold for equivalence operator in second round of discovery tasks

Fold	Best predicate	Predicate
1	1	(AND "age_B" "serum-insulin_B")
2	1	(AND "pedigree-function_B" "times-pregnant_B")
3	1	(AND "pedigree-function_B" "bmi_B")
4	2	(AND "pedigree-function_B" "age_B" "serum-insulin_B")
5	1	(AND "times-pregnant_B" "age_B" "pedigree-function_B" "serum-insulin_B" "triceps-skin-thickness_B")
6	7	(AND "times-pregnant_B" "age_B" "bmi_B" "serum-insulin_B" "pedigree-function_B")
7	1	(AND "bmi_B" "age_B" "pedigree-function_B")
8	1	(AND "age_B" "serum-insulin_B")
9	1	(AND "bmi_B" "times-pregnant_B" "pedigree-function_B" "age_B" "serum-insulin_B")
10	1	(AND "pedigree-function_B" "age_B")

In the application of the model, the predicates discovered in the first and second rounds were evaluated, in addition to the conjunction and disjunction of the predicates found in both rounds.

Finally, the truth value from the predicate evaluation was scaled according to the maximum truth value calculated in each instance for each strategy in the test data sets.

Table 13 Accuracy percentage for each threshold on each strategy for implication operator

Threshold	Pa (%)	Pb (%)	Pa AND Pb (%)	Pa OR Pb (%)	S(Pa) (%)	S(Pb) (%)	S(Pa AND Pb) (%)	S(Pa OR Pb) (%)
0.05	58.3	33.0	34.5	33.6	33.8	33.6	33.4	33.4
0.1	73.4	35.7	49.2	40.6	44.9	34.7	33.4	34.3
0.15	71.5	39.4	65.8	49.4	54.2	35.7	34.3	36.4
0.2	69.2	46.8	73.4	55.8	60.9	37.9	35.7	39.6
0.25	67.9	49.4	70.2	62.3	65.3	40.0	39.1	42.8
0.3	67.5	52.5	69.8	64.5	69.1	44.2	45.3	45.5
0.35	67.5	54.9	67.9	66.2	70.4	46.6	51.3	48.7
0.4	67.4	59.6	67.4	67.5	71.7	48.5	57.7	52.6
0.45	67.0	63.8	67.0	67.9	72.5	52.5	60.9	57.5
0.5	67.0	64.5	67.0	69.6	72.1	55.1	65.7	61.9
0.55	67.0	65.8	67.0	69.2	71.5	58.7	67.2	64.0
0.6	67.0	67.2	67.0	68.1	71.1	61.1	70.2	66.6
0.65	67.0	67.4	67.0	68.1	71.5	63.2	72.3	68.7
0.7	67.0	68.7	67.0	67.4	71.1	65.5	71.9	69.2
0.75	67.0	68.9	67.0	67.4	69.6	68.3	72.5	69.8
0.8	67.0	68.5	67.0	67.0	69.1	70.0	71.9	69.2
0.85	67.0	67.7	67.0	67.0	68.9	70.2	71.5	68.7
0.9	67.0	67.2	67.0	67.0	68.7	69.4	70.2	68.9
0.95	67.0	67.2	67.0	67.0	68.5	68.5	69.1	68.7

Of the proposed strategies, combining the predicate discovery using the equivalence operator with the change in the threshold value increased the classification accuracy from 67 to 76.6%.

This methodology is capable of obtaining results in different fields. There is a scientific community working on this issue with the aim of demonstrating that it is a technique with a high level of generality, interpretable according to language.

These methods of solution are general, not only of classification, general method, the generality reduces precision, this leads us to seek improvements, for example, evaluation of a greater number of predicates that collect in a differentiated way the knowledge that is contained in the data.

The accuracy obtained indicates good performance, and that this methodology with some changes can improve the accuracy of the classification, proposing in future work using several compensatory fuzzy logics to contrast results and using a higher number of predicates per fold.

Table 14 Accuracy percentage for each threshold on each strategy for equivalence operator

Threshold	Pa (%)	Pb (%)	Pa AND Pb (%)	Pa OR Pb (%)	S(Pa) (%)	S(Pb) (%)	S(Pa AND Pb) (%)	S(Pa OR Pb) (%)
0.05	37.0	59.1	43.2	39.6	35.7	40.8	33.8	36.6
0.1	48.7	68.1	66.4	58.9	44.2	51.3	41.1	47.4
0.15	59.6	70.8	74.0	73.0	53.0	60.9	52.5	60.6
0.2	67.2	69.4	72.1	73.0	62.3	64.9	60.4	67.0
0.25	71.7	69.6	72.1	76.4	67.9	67.2	69.6	71.5
0.3	72.3	68.7	71.5	76.8	70.0	67.4	71.5	73.8
0.35	73.6	67.9	69.6	75.7	72.6	67.7	74.2	75.7
0.4	75.7	68.1	69.1	75.5	73.2	68.7	73.8	75.8
0.45	76.6	67.9	68.1	74.3	74.0	69.2	72.8	75.8
0.5	75.7	67.9	67.5	73.0	74.5	69.6	71.1	74.3
0.55	74.9	67.5	67.5	72.5	74.7	70.0	71.3	73.2
0.6	74.5	67.5	67.2	71.3	73.8	69.1	70.2	73.6
0.65	74.2	67.4	67.2	69.6	73.6	68.9	70.2	73.2
0.7	72.3	67.4	67.0	68.1	74.0	68.9	71.1	72.5
0.75	72.1	67.4	67.0	67.7	73.6	68.7	71.3	72.1
0.8	71.7	67.4	67.0	67.4	72.8	69.1	70.6	71.5
0.85	69.4	67.2	67.0	67.5	71.5	69.1	69.6	70.4
0.9	67.7	67.2	67.0	67.2	71.1	69.1	69.6	70.2
0.95	67.4	67.2	67.0	67.2	69.6	68.9	69.4	69.1

Table 15 Comparison of explored strategies listed from worst to best one

Place	Strategy	Threshold	Operator	Scaling	Accuracy (%)
10	First round discovery task (P_A)	0.50	IMP	No	67.0
9	Second round discovery tasks (P_B)	0.50	EQV	No	67.9
8	Second round discovery tasks (P_B)	0.45	EQV	No	67.9
7	Conjunction of P_A and P_B	0.45	EQV	No	68.1
6	Disjunction of P_A and P_B	0.50	IMP	No	69.6
5	Conjunction of P_A and P_B	0.45	EQV	Yes	72.8
4	First round discovery task (P_A)	0.10	IMP	No	73.4
3	First round discovery task (P_A)	0.50	EQV	Yes	74.5
2	Disjunction of P_A and P_B	0.45	EQV	Yes	75.8
1	First round discovery task (P_A)	0.45	EQV	No	76.6

Table 16 Parameters discovered for the GMF with predicate 1

Glucose_A	{β = 145.200186101007, γ = 71.7181171659611, m = 0.979345454206472}
Triceps-skin-thickness_A	{β = 33.261859875741, γ = 11.7243054259065, m = 0.87768971321616}

Table 17 Parameters discovered for the GMF with predicate 2

Times-pregnant_A	{β = 11.6339803493541, γ = 0.763007771827193, m = 0.604234750586012}
Glucose_A	{β = 158.393398962437, γ = 67.1073981559019, m = 0.665630549307261}
Triceps-skin-thickness_A	{β = 62.6304094077607, γ = 9.44085412698131, m = 0.775204461929807}
Pedigree-function_A	{β = 1.90213112218371, γ = 0.153293766485169, m = 0.606792830721345}
age_A	{β = 67.2938878784107, γ = 22.2657802480769, m = 0.712357532777561}

Table 18 Parameters discovered for the GMF with predicate 3

Glucose_A	{β = 174.481479158092, γ = 68.9422648427739, m = 0.828662591387173}
Bmi_A	{β = 31.9248200180401, γ = 19.2452950031731, m = 0.917779321818711}
Age_A	{β = 35.3077979760602, γ = 23.5243577363439, m = 0.981594058514275}

Table 19 Parameters discovered for the GMF with predicate 4

Glucose_A	{β = 172.562537015028, γ = 58.3420180740638, m = 0.854507586338747}
Blood-pressure_A	{β = 79.7915861051315, γ = 27.5181219688206, m = 0.987886050002784}
Serum-insulin_A	{β = 638.959814611226, γ = 7.90937904770252, m = 0.324251199077224}
Bmi_A	{β = 33.6914332687054, γ = 21.5877520943198, m = 0.979344221009564}
Age_A	{β = 39.0377866610317, γ = 21.970146358113, m = 0.821541960287706}

Table 20 Parameters discovered for the GMF with predicate 5

Glucose_A	{β = 154.759798936423, γ = 69.4641369024401, m = 0.810542025059508}
Triceps-skin-thickness_A	{β = 58.8774642887287, γ = 11.4807248795169, m = 0.823101877965104}
Pedigree-function_A	{β = 2.39436887331231, γ = 0.09421240902852, m = 0.814572761051348}
Age_A	{β = 59.3451059377436, γ = 21.3960058475121, m = 0.727999111323469}

Table 21 Parameters discovered for the GMF with predicate 6

Times-pregnant_A	{β = 11.9859355412658, γ = 0.590325391320615, m = 0.796725919108146}
Glucose_A	{β = 130.381908880445, γ = 59.9105560286202, m = 0.955864522131949}
Triceps-skin-thickness_A	{β = 63.0719466305994, γ = 7.7656334279512, m = 0.557938771217536}
Age_A	{β = 44.8991859344449, γ = 22.3684498560264, m = 0.706962742918363}

Table 22 Parameters discovered for the GMF with predicate 7

Glucose_A	{β = 161.552985004022, γ = 67.4799844659697, m = 0.845176090922227}
Bmi_A	{β = 42.1630984984678, γ = 20.1159338727802, m = 0.984094452846603}

Table 23 Parameters discovered for the GMF with predicate 8

Glucose_A	{β = 156.934287225816, γ = 65.5870972124601, m = 0.989090760282475}
Triceps-skin-thickness_A	{β = 33.0797243502871, γ = 10.6927641464043, m = 0.929925945831367}

Table 24 Parameters discovered for the GMF with predicate 9

Times-pregnant_A	{β = 9.70613675858576, γ = 0.574090095458934, m = 0.999840467958595}
Glucose_A	{β = 152.889276364102, γ = 65.3663254444622, m = 0.844086938310618}

Table 25 Parameters discovered for the GMF with predicate 10

Glucose_A	{β = 168.462672858977, γ = 62.6522478589042, m = 0.629415275955952}
Triceps-skin-thickness_A	{β = 84.7893493826136, γ = 8.09753315551548, m = 0.927494051825464}
Serum-insulin_A	{β = 790.95667131317, γ = 7.63289486932767, m = 0.651771567986116}

Table 26 Comparison with the works of the state of the art

Algorithm	Method	Number of rules or predicates (10-FCV)	Precision (%)
SFPC [13]	CFL	N.A	72
SM-RuleMiner [14]	Optimization-based rule miner	41 rules	89.97
Re-Rx with J48Graft [15]	Sampling Re-Rx with J48Graft	8 rules	83.83
This work	CFL Predicates	10 predicates	76.6

6 Annex: Algorithms

This section shows the algorithms used to guide the process for the knowledge discovery.

Algorithm 1. Discovery task to generate the best predicates list per fold

1. $for\ f = 1\ to\ 10$
   a. $predicates\_list_f = discovery(simbolic\_predicate, instance_f)$

where
$simbolic\_predicate = (operator * diabetic)$ with $operator = \{IMP, EQV\}$

Algorithm 2. Accuracy calculation for each threshold value in the thresholds list

$\boldsymbol{threshold\_list = [0.05, 0.95]}$ **with leaps of** $\boldsymbol{0.05}$

1. $\boldsymbol{total\ records = \sum_{f=1}^{10} n_f m_f}$
2. $\boldsymbol{for\ each\ t = 1\ to\ 19}$
   a. $\boldsymbol{hits_{t,f,p} = \sum_{f=1}^{10} \sum_{p=1}^{n_f} \sum_{i=1}^{m_f} x_{tfpi}}$
   b. $\boldsymbol{A_t = \frac{hits_t}{total\ records}}$

where

$$x_{tfpi} = \begin{cases} 1 & if\ record_{i,class} = 1\ AND\ evaluate(predicates\_list_{f\,p}, record_i) \geq th \\ 1 & if\ record_{i,class} = 0\ AND\ evaluate(predicates\_list_{f\,p}, record_i) < th \\ 0 & otherwise \end{cases}$$

$\boldsymbol{n_f}$= total discovered predicates in $\boldsymbol{predicates\_list_f}$
$\boldsymbol{m_f}$= total records in $\boldsymbol{instance_f}$
$\boldsymbol{hits_t = \sum_{f=1}^{10} \sum_{p=1}^{n_f} hits_{tfp}}$ (total hits for threshold$\boldsymbol{t}$)
$\boldsymbol{A_t}$ = Accuracy for threshold $\boldsymbol{t}$

Algorithm 3. List best predicates for each fold

1. $for\ fold = 1\ to\ 10$
   a. $best\_predicate_f = predicates\_list_{f,1}$
   b. $for\ p = 2\ to\ n_f$
      i. $if\ (hits_{max_t fold\ p} > hits_{max_t fold\ max_p})$
      ii. $best\_predicate_f = predicates\_list_{f,p}$

where
$\boldsymbol{max_t}$ = threshold with best accuracy in $\boldsymbol{threshold\_list}$
$\boldsymbol{n_f}$ = total predicates in $\boldsymbol{predicates\_list_f}$

References

1. American Diabetes Association. (2010). Diagnosis and classification of diabetes mellitus. *Diabetes Care, 33*(Supplement 1), S62–S69.
2. Tabla de los rangos de los niveles de azucar en la sangre (ND) Diabetes mellitus MX, recovered from: https://www.diabetesmellitus.mx/rangos-los-niveles-azucar-en-la-sangre/
3. Ruiz-Ramos, M., Escolar-Pujolar, A., Mayoral-Sánchez, E., Corral-San Laureano, F., & Fernández-Fernández, I. (2006). La diabetes mellitus en España: mortalidad, prevalencia, incidencia, costes económicos y desigualdades. *Gaceta Sanitaria, 20*, 15–24.
4. Atkinson, M. A., Eisenbarth, G. S., & Michels, A. W. (2014). Type 1 diabetes. *The Lancet, 383*(9911), 69–82.
5. Type 2 Diabetes (ND), International Diabetes Federation, recovered from: https://www.idf.org/aboutdiabetes/what-is-diabetes/type-2-diabetes.html
6. Symptoms & Causes of Diabetes. (2016). National Institute of Diabetes and Digestive and Kidney Diseases, recovered of https://www.niddk.nih.gov/health-information/diabetes/overview/symptoms-causes
7. Predict the onset of diabetes based on diagnostic measures (ND) Pima Indians Diabetes Database, recovered from: https://www.kaggle.com/uciml/pima-indians-diabetes-database
8. Jayaraman, P., Subrahmanya, R., & Parthasarathy, S. (2016). Prediction of Diabetes in PIMA Women, recovered from: https://rpubs.com/jayarapm/PIMAIndianWomenDiabetes
9. Andrade, R. A. E., González, E. F., & Caballero, E. G. (2014). Un sistema lógico para el razonamiento y la toma de decisiones: La lógica difusa compensatoria basada en la media geométrica. *Investigación operacional, 32*(3), 230–245.
10. Cejas-Montero, J. (2011). La lógica difusa compensatoria/The compensatory fuzzy logic. *Ingeniería Industrial, 32*(2).
11. González-Caballeroa, E., Espín-Andrade, R. A., Pedryczc, W., Ramos, L. G. (En proceso de publicación). Continuous linguistic variables for data mining.
12. Bouchet, A., Pastore, J., Brun, M., & Ballarin, V. (2010). *Lógica Difusa Compensatoria basada en la media aritmética y su aplicación en la Morfología Matemática Difusa*. Universidad Nacional de Mar del Plata, Argentina.
13. Meschino, G. J., Comas, D. S., Ballarin, V. L., Scandurra, A. G., & Passoni, L. I. (2015). Automatic design of interpretable fuzzy predicate systems for clustering using self-organizing maps. *Neurocomputing, 147*, 47–59.

14. Cheruku, R., Edla, D. R., & Kuppili, V. (2017). SM-RuleMiner: Spider monkey based rule miner using novel fitness function for diabetes classification. *Computers in Biology and Medicine, 81*, 79–92.
15. Hayashi, Y., & Yukita, S. (2016). Rule extraction using recursive-rule extraction algorithm with J48graft combined with sampling selection techniques for the diagnosis of type 2 diabetes mellitus in the Pima Indian dataset. *Informatics in Medicine Unlocked, 2*, 92–104.

Comparison of the Effect of Parameter Adaptation in Bio-inspired CS Algorithm Using Type-2 Fuzzy Logic

Maribel Guerrero, Fevrier Valdez, and Oscar Castillo

1 Introduction

When we use the word optimize or optimization, "the best way to do something" should come to mind [1]. Without a doubt, optimization is found in our daily lives, every day we live with technological systems, mobile applications, electronic devices, a traffic light on the street of the city where we live, when using an elevator to go from one point to another, driving to our car or using public transport, taking the next flight, but what exactly is it when we say that we are solving an optimization problem in mathematical, computational, statistical or economic terms.

It is defined as the problem that tries to find a better solution, from a set of feasible solutions, in many cases that problem can become difficult to solve with some traditional technique, and for this, it will be necessary to resort to techniques that apply optimization algorithms that are inspired by nature, by species of birds, fish, the behavior of ants, bees, genetics, evolution, plants, etc. the best solution, lower costs, reduce time, etc. [1, 2].

Bio-inspired algorithms resemble the natural evolution of species, their behavior, and the response of social systems to any challenge that arises [3, 4].

In administration, optimization takes place in managerial areas in order to plan and manage, with the intention of improving work processes to increase productivity and performance.

M. Guerrero · F. Valdez (✉) · O. Castillo
Tijuana Institute of Technology, Tijuana, Mexico
e-mail: fevrier@tectijuana.mx

M. Guerrero
e-mail: maribel.guerrero@tectijuana.edu.mx

O. Castillo
e-mail: ocastillo@tectijuana.mx

O. Castillo and P. Melin (eds.), *Hybrid Intelligent Systems Based on Extensions of Fuzzy Logic, Neural Networks and Metaheuristics*, Studies in Computational Intelligence 1096,
https://doi.org/10.1007/978-3-031-28999-6_14

In terms of economy, optimization makes sense in the processes to obtain the increase in performance with the minimum quantality of resources, or by reducing unnecessary costs.

For all of the above, optimization is a topic of research interest for any area that requires improving a problem [5–8], and for these multiple solutions are presented, but the indicated one will always be to improve times, and costs or reduce effort [9].

The purpose of this work is to analyze the different parameters that are immersed in the bio-inspired cuckoo search algorithm, with the aim of finding the best parameter to be controlled in the algorithm using type 2 fuzzy logic.

The contribution of this article is the development of parameter variation using interval type-2 fuzzy logic to identify dynamic Cuckoo search algorithms and verify the behavior of the parameters when adjusted.

The second section will find the main characteristics of bio-inspired algorithms, in Sect. 3 Cuckoo search algorithm and the next section Levy Flights, the Sect. 5 flowchart, the next section proposes and the next results and conclusions.

2 Main Characteristics of Bio-inspired Algorithms

When talking about biological evolution, we must understand the different distinguishing properties between one organism to another, its adaptation to survival, and the relationship between one and other species.

Charles Darwin [10], in the development of the theme regarding the origin of species, explains that the different species of living beings have been organized through natural selection, the species that survive the following generations are not the strongest, the fastest, nor the most intelligent, has to do with the best adaptation to changes.

Bio-inspired algorithms include the following aspects [11]:

(1) **Population:** The population is the set of candidates to be solved.
(2) **Mechanism of Evolution:** It is the set of methods by which the population is modified.
(3) **Rating Mechanism:** the mechanism by which value is assigned to each element of the population. At this point, we specify the function that we want to be optimized with either a minimum or maximum value.
(4) **Stop criterion:** Procedure that allows control and evaluation if the solution was found.

3 Cuckoo Search Algorithm

The Cuckoo Search algorithm is an algorithm inspired by nature to solve optimization problems of different engineering [12]. It is a promising algorithm it was created by Xin-she Yang and Suash Deb in 2009.

This algorithm was inspired by the reproduction of the species of birds called cuckoos, which consists of placing their eggs in the nests of other species [13], if the birds discover the unsafe egg, it will eliminate it, otherwise, it will reproduce and feed as if it were of said species, this species evolved to such a degree that it can mimic the colors or patterns of host eggs [9].

Algorithm features

(1) use Levy flights.
(2) uses randomness to select new solutions and also to explore the solution space.

Parameters used by the algorithm

(1) The parameter Pa is the discovery factor; it allows to replacement of the nests that do not contain good solutions [14].
(2) The Beta parameter is a scale factor, it is implicit in the equivalence of Levy Flights, allowing the exploration and exploitation of the search [15].

4 Lévy Flights

The search for food for any animal is random [12], therefore, the route that an animal in search of food can select will be through a random walk [16], based on the current location or state and the transition from probability to probability [17], the next location [18].

So, the next location will come from probability, which can be modeled mathematically [19, 20].

Many animals and insects follow Lévy's flights. The flights are characterized by random walks with step lengths that have a distribution according to the probability of heavy tails [21].

5 Algorithm

Below we show the cuckoo search algorithm.

In Fig. 1 most of the flowchart consists of declaring the necessary variables, initializing the parameters [22], dimensions as well as the optimization function, the number of iterations, and the population, later the cuckoo is generated randomly, and the quality of the solution must be evaluated and keep the best solution found so far if the new solution is better than the one we have found so far, it should be replaced and the previous one forgotten, a fraction of nests will be eliminated and the new positions will be known with levy flights, the best solutions will be kept and if the stopping criterion is found then the results will be displayed, otherwise, this process will continue until the stopping criterion is satisfied [13, 23].

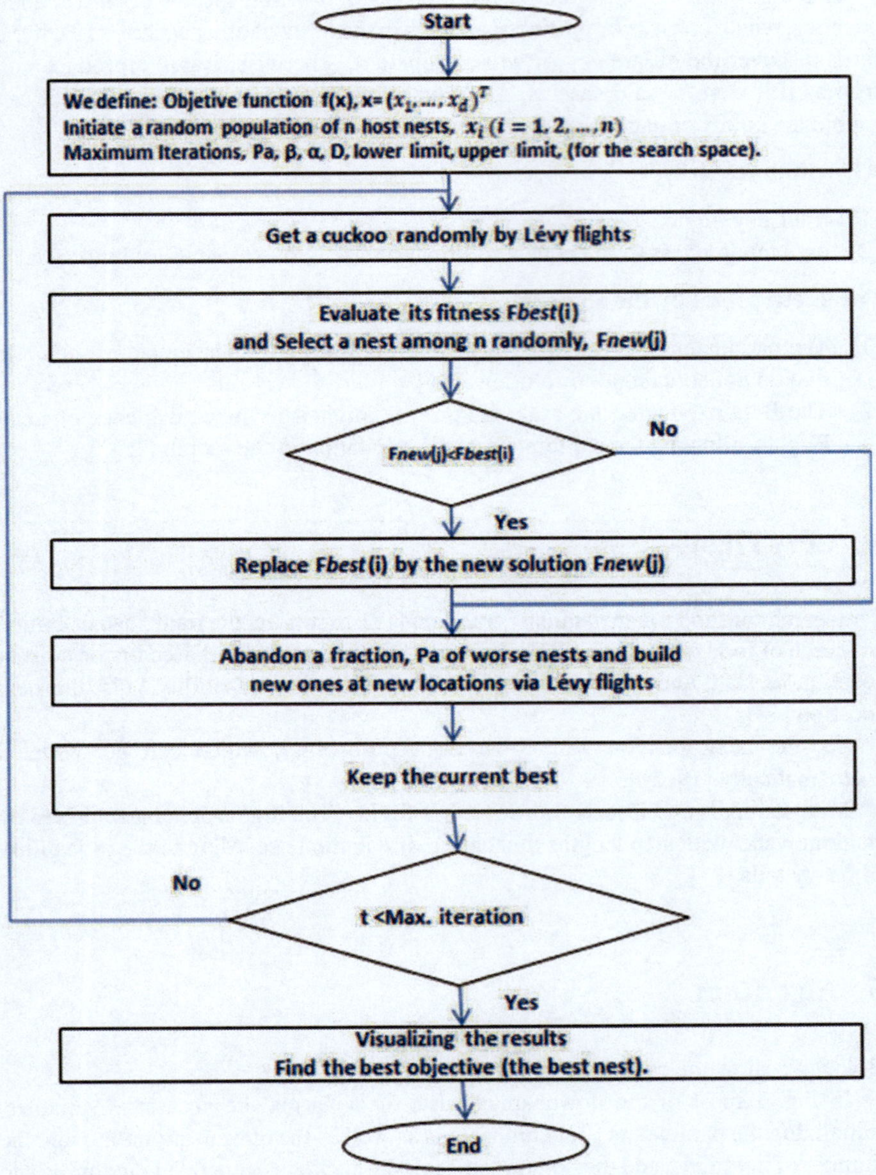

Fig. 1 Flowchart of the CS algorithm

For this article, we have implemented type 2 fuzzy logic by intervals, and we have analyzed the parameters pa, beta, and alpha in order to measure the performance of the algorithm by varying these parameters in order to decide which is the most optimal behavior of the algorithm.

6 Proposed Approach

In the following images we have represented the proposal of our algorithms.

In Fig. 2 shows be proposal approach, in this complementing CS algorithm with interval type 2 fuzzy sets using a variation of parameters and benchmark mathematical functions to review the efficiency,

In Fig. 3 shows the variation of parameters in CS algorithm with fuzzy logic type 2, Input is iteration, Mamdani system, and output Pa or beta and alpha.

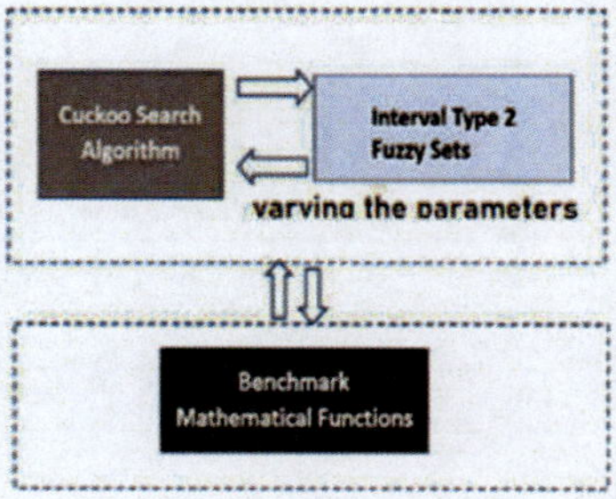

Fig. 2 Propose

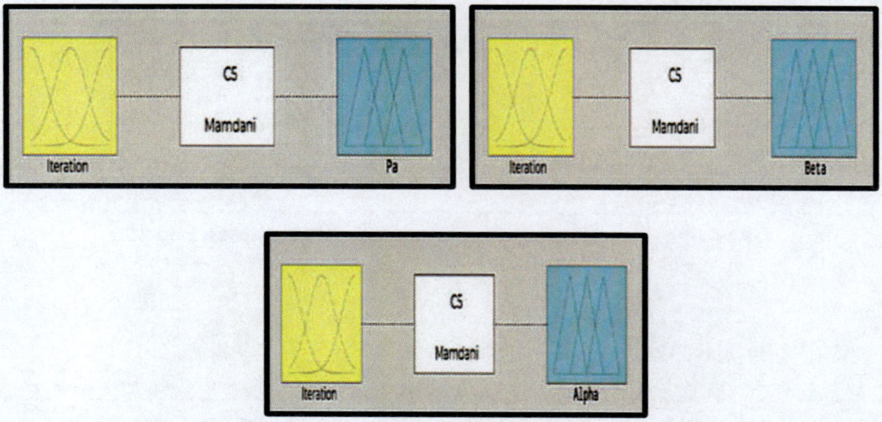

Fig. 3 Variation of parameters in CS algorithm with fuzzy logic type 2

7 Simulations Results

In order to carry out the set of simulations, it is necessary to establish values for each parameter, which allows the dynamic adjustment of parameters in the cuckoo search algorithm. For this, the following results are presented:

In Table 1 shows the parameters and respectively values, in this case, the population of 100 nests, variation of dimensions 8, 16, 32, 64 and 128.

In Fig. 4 shows the benchmark functions used in simulation results, in this case, shows the figure and mathematical function.

In Table 2 we show the results with sphere function, in this case the best result obtained in FCS (Alpha).

In Table 3 we show the results with Ackley function, the best average is FCS(Alpha).

In Table 4 we show the results with Rastrigin function, in this case the best solution is FCS (Alpha).

In Table 5 we show the best solution with for the Griewank function, the best average is FCS (Alpha).

Table 1 Parameters and values

Parameters	FSC2 (Alpha)	FSC2(Beta)	FCS2(Pa)
Alpha	Dynamic	5%	5%
Beta	1.5	Dynamic	1.5
Pa	75%	75%	Dynamic

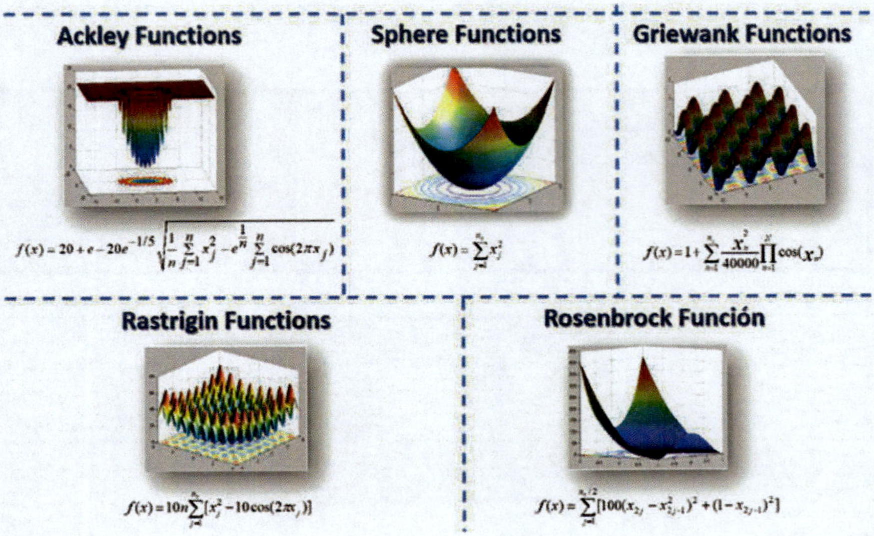

Fig. 4 Benchmark functions

Table 2 Simulation results for Sphere function

	Sphere function					
Dimensions	Average			Standard deviation		
	FCS(PA)	FCS(ALPHA)	FCS(BETA)	FCS(PA)	FCS(ALPHA)	FCS(BETA)
8	7.91E-48	9.21E-87	5.17E-92	1.16E-47	1.77E-86	1.61E-91
16	1.36E-28	5.30E-50	4.44E-48	3.13E-29	2.78E-50	5.52E-50
32	1.40E-16	4.72E-25	1.31E-20	2.88E-17	2.36E-25	1.31E-22
64	1.29E-09	1.04E-11	1.67E-07	2.01E-09	3.88E-12	8.76E-10
128	1.67E-05	3.08E-05	1.54E-03	1.54E-04	6.38E-06	1.16E-04

Table 3 Simulation results for the Ackley function

	Ackley function					
Dimensions	Average			Standard deviation		
	FCS(PA)	FCS(ALPHA)	FCS(BETA)	FCS(PA)	FCS(ALPHA)	FCS(BETA)
8	2.30E-13	0.00E + 00	3.02E-04	3.84E-12	0.00E + 00	1.14E-03
16	3.08E-01	2.56E-01	4.56E-01	4.11E-01	9.49E-02	6.58E-02
32	1.73E + 01	2.14E + 00	2.19E + 00	2.46E + 00	4.76E-01	2.83E + 00
64	6.37E + 00	8.54E + 00	7.48E + 00	7.19E + 00	1.34E + 00	6.52E + 00
128	1.85 + 01	1.04E + 01	2.49E + 01	2.20E + 01	1.16E + 00	1.93E + 00

Table 4 Simulation results for the Rastrigin function

	Rastrigin function					
Dimensions	Average			Standard deviation		
	FCS(PA)	FCS(ALPHA)	FCS(BETA)	FCS(PA)	FCS(ALPHA)	FCS(BETA)
8	4.14E-01	1.03E + 00	5.74E-05	3.07E-13	0.00E + 00	1.71E-05
16	5.24E-02	2.74E-02	8.21E-02	7.81E-02	8.79E-02	8.60E-02
32	1.90E + 00	2.14E + 00	4.16E-01	4.18E-01	7.69E + 05	5.66E + 00
64	1.15E + 00	1.54E + 00	1.35E + 00	1.44E + 00	6.77E + 00	8.62E + 00
128	3.33E + 00	1.35E + 00	4.73E + 00	3.96E + 00	9.18E + 00	7.93E + 00

In Table 6 we show the results with Rosenbrock function. The best average is with FSC (Alpha).

In Tables 2, 3, 4, 5 and 6 we show the results obtained, implementing the benchmark functions, dynamically adjusting the parameters, comparing their results FSC

Table 5 Simulation results for the Griewank function

	Griewank Function					
Dimensions	Average			Standard deviation		
	FCS(PA)	FCS(ALPHA)	FCS(BETA)	FCS(PA)	FCS(ALPHA)	FCS(BETA)
8	9.15E-08	9.15E-10	9.15E-09	3.00E-03	3.00E-05	3.00E-04
16	6.50E-16	6.50E-18	6.50E-17	4.64E-12	4.64E-12	4.64E-13
32	4.67E-13	4.67E-15	4.67E-14	8.03E-13	8.03E-15	8.03E-14
64	1.65E-09	1.65E-11	1.65E-10	1.71E-07	1.71E-09	2.72E-08
128	2.00E-04	2.00E-06	2.00E-05	1.00E-03	1.00E-05	2.00E-04

Table 6 Simulation results for the Rosenbrock function

	Rosenbrock function					
Dimensions	Average			Standard deviation		
	FCS(PA)	FCS(ALPHA)	FCS(BETA)	FCS(PA)	FCS(ALPHA)	FCS(BETA)
8	2.05E-11	1.85E-09	9.67E-12	1.17E-07	2.40E-11	4.94E-11
16	7.79E-02	1.61E-02	5.68E-02	6.05E-01	6.65E-02	7.55E-02
32	4.81E + 00	5.35E-01	4.81E + 00	1.75E + 01	3.45E-01	6.68E-01
64	3.87E + 01	1.71E + 00	3.87E + 01	5.78E + 01	2.21E-01	3.00E-01
128	4.23E + 01	1.02E + 01	4.23E + 01	1.25E + 02	9.34E-01	3.47E-01

(alpha), FSC(Beta), FSC(Pa), and showing the averages and their standard deviations; perceiving the algorithm which represents the optimal solutions with 8, 16, 32, 64 and 128 dimensions.

8 Conclusions

The results of the tests with benchmark mathematical functions show that for all the dimensions of our proposals with respect to the results of the average, they show that the algorithm FCS2(Alpha) in the functions Rosenbrock, Sphere, Griewank, and Ackley. The Rastrigin function is a complex function for the proposal, for which it is required to continue studying to find alternatives for the search space range, and iterations to analyze its behavior.

Alpha is a parameter that accelerates convergence, Pa is used for nest discovery and beta for exploration and exploitation. That is why in the results we can see that Alpha obtains good results; however, this parameter is recommended to have a balance, its value should not be very large, it could cause opposite effects.

In future work we will consider applying the fuzzy cuckoo search approach to designing fuzzy controllers for different cases.

References

1. Martínez-Álvarez, F., Asencio-Cortés, G., Torres, J. F., Gutiérrez-Avilés, D., Melgar-García, L., Pérez-Chacón, R., Rubio-Escudero, C., Riquelme, J. C., & Troncoso, A. (2020). Coronavirus optimization algorithm: A bioinspired metaheuristic based on the COVID-19 propagation model. *Big Data, 8*(4), 308–322.
2. Oyelade, O. N., Ezugwu, A.E.-S., Mohamed, T. I., & Abualigah, L. (2022). Ebola optimization search algorithm: A new nature-inspired metaheuristic optimization algorithm. *IEEE Access, 10*, 16150–16177.
3. Su, B., Lin, Y., Wang, J., Quan, X., Chang, Z., & Rui, C. (2022). Sewage treatment system for improving energy efficiency based on particle swarm optimization algorithm. *Energy Reports, 8*, 8701–8708.
4. Chang, C. C. W., Ding, T. J., Bhuiyan, M. A. S., Chao, K. C., Ariannejad, M., & Yian, H. C. (2022). Nature-inspired optimization algorithms in solving partial shading problems: a systematic review. In *Archives of computational methods in engineering, 2022* (pp. 1–27).
5. Trojovský, P., & Dehghani, M. (2022). Pelican optimization algorithm: A novel nature-inspired algorithm for engineering applications. *Sensors, 22*(3), 855.
6. Cuevas, F., Castillo, O., & Cortes, P. (2022). Optimal setting of membership functions for interval type-2 fuzzy tracking controllers using a shark smell metaheuristic algorithm. *International Journal of Fuzzy Systems, 24*(2), 799–822.
7. Cuevas, F., Castillo, O., & Cortés-Antonio, P. (2022). Generalized type-2 fuzzy parameter adaptation in the marine predator algorithm for fuzzy controller parameterization in mobile robots. *Symmetry, 14* (5), 859.
8. Castillo, O., Castro, J. R., & Melin, P. (2022). *Type-2 fuzzy logic systems: 'Interval type-3 fuzzy systems: theory and design'* (pp. 5–11). Springer.
9. Fister, I., Yang, X.-S., & Fister, D. (2014) Cuckoo search: A brief literature review. In: *Cuckoo search and firefly algorithm, 2014* (pp. 49–62).
10. Darwin, C. (2004). *On the origin of species, 1859*. Routledge.
11. Kureichik, V. V., & Zaruba, D. V. (2015). The bioinspired algorithm of electronic computing equipment schemes elements placement. In *Artificial intelligence perspectives and applications*. (pp. 51–58). Springer.
12. Fan, J., Xu, W., Huang, Y., & Dinesh Jackson Samuel, R. (2021). Application of Chaos Cuckoo search algorithm in computer vision technology. *Soft Computing, 25*(18), 12373–12387.
13. Mareli, M., & Twala, B. (2018). An adaptive Cuckoo search algorithm for optimisation. *Applied Computing and Informatics, 14*(2), 107–115.
14. Ochoa-Zezzatti, A., Castillo, O., Melin, P., Castillo, N., Bustillos, S., & Arreola, J. (2014). Shipwrecked on fear: selection of electives in school minorities in a university using cuckoo search algorithm. In *Recent advances on hybrid approaches for designing intelligent systems* (pp. 139–150). Springer.
15. Chakraborty, S., & Mali, K. (2022). Biomedical image segmentation using fuzzy multilevel soft thresholding system coupled modified cuckoo search. *Biomedical Signal Processing and Control, 72*, 103324.
16. Viswanathan, G., Afanasyev, V., Buldyrev, S. V., Havlin, S., Da Luz, M., Raposo, E., & Stanley, H. E. (2000). Lévy flights in random searches. *Physica A: Statistical Mechanics and its Applications, 282*(1–2), 1–12.
17. Kishnani, M., Pareek, S., & Gupta, R. (2014). Optimal tuning of PID controller by cuckoo search via Lévy flights. In *Book optimal tuning of PID controller by cuckoo search via Lévy flights* (pp. 1–5). IEEE.

18. Guerrero, M., Castillo, O., & García, M. (2015). Cuckoo search via lévy flights and a comparison with genetic algorithms. In *Fuzzy logic augmentation of nature-inspired optimization metaheuristics*. (pp. 91–103). Springer.
19. Guerrero, M., Castillo, O., & García, M. (2015). Fuzzy dynamic parameters adaptation in the Cuckoo Search Algorithm using fuzzy logic. In *IEEE congress on evolutionary computation (CEC), 2015* (pp. 441–448).
20. Chechkin, A. V., Metzler, R., Klafter, J., & Gonchar, V. Y. (2008). Introduction to the theory of Lévy flights. In *Anomalous transport: Foundations and applications, 2008* (pp. 129–162).
21. Guerrero, M., Valdez, F., and Castillo, O. (2022). A new Cuckoo search algorithm using interval type-2 fuzzy logic for dynamic parameter adaptation. In *Book a new cuckoo search algorithm using interval type-2 fuzzy logic for dynamic parameter adaptation* (pp. 853–860). Springer International Publishing.
22. Santillan, J. H., Tapucar, S., Manliguez, C., & Calag, V. (2018). Cuckoo search via Lévy flights for the capacitated vehicle routing problem. *Journal of Industrial Engineering International, 14*(2), 293–304.
23. Gandomi, A. H., Yang, X.-S., & Alavi, A. H. (2013). Cuckoo search algorithm: A metaheuristic approach to solve structural optimization problems. *Engineering with Computers, 29*(1), 17–35.

Interpretability of an Archimedean Compensatory Fuzzy Logic in Data Analytics: Some Case Studies

Carlos Eric Llorente-Peralta, Laura Cruz-Reyes, Rafael Alejandro Espín-Andrade, and José Fernando Padron-Tristan

1 Introduction

Compensatory Archimedean fuzzy logic is presented in the following paper as an interpretable logical theory. Previously it was proved that compensatory fuzzy logic is a transdisciplinary and interpretable logical theory.

On this occasion, through the solution of data analytics problems, it is explained how this theory extends the results previously achieved with compensatory fuzzy logic since, in this case, compensatory Archimedean fuzzy logic includes characteristics of logical pluralism, which is closely related to modal logic, as well as through the inclusion of a family of membership functions called generalized continuous linguistic variable, logic is given the sensitivity to data.

The best membership function is generated based on the modeled data. Also, under the parameters, each of these variables is interpretable naturally.

Therefore, it is shown that compensatory Archimedean fuzzy logic is an interpretable logic under its concepts, operators, and membership functions, which is why the results can be presented in natural language.

C. E. Llorente-Peralta · J. F. Padron-Tristan
Tecnológico Nacional de México, Instituto Tecnológico de Tijuana, Tijuana, Mexico

L. Cruz-Reyes (✉)
Tecnológico Nacional de México, Instituto Tecnológico de Ciudad Madero, Ciudad Madero, Mexico
e-mail: lauracruzreyes@itcm.edu.mx

R. A. Espín-Andrade
Universidad Autónoma de Coahuila, Saltillo, Mexico
e-mail: espinr@uadec.edu.mx

O. Castillo and P. Melin (eds.), *Hybrid Intelligent Systems Based on Extensions of Fuzzy Logic, Neural Networks and Metaheuristics*, Studies in Computational Intelligence 1096,
https://doi.org/10.1007/978-3-031-28999-6_15

2 Data Analytics

Data analytics is described in various ways; sometimes it is said that it is a data extraction process that allows the acquisition of knowledge that will be useful in the decision-making process, sometimes it is said that this process is linked to the use of algorithms that allow finding relationships between variables [7, 13].

Data analytics can be described in three large groups: descriptive analytics, which seeks to find patterns of behavior in variables through historical data by performing regression processes and correlation analysis. Predictive analytics where historical events seek to understand past events and events through data, in this way, models are generated that allow forecasting future events and prescriptive analytics, where optimization and modeling models are built-in variables of various interests [11].

One of the processes of data analytics is knowledge discovery in databases (KDD), which is defined as the automatic process by which data discovery and analysis are combined [12].

To this end, tools have been developed that use statistical methods, databases, artificial intelligence, among others, in order to obtain information from massive data sets. The main objective of this process is to convert the data into valuable knowledge that can be presented to the end-user [10].

Regarding the compensatory fuzzy logic (CFL) as a KDD method, various investigations have been carried out that prove that the results are exciting and provide relevant information; among them is the analysis of the image of magnetic resonance, in which It is possible to identify different elements selected efficiently [8], in another article through semantic models the process of knowledge discovery and decision making is carried out these processes are very close to the process of human reasoning [3].

2.1 Interpretability

In Espín-Andrade et al. [4], compensatory fuzzy logic (CFL) is spoken of as a transdisciplinary discipline, that is, a new discipline that combines variables, concepts, results, and methods from various disciplines and obtains new variables, concepts, results and methods of a new theory, a transdisciplinary theory.

One of the advantages of using CFL as a transdisciplinary logic applied to knowledge discovery is the use of fuzzy predicates. Some reasons for this are [4]:

- The generality of the logical predicate structures allows dealing with various data mining problems and support and analysis problems in decision making
- The possibility of interpreting predicates in natural language.
- The possibility of representation through trees, graphs, neural networks, and maps, among others.

Using a CFL, an interpretable logical theory is defined, that is, a fuzzy model that allows understanding a model's behavior. Even so, interpretability is a personal property that depends on several factors such as the structure of the model, the fuzzy rules, the linguistic terms, and the shape of the fuzzy functions, among others [1, 4].

It is such that in order to prove that the CFL meets the characteristics of an interpretable logical theory, a series of points are demonstrated, which are:

1. Create a new theoretical approach that is interpretable by bivalent logic and compatible with selected decision-making theories and mathematical statistics elements.
2. Development of computational tolos.
3. Applications to different fields.
4. Preparation and application of experimental tests the following elements:
    1. Compatibility with human behavior.
    2. Support for knowledge discovery methods for particular problems.
    3. Compatibility with the Mamdani approach in simple cases.
    4. Improved accuracy in complex and high-dimensional cases of Fuzzy Control.
    5. Compatibility with protoforms used in CWW and SUD.
    6. Robustness.

2.2 *Archimedean Compensatory Fuzzy Logic*

Archimedean compensatory fuzzy logic (ACFL) is a type of logic defined from the unification of two different approaches, and these are Archimedean fuzzy logic and compensatory fuzzy logic [5].

Archimedean fuzzy logic (AFL) is a t-norm and t-conorm logical system and is composed of "a trio (c, d, n) of operators, where c is an Archimedean continuous t-norm, d is its corresponding t-Archimedean continuous t-conorm, and n is the negation operator [5]".

An AFL has the following properties [5]:

- Every t-norm $T : [0, 1]^2 \to [0, 1]$, satisfies $T(x, y) \leq min\{x, y\}$
- Every t-conorma $S : [0, 1]^2 \to [0, 1]$ satisfies $S(x, y) \geq max\{x, y\}$
- $min(x, y)$ is the only idempotent continuous t-norm and the maximum function of the family of t-norm operators.
- $max(x, y)$ is the only idempotent continuous t-conorm and the minimum function of the family of operators t-conorm.
- As a consequence of the preceding properties, each t-norm (t-conorm) belongs to only one of two subsets, the singleton of the minimum t-norm (the maximum t-conorm) or the set of non-idempotent t-norms (t-conorms).
- For each archimedean continuous t-norm, there is a non-increasing continuous function $f : [0, 1] \to [0, +\infty]$ satisfaciendo $f(1) = 0$, tal que:

$$T(x_1, x_2, \ldots, x_n) = f^{(-1)}\left(\sum_{i=1}^{n} f(x_i)\right) \tag{1}$$

where:

$$f^{(-1)}(z) = \begin{Bmatrix} f^{-1}(z), & if \quad z \in [0, f(0)] \\ 0 \ if & [f(0), +\infty] \end{Bmatrix} \tag{2}$$

- There is an equivalent property for t-conorm.

A compensatory fuzzy logic (CFL) is composed of a quartet of continuous operators (c, d, o, n), c and d of $[0, 1]^n$ in $[0, 1]$, o of $[0, 1]^2$ in $[0, 1]$ and n a negation operator, which satisfies the following group of axioms [4, 5, 6, 9]:

I. **Compensation axiom:** $min(x_1, x_2, \ldots, x_n) \leq c(x_1, x_2, \ldots, x_n) \leq max(x_1, x_2, \ldots, x_n)$
II. **Commutativity or symmetry axiom:** $c(x_1, x_2, \ldots, x_n) = c(x_1, x_2, \ldots, x_n)$
III. **Strict growth axiom:** if $x_1 = y_1, x_2 = y_2, \ldots, x_{i-1} = y_{i-1} = x_{i+1} = y_{i+1}, \ldots, x_n = y_n$ are different from, and $x_i > y_j$then $c(x_1, x_2, \ldots, x_n) > c(x_1, x_2, \ldots, x_n)$
IV. **Veto axiom: if** $x_i = 0$ for some i then $c(x) = 0$
V. **Fuzzy reciprocity axiom:** $o(x, y) = n[o(y, x)]$
VI. **Fuzzy transitivity axiom:** if $o(x, y) \geq 0.5$ y $o(y, z) \geq 0.5$ then $o(x, y) \geq \max(o(x, y), o(y, z))$
VII. **De Morgan's Laws:** $n(c(x_1, x_2, \ldots, x_n)) = d(n(x_1), n(x_2), \ldots, n(x_n))$

$$n(d(x_1, x_2, \ldots, x_n)) = c(n(x_1), n(x_2), \ldots, n(x_n))$$

Having said the above, an ACFL is defined as "The sextet of operators $(c_t, c_c, d_t, d_c, o, n)$ such that (c_t, d_t, n) is an AFL and (c_c, d_c, o, n) it is a CFL, so the AFL and the CFL are compatible". In addition, it is mentioned that "all AFL and its corresponding CFL are compatible for each propositional formula recursively formed by simple predicates, conjunction, disjunction, negation, implication and equivalence operators" [5].

In the ACFL, what they call a new approach to the universal and existential quantifier is observed. Since a quartet of quantifiers $(\forall_T, \forall_c, \exists_T, \exists_c)$ i s defined as follows:

Let X = $\{x_1, x_2, \ldots, x_n\} \subset [0, 1]$ be the universal quantifier for the finite set X and the Archimedean continuous T-norm, defined by: $\forall_T x_i \in X = c_T(x_1, x_2, \ldots, x_n)$. Let's call it a universal quantifier of non-refutation [5].

Let $X = \{x_1, x_2, \ldots, x_n\} \subset [0, 1]$ be the universal quantifier for the finite set X and the conjunction operator of a CFL based on quasi-arithmetic equations, defined by: $\forall_c x_i \in X = c_c(x_1, x_2, \ldots, x_n)$. Let's call it a universal quantifier of the statement [5].

Let $X = \{x_1, x_2, \ldots, x_n\} \subset [0, 1]$ be the existential quantifier for the finite set X and the Archimedean continuous T-norm, defined by $\exists_T x_i \in X = d_T(x_1, x_2, \ldots, x_n)$. let's call it existential quantifier of non-refutation [5].

Let $X = \{x_1, x_2, \ldots, x_n\} \subset [0, 1]$ e the existential quantifier for the finite set X and the CFL, defined by: $\exists_c x_i \in X = d_c(x_1, x_2, \ldots, x_n)$ Let's call it the existential quantifier of the statement [5].

Applying the $\forall_T$ operator transforms fuzzy predicates into neat predicates because if n increases and $\forall_T x_i \in X$ is not completely true, the predicate tends to become false. Furthermore, since the simple predicates are true, then $\forall_T x_i \in X$ will always be true, and the system tends to be a bivalent logical system.

And by applying the $\forall_c$ operator, it is possible to maintain the regularity of the truth values of the simple predicates; this is due to the property of idempotency. The tendency of this operator is to affirm the proposition.

The non-refutation operator receives that name because it has the tendency to reject the acceptance that the propositions are true, and the affirmation operator tends to confirm the true value of each proposition [5].

Furthermore, the quatrain of predicate quantifiers can be defined as follows [5]:

$\boldsymbol{\forall_T x\, p_T(x)}$: It is called the value of the need for non-refutation of $p(x)$.

$\boldsymbol{\forall_c x\, p_c(x)}$: It is called the value of the necessity of the statement of $p(x)$.

$\boldsymbol{\exists_c x\, p_c(x)}$: It is called the value of the possibility of the statement of $p(x)$.

$\boldsymbol{\exists_T x\, p_T(x)}$: It is called the value of the possibility of non-refutation of $p(x)$.

These four operators are approaches to the concepts of necessity and possibility observed in modal logic and included in the fuzzy logic theory of necessity.

2.3 *Generalization of Concepts of an Archimedean Compensatory Fuzzy Logic*

This section observes how an Archimedean compensatory fuzzy logic (ACFL) is defined through a generator function f that allows the generalization of some fuzzy concepts [2].

Then through an Archimedean compensatory fuzzy logic that is denoted by $L = (c_c, c_t, d_c, d_t, o, n)$ and a generator function f, a generalized linguistic modifier $m(x, L)$ is defined through the following equation:

$$x_L^a = f^{(-1)}(af(x)) \tag{3}$$

where $a \in \mathbb{R}^+ y x_L^a$ denotes $m(x, L)$ and substituting $f(x)$ por $-g(\ln(x))$ in Eq. (1), we call g as a secondary generating function of L (SGF). Also, when $a = 1, f(x) = 0$ and g is an odd function, then $x_L^a = f^{(-1)}(0) = e^{-g^{-1}(0)}$.

And from a logic L and a secondary generator function g, a generalized sigmoidal function (GSF) is obtained expressed in Eq. (2):

$$S_g(x) = \frac{1}{1 + e^{-g^{(-1)}(x)}} \tag{4}$$

And implementing the parameters $\alpha, \gamma \in \mathbb{R}$, $\alpha > 0$ in Eq. (2) a generalized parameterized sigmoidal function (GSFP) is obtained defined by Eq. (3) [2].

$$S_g(x; \alpha, \gamma) = \frac{1}{1 + e^{-g^{(-1)}(\alpha(x-\gamma))}} \tag{5}$$

Thus, by means of an SGF g of an ACFL L and the parameters $\alpha, \gamma \in \mathbb{R}$, $\alpha > 0$, and $m \in [0, 1]$ we have a family of parameterized functions such as increasing sigmoid, decreasing sigmoid and convex functions. This family is called a generalized linguistic continuous variable (GCLV).

Defined by the following equation:

$$GCLV_L(x; \alpha, \gamma, m) = \frac{C_T\left(S_g(x; \alpha, \gamma)_L^m, \left(1 - S_g(x; \alpha, \gamma)\right)_L^{1-m}\right)}{M} \tag{6}$$

where c_t is an Archimedean continuous t-norm in L and M is the maximum of $c_t\left(S_g(x; \alpha, \gamma)_L^m, \left(1 - S_g(x; \alpha, \gamma)_L^{1-m}\right)\right)$ in $\mathbb{R}$ [2].

Also, using a GSF g of an ACFL L, the quartet of operators (c_t, c_c, d_t, d_c) is calculated using the following equations:

$$c_t(x_1, x_2) = e^{-g^{(-1)}(-g(ln(x_1)) - g(ln(x_2)))} \tag{7}$$

$$c_c(x_1, x_2, \ldots, x_n) = e^{-g^{(-1)}\left(\frac{-\sum_{i=1}^{n} g(ln(x_i))}{n}\right)} \tag{8}$$

$$d_T(x_1, x_2) = 1 - e^{-g^{(-1)}(-g(ln(1-x_1)) - g(ln(1-x_2)))} \tag{9}$$

$$d_c(x_1, x_2, \ldots, x_n) = 1 - e^{-g^{(-1)}\left(\frac{-\sum_{i=1}^{n} g(ln(1-x_i))}{n}\right)} \tag{10}$$

his series of operators allows us to evaluate an element of a record for a given logical predicate. However, through the universality and existentiality quantifiers, it is possible to evaluate the entire data set.

The compensatory and Archimedean universality quantifier ($\forall_c y \forall_t$) and the compensatory and Archimedean existentiality quantifier ($\exists_c y \exists_t$) that for a predicate p of a logic L are calculated by means of Eqs. (11), (12), (13) and (14), respectively.

$$\forall_{cx x\in U} p(x) = \begin{cases} f^{-1}\left(\frac{1}{n}\sum_{x\in U} f(p(x))\right) if\ p(x) \neq 0\ for\ all\ x \in U \\ 0\ if\ xp(x) = 0 \end{cases} \tag{11}$$

$$\forall_{tx x\in U} p(x) = \begin{cases} f^{-1}\left(\sum_{x\in U} f(p(x))\right) if\ p(x) \neq 0\ for\ all\ x \in U \\ 0\ if\ xp(x) = 0 \end{cases} \tag{12}$$

$$\exists_{tx x\in U} p(x) = \begin{cases} 1 - f^{-1}\left(\frac{1}{n}\sum_{x\in U} f(1 - p(x))\right) if\ p(x) \neq 0\ for\ all\ x \in U \\ 0\ if\ xp(x) = 0 \end{cases} \tag{13}$$

$$\exists_{tx x\in U} p(x) = \begin{cases} 1 - f^{-1}\left(\sum_{x\in U} f(1 - p(x))\right) if\ p(x) \neq 0\ for\ all\ x \in U \\ 0\ if\ xp(x) = 0 \end{cases} \tag{14}$$

As mentioned above, it is necessary to provide a generating function that defines the ACFL, in Espín-Andrade et al. [2] the use of an exponential logarithmic function $f(x) = -\left(log_b(x)\right)^{ex}$ with b representing the logarithmic base and ex he exponential value and from this its inverse function $f^{-1}(x) = b^{-\sqrt[ex]{x}}$.

Based on these, the following concepts are defined:

- The linguistic modifier:

$$x_L^a = b^{-\sqrt[ex]{(a-(log_b(x))^{ex})}} \tag{15}$$

- The generalized sigmoidal function:

$$S_g(x; \alpha, \gamma) = \frac{1}{1 + b^{-\sqrt[ex]{(\alpha(x-\gamma)}}} \tag{16}$$

- The generalized continuous linguistic variable:

$$GCLV_L(x; \alpha, \gamma, m) = \frac{b^{\sqrt[ex]{log_b^{ex}\left(s_g(x;\alpha,\gamma)_L^m\right) + log_b^{ex}\left(1 - s_g(x;\alpha,\gamma)_L^{1-m}\right)}}}{max_{x\in\mathbb{R}}\left[b^{\sqrt[ex]{log_b^{ex}\left(s_g(x;\alpha,\gamma)_L^m\right) + log_b^{ex}\left(1 - s_g(x;\alpha,\gamma)_L^{1-m}\right)}}\right]} \tag{17}$$

Thus, also the operators c_c, c_t, d_c, d_t are defined as follows:

$$c_c\left(x_1, x_{2,\ldots,} x_n\right) = b^{\sqrt[ex]{\frac{\sum_{i=1}^{n} log_b^{ex}(x_i)}{n}}} \tag{18}$$

$$c_T(x_1, x_2) = b^{\sqrt[ex]{log_b^{ex}(x_1) + log_b^{ex}(x_2)}} \tag{19}$$

$$d_c(x_1, x_{2,\cdots}, x_n) = 1 - b^{\sqrt[ex]{\frac{\sum_{i=1}^{n} log_b^{ex}(1-x_i)}{n}}} \quad (20)$$

$$d_T(x_1, x_2) = 1 - b^{\sqrt[ex]{log_b^{ex}(1-x_1)+log_b^{ex}(1-x_2)}} \quad (21)$$

And the quartet of quantifiers as follows:

$$\forall_{cx\in U} p(x) = b^{-\sqrt[ex]{\frac{1}{n}\sum_{i=1}^{n} log_b^{ex}(p(x_i))}} \quad (22)$$

$$\forall_{tx\in U} p(x) = b^{-\sqrt[ex]{\sum_{i=1}^{n} log_b^{ex}(p(x_i))}} \quad (23)$$

$$\exists_{cx\in U} p(x) = 1 - b^{-\sqrt[ex]{\frac{1}{n}\sum_{i=1}^{n} log_b^{ex}(1-p(x_i))}} \quad (24)$$

$$\exists_{tx\in U} p(x) = 1 - b^{-\sqrt[ex]{\sum_{i=1}^{n} log_b^{ex}(1-p(x_i))}} \quad (25)$$

Thus, a compensatory Archimedean fuzzy logic based on an exponential logarithmic function (ACFL-ELF) is defined.

3 Interpretability of the ACFL in the Context of Decision-Making Theories

Example 1 Decision-making process, where the interpretability is observed in a competitiveness evaluation model in companies.

According to a model for evaluating the competitiveness of companies in the market prepared by a group of experts: a company is competitive in a product line in a market, if its economy is solid, its technological position is advanced, and it has a significant presence in this product line on the market.

The attributes expressed in the model are represented as follows.

$C(x)$: Company x is competitive.

$s(x)$: Company x has a solid economy.

$T(x)$: Company x has an advanced technological position.

$l(x)$: Company x is powerful in the product line of its market.

Employing this information, the competitiveness evaluation model is expressed by the following logical predicate $C(x) = s(x) \wedge T(x) \wedge l^2(x)$ can be represented graphically through the tree observed in Fig. 1.

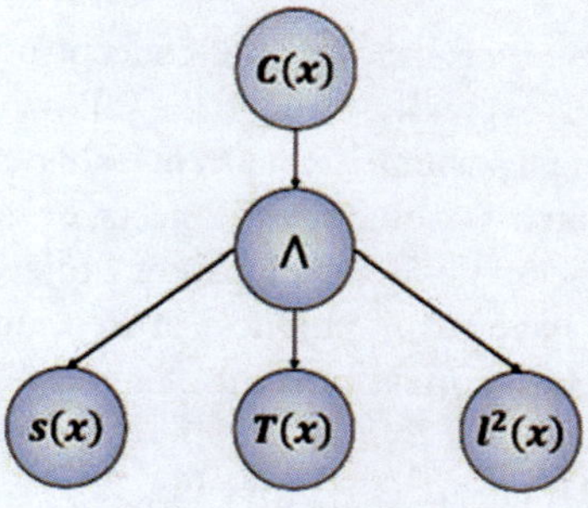

Fig. 1 Graphic representation of the evaluation model

Also, the attribute solid economy $s(x)$ is defined as follows: A company has a stable economy if it has a good financial condition and a good amount of sales. If the financial status is something negative, it must be offset by an outstanding amount of sales. Which is expressed by the logical predicate: $s(x) = f(x) \wedge V(x) \wedge \left((\neg f^{0.5}(x)) \rightarrow V^2(x)\right)$ (See Fig. 2a).

Where each attribute is expressed through the following propositions:

$f(x)$: Company x has an excellent financial condition.

$V(x)$: Company x has a good amount of sales.

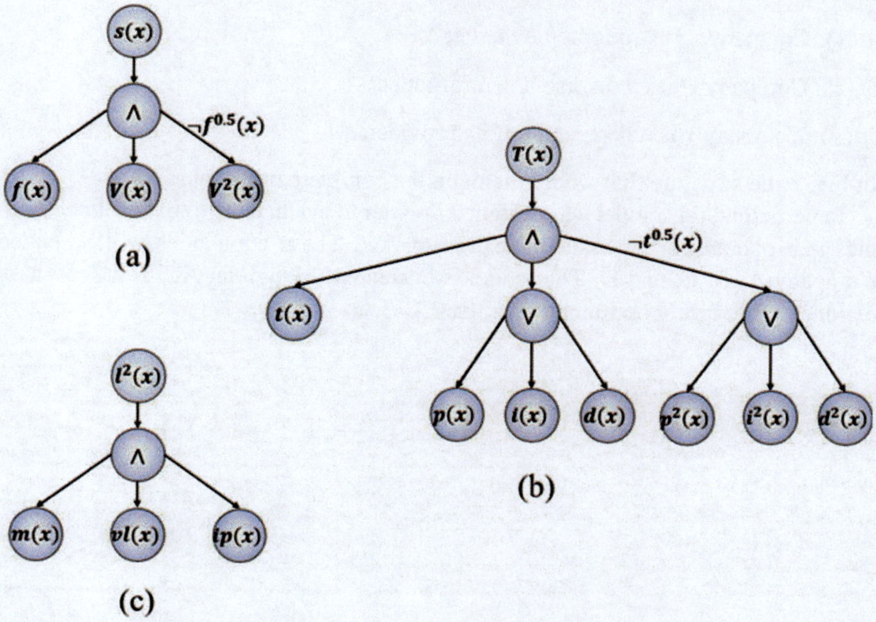

Fig. 2 Graphic representation of evaluation sub models

In order to obtain a more precise model, the concept of advanced technological position $T(x)$ is introduced, expressed through the following statement: A company has an advanced technological position if its current technology is good; owns patents, or many products in Research-Development Projects, or is used to use a significant amount of money for Research and Development Projects. If your technology is not good, then you must have many products in Research-Development Projects or it is used to use a very important amount of money for this type of projects, where this model is expressed as follows: $T(x) = t(x) \wedge (p(x) \vee i(x) \vee d(x)) \wedge \left(\left(\neg t^{0.5}(x)\right) \rightarrow \left(p^2(x) \vee i^2(x) \vee d^2(x)\right)\right)$ (See Fig. 2b).

Where for each attribute, we have the following propositions:

$t(x)$: Company x has good current technology.

$p(x)$:Companies x have patents.

$i(x)$: Companies x have products in research-development projects.

$d(x)$: Companies x is used to use a significant amount of money for research and development projects.

Finally, the concept of good in a product line in the market $l^2(x)$ is defined as A company that is good in a product line in a market, if it has a solid position in that market, a diverse line of products and is independent. of suppliers, which is expressed as follows: $l(x) = m(x) \wedge vl(x) \wedge ip(x)$ (Ver Fig. 2c).

These attributes are expressed through the following propositions:

$m(x)$: Company x is strong in the market.

$vl(x)$: Company x has a diverse line of products.

$ip(x)$: Company x is independent of its providers.

In Fig. 3 the structure that would maintain the complete tree is observed.

In particular, this model was designed in order to evaluate a group of four companies that operate in the tissue adhesives market. These companies will be called Company A, B, C, and D. The model is evaluated through an ACFL based on an exponential-logarithmic function (ACFL-ELF).

Fig. 3 Evaluation tree that represents the business competitiveness evaluation model

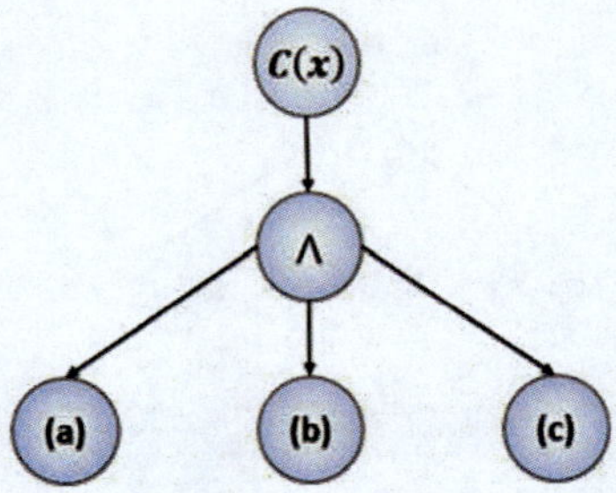

Table 1 Table of attribute values

Company	A	B	C	D
f(x)	0.5	0.6	0.9	0
V(x)	0.47	0.63	0.75	0.99
t(x)	0.3	0.5	0.7	0.8
P(x)	0.93	0.41	0.62	0.81
i(x)	0.81	1	0.55	0.79
d(x)	0.61	0.95	0	0.7
Ip(x)	0.6	0.8	1	0.5
vl(x)	0.23	0.77	0.92	0.39
m(x)	0.1	0.4	0.8	1

Table 1 shows the fuzzy values for each attribute of each company.

To demonstrate how the evaluation process is carried out by using an ACFL-ELF to which we will give an exponential value of one and with a logarithmic base e, we proceed to evaluate the predicate $s(x)$ observed in Fig. 2a.

To evaluate $\neg f^{0.5}(x)$ the linguistic modifier x_L^a (3) is used to evaluate the negate of $f(x)$. Which is expressed by the following equation: $x_L^{0.5} = f^{(-1)}(|0.5|f(1-0.5)) = 0.7071$. Using the same equation to calculate V^2 we have that it is equal to 0.2208, en The following expression evaluates $\neg f^{0.5}(x) \to V^2(x)$ process for which the implication of Zadeh is used, which is expressed as follows: $i(x, y) = d_c(\neg 0.7071, c_c(0.7071, 0.2208)) = 0.3460$. Where d_c corresponds to (20) and c_c to (18).

Once these values have been calculated, we proceed to evaluate $s(x) = f(x) \wedge V(x) \wedge \left(\left(\neg f^{0.5}(x)\right) \to V^2(x)\right)$, for this case we proceed from the In the following way, if $f(x) < 0.5$, then it will be calculated as follows: $c_c(0.25, 0.47, 0.3460)(18) = 0.4332$. Otherwise, it will be evaluated as follows: $c_c(0.25, 0.47)(18) = 0.485$, this expression must also be evaluated using the normal t operators, which is achieved by changing the c_c and d_c operators for the c_t and d_t.

Table 2 shows the results of evaluating the companies table using a logarithmic base *e* and exponential value of one.

Table 2 Results of the evaluation of the model through an ACFL-ELF

Company	A		B		C		D	
Logic	AFL	CFL	AFL	CFL	AFL	CFL	AFL	CFL
s(x)	0.235	0.485	0.378	0.615	0.675	0.822	0.000	0.000
T(x)	0.252	0.516	0.500	0.696	0.580	0.558	0.790	0.785
l(x)	0.014	0.240	0.246	0.627	0.735	0.903	0.195	0.580
l^2(x)	1.9E-04	0.058	0.061	0.393	0.541	0.815	0.038	0.336
C(x)	1.1E-05	0.243	0.011	0.552	0.212	0.720	0.000	0.000

Table 3 Results of evaluating the model through the quantifiers of an ACFL

Cuantifiers	$\forall_c$	$\exists_c$	$\forall_t$	$\exists_t$
True value	0.5558	0.4444	3.1661E-08	0.2208

As can be seen in the table, company C is the only one that meets the requirements of the model. In Espin-Andrade et al. [4], it is said that in terms of natural language, this result is expressed as "competitiveness in company C it is something true". For this case, it is added that this expression has an affirmation value of 0.720 degrees of truth and a non-refutation value of 0.212 degrees of truth.

Likewise, the ACFL allows us to evaluate the model above all companies through a quartet of quantifiers that indicate the true value of the need for the statement ($\forall_c$)(22), the truth value of the possibility of the statement ($\exists_c$)(24), the truth value of the need for non-refutation ($\forall_t$)(23) and the truth value of the possibility of non-refutation ($\exists_t$)(25). These values are observed in Table 3.

4 Interpretability of an ACFL According to Bivalent Logic and Statistics

Example 2 Hypothesis of the Mexican economy in order to illustrate the knowledge discovery process.

For this example, the following hypotheses are proposed to describe the dynamics of the Mexican economy in the natural language.

- **H1**: If after time t from time t_0 is short, GDP in t_0 is high, and the peso-dollar exchange rate is reasonable, and inflation is good, then inflation at $t_0 + t$ s will be good. (sufficient condition to have good future inflation).

$$if short\_time_t \wedge hight\_GDP_t \wedge reasonable\_exchange_t \\ \wedge\, good\_inflation_t \rightarrow good\_inflation_{t_0+t}$$

- **H2**: If after time t from time t_0 is short, GDP in t_0 is high, and the peso-dollar exchange rate is reasonable, and inflation is good, then the exchange rate at $t_0 + t$ will be good (sufficient condition to have a good exchange rate future).

$$if short\_time_t \wedge hight\_GDP_t \wedge reasonable\_exchange_t \\ \wedge\, good\_inflation_t \rightarrow good\_exchange_{t_0+t}$$

- **H3**: If after time t from time t_0 is short, GDP in t_0 is high, and the peso-dollar exchange rate is reasonable, and inflation is good, then GDP at $t_0 + t$ will be good. (sufficient condition to have an excellent future GDP).

$$
\begin{aligned}
if short\_time_t \wedge hight\_GDP_t \wedge reasonable\_exchange_t \\
\wedge good\_inflation_t \rightarrow good\_GDP_{t_0+t}
\end{aligned}
$$

To optimize the decision variables of the proposed hypotheses, use is made of the generalized continuous linguistic variable (GCLV), which is observed in Eq. (6).

The optimization is carried out by varying the parameters α, γ and$m$, which allows the modeling of the information of each variable. In Fig. 4 it is observed how these parameters model the membership function.

Through an ACFL-ELF that makes use of a GCLV to model the data and compensating operators for evaluation. For this example, two cases are proposed. In the first for hypotheses 1, 2, and 3, the values of α, γ and$m$ are defined by a human expert, and the logarithmic base (b) is established with values e and exponent (exp) one. For the second case, for the hypothesis, called **H1'**, **H2'** and **H3'**, the parameter $\alpha, \gamma m, b$andexp are discovered that allow improving the results of the first case. The results will be evaluated through the quantifiers $\forall_c$ y $\exists_c$. In Table 4. the values of α, γ, m are observed for the first and second cases.

Through the optimization process, it was obtained that the best values for b and *exp* are 1,708 and 3, respectively. Thus, also in Table 5, the results of evaluating the hypotheses for each record of the data set are observed, as well as in the last four rows, the value of the quantifiers is observed for all, and there are compensations for the entire data set and the value for all and exists from the last six records in the dataset.

According to the observed results, a noticeable improvement is observed in the hypotheses raised at the beginning of the experiment. Of these hypotheses, hypothesis

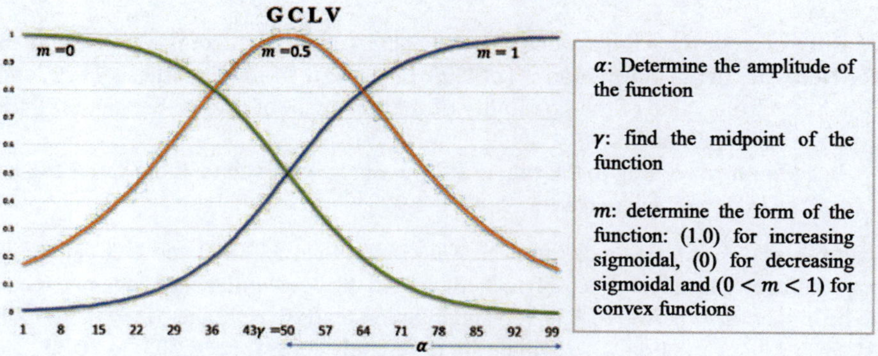

Fig. 4 Figure that shows the modeling of the data through the parameters of a GCLV

Table 4 Resulting values of α, γ and *m* for the first and second case

	α	γ	m	α	γ	m
$Time$	0.5	2.0	0.0	1.946	3.820	0.109
$Inflation_{t_0}$	0.5	11.0	0.0	1.990	12.550	0.860
GDP_{t_0}	0.5	2.0	1.0	0.796	3.130	0.522
$Exchangerate_{t_0}$	0.5	7.0	0.0	0.912	4.160	0.769
$inflation_{t_0+t}$	0.5	11.0	0.0	0.055	10.330	0.533
GDP_{t_0+t}	0.5	2.0	1.0	0.608	1.700	0.405
$Exchangerate_{t_0+t}$	0.5	7.0	0.0	2.027	2.750	0.646

Table 5 Results of the hypothesis evaluation process through a GCLV of an ACFL-ELF

	H1	*H2*	*H3*	*H1'*	*H2'*	*H3'*
1	1.00	0.66	1.00	0.99	0.88	1.00
2	0.16	0.17	0.72	0.81	0.85	0.91
3	0.19	0.74	0.59	0.90	0.85	0.86
4	0.26	1.00	0.56	1.00	0.79	0.83
...						
17	0.27	0.67	0.35	0.96	0.90	0.72
18	0.70	0.82	0.71	0.98	1.00	0.68
19	0.31	0.69	0.39	0.96	0.90	0.72
20	0.34	0.61	0.37	0.98	1.00	0.68
21	0.52	0.71	0.54	0.98	1.00	0.68
$\forall_c$	**0.34**	**0.64**	**0.57**	**0.90**	**0.85**	**0.74**
$\exists_c$	**0.63**	**0.77**	**0.74**	**0.95**	**0.83**	**0.81**
$\forall_c$	**0.59**	**0.90**	**0.64**	**0.98**	**0.91**	**0.80**
$\exists_c$	**0.33**	**0.71**	**0.46**	**0.89**	**0.67**	**0.58**

1' is the closest to having a value close to one; in addition, there is a truth value of the Necessity of the affirmation ($\forall$) of 0.90, 0.85 and 0.74 for Hypothesis 1 ', 2' and 3 '. And a truth value of the possibility of the statement of 0.95, 0.83, and 0.81 for Hypothesis 1 ', 2', and 3.

In addition, according to the values of α, γ and *m* observed in Table 4.1, it can be expressed in natural language for:

Hypothesis 1': if time tends to be less ($m = 109$) than 3.82 (γ) and GDP at t_0 e is approximately equal ($m = 0.522$) to 3.13 (γ) and the peso-dollar exchange rate tends to be higher ($m = 0.769$) to 4.16 (γ) and inflation tends to be higher ($m = 0.86$) than 12.55 (γ), then inflation $t_0 + t$ will be approximately equal ($m = 0.533$) to 10.33 (γ).

Hypothesis 2': if time tends to be less ($m = 109$) than 3.82 (γ) and GDP at t_0 e is approximately equal ($m = 0.522$) to 3.13 (γ) and the peso-dollar exchange rate tends to be higher ($m = 0.769$) to 4.16 (γ) and inflation tends to be higher ($m = 0.86$) than 12.55 (γ), so the exchange rate in $t_0 + t$ tends to be higher ($m = 0.646$) to 2.75 (γ).

Hypothesis 3': if time tends to be less ($m = 109$) than 3.82 (γ) and GDP at t_0 e is approximately equal ($m = 0.522$) to 3.13 (γ) and the peso-dollar exchange rate tends to be higher ($m = 0.769$) to 4.16 (γ) and inflation tends to be higher ($m = 0.86$) than 12.55 (γ), then GDP at $t_0 + t$ tends to be lower ($m = 0.405$) to 1.7 (γ).

5 Conclusions

In the following document, data analytics cases prove that compensatory Archimedean fuzzy logic is an interpretable logical theory that allows expressing the results in the natural language under its components.

Thus, it is also observed how the inclusion of a logical pluralism formalizes the results from the concept of a double interpretation. Thus, it is also observed how it improves the results previously achieved by compensatory fuzzy logic and expands the possibilities concerning interpretable data analytics processes.

References

1. Casillas, J., Cordón, O., Herrera, F., & Magdalena, L. (2003). Accuracy improvements to find the balance interpretability-accuracy in linguistic fuzzy modeling: An overview. In J. Casillas, O. Cordón, F. Herrera (Eds.), *Studies in fuzziness and soft computing* (129th ed., pp. 3–24). Berlin: Springer. https://doi.org/10.1007/978-3-540-37058-1_1
2. Espín-Andrade, R. A., Cruz-Reyes, L., Llorente-Peralta, C., González-Caballero, E., Pedrycz, W., & Ruiz, S. (2021). Archimedean compensatory fuzzy logic as a pluralist contextual theory useful for knowledge discovery. *International Journal of Fuzzy Systems*, 1–21. https://doi.org/10.1007/s40815-021-01150-6
3. Espin-Andrade, R. A., Fernández Gonzalez, E. R., & González, E. (2014). Compensatory fuzzy logic: A frame for reasoning and modeling preference knowledge in intelligent systems. *Soft Computing for Business Intelligence. Studies in Computational Intelligence. Springer, Berlin., 537*, 3–23. https://doi.org/10.1007/978-3-642-53737-0_1
4. Espin-Andrade, R. A., Gonzalez, E., Pedrycz, W., & Fernandez, E. (2016). An interpretable logical theory: The case of compensatory fuzzy logic. *International Journal of Computational Intelligence Systems, 9*(4), 612–626. https://doi.org/10.1080/18756891.2016.1204111
5. Espin-Andrade, R. A., Gonzalez, E., Pedrycz, W., & Fernández González, E. R. (2015). Archimedean-compensatory fuzzy logic systems. *International Journal of Computational Intelligence Systems, 8*(2), 54–62. https://doi.org/10.1080/18756891.2015.1129591
6. Espín Andrade, R., & a., Fernándes, E., & Gonzaléz, E. (2011). Un Sistema Lógico Para El Razonamiento Y La Toma De Decisiones : La Lógica Difusa Compensatoria Basada En La Media Geométrica. *Revista Investigación Operacional, 3*, 230–245.
7. Gandomi, A., & Haider, M. (2015). Beyond the hype: Big data concepts, methods, and analytics. *International Journal of Information Management, 35*(2), 137–144. https://doi.org/10.1016/j.ijinfomgt.2014.10.007

8. Meschino, G. J., Espín-Andrade, R. A., & Ballarin, V. L. (2008). A framework for tissue discrimination in magnetic resonance brain images based on predicates analysis and compensatory fuzzy logic. *IC-MED International Journal of Intelligent Computing in Medical Sciences and Image Processing, 2*(3), 207–222. https://doi.org/10.1080/1931308X.2008.10644165
9. Montero, J. C., A. (2011). La Lógica Difusa Compensatoria. *Ingeniería Industrial, XXXII*(2), 157–161
10. Pazzani, M. J. (2000). Knowledge discovery from data? *IEEE Intelligent Systems, 15*(2), 10–12. https://doi.org/10.1109/5254.850821
11. Pusala, M. K., Amini Salehi, M., Katukuri, J. R., Xie, Y., & Raghavan, V. (2016). Massive data analysis: Tasks, tools, applications, and challenges. In *Big data analytics* (pp. 11–40). India: Springer. https://doi.org/10.1007/978-81-322-3628-3_2
12. Timarán-Pereira, S. R., Hernández-Arteaga, I., Caicedo-Zambrano, S. J., Hidalgo-Troya, A. y Alvarado- Pérez, J. C. (2016). El proceso de descubrimiento de conocimiento en bases de datos. *Ediciones Universidad Cooperativa de Colombia.*, *8*(26), 63–86. https://doi.org/10.16925/9789587600490
13. Watson, H. J. (2014). Tutorial: Big data analytics: Concepts, technologies, and applications. *Communications of the Association for Information Systems, 34*(1), 1247–1268. https://doi.org/10.17705/1cais.03465

A New Selection and Class Prediction Using Type-1 Fuzzy Logic Applied to a Convolutional Neural Network

Yutzil Poma and Patricia Melin

1 Introduction

Neural networks have been of great help for the development of new technologies as in [1] where a human emotion is predicted with a neural network, other uses of the NN is in the acceleration of hardware using deep neural networks as in [2], also in the development of effective classification systems as in [3], it has been demonstrated that they are very effective for classification with the help of type-2 fuzzy systems in conjunction with genetic algorithms [4].

Convolutional neural networks these days have brought great advances to science and technology as they have been applied in conjunction with different methods to take advantage of their virtues as for example in medicine where breast classification and histopathological analysis on tumor detection [5] or in [6] where tuberculosis is detected using these networks also in [7], or in [8] where diabetic rhinoplasty is classified using cnn's, as well as in [9, 10] or in [11] where use is made of modular neural networks for blood pressure classification, where these networks are used for the medical area.

Fuzzy systems in conjunction with neural networks have shown that considerable improvements are obtained in their results as in [12] where a type-2 fuzzy system was used for pressure classification, or in optimization using type-1 and type-2 interval fuzzy systems in conjunction with genetic algorithms for blood pressure classification [13], or in various applications where fuzzy systems have improved many results and developments for classification and recognition [14]. In [15] the type-2 interval fuzzy system was used to classify the heart rate using the bird swarm algorithm, the type-2 fuzzy system uses for recognition systems which helps to simplify the analysis of images [16].

Y. Poma · P. Melin (✉)
Tijuana Institute of Technology, Tijuana, Mexico
e-mail: pmelin@tectijuana.mx

O. Castillo and P. Melin (eds.), *Hybrid Intelligent Systems Based on Extensions of Fuzzy Logic, Neural Networks and Metaheuristics*, Studies in Computational Intelligence 1096,
https://doi.org/10.1007/978-3-031-28999-6_16

The main contribution of this work is the classification using a type-1 fuzzy system which is limited to only 4 classes of the selected case study which helps for analysis and comparison of results and to see what improvements can be made for future experimentation.

This work is composed as follows: In Sect. 2 we see the relative words where basic concepts of this work are addressed, in Sect. 3 the detailed method proposal, in Sect. 4 we can appreciate the Results and discussion, later in Sect. 5 we can observe the statistical test of the results obtained, finally we end with the conclusions of the work presented.

2 Literature Review

This section we presents the relative words to understand the proposed method:

2.1 Convolutional Neural Networks

Also called CNN's are deep networks which one of its main advantages is that they are used for classification and image recognition as it is responsible for extracting the main features of the images [17] these neural networks have a basic architecture, which is made up of layers. In the convolution layer is responsible for extracting the main features of the image through filters that the developer determines creating a map of main features [18] which passes to the next layer which is the pooling layer also called reduction layer which uses the feature map and reduces these features by means of a mask that goes through the image and takes the values of this and either by the average or the highest value takes the value and stores it thus making the extraction of features [19]. Finally, there is the classification layer which is responsible for classifying the images to their respective classes using the activation function softmax generally, which uses the last weights of the last hidden layer and exponentiates that weight and divides it by the sum of the exponentiation of all the weights of the last outputs of the last hidden layer and at the end the highest value of this operation classifies it to its respective class [20].

3 Proposed Method

The architecture used in the convolutional neural network is: convolution layers which result in the ReLU activation function and then 2 pooling layers, ending with a classification layer.

The type-1 fuzzy system designed for this proposal consists of 4 inputs, each of which represents the last weight of the last hidden layer of the full connected.

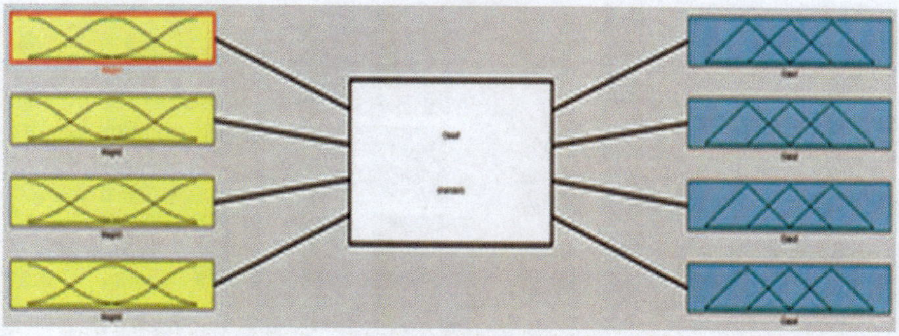

Fig. 1 Proposed type-1 fuzzy system

These inputs (weights) enter the fuzzy system which classifies each of the fuzzy rules described below based on the fuzzy rules:

1. If (Weight1 is Down) and (Weight2 is Down) and (Weight3 is Down) and (Weight4 is Down) then (Class1 is MinClasif)(Class2 is MinClasif)(Class3 is MinClasif)(Class4 is MinClasif)
2. If (Weight1 is Medium) and (Weight2 is Medium) and (Weight3 is Medium) and (Weight4 is Medium) then (Class1 is MedClasif)(Class2 is MedClasif)(Class3 is MedClasif)(Class4 is MedClasif)
3. If (Weight1 is High) and (Weight2 is High) and (Weight3 is High) and (Weight4 is High) then (Class1 is MaxClasif)(Class2 is MaxClasif)(Class3 is MaxClasif)(Class4 is MaxClasif).

The output of the fuzzy system consists of 4 outputs which represent the percentage of classification which is the output of first four class of monkeys. The outputs and inputs have Gaussian type membership functions. The type-1 fuzzy system can be seen in Fig. 1. The FGSA method is used to find the adaptation of the points that build the membership functions of the fuzzy system (inputs and outputs).

4 Results and Discussion

The case study used for the implementation of the proposal was the database of 10 Monkey Species [21] which consists of a total of 1360 images which has 272 images to test and 1088 to train each image, originally it has a size of 400 * 300 or more pixels, we use 1000 images in total for testing which represents a 73.53% sample of the total original size of images in this database. 600 images were used to train and to test 400 images these images were originally in color but they were changed to grayscale and their size was reduced leaving a width of 90 pixels by 100 pixels in height and each image is in a.jpg format.

Fig. 2 Images from the 10 Monkeys species database

Next in Fig. 2 we can see the original images on the left side while on the right side we can see the end of the preprocessing of the final images that was used to test the proposal.

Each of the experiments was performed 30 times, the epochs were increased from 50 to 2000 epochs of training.

In the following table we can observe the results of the experimentation of the method using a type-1 fuzzy system using the FGSA to adapt the points that form the membership functions, using 60% of the data to train.

In Table 1 we can see that the best classification percentage was when the convolutional neural network has been trained at 2000 epochs with 38.23% classification, training with 60% of the data and having 40% of the data to test.

Based on past experiments and noticing that the classification percentage is low, it was decided to increase the number of images to train and reduce the number of images to test, taking 80% of the data for training while 20% has been test using type-1 fuzzy logic. In Table 2 we can observe the results of this experiment in which it is shown that the best result was when the cnn has been trained at 2000 epochs, having

Table 1 Results the experiments using FGSA for built the membership functions in a type-1 fuzzy system

Epochs	Max recognition rate	Number of filter 1	Number of filter 2	Average	σ	Train (%)
50	26.31	80	65	26.57	0.98	60
70	28.88	65	46	29.26	0.52	60
100	30.74	71	70	29.31	0.82	60
200	30.89	63	45	31.36	0.84	60
500	32.14	35	52	32.51	0.78	60
1000	33.12	21	75	35.65	0.71	60
2000	**38.01**	**56**	**45**	**38.23**	**0.67**	**60**

Table 2 Results the experiments with 80% of data to training and 20% to tester using type-1 fuzzy logic

Epochs	Max recognition rate	Number of filter 1	Number of filter 2	Average	σ	Train (%)
50	35.02	56	55	35.45	0.48	80
70	36.52	75	48	37.86	0.52	80
100	37.71	74	80	37.25	0.41	80
200	38.89	63	85	38.36	0.64	80
500	42.24	85	62	43.56	0.88	80
1000	53.11	61	55	54.52	0.91	80
2000	**68.03**	**76**	**65**	**67.65**	**0.77**	**80**

a maximum in the classification average of 67.65% and with a standard deviation of 0.77.

Based on past experiments and the results that were still low for the classification percentage, it was decided to add a convolution layer and a pooling layer to the existing architecture. The architecture used has 3 convolution layers and 3 polling layers using the type-1 fuzzy system. The network was trained with 80% of the data to improve the classification percentage, obtaining a maximum of 71.65% with 2000 training epochs. We can see in Table 3.

In Table 4 we can see the results obtained from the experiments carried out with the proposed method compared to other methodologies and neural networks for classification, which we can see that the convolutional neural network and the proposed method are not as optimal as the others.

Table 3 Results of experiments using the architecture with 3 convolutional layers and 3 pooling layer and type-1 fuzzy logic

Epochs	Max recognition rate	Number of filter 1	Number of filter 2	Average	σ	Train (%)
50	36.51	57	56	35.15	0.65	80
70	36.89	65	49	37.46	0.52	80
100	38.97	78	81	38.25	0.62	80
200	39.45	56	75	39.38	0.55	80
500	43.04	78	67	44.02	0.63	80
1000	63.10	57	85	66.12	0.45	80
2000	**71.13**	**81**	**23**	**71.65**	**0.62**	**80**

Table 4 Comparison of results with the proposed method with other methods

Neural Network	Epochs	Separation method	Method	Train (%)	Test (%)	Average (%)	σ
CNN	200	OvO [25]	Scratch -Inception-V3	50	50	86.28	0.63
CNN	200	OvO [25]	Scratch -Inception-V3	80	20	93	1.73
CNN	200	OvA [25]	Scratch -Inception-V3	50	50	84.1	1.95
CNN	200	OvA [25]	Scratch -Inception-V3	80	20	90.94	1.94
CNN	200	OvO [25]	Scratch-Resnet-50	50	50	80.81	2.83
CNN	200	OvO [25]	Scratch-Resnet-50	80	20	89.56	2.07
CNN	200	OvA [25]	Scratch-Resnet-50	50	50	81.46	1.19
CNN	200	OvA [25]	Scratch-Resnet-50	80	20	89.64	0.71
CNN	200	OvO [25]	Fine-tuned-Inception-V3	50	50	97.52	0.73
CNN	200	OvO [25]	Fine-tuned-Inception-V3	80	20	97.67	1.15
CNN	200	OvA [25]	Fine-tuned-Inception-V3	50	50	98.17	0.94
CNN	200	OvA [25]	Fine-tuned-Inception-V3	80	20	99.13	0.41
CNN	200	OvO [25]	Fine-tuned-Resnet-50	50	50	95.77	0.97
CNN	200	OvO [25]	Fine-tuned-Resnet-50	80	20	97.37	0.64
CNN	200	OvA [25]	Fine-tuned-Resnet-50	50	50	96.79	1.65
CNN	200	OvA [25]	Fine-tuned-Resnet-50	80	20	97.37	1.4
CNN	**2000**	**–**	**Optimized with fuzzy logic**	**60**	**40**	**38.23**	**0.67**
CNN	**2000**	**–**	**Optimized with fuzzy logic**	**80**	**20**	**67.65**	**0.77**
CNN	**2000**	**–**	**Optimized with fuzzy logic**	**80**	**20**	**71.65**	**0.62**

5 Statistical Test

The statistical test was carried out to verify that the results obtained are correct and formally determine the conclusions.

In Table 5 we can see the parameters used for the statistical test.

The following null and alternative hypotheses need to be tested:

$$H_0:\mu_1 \geq \mu_2$$

$$H_a:\mu_1 < \mu_2$$

In the Fig. 3 we can see the graphical Z test Results.

It is concluded that the null hypothesis H_0 is not rejected. Therefore, there is not enough evidence to claim that the population mean μ_1 is less than μ_2 at the α=0.05 significance level.

Table 5 Concepts and values for statistical test

Concept	Value
Sample mean 1 ($\overline{X}_1$)	71.65
Population standard deviation 1 (σ_1)	0.62
Sample size 1 (n_1)	30
Sample mean 2 ($\overline{X}_2$)	67.65
Population standard deviation 2 (σ_2)	0.77
Sample size 2 (n_2)	30
Significance level (α)	0.05

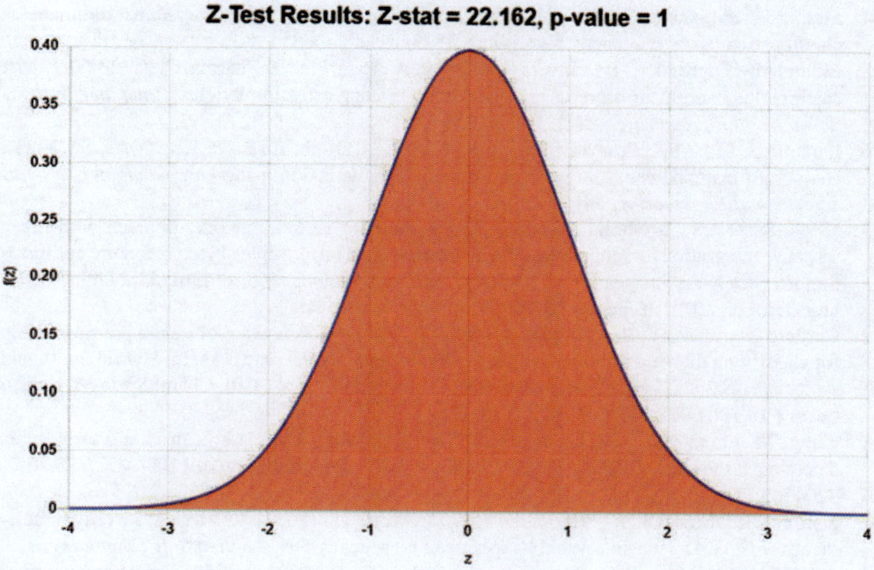

Fig. 3 Graphical Z-test results

6 Conclusions

We conclude that the initial version of the classifier for the replacement of the softmax function using type-1 fuzzy logic, it is still not the most optimal since the results in the classification with the comparison with other methods there is a notable difference in the results. However, we are still trying to modify the fuzzy system and adjust it, expecting better results in the future. In addition, we plan to consider other application areas, such as the ones presented in [22, 23].

Acknowledgements We thank sour sponsor CONACYT & the Tijuana Institute of Technology for the financial support provided with the scholarship number 816488.

References

1. Antonucci, A., Papini, G. P. R., Bevilacqua, P., Palopoli, L., & Fontanelli, D. (2022). Efficient Prediction of Human Motion for Real-Time Robotics Applications With Physics-Inspired Neural Networks. *IEEE Access, 10*, 144–157. https://doi.org/10.1109/ACCESS.2021.3138614
2. Ansari, A., & Ogunfunmi, T. (2022). Hardware Acceleration of a Generalized Fast 2-D Convolution Method for Deep Neural Networks. *IEEE Access, 10*, 16843–16858. https://doi.org/10.1109/ACCESS.2022.3149505
3. Espejel-Cabrera, J., Cervantes, J., García-Lamont, F., Castilla, J. S. R., & Jalili, L. D. (2021). Mexican sign language segmentation using color based neuronal networks to detect the individual skin color. *Expert Systems with Applications, 183*, 2021, 115295. ISSN 0957-4174. https://doi.org/10.1016/j.eswa.2021.115295
4. Melin, P., & Sánchez, D. (2021). Optimal design of type-2 fuzzy systems for diabetes classification based on genetic algorithms, 15–32.
5. Banumathy, D., Khalaf, O. I., Tavera Romero, C. A., Raja, P. V., & Sharma, D. K. (2023). Breast calcifications and histopathological analysis on tumour detection by cnn. *Computer Systems Science and Engineering, 44*(1), 595–612.
6. Ignatius, J. L. P., Selvakumar, S., Paul, K. G. J. L., Kailash, A. B., & Keertivaas, S. (2023). Histogram matched chest x-rays based tuberculosis detection using cnn. *Computer Systems Science and Engineering, 44*(1), 81–97.
7. Varela-Santos, S., & Melin, P. (2021). A new modular neural network approach with fuzzy response integration for lung disease classification based on multiple objective feature optimization in chest X-ray images. *Expert Systems with Applications, 168*, 114361. ISSN 0957-4174. https://doi.org/10.1016/j.eswa.2020.114361
8. Cordero-Martínez, R., Sánchez, D., & Melin, P. (2022). Comparison of image pre-processing for classifying diabetic retinopathy using convolutional neural networks. In: Hybrid intelligent systems. HIS 2021. Lecture notes in networks and systems (vol. 420). Cham: Springer. https://doi.org/10.1007/978-3-030-96305-7_18
9. Farag, M. M. (2022). A Self-contained STFT CNN for ECG classification and arrhythmia detection at the edge. *IEEE Access, 10*, 94469–94486. https://doi.org/10.1109/ACCESS.2022.3204703
10. Karthik, R., Menaka, R., Hariharan, M., & Won, D. (2022). CT-based severity assessment for COVID-19 using weakly supervised non-local CNN. *Applied Soft Computing, 121*, 108765. ISSN 1568–4946, https://doi.org/10.1016/j.asoc.2022.108765. (https://www.sciencedirect.com/science/article/pii/S156849462200196X)
11. Miramontes, I., Melin, P., & Prado-Arechiga, G. (2020). Particle Swarm optimization of modular neural networks for obtaining the trend of blood pressure. In: Castillo, O., Melin, P., Kacprzyk, J. (Eds.), *Intuitionistic and type-2 fuzzy logic enhancements in neural and optimization algorithms: theory and applications*. Studies in computational intelligence (Vol. 862). Cham: Springer. https://doi.org/10.1007/978-3-030-35445-9_19
12. Guzmán, J. C., Melin, P., & Prado-Arechiga, G. (2020). Design of interval type-2 fuzzy systems for classification of blood pressure load. In: Castillo, O., Melin, P. (Eds.), *Hybrid intelligent systems in control, pattern recognition and medicine*. Studies in computational intelligence (Vol. 827). Cham: Springer. https://doi.org/10.1007/978-3-030-34135-0_16
13. Guzmán, J. C., Melin, P., & Prado-Arechiga, G. (2020). Optimization for type-1 and interval type-2 fuzzy systems for the classification of blood pressure load using genetic algorithms. In: Castillo, O., Melin, P., Kacprzyk, J. (Eds.), Intuitionistic and type-2 fuzzy logic enhancements in neural and optimization algorithms: theory and applications. Studies in computational intelligence (Vol. 862). Cham: Springer. https://doi.org/10.1007/978-3-030-35445-9_5
14. Melin, P., & Castillo, O. (2013). A review on the applications of type-2 fuzzy logic in classification and pattern recognition. *Expert Systems with Applications, 40*(13), 5413–5423. ISSN 0957–4174. https://doi.org/10.1016/j.eswa.2013.03.020

15. Miramontes, I., Guzman, J. C., Melin, P., & Prado-Arechiga, G. (2018). Optimal design of interval type-2 fuzzy heart rate level classification systems using the bird swarm algorithm. *Algorithms, 11*(12), 206. https://doi.org/10.3390/a11120206
16. Melin, P., Castillo, O., Gonzalez, C. I., Castro, J. R., & Mendoza, O. (2016).General Type-2 fuzzy edge detectors applied to face recognition systems. In: 2016 annual conference of the North American fuzzy information processing society (NAFIPS), 2016 (pp. 1–6). https://doi.org/10.1109/NAFIPS.2016.7851625
17. LeCun, Y., & Bengio, Y. (1998). Convolution networks for images, speech, and time-series. *Igarss, 2014*(1), 1–5.
18. Wang, Y., Li, J., Zhang, Y., & Sinnott, R. (2021). Identifying lameness in horses through deep learning. In *Proceedings of the 36th annual ACM symposium on applied computing* (pp. 976–985)
19. Yang, J., Yu, K., Gong, Y., & Beckman, T. H. (2009). Linear spatial pyramid matching using sparse coding for image classification. In *IEEE computational social conference computer visual pattern recognition* (pp. 1794–1801).
20. Venkatesan, R., & Li, B. (2017). *Convolutional neural networks in visual computing: A concise guide.* CRC Press.
21. Image dataset for fine-grain classification - 10 Monkey Species. https://www.kaggle.com/slothkong/10-monkey-species. [Retrieved 4 Oct, 2021].
22. Castillo, O., Castro, J. R., Melin, P., & Rodriguez-Diaz, A. (2014). Application of interval type-2 fuzzy neural networks in non-linear identification and time series prediction. *Soft Computing, 18*(6), 1213–1224.
23. Castillo, O., & Melin, P. (2003). *Soft computing and fractal theory for intelligent manufacturing*. Heidelberg: Springer.

Relaxed Differential Evolution Algorithm

Prometeo Cortés-Antonio, Arturo Téllez-Velázquez, Raúl Cruz-Barbosa, and Oscar Castillo

1 Introduction

Metaheuristic optimization algorithms have been widely used in many real-life applications. Moreover, they have been used in the system modeling and optimization. For example, in the design of neural networks and fuzzy systems used for medical and financial diagnostic systems [1, 2], control and manipulation of robots or automobiles [3, 4, 5], where the optimization algorithms play an important role to fine-tune the parameters mainly in non-linear or complex problems.

There is a wide variety of metaheuristic optimization methods, and one way to classify them is according to their conceptual inspiration. Evolutionary Algorithms such as Genetic Algorithms, Evolutionary Strategies, and Differential Evolution Algorithms, among others [6] are based on the natural evolutionary process. Other kinds of metaheuristics have been inspired by intelligent behavior that exhibits physical and natural phenomena; to mention a few, we can state the Particle Swarm Optimization, Artificial Bee Colony algorithm, Whale Optimization Algorithm, and Marine Predator Algorithm [7].

P. Cortés-Antonio (✉) · O. Castillo
Tijuana Institute of Technology, Tijuana, México
e-mail: prometeo.cortes@tectijuana.edu.mx

O. Castillo
e-mail: ocastillo@tectijuana.mx

A. Téllez-Velázquez · R. Cruz-Barbosa
Applied Artificial Intelligence Laboratory, Computer Science Institute, Universidad, Tecnológica de la Mixteca, Huajuapan de León 69000, Oaxaca, México
e-mail: atellezv@mixteco.utm.mx

R. Cruz-Barbosa
e-mail: rcruz@mixteco.utm.mx

O. Castillo and P. Melin (eds.), *Hybrid Intelligent Systems Based on Extensions of Fuzzy Logic, Neural Networks and Metaheuristics*, Studies in Computational Intelligence 1096,
https://doi.org/10.1007/978-3-031-28999-6_17

Among the Evolutionary Algorithms, the Differential Evolution Algorithm has shown to be simple and efficient in terms of convergence and certainty to get the optimal solutions when it is compared to other optimization techniques. Generally, it contains only two control parameters in its evolutionary process, the Cr parameter, to control the Crossover mechanism, and the F parameter to scale the difference between two solutions of the current population.

Due to the simplicity of the DEA, a number of variants of this algorithm have been proposed with the aim of getting more robustness while maintaining its simplicity. Reference [8] showed some variants defined for the generation of new solutions, called DE/x/y/z, to specify some alternatives such as, whether or not the best solution from the population will be used, the number of pairs vector differences and, the way to perform the Crossover operation respectively. A comparative analysis of these variants is presented in [9].

Other studies about the DEA focused on some combinations of the most prominent DE/x/y/z versions [10]; for example in [11] are randomly selected some versions or, [12] uses a probabilistic approach for the selection of versions, the use of external files [13]. In [14], a comparative analysis of some of these approaches was performed. On the other hand, there are studies where the control parameters Cr and F are adapted dynamically during the evolutionary process, e.g. [15] proposes dynamic parameters based on fuzzy systems to handle uncertainty and robustness.

This paper proposes a new variant of the DEA focused on relaxing the condition for the individuals' selections in the creation of mutated vectors, which states that the selected individuals are different from each other, which implicitly involves a sequential process. By relaxing (delete) the selection condition of the DEA, it can be supposed that the convergence rate of the algorithm may be slower. Therefore in this research work, it is proposed to carry out an experimental analysis that shows the execution time and the optimal solution of relaxed DEA compared to DEA without relaxing.

Although the DEA provides good convergence and efficiency compared to other optimization algorithms, there are research works focused on reducing its execution time, by using either parallel or distributed computing techniques [16], and who could benefit from the proposed modification. Some implementations of the DEA using Graphical Processing Unit (GPUs) can be found in the literature. For instance [17] a simple implementation of DEA in GPU using CUDA-C is presented, while in [18] a design is made to optimize memory and maximize the use of GPU resources. On the other hand, custom hardware designs of the DEA on Field-Programmable Gate Arrays (FPGA) have been developed, i.e., an architecture for a floating-point numerical representation was designed [19], whereas other FPGA designs are focused on implementations with low resource consumption using micro or compact versions of the DEA [20, 21]. Additionally, there are research works for implementing DEA, which promote heterogeneous computing systems to exploit the capabilities of different kinds of processors and work in a distributed way [22].

Paper structure describes the underlying search strategy of the Differential Evolution Algorithm and details the proposed variant in Sect. 2. Section 3 presents the case study for evaluating the proposed version of the DEA. Section 4 shows the results of the DEA when it is evaluated for different dimensionality of the benchmark, and it presents a comparative analysis using Hypothesis Testing. Finally, this article presents the conclusions and future work of this research in Sect. 5.

2 Differential Evolution Algorithm and Proposed Version

The DEA belongs to the wide range of algorithms and methods of optimization and direct search approaches, which can be classified as metaheuristics or evolutionary algorithms due to it is based on the postulates of biological evolution. Among the advantages that the DEA exhibits when it is compared with other metaheuristics are: it is simple from a computational perspective; it contains only two configuration parameters: F, Cr and; it has been shown to have faster convergence to the solution for many situations at different applications.

The DEA search strategy for finding optimal solutions can be summarized by three simple subprocess: Mutation, Crossover and Selection, which are repeated during an evolution process for each D-dimensional individual for a given population with a predefined size.

Mutation operation requires the selection of three individuals from the current population, and they should meet the condition that says; they should be mutually different and different from the individual to be evolved. Equation (1) indicates the mutation using three solution vectors.

$$v = x_{r_1} + F(x_{r_2} - x_{r_3}) \tag{1}$$

where F is recommended to be in [0, 2], and it is used as the attenuator/amplifier of the differential term.

The Crossover process uses the vector generated from Equation (1) and the individual that is intended to evolve, i.e. x_i, for generating a temporal solution with the components of these two vectors according to independent binomial experiments of selection. Equation (2) shows the crossover operation.

$$u_j = \begin{cases} v_j & randb(j) \leq Crorjrand \\ x_{j,i} & otherwise \end{cases} \tag{2}$$

where Cr belong to [0,1], and his value is predetermined by the user, whereas *jrand* is an integer random number that belongs to [1:D] to ensure that the u vector contains at least one component of v.

Finally, the Selection operation decides if the vector generated from the Mutation and Crossover operations replaces the individual x_i, according to the cost generated by the fitness function.

The pseudocode, in Fig. 1, shows the whole process of the Differential Evolution Algorithm, which can be implemented in different programming languages. In

```
1 Set up parameters: NP, Cr, F; and stop conditions: Gmax, ErrorMin
2 Create an randomly initial population, P, with NP individuals, x_i, each x_i has D-dimensions
3 Compute the fitness, f(x_i), for each individual of the population, and
4 find the best individual x_best
5 while stop condition is not satisfied
6     for each individual x_i
7         select randomly three individuals from P with condition: x_r1 ≠ x_r2 ≠ x_r3 ≠ x_i
8         generate a mutated vector: v = x_r1 + F (x_r2 − x_r3)
9         generate a default element: jrand = randInt[1: D]
10        for each dimension of the test vector do
11            if (rand[0,1] > CR or dimension ≠ jrand) then
12                v_d = x_i,d
13            end if
14        end for
15            if (f(v) < f(x_i)) then
16            Replace x_i, f(x_i) by v and f(v) respectively end if
17    end for
18 end while
```

Fig. 1 Pseudocode of differential evolution algorithm

order to make more computational efficient the DEA, the pseudocode contains small variations of the above-described process, but the three operations remain the same. Code lines 7–8 perform the Mutation operation, lines 9–14 perform the Crossover operation, whereas lines 15 and 16 perform the Selection.

The DEA is very flexible and many variants have been proposed, the described version above is one of the most efficient and used in the literature. The name and notation to distinguish it from other variants is DE/rand/1/bin.

2.1 Relaxed Differential Evolution Algorithm

One of the main advantages offered by Optimization Algorithms based on heuristics is the high parallelization of the methods and operations performed by them. Therefore, many current applications that involve the optimization of high-dimensional problems and/or large amounts of data are developed with parallel processing designs.

Under this direction, in this paper, we propose an improvement to the DEA that consists of eliminating the condition that dictates that the individuals selected in the Mutation operation be mutually different (see line 7 of Fig. 1). Since this condition involves a purely sequential process. Note that this condition is required to perform the Mutation operation and it has nothing to do with the Selection operation. The proposed version of the DEA may be referred to as relaxed DEA, rDEA for short.

The modification to the DEA shows clearly that the processing of the Mutation operation will be faster and it leads to the whole processing time of the DEA being less than the original DEA, when they are executed with the same number of generations. However, this modification makes us suspect that the performance of the convergence rate will be slower.

For this reason, the rest of this article presents a list of mathematical functions that are used as a case study to measure the performance of the rDEA compared to the original DEA. It should be noted that the implementation of parallel processing techniques for the execution of the DEA is outside the scope of this analysis, and therefore the performed tests are over sequential executions according to the pseudocode in Fig. 1. The following Section presents a set of mathematical functions that are used as a case study to measure the performance of the algorithm.

3 Case Study: Mathematical Benchmark Functions

Comparisons to measure the performance of the Optimization Algorithms can be carried out using different applications, where a model of some system should be optimized. However, one of the common comparisons to evaluate algorithms is through a set of mathematical functions, also called benchmark, in which each one presents a different complexity. In the literature, you can find some different proposals of benchmarks that have been used for different Competitions, Special Sessions, and scientific papers for the evaluation of Optimization Algorithms. The analysis of the proposed algorithm uses part of the set of functions presented in the Test Suite for the Special Issue of Soft Computing on Scalability of Evolutionary Algorithms and other Metaheuristics for Large Scale Continuous Optimization Problems [23]. Table 1 presents the set of functions used in this research and shows the name function, mathematical equation, and the range of the D-variables. The parameter D indicates that the functions dimensionality is set to a desired value.

4 Results and Comparative Analysis

The study carried out in this article only considers the comparison of the conventional DEA and the rDEA based on the DEA/rand/1/bin variant. The reasons for carrying out this simple analysis are: (1) because different articles have compared the performance between different variants DEA or the DEA against other metaheuristics, showing DEA/rand/1/bin is one of the variants that have shown the best performance, and because it is also the most used variant in system optimization applications; and (2) because the authors prefer to carry out a simple and clear analysis of problems of different dimensionality.

To measure and compare the performance of the DEA and rDEA algorithms, they are evaluated using the benchmark shown in Table 1 for three cases, where $D = 10$, 30 and 50.

Table 1 Benchmark mathematical functions

Fun	Name	Definition	Range
F1	sphere	$\sum_{i=1}^{D} {x_i}^2$	$x_i \in [-100, 100]$
F2	Schwefel 2.21	$max_i\{\|x_i\|, 1 \le i \le D\}$	$x_i \in [-100, 100]$
F3	Rosenbrock	$\sum_{i=1}^{D-1}\left(100({x_i}^2 + x_i)^2 + (x_i - 1)^2\right)$	$x_i \in [-100, 100]$
F4	Rastigirin	$\sum_{i=1}^{D}({x_i}^2 - 10\cos(2\pi x_i) + 10)$	$x_i \in [-5, 5]$
F5	Greiwank	$\sum_{i=1}^{D} \frac{{x_i}^2}{4000} - \prod_{i=1}^{D} \cos\frac{x_i}{\sqrt{i}} + 1$	$x_i \in [-600, 600]$
F6	Ackley	$-20\exp\left(-0.2\sqrt{\frac{\sum_{i=1}^{D} {x_i}^2}{D}}\right) - \exp\left(\frac{1}{D}\sum_{i=1}^{D}\cos(2\pi x_i)\right) + 20$	$x_i \in [-32, 32]$
F7	Schwefel 2.22	$\sum_{i=1}^{D}\|x_i\| - \prod_{i=1}^{D}\|x_i\|$	$x_i \in [-10, 10]$
F8	Schwefel 1.2	$\sum_{i=1}^{D}\left(\sum_{j=1}^{i} x_j\right)^2$	$x_i \in [-65.536, 65.536]$
F9	Extended	$\sum_{i=1}^{D-1} f_{10}(x_i, x_{i+1}) + f_{10}(x_D, x_1)$ $f_{10} = (x^2 + y^2)^{0.25}\left(\sin^2\left(50(x^2 + y^2)^{0.1}\right) + 1\right)$	$x_i \in [-100, 100]$
F10	Bohachevsky	$\sum_{i=1}^{D-1}({x_i}^2 + x_{i+1}^2 - 0.3\cos(3\pi x_i) - 0.4\cos(4\pi x_{i+1}) + 0.7)$	$x_i \in [-15, 15]$
F11	Schaffer	$\sum_{i=1}^{D-1}({x_i}^2 + x_{i+1}^2)^{0.25}\left(\sin^2\left(50({x_i}^2 + x_{i+1}^2)^{0.1}\right) + 1\right)$	$x_i \in [-100, 100]$

Note That the functions F3–F6 are multimodal, whereas the remainder are monomodal

4.1 Configuration and Control parameters values

All tests performed in this analysis use the same values for the control parameters F and CR, independently of the dimensionality of the functions, with F = Cr = 0.7. This values are taken from results showed in [9, 10]. For the configuration parameters, the population size, NP = 50, the number of generations, Gmax = 10*D. Note, in order to perform a time and convergence comparison, the only stop condition of the DEA is the total number of evolutions, Gmax.

4.2 Results and Analysis

In order to perform a comparative analysis, the DEA and rDEA algorithms were executed 30 times. Table 2 shows the results obtained from the execution of the 11 mathematical functions presented in Table 1 when its dimensionality equals 10. Table 2 shows in columns 2–4 the corresponding values to the average value of the final solution, the standard deviation and the average execution time, generated by executing the DEA according to the objective function indicated in each row. Whereas columns 5–7 correspond to the rDEA values.

Also from Table 2 it can be seen, in all functions the average execution time of each function was minor when executing the rDEA (comparison of columns 4 and 7), a result that authors expected, and values indicate that by removing the sequential process in the Mutation operator, the execution time is reduced by at least 10%. Regarding to convergence, it is observed that the mean of the solution of each function was also better in the case of rDEA, i.e., it was closer to the local minimum (except in F10, where both algorithms obtained the global minimum), which is the

Table 2 Results of DEA and rDEA runs in the benchmark with functions of 10 variables

	DEA			rDEA		
Fun	average	SD	Time	average	SD	Time
F1	2.39×10^{-23}	1.73×10^{-23}	1.12	2.31×10^{-29}	2.37×10^{-29}	0.95
F2	3.15×10^{-6}	1.17×10^{-6}	1.16	1.08×10^{-8}	7.90×10^{-9}	0.89
F3	1.05×10^{-1}	6.28×10^{-2}	1.10	1.35×10^{-2}	1.23×10^{-2}	0.87
F4	1.69×10^{1}	3.78	1.09	8.22×10^{-1}	1.28	0.90
F5	2.95×10^{-1}	4.50×10^{-2}	1.15	9.03×10^{-3}	5.90×10^{-3}	0.92
F6	1.47×10^{-11}	8.39×10^{-12}	1.30	1.05×10^{-11}	6.76×10^{-12}	0.84
F7	4.22×10^{-13}	3.86×10^{-13}	1.20	2.71×10^{-16}	1.14×10^{-16}	0.93
F8	8.36×10^{-23}	9.57×10^{-23}	1.22	1.15×10^{-28}	7.81×10^{-29}	0.93
F9	2.05×10^{-2}	7.90×10^{-3}	1.56	1.07×10^{-4}	4.86×10^{-5}	1.10
F10	0	0	1.29	0	0	0.93
F11	1.95×10^{-2}	8.60×10^{-3}	1.46	9.27×10^{-5}	3.86×10^{-5}	1.07

opposite to we initially belief, that the convergence rate of rDEA could be slower than the original DEA.

Format in Tables 3 and 4 is the same as shown in Table 2, but now the results are shown when using the Benchmark for the cases where D=30, 50 are shown. Here it can see that the greater the dimensionality of the problem, the rDEA variant has the possibility to get a greater difference in the execution time with respect to DEA while maintains the relationship that the execution time decreases at least 10%.

Regarding to convergence, it is also observed that, for different dimensionality, the rDEA was closer to the local minimum for all cases. These final results were not

Table 3 Results of DEA and rDEA runs in the benchmark with functions of 30 variables

Algorithm	DEA			rDEA		
Fun	Average	SD	Time	Average	SD	Time
F1	4.74×10^{-6}	1.66×10^{-6}	3.35	3.09×10^{-26}	2.25E-26	2.72
F2	9.53	4.42	3.38	1.58×10^{-2}	1.11E-02	2.75
F3	4.72×10^{1}	1.11×10^{1}	3.35	2.67×10^{1}	2.78×10^{1}	2.72
F4	1.82×10^{2}	7.26	3.46	2.14×10^{1}	4.44	2.81
F5	4.88×10^{-2}	1.09×10^{-1}	3.84	0.00	0.00	2.92
F6	2.01×10^{-11}	9.43×10^{-12}	2.93	1.52×10^{-11}	1.09×10^{-11}	2.35
F7	1.36×10^{-3}	6.66×10^{-4}	3.58	2.11×10^{-14}	1.07×10^{-14}	2.87
F8	2.09×10^{-5}	1.02×10^{-5}	3.75	3.09×10^{-25}	2.21×10^{-25}	2.90
F9	2.92×10^{1}	1.93	5.36	1.08×10^{-3}	2.94×10^{-4}	4.64
F10	1.18×10^{-4}	8.30×10^{-5}	4.09	0.00	0.00	2.97
F11	2.57×10^{1}	2.50	5.38	1.36×10^{-3}	6.36×10^{-4}	4.15

Table 4 Results of DEA and rDEA runs in the benchmark with functions of 50 variables

	DEA			rDEA		
Fun	average	SD	Time	average	SD	Time
F1	2.78×10^{-1}	1.12×10^{-1}	5.80	4.96×10^{-29}	4.21×10^{-29}	4.74
F2	8.05×10^{1}	5.19	5.83	9.50	3.81	4.78
F3	3.30×10^{4}	1.67×10^{4}	5.79	3.35×10^{1}	4.93	4.75
F4	4.18×10^{2}	1.67×10^{1}	6.15	5.00×10^{1}	6.96	4.99
F5	3.17×10^{-1}	6.30×10^{-2}	9.19	0	0	5.15
F6	3.41×10^{-11}	1.27×10^{-11}	4.15	2.70×10^{-11}	1.63×10^{-11}	4.02
F7	7.28×10^{-1}	2.52×10^{-1}	6.18	2.11×10^{-16}	1.11×10^{-16}	4.90
F8	2.13	7.15×10^{-1}	6.67	1.15×10^{-28}	5.18×10^{-29}	5.33
F9	1.53×10^{2}	1.74×10^{1}	10.70	2.08×10^{-2}	1.42×10^{-2}	9.03
F10	3.61	8.62×10^{-1}	7.48	8.26×10^{-2}	1.85×10^{-1}	5.10
F11	1.46×10^{2}	9.19	10.19	5.02×10^{-2}	5.41×10^{-2}	8.65

Table 5 Hypothesis test for comparison of DEA and rDEA

Fun	D = 10	D = 30	D = 50
F1	−7.57	−15.64	−13.63
F2	−14.71	−11.79	−60.32
F3	−7.85	−3.75	−10.81
F4	−22.06	−103.33	−111.21
F5	−34.46	−2.45	−27.60
F6	−2.18	−1.84	−1.88
F7	−5.99	−11.21	−15.81
F8	−4.79	−11.24	−16.36
F9	−14.14	−83.06	−48.35
F10	–	−7.79	−21.91
F11	−12.39	−56.19	−87.21

expected in an initial hypothesis and they lead to the conclusion that: the proposed variant is not only more efficient from the computational point of view, but also that it converges to the global solution more efficiently.

The authors interpret that the reason why the rDEA converges faster than the DEA is due to the fact that of not fulfilling the condition of mutually different individuals, the mutation can be considered as an intermediate variant between the DEA/rand/1/ and DEA/best/1/bin, and it helps to approach to the optimal solution faster.

Finally, in order to give a mathematical support to the statement that the rDEA converges faster than DEA to the global solution for all cases of this study, Table 5 shows the z values of the hypothesis test defined as: $H_0 : \mu_{DEA} \leq \mu_{rDEA}$ and $H_1 : \mu_{DEA} > \mu_{rDEA}$, given a confidence level of 95% and therefore $z_c = -1.645$. As can be seen from the z values in Table 5, in all cases, the results are located in the rejection region of the null hypothesis.

5 Conclusions and Future Work

This article presents a new variant to the DEA, which is based on the modification in Mutation process of the algorithm that involves a sequential process. To test the performance of the proposed, the DEA y rDEA were tested in a benchmark for functions with 10, 30 and 50 dimensions. The rDEA variant showed not only a decrease in the execution time of the DEA but also a higher convergence rate than DEA to reach the global solution.

With the analysis presented, doors can be opened to different studies of the algorithm: (1) Parallel and/or Distributed Computing Designs that obtain benefits from not having to carry out the sequential selection of the population in the Mutation operation. (2) Extend the study of rDEA by the modification of different variants of DEA and extend the performance analysis against other metaheuristics. (3) Use rDEA for optimization applications of parametric models such as Fuzzy Systems and Neural Networks among others.

References

1. Nagib, A., Saeed, M., El-Feky, S., & Mohamed, A. (2022). Hyperparameters optimization of deep convolutional neural network for detecting COVID-19 using differential evolution. In Decision sciences for COVID-19 (pp. 305–325). Berlin: Springer.
2. Hu, Z. (2020). Statistical optimization of supply chain financial credit based on deep learning and fuzzy algorithm. *Journal of Intelligent & Fuzzy Systems, 38*, 7191–7202.
3. Valdez, F., Castillo, O., & Peraza, C. (2020). Fuzzy logic in dynamic parameter adaptation of harmony search optimization for benchmark functions and fuzzy controllers. *International Journal of Fuzzy Systems, 22*, 1198–1211.
4. Cuevas, F., Castillo, O., & Cortés, P. (2021). Optimal setting of membership functions for interval type-2 fuzzy tracking controllers using a shark smell metaheuristic algorithm. *International Journal of Fuzzy Systems,* 1–24.
5. Bejar, E., & Morán, A. (2018). Backing up control of a self-driving truck-trailer vehicle with deep reinforcement learning and fuzzy logic. In *2018 IEEE International Symposium on Signal Processing and Information Technology (ISSPIT)* (pp. 202–207).
6. Greiner, D., Periaux, J., Quagliarella, D., Magalhaes-Mendes, J., & Galván, B. (2018). Evolutionary algorithms and metaheuristics: applications in engineering design and optimization. Hindawi
7. Osaba, E., Villar-Rodriguez, E., Del Ser, J., Nebro, A., Molina, D., LaTorre, A., Suganthan, P., Coello, C., & Herrera, F. (2021). A tutorial on the design, experimentation and application of metaheuristic algorithms to real-world optimization problems. *Swarm and Evolutionary Computation, 64*, 100888.
8. Storn, R., & Price, K. (1997). Differential evolution–a simple and efficient heuristic for global optimization over continuous spaces. *Journal of global optimization, 11*, 341–359.
9. Mezura-Montes, E., Velázquez-Reyes, J., & Coello Coello, C. (2006). A comparative study of differential evolution variants for global optimization. In *Proceedings of the 8th annual conference on Genetic and evolutionary computation* (pp. 485–492).
10. Wang, Y., Cai, Z., & Zhang, Q. (2011). Differential evolution with composite trial vector generation strategies and control parameters. *IEEE Transactions on Evolutionary Computation, 15*, 55–66.
11. Brest, J., Bošković, B., Greiner, S., Žumer, V., & Maučec, M. (2007). Performance comparison of self-adaptive and adaptive differential evolution algorithms. *Soft Computing, 11*, 617–629.
12. Qin, A., Li, X., Pan, H., & Xia, S. (2013). Investigation of self-adaptive differential evolution on the CEC-2013 real-parameter single-objective optimization testbed. In *2013 IEEE Congress on Evolutionary Computation* (pp. 1107–1114)
13. Zhang, J., & Sanderson, A. (2009). JADE: adaptive differential evolution with optional external archive. *IEEE Transactions on Evolutionary Computation, 13*, 945–958.
14. Georgioudakis, M., & Plevris, V. (2020). A comparative study of differential evolution variants in constrained structural optimization. *Frontiers in Built Environment, 6*, 102.
15. Ochoa, P., Castillo, O., Soria, J., & Cortés-Antonio, P. (2018). Differential evolution algorithm using a dynamic crossover parameter with high-speed interval type 2 fuzzy system. In *Mexican International Conference on Artificial Intelligence* (pp. 369–378).
16. Park, S., Shires, D., & Henz, B. (2008). Coprocessor computing with FPGA and GPU. In *2008 DoD HPCMP Users Group Conference* (pp. 366–370).
17. Veronese, L., & Krohling, R. (2010). Differential evolution algorithm on the GPU with C-CUDA. In *IEEE Congress on Evolutionary Computation* (pp. 1–7).

18. Qin, A., Raimondo, F., Forbes, F., & Ong, Y. (2012). An improved CUDA-based implementation of differential evolution on GPU (pp. 991–998).
19. Cortés-Antonio, P., Rangel-González, J., Villa-Vargas, L., Ramírez-Salinas, M., Molina-Lozano, H., & Batyrshin, I. (2014). Design and implementation of differential evolution algorithm on FPGA for double-precision floating-point representation. *Acta Polytechnica Hungarica, 11*, 139–153.
20. León-Javier, A., Moreno-Armendáriz, M., & Cruz-Cortés, N. (2009). Designing a compact genetic algorithm with minimal fpga resources. In *Advances in Computational Intelligence* (pp. 349–357). Berlin: Springer.
21. Hidalgo, J., Baraglia, R., Perego, R., Lanchares, J., & Tirado, F. (2001). A parallel compact genetic algorithm for multi-FPGA partitioning. In *Proceedings Ninth Euromicro Workshop on Parallel and Distributed Processing* (pp. 113–120).
22. Mancilla, A., Castillo, O., & Valdez, M. (2022). Mixing population-based metaheuristics: An approach based on a distributed-queue for the optimal design of fuzzy controllers. In *International Conference on Intelligent and Fuzzy Systems* (pp. 839–846).
23. Lozano, M., Molina, D., & Herrera, F. (2011). Editorial scalability of evolutionary algorithms and other metaheuristics for large-scale continuous optimization problems. *Soft Computing, 15*, 2085–2087.

Optimization: Theory and Applications

Automatic Characterization of Time Series Using Metaheuristic Algorithms for Epidemics Spread Analysis

Valentín Calzada-Ledesma and Andrés Espinal

1 Introduction

Throughout history, humanity has faced different epidemics that have claimed the lives of millions of people. For example, in the fourteenth century, the "Black Death" spread through Europe and Africa, causing an estimated 75–200 million deaths [1]. In January 2020, a new virus called Coronavirus Type 2 Caused by Severe Acute Respiratory Syndrome (SARS-CoV2) arrived in Mexico, causing Coronavirus disease 2019 (COVID-19), which, like Influenza and Dengue, it became a disease of national concern. Currently, these epidemics have become seasonal, there are vaccines available and there is a human culture regarding certain health protocols. However, it is important not to lower our guard and continue developing new research that provides methodologies capable of analyzing the information available on epidemics in order to be better prepared to face them in the future.

In the state of the art, different epidemiological approaches have been proposed to understand the epidemiological behavior of a disease; one of them is the SIR model, which tries to describe epidemics through the analysis of three main factors: the Susceptible population, the Infected population and the Recovered population [7, 10]. However, it is well known that highly specialized knowledge is needed to properly consider and set all factors and parameters of that model [4, 15, 17]. This is why several researchers outside the area of epidemiology have chosen to use non-deterministic artificial intelligence tools to provide a quick and approximate

V. Calzada-Ledesma (✉)
Tecnológico Nacional de México/Instituto Tecnológico Superior de Purísima del Rincón, Purísima del Rincón, Guanajuato, México
e-mail: valentin.cl@purisima.tecnm.mx

A. Espinal
Departamento de Estudios Organizacionales, DCEA – Universidad de Guanajuato, Guanajuato, Mexico

O. Castillo and P. Melin (eds.), *Hybrid Intelligent Systems Based on Extensions of Fuzzy Logic, Neural Networks and Metaheuristics*, Studies in Computational Intelligence 1096,
https://doi.org/10.1007/978-3-031-28999-6_18

insight into the behavior of an epidemic since these types of approaches have shown adequate performance to model the behavior of diseases such as Ebola, Zika and Dengue [2, 5, 11, 12].

Characterization is conceptualized as a way of representing something through a model from which features can be derived. In this article, we address the problem of characterizing the spread of three epidemics: COVID-19, Influenza, and Dengue, considering Mexico as a test case, since there is historical data on a sentinel epidemiological model, which documents probable cases, confirmed cases, and deaths for each of the epidemics. To achieve this, we propose the use of META-COVID19 [6], which is an algorithm proposed by our research team initially conceived to describe COVID-19 time series, however, it can be used with other epidemics, the only restriction is that the input data be time series.

META-COVID19 implements computational optimization techniques to automatically fit an epidemic time series using orthogonal polynomials, without prior knowledge of the data and without involving a human expert. In general, the system input is a time series of data collected on some epidemic. Subsequently, an iterative process guided by two Boltzmann-based selection operators is started, in such a way that the system automatically searches for an explicit mathematical model in a characterization space based on the family of orthogonal Jacobi polynomials, which fits the behavior of the time series. Finally, these models can be analyzed to obtain mathematical and statistical features of epidemic behavior.

The mathematical models found in this research work can be a starting point for other researchers interested in studying the spread of different epidemics. This is highly relevant since it is possible that new diseases will soon reach the world, so it is important to have methodologies capable of providing useful information on the behavior of diseases, which will allow the design of better strategies to face them in the future.

2 Related Work

Nowadays, there is a growing trend in the use of artificial intelligence to address problems related to the health sector, focusing computer science research efforts on real-time forecasting, computer simulation of the diseases spread, and regions of study of possible outbreaks of new viruses [14, 18, 22, 23].

In [8], an in-depth discussion was made on why various methods that attempt to forecast diseases have not worked well. Likewise, the journal Nature, published a manifesto to avoid bad practices of prediction approaches [19]. It is for these reasons that in [6], the META-COVID 19 algorithm was proposed to address the problem of characterization instead of prediction/forecasting of epidemics.

As stated above, to characterize the spread of the epidemic, the time series must be represented mathematically by polynomials, these must be compact, robust, and easy to handle. However, there is a great variety of polynomial families in the literature, and choosing the appropriate family, as well as its respective parameter settings, is an extremely difficult task. It is for this reason that META-COVID19 focuses on the use of the Jacobi family of polynomials since several polynomial families, such as Gegenbauer, Legendre, Zernike, and Chebyshev, are particular cases of Jacobi polynomials, thus solving most of the aforementioned problems. Likewise, the Jacobi polynomials have a large number of mathematical properties [21], and this is extremely important since from a single polynomial model it is possible to obtain multiple information that can have an important impact from the point of view of statistical inference in epidemiology [9].

In Mexico, one of the main deterministic models developed by the Centro de Investigación en Matemáticas (CIMAT) and the Universidad Autónoma de México (UNAM), to address problems related to COVID-19, is the AMMA model [3]; which is a SEIRD type dynamic model which considers people: Susceptible, Exposed, Infected, Recovered and Deaths, based on Bayesian inference of sequential data assimilation. Currently, this model allows estimating the increase or decrease in the number of cases of COVID-19, at the federal entity level and for the Metropolitan Area of the Valley of Mexico, based on the records of daily confirmed cases, accumulated deaths, and daily hospital demand (beds normal and intensive care units), published by the Ministry of Health.

From the mathematical characterization of diseases that is proposed in this work, the assumptions could be deepened to generate more information and support assertive decision-making with mathematical foundations, which could be combined and strengthened with other epidemiological models such as the AMMA, in favor of Mexican society.

3 Material and Methods

3.1 Current Context of Health in Mexico

The global health crisis unleashed at the end of 2019 has put the health sector at the center of political and social discussion around the world. With 5.5% of the national Gross Domestic Product (GDP) allocated to the health sector, Mexico is currently positioned as one of the countries of the Organization for Economic Cooperation and Development (OECD) that does not invest as much in this sector in proportion to its economics [13]. Of the total GDP in health registered in the country in 2020, one-fifth corresponds to hospital services (the segment with the highest percentage of spending) followed by outpatient medical services. Regarding the sectoral division, 38% of the value is attributed to the public sector, while the rest is shared between unpaid work and the private sector [13].

In 2019, acute respiratory infections were the most common type of illness in Mexico. That year, a total of close to 24 million cases of this type of infection was estimated in the country. Intestinal infections were the second most common type of illness in the Mexican population, with more than 5 million estimated cases.

Conjunctivitis and obesity are also among the ten most frequent diseases in Mexico [13].

Finally, the disease caused by the SARS-CoV-2 virus was the second most common cause. In 2020, with approximately 218,885 deaths, heart disease (excluding cardiac arrest) was the leading cause of death in Mexico. Influenza, and pneumonia were also among the ten leading causes of mortality in the country [13].

3.2 META-COVID19

This section describes in detail the elements of the META-COVID19 metaheuristic approach, whose general scheme is shown in Fig. 1.

In this methodology, a time series can be fitted through the linear combination of multiple Jacobi polynomials, as shown in Eq. 1:

$$F(t; x) = F(t; \alpha, \beta, d) = \sum_{j=1}^{d} a_j P_j^{\alpha,\beta}(t) \tag{1}$$

where $P_j^{\alpha,\beta}(t)$ is the real part of a Jacobi polynomial used to fit the time series, a_j are modulation coefficients calculated mathematically (for more details see [6]), and $t \in [-1, 1]$ is the domain of the time series. In this way, a characterization space is established (blue cube) restricted by the fundamental parameters $x = [\alpha, \beta, d]$ where $\alpha \in [-1, 1]$, $\beta > -1$, and $d > 1$ is the polynomial degree [6]. It is worth

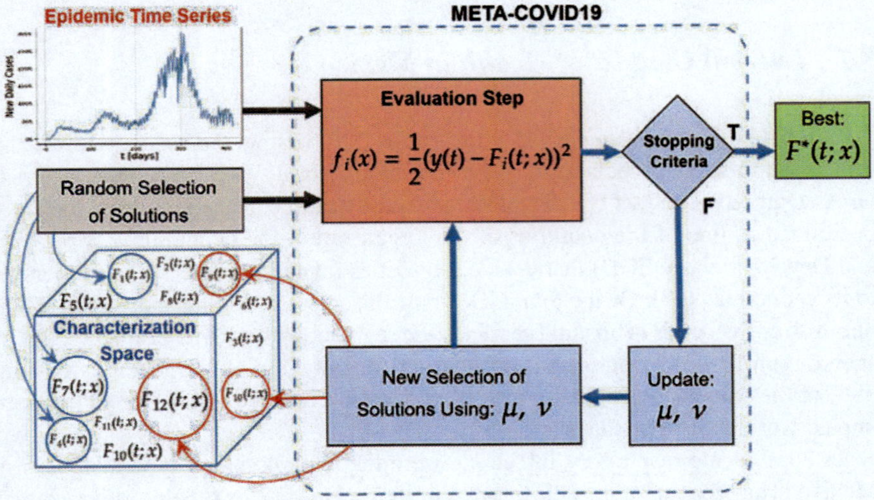

Fig. 1 META-COVID19 general scheme [6]

mentioning that in Eq. 1, which from now on will be called "solution", the fundamental parameters x are set with random numbers sampled from a distribution with μ and ν parameters in order to select a solution from the characterization space. Initially, a population J_T of n solutions is randomly set at iteration T using a uniform distribution. To find out how good the fit is for each solution $F_i(t; x)$ (where $i \in [1, n]$) regarding the epidemic time series $y(t)$, , an evaluation step that uses the mean square error (Eq. 2) as fitness function, is performed:

$$f_i(x) = \frac{1}{2}(y(t) - F_i(t; x))^2. \tag{2}$$

In case the established stopping criterion is not met, META-COVID19 uses two Boltzmann-based operators μ and υ to select a new population of solutions to guide the evolutionary process towards the best solution $F^*(t; x)$ that minimizes the fitness function. The parameter μ positions the population on a trajectory towards the optimum, and the parameter ν provides the necessary diversity to the population, so that the algorithm can perform a guided exploration in the characterization space. The selection operators are updated at each iteration using Eqs. 3 and 4 respectively, as shown below:

$$\mu \approx \frac{\frac{b-a}{m}\sum_{i=1}^{m}\left[\frac{1}{Z}g(f_i(x))\exp(\beta f_i(x))x_i + \log[g(f_i(x))]x_i + \beta f_i(x)x_i\right]}{1 + \frac{b-a}{m}\sum_{i=1}^{m}\left[\log[g(f_i(x))] + \beta f_i(x)\right]} \tag{3}$$

$$\nu \approx \frac{\sum_{i=1}^{m}\left[\frac{1}{Z}g(f_i(x))\exp(\beta f_i(x))(x_i - \mu)^2 + \log[g(f_i(x))](x_i - \mu)^2 + \beta f_i(x)(x_i - \mu)^2\right]}{\sum_{i=1}^{m}\left[\log[g(f_i(x))] + \beta f_i(x)\right]} \tag{4}$$

where x_i is the set of fundamental parameters of the i th solution, $g(\cdot)$ is a function that depends on the i th solution's performance $f_i(x)$ (Eq. 2), and $\beta = \frac{1}{f^*(x)}$; in the latter, $f^*(x)$ is the performance of the best solution found so far in the evolutionary process. Note that to perform this update, m solutions are needed, they must be the ones with the best performance at each iteration to guarantee the convergence of the algorithm, allowing the selection of new solutions (red circles) whose adjustment will be closer to the behavior of the analyzed time series. The algorithm of META-COVID19 is shown below.

Algorithm 1 META-COVID19 Pseudocode (De Anda-Suárez et al. 2022).

Require: n: population size, $y(t)$: time series.
Ensure: $F^*(t; x)$: Best solution.
$T \leftarrow 0$: Current iteration.
$J_T \leftarrow \text{initialPopulation}(n, \mu_0 = 0, \nu_0 = 1)$.
Evaluate$(J_T, y(t))$.
$F^*(t; x) \leftarrow \text{getBestSolution}(J_T)$.
while $f^*(x) \leq \epsilon$ **do**
    $S \leftarrow \text{getTopSolutions}(J_T, m)$.
    $\mu_{T+1}, \nu_{T+1} \leftarrow \text{updateSelectionOperators}(S)$.
    $J_{T+1} \leftarrow \text{NewPopulation}(n, \mu_{T+1}, \nu_{T+1})$.
    Evaluate$(J_{T+1}, y(t))$.
    $F^*(t; x) \leftarrow \text{getBestSolution}(J_T)$.
    $T \leftarrow T + 1$.
end while
return $F^*(t; x)$.

4 Experimental Setup

4.1 Datasets

In this work, we focus on the time series of confirmed positive cases of COVID-19, Influenza, and Dengue reported in Mexico. Open data available from the Mexican Ministry of Health [20] and Our World in Data [16] were used. The information is periodically updated on these platforms and includes data based on a sentinel epidemiological model.

However, this data was not in a suitable format for analysis by META-COVID19. Therefore, a data engineering process was necessary to format the information before carrying out the characterization proposed in this work. It is worth mentioning that the experimentation was carried out using the data available at the date of creation of this article (August 2022), but it can be applied with the data collected at any time. Figures 2, 3 and 4 graphically show the time series of the epidemics.

4.2 Parameter Settings

The settings of the META-COVID19 parameters are shown below, to carry out the corresponding experimentation. The population of solutions is set to $n = 50$. After a stage of meticulous experimentation, it was observed that to calculate the selection

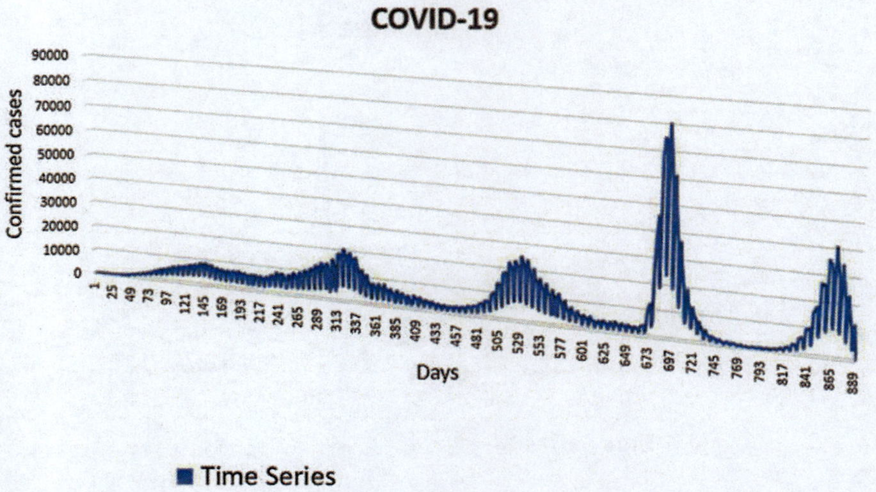

Fig. 2 COVID-19 confirmed cases time series

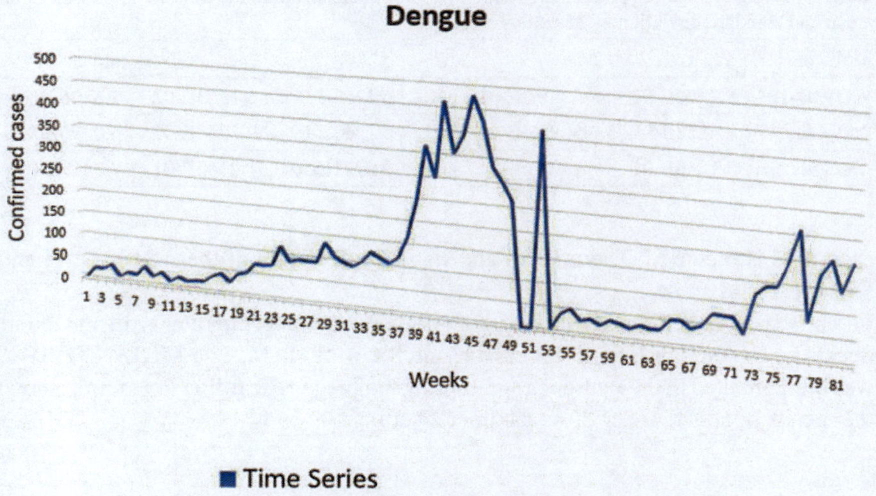

Fig. 3 Dengue confirmed cases time series

operators μ and ν, $m = 20$ (i.e., 40% of the population) top-ranked solutions allow conducting the search for the best polynomial in an adequate way. Using 100% of the population is possible, but this leads to higher computational cost. For the characterization space, as mentioned in Sect. 3.2, the domains are established as follows: $\alpha \in [-1, 1]$, $\beta \in [-1, 10]$, and $d \in [1, 60]$.

In every experiment, the convergence criterion is $\epsilon \leq 1 \times 10^{-3}$, and the number of iterations is set to 50. Finally, to ensure the consistency of the experimental results, each experiment is executed 31 times independently, and the best solution found

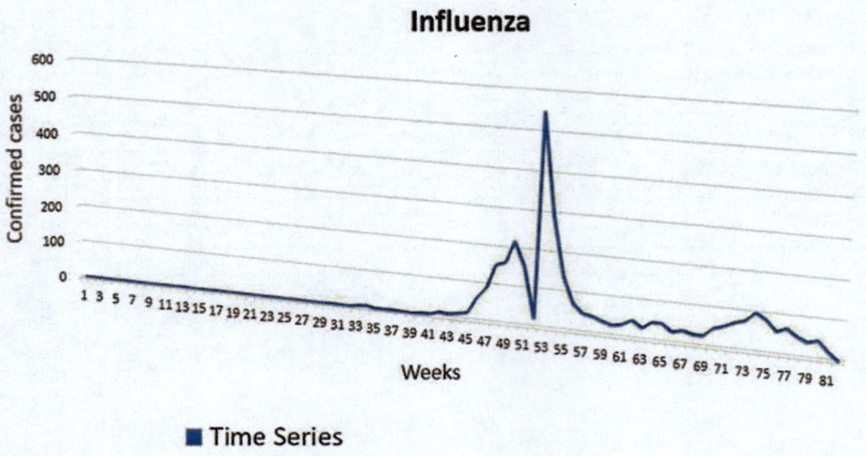

Fig. 4 Influenza confirmed cases time series

Table 1 Best fundamental parameters found to approximate an epidemic time series as well as the mean and standard deviation of 31 experiments

Epidemic	α	β	d	$f^*(x)$	Median	Std
COVID-19	8.1260E-01	−5.4060E-01	60	6.1263E-02	6.1503E-02	7.6706E-05
Influenza	8.4838E-01	−5.6994E-01	59	6.7363E-02	9.5743E-02	6.0096E-03
Dengue	5.1360E-01	−1.7626E-01	60	9.0721E-02	1.2345E-01	7.3413E-03

each time is recorded. The median and standard deviation of these 31 experiments are reported in Table 1.

The experimentation was carried out on a conventional computer with an Ubuntu operating system, an Intel i5 processor, and 8 GB of RAM. The META-COVID19 was implemented in the Python programming language. The following section shows the results obtained under these parameter settings.

4.3 Results

Table 1 shows the best fundamental parameters for the linear combination of Jacobi Polynomials (i.e., solution), as well as their corresponding fitness value $f^*(x)$ for each epidemic. In addition, the Median and the Standard deviation (Std) of the 31 experiments are shown. Full results are reported in Appendix A.

As can be seen, for these experiments, the best polynomial found was for COVID-10 with a median fitness value of 6.1503E-02 followed by Influenza and Dengue. This may be due to the amount of data available, since the COVID-19 time series is made up of daily data, on the contrary, Influenza and Dengue time series by weekly data.

Table 2 reports the coefficients a_j of the linear combination of Jacobi polynomials to fit the epidemics time series shown in Figs. 2, 3, and 4; the fundamental parameters reported in Table 1 were also used to acomplish this task, according to Eq. 1.

To graphically exemplify the performance of the best solutions found by META-COVID19, we show in Figs. 5, 6, and 7, the fit achieved using the parameters reported in Table 1.

Table 2 Coefficients a_j of the solutions to fit epidemic time series

a_j	COVID-19	Influenza	Dengue	a_j	COVID-19	Influenza	Dengue
1	0.05	442.39	−81.56	31	−0.10	−1728.71	98.60
2	0.08	−757.83	268.88	32	-0.05	1245.78	−1065.05
3	−0.04	989.35	−188.22	33	0.02	−1539.89	−19.77
4	0.02	−1024.16	578.69	34	0.07	918.70	−1020.45
5	−0.02	1219.38	−190.61	35	0.06	−1292.86	−101.56
6	−0.03	−1031.10	862.85	36	−0.01	622.54	−897.14
7	0.04	1270.96	−122.65	37	−0.07	−1019.70	−143.32
8	0.05	−848.48	1092.12	38	−0.05	381.65	−725.50
9	0.05	1171.67	−2.01	39	0.01	−751.89	−149.19
10	−0.01	−515.50	1242.54	40	0.03	206.08	−537.30
11	0.03	942.20	150.02	41	0.05	−513.49	−129.26
12	−0.02	−74.94	1293.69	42	0.04	93.40	−361.87
13	−0.15	606.03	312.89	43	−0.03	−321.31	−96.55
14	−0.02	424.07	1237.21	44	−0.08	30.94	−218.18
15	0.13	194.47	463.66	45	−0.04	−180.56	−61.90
16	0.07	928.98	1072.11	46	0.04	3.43	−115.07
17	−0.05	−256.28	582.74	47	0.07	−88.56	−33.71
18	−0.09	1386.38	812.08	48	0.02	−4.68	−50.88
19	−0.07	−707.39	653.21	49	−0.03	−35.98	−14.61
20	0.01	1750.38	479.79	50	−0.05	−4.58	−17.09
21	0.09	−1119.57	666.26	51	−0.05	−10.57	−4.43
22	0.07	1986.13	109.22	52	0.02	−2.22	−3.42
23	−0.01	−1459.06	620.16	53	0.07	−1.35	−0.55
24	−0.04	2075.71	−261.55	54	0.02	−1.03	0.51
25	−0.03	−1698.94	523.04	55	−0.02	0.64	0.53
26	−0.04	2020.39	−592.54	56	−0.02	−0.15	0.40
27	−0.01	−1823.63	389.88	57	−0.05	0.38	0.20
28	0.09	1839.40	−850.19	58	0.02	−0.24	0.06
29	0.09	−1830.40	241.28	59	0.02	−0.09	0.10
30	−0.04	1567.01	−1010.61	60	0.01	–	–

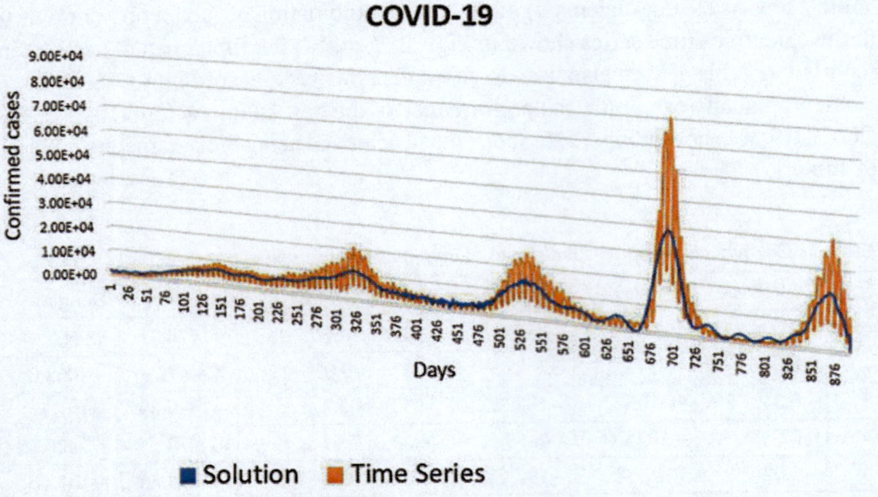

Fig. 5 COVID-19 time series fitting achieved by the best solution

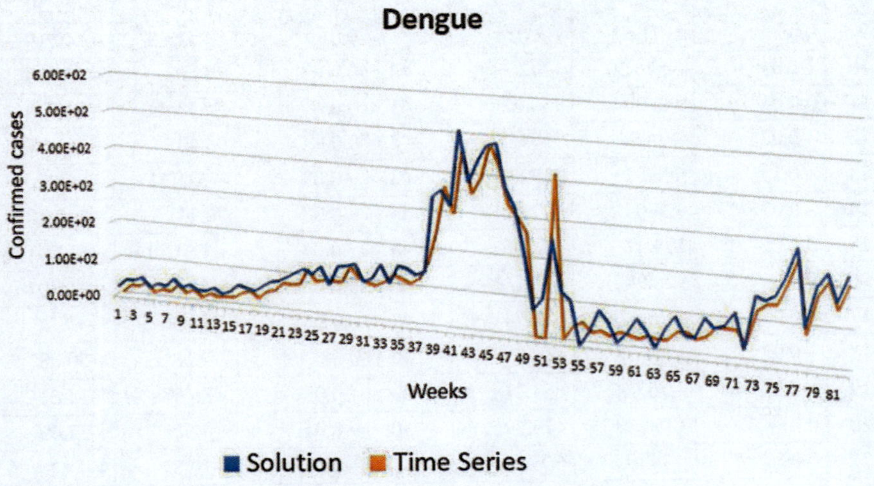

Fig. 6 Dengue time series fitting achieved by the best solution

In general, the graphical results show that META-COVID19 is capable of achieving a fit in each of the time series used in this study. It is worth mentioning that, as stated in Sect. 4.1, the information provided by the health department in Mexico directly influences the result of the approximation.

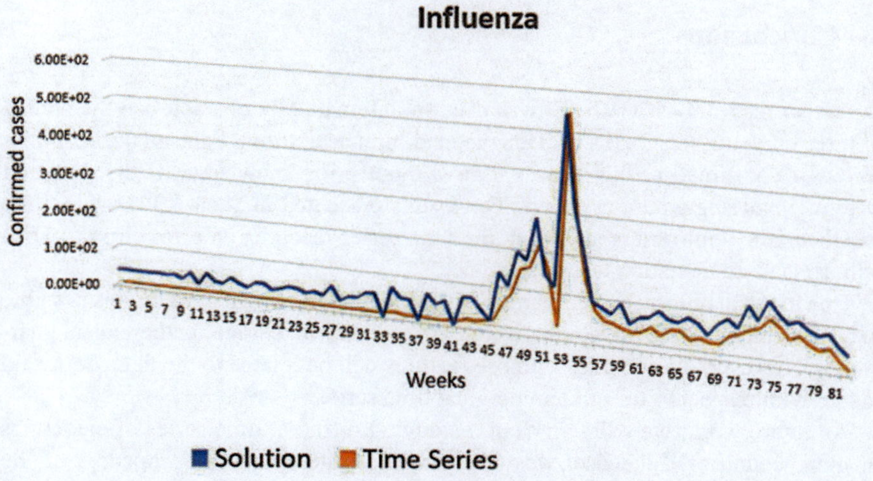

Fig. 7 Influenza time series fitting achieved by the best solution

4.4 Features Calculation Examples from Polynomial Models

One of the elementary features that can be computed from polynomial models is the derivative with respect to time, using the derivative property of Jacobi polynomials shown in Eq. 5.

$$\frac{d}{dt}\left[P_d^{\alpha,\beta}(t)\right] = \frac{1}{2}(\alpha+\beta+d+1)P_{d-1}^{\alpha+1,\beta+1}(t). \tag{5}$$

The derivative of a solution provides descriptive information on the spread change (either positive or negative) of new cases with respect to elapsed time.

Another important feature is the cumulative probability $\widehat{P}$ in the domain $g, h \in [e, f]$, providing probabilistic information of infections in a time window, thus generating quantitative information from the model.

$$\widehat{P}(g \le t \le h) = \frac{\sum_j a_j P_d^{\alpha,\beta}(t)\Big|_g^h}{\sum_j a_j P_{d+1}^{\alpha-1,\beta-1}(t)\Big|_e^f}. \tag{6}$$

A valuable statistical element in epidemiology is the expected value. It can be calculated for a time window $t \in [e, f]$ of the epidemic spread as shown below:

$$E[t] = \frac{\sum_j a_j t_j P_{d+1}^{\alpha-1,\beta-1}(t)\Big|_e^f - 2\frac{P_{d+2}^{\alpha-2,\beta-2}(t)\Big|_e^f}{\alpha+\beta+d-1}}{\sum_j a_j P_{d+1}^{\alpha-1,\beta-1}(t)\Big|_e^f}. \tag{7}$$

5 Conclusions

In this work, META-COVID19 was used to automatically characterize the spread of three epidemics: COVID-19, Dengue, and Influenza, through the analysis of time series of confirmed positive cases, this without prior knowledge of the data and without involving a human expert. The results presented in Sect. 4.3 show that the metaheuristic approach is able to fit the time series reaching an error close to 1E-3 with a standard deviation less than 8E-5.

Due to the mathematical properties of the Jacobi polynomial models, it is possible to obtain epidemiological features of the time series with certain mathematical guarantees. However, the accuracy of those features will be related to the fit error of said model with respect to the epidemiological time series.

As future work, we will carry out a study on different time series of epidemics in more countries. Likewise, we will study in more detail the properties of the polynomial models found.

Acknowledgements We thank the Tecnológico Nacional de México and the Instituto Tecnológico Superior de Purísima del Rincón for the support provided for the development of this research.

Appendix A: Full Results of the Evolutionary Process

The complete results of the 31 experiments carried out independently to fit the time series of each of the epidemics analyzed in this work are shown below (Tables 3, 4 and 5).

Table 3 Full results for COVID-19 time series fitting

Experiment	α	β	d	$f^*(x)$
1	8.1260E-01	−5.4060E-01	60	6.1263E-02
2	−6.9403E-03	−3.4019E-01	59	6.1315E-02
3	7.6225E-02	1.7762E-01	59	6.1347E-02
4	6.9265E-03	2.3384E-01	58	6.1387E-02
5	2.7295E-02	8.2915E-01	58	6.1388E-02
6	−4.9190E-03	−3.9003E-02	58	6.1391E-02
7	1.7801E-01	−6.0675E-02	59	6.1418E-02
8	−3.7202E-03	1.9409E-01	59	6.1437E-02
9	7.5695E-01	4.0055E-02	58	6.1469E-02
10	−3.2709E-02	1.5138E-01	60	6.1474E-02
11	3.8821E-03	1.7623E-03	59	6.1474E-02
12	−2.3249E-04	1.0980E-01	59	6.1475E-02
13	−2.0915E-03	−1.8362E-02	59	6.1480E-02
14	1.0737E-01	7.9252E-02	60	6.1483E-02
15	−2.4474E-02	4.7542E-01	59	6.1495E-02
16	4.7928E-02	−3.9262E-02	58	6.1503E-02
17	−2.3242E-01	−1.3136E-01	59	6.1507E-02
18	1.3797E-01	−4.7296E-02	60	6.1508E-02
19	−1.1074E-01	4.4323E-02	59	6.1509E-02
20	3.7600E-01	6.3902E-02	58	6.1513E-02
21	2.2202E-02	1.1082E-01	58	6.1519E-02
22	8.8068E-01	−2.5098E-01	59	6.1522E-02
23	−1.8557E-02	−9.0628E-02	59	6.1540E-02
24	−1.0060E-01	7.9679E-02	59	6.1540E-02
25	1.7334E-03	−8.4682E-04	58	6.1543E-02
26	5.2267E-02	−3.4513E-02	59	6.1546E-02
27	9.7821E-03	3.8568E-02	57	6.1547E-02
28	5.6211E-01	2.1058E-01	60	6.1547E-02
29	2.1390E-03	−1.1735E-03	58	6.1555E-02
30	9.1721E-02	−4.2112E-02	57	6.1564E-02
31	5.6344E-03	−4.2419E-03	58	6.1575E-02

Table 4 Full results for Influenza time series fitting

Experiment	α	β	d	$f^*(x)$
1	8.4838E-01	−5.6994E-01	59	6.7363E-02
2	5.8781E-02	−6.6349E-02	59	8.4100E-02
3	2.4000E-01	−5.0672E-02	60	8.6164E-02
4	1.9016E-03	1.4764E-01	59	8.6943E-02
5	5.2840E-01	3.8580E-01	59	8.9235E-02
6	2.7085E-01	1.7122E-01	59	9.0390E-02
7	−1.7290E-02	2.5582E-01	60	9.0614E-02
8	−2.7011E-02	6.9444E-03	60	9.0772E-02
9	5.0982E-03	4.0160E-01	60	9.1628E-02
10	6.3226E-01	5.1204E-02	60	9.2900E-02
11	−1.6113E-02	4.6209E-02	60	9.4011E-02
12	−3.2293E-01	5.0641E-01	59	9.4160E-02
13	−2.0584E-02	3.9327E-01	59	9.4784E-02
14	−1.2646E-02	1.0757E-01	60	9.4850E-02
15	−2.2484E-01	5.9621E-01	60	9.5435E-02
16	9.0968E-02	−3.4085E-02	59	9.5743E-02
17	3.7884E-01	1.0236E-01	60	9.5942E-02
18	−1.9208E-01	−9.6376E-02	57	9.5951E-02
19	−6.6023E-01	5.4990E-01	21	9.6798E-02
20	−9.1951E-01	1.3987E+00	20	9.6798E-02
21	−3.7141E-03	8.2386E-02	20	9.6798E-02
22	−9.7102E-04	−8.4625E-03	20	9.6798E-02
23	5.2415E-02	2.8087E-04	20	9.6798E-02
24	4.7487E-01	1.0868E+00	21	9.6798E-02
25	−3.7248E-01	6.4116E-01	20	9.6798E-02
26	2.0620E-01	1.5777E-01	21	9.6798E-02
27	−2.9789E-03	1.4101E-01	21	9.6798E-02
28	5.2377E-01	2.4734E+00	20	9.6798E-02
29	−8.5019E-01	−8.6898E-02	20	9.6798E-02
30	−2.7166E-01	4.1158E-02	20	9.6798E-02
31	−6.1453E-03	−2.0267E-01	22	9.7321E-02

Table 5 Full results for Dengue time series fitting

Experiment	α	β	d	$f^*(x)$
1	5.1360E-01	−1.7626E-01	60	9.0721E-02
2	2.5975E-01	−1.1904E-01	60	9.9808E-02
3	3.9105E-03	−1.9518E-03	59	1.1451E-01
4	−2.6351E-02	8.7030E-02	59	1.1568E-01
5	−5.9696E-03	3.2948E-05	59	1.1571E-01
6	2.6769E-03	1.7051E-01	60	1.1626E-01
7	3.3806E-02	1.5299E-01	60	1.2057E-01
8	1.6077E-01	6.4142E-01	58	1.2081E-01
9	−1.5414E-02	−3.5862E-01	60	1.2309E-01
10	2.1397E-01	3.9529E-01	59	1.2333E-01
11	−3.8959E-01	1.9964E+00	22	1.2345E-01
12	−5.0625E-02	6.1332E-01	21	1.2345E-01
13	−2.6034E-02	−2.7918E-01	21	1.2345E-01
14	5.4194E-03	1.1548E-01	22	1.2345E-01
15	−1.7640E-01	3.4540E-01	22	1.2345E-01
16	−1.0251E-03	−3.4520E-02	21	1.2345E-01
17	7.5168E-02	1.5025E-01	22	1.2345E-01
18	3.1229E-01	5.8538E-02	21	1.2345E-01
19	9.0026E-01	−9.8015E-01	22	1.2345E-01
20	−2.0655E-01	6.7970E-02	21	1.2345E-01
21	−6.9117E-01	1.6085E+00	21	1.2345E-01
22	3.2106E-03	7.4319E-02	21	1.2345E-01
23	−6.3980E-02	−2.3294E-01	21	1.2345E-01
24	−4.3106E-02	6.6467E-04	21	1.2345E-01
25	4.0287E-01	8.7269E-01	22	1.2345E-01
26	3.3154E-02	1.1776E+00	21	1.2345E-01
27	−1.7930E-02	1.5733E+00	21	1.2345E-01
28	1.1777E-02	8.1956E-01	22	1.2345E-01
29	2.4731E-01	2.0408E+00	21	1.2345E-01
30	−2.7264E-04	1.3756E-03	22	1.2345E-01
31	5.6332E-02	1.2349E-01	20	1.2412E-01

References

1. Alchon, S. A. (2003). *A pest in the land: New world epidemics in a global perspective* (pp. 109–143). UNM Press.
2. Anno, S., Hara, T., Kai, H., Lee, M. A., Chang, Y., Oyoshi, K., ..., & Tadono, T. (2019). Spatiotemporal dengue fever hotspots associated with climatic factors in Taiwan including

outbreak predictions based on machine-learning. *Geospatial Health, 14*(2).
3. Capistran, M. A., Capella, A., & Christen, J. A. (2021). Forecasting hospital demand in metropolitan areas during the current COVID-19 pandemic and estimates of lockdown-induced 2nd waves. *PLoS ONE, 16*(1), e0245669.
4. Carletti, T., Fanelli, D., & Piazza, F. (2020). COVID-19: The unreasonable effectiveness of simple models. *Chaos, Solitons & Fractals: X, 5*, 100034.
5. Chenar, S. S., & Deng, Z. (2018). Development of genetic programming-based model for predicting oyster norovirus outbreak risks. *Water Research, 128*, 20–37.
6. De Anda-Suárez, J., Calzada-Ledesma, V., Gutiérrez-Hernández, D. A., Santiago-Montero, R., Villanueva-Jiménez, L. F., & Rodríguez-Miranda, S. (2022). A novel metaheuristic framework based on the generalized Boltzmann distribution for COVID-19 spread characterization. *IEEE Access, 10*, 7326–7340.
7. Huppert, A., & Katriel, G. (2013). Mathematical modelling and prediction in infectious disease epidemiology. *Clinical Microbiology and Infection, 19*(11), 999–1005.
8. Ioannidis, J. P., Cripps, S., & Tanner, M. A. (2020). Forecasting for COVID-19 has failed. *International Journal of Forecasting*.
9. Jewell, N. P. (2003). *Statistics for epidemiology*. Chapman and Hall/CRC.
10. Kermack, W. O., & McKendrick, A. G. (1927). A contribution to the mathematical theory of epidemics. In *Proceedings of the Royal Society of London. Series A, Containing Papers of a Mathematical and Physical Character*, *115*(772), 700–721.
11. Koike, F., & Morimoto, N. (2018). Supervised forecasting of the range expansion of novel non-indigenous organisms: Alien pest organisms and the 2009 H1N1 flu pandemic. *Global Ecology and Biogeography, 27*(8), 991–1000.
12. Liang, R., Lu, Y., Qu, X., Su, Q., Li, C., Xia, S., ..., Niu, B. (2020). Prediction for global African swine fever outbreaks based on a combination of random forest algorithms and meteorological data. *Transboundary and Emerging Diseases, 67*(2), 935–946.
13. Martha, C. (2022). Statista research department. *Statista*. https://es.statista.com/temas/7646/el-sector-de-la-salud-en-mexico/. (4 Jan 2022).
14. Melin, P., Monica, J. C., Sanchez, D., & Castillo, O. (2020). Multiple ensemble neural network models with fuzzy response aggregation for predicting COVID-19 time series: The case of Mexico. In *Healthcare* (Vol. 8, no. 2, p. 181). MDPI. (June 2020).
15. Postnikov, E. B. (2020). Estimation of COVID-19 dynamics "on a back-of-envelope": Does the simplest SIR model provide quantitative parameters and predictions? *Chaos Solitons & Fractals, 135*, 109841.
16. Ritchie, H., Mathieu, E., Rodés-Guirao, L., Appel, C., Giattino, C., Ortiz-Ospina, E., ... & Roser, M. (2020). Coronavirus pandemic (COVID-19). Our world in data.
17. Roda, W. C., Varughese, M. B., Han, D., & Li, M. Y. (2020). Why is it difficult to accurately predict the COVID-19 epidemic? *Infectious Disease Modelling, 5*, 271–281.
18. Saba, A. I., & Elsheikh, A. H. (2020). Forecasting the prevalence of COVID-19 outbreak in Egypt using nonlinear autoregressive artificial neural networks. *Process Safety and Environmental Protection, 141*, 1–8.
19. Saltelli, A., Bammer, G., Bruno, I., Charters, E., Di Fiore, M., Didier, E., ..., Vineis, P. (2020). Five ways to ensure that models serve society: A manifesto.
20. Secretaría de Salud (2022). Datos Abiertos Dirección General de Epidemiología. Gobierno de México. https://www.gob.mx/salud/documentos/datos-abiertos-152127. (01 Sept 2022).
21. Szeg, G. (1939). *Orthogonal polynomials* (Vol. 23). American Mathematical Society.
22. Tuli, S., Tuli, S., Tuli, R., & Gill, S. S. (2020). Predicting the growth and trend of COVID-19 pandemic using machine learning and cloud computing. *Internet of Things, 11*, 100222.
23. Wang, P., Zheng, X., Li, J., & Zhu, B. (2020). Prediction of epidemic trends in COVID-19 with logistic model and machine learning technics. *Chaos, Solitons & Fractals, 139*, 110058.

Comparative Study of Heuristics for the One-Dimensional Bin Packing Problem

Jessica González-San-Martín, Laura Cruz-Reyes, Claudia Gómez-Santillán, Héctor Fraire, Nelson Rangel-Valdez, Bernabé Dorronsoro, and Marcela Quiroz-Castellanos

1 Introduction

The One-dimensional Bin Packing Problem (1D-BPP) is a challenging NP-Hard combinatorial optimization problem [1] where each object has a unique weight, and the objective is to pack the greatest number of objects in the least number of possible bins. Martello and Toth [2] formally defined the 1D-BPP as follows:

$$\text{minimize} \sum_{k=1}^{K} y_k \tag{1}$$

subject to

$$\sum_{k=1}^{K} x_{ik} = 1, \qquad \forall i \in I \tag{2}$$

J. González-San-Martín (✉) · L. Cruz-Reyes · C. Gómez-Santillán · H. Fraire
Tecnológico Nacional de México/Instituto Tecnológico de Ciudad Madero, Cd. Madero, Tamaulipas, México
e-mail: jessica_san_martin@hotmail.com

N. Rangel-Valdez
CONACyT Research Fellow at Graduate Program Division, Tecnológico Nacional de México, Instituto Tecnológico de Ciudad Madero, Cd. Madero, Tamaulipas, México

B. Dorronsoro
Computer Science Engineering Department, University of Cadiz, Puerto Real, Cadiz, Spain

M. Quiroz-Castellanos
Artificial Intelligence Research Center, Universidad Veracruzana, Xalapa-Enriquez, Veracruz, México

O. Castillo and P. Melin (eds.), *Hybrid Intelligent Systems Based on Extensions of Fuzzy Logic, Neural Networks and Metaheuristics*, Studies in Computational Intelligence 1096,
https://doi.org/10.1007/978-3-031-28999-6_19

$$\sum_{i=1}^{n} w_i x_{ik} \le w_{yk}, \qquad \forall k \in K \tag{3}$$

$$y_k \in \{0, 1\}, \qquad \forall k \in K \tag{4}$$

$$x_{ik} \in \{0, 1\}, \qquad \forall i \in I, \forall k \in K \tag{5}$$

where n is the number of items, w_i is the weight of the item i, W is the bin capacity, $I = \{1, ..., n\}$ is the set of items, $K = \{1, ..., K\}$ is the set of bins, the variables x_{ik} and y_{ik} are defined as:

$$x_{ik} = \begin{cases} 1 \text{ if item } i \text{ is assigned to bin } k, \\ 0 \text{ otherwise} \end{cases} \qquad y_k = \begin{cases} 1 \text{ if bin } k \text{ is used,} \\ 0 \text{ otherwise} \end{cases}$$

This problem has been preserved as a current case study due to its complexity and the various applications it offers. In recent decades, different works related to 1D-BPP have been published. In the state-of-the-art we can find different types of algorithms, mainly heuristics and metaheuristics that focus on solving One-dimensional Bin Packing, however, some of the exact algorithms proposed in the literature are also contemplated.

In this study, we present an empirical comparison of the algorithms and the strategies that have been used to solve this problem, especially the algorithms that report the best results. Our paper contemplates the most recent works up to those presented twenty years ago, to show the evolution of the strategies and approaches used in this line of research.

2 One-Dimensional Bin Packing Techniques

This section review 1D-BPP techniques proposed in the past twenty years and report their strategies and strengths. The selection method used was based on the results reported by its authors and the number of quotes in the articles. We present twenty-one works among which we can find heuristics, metaheuristics, and some exact algorithms.

2.1 *Heuristic Techniques*

Over time, different heuristics have been proposed to solve the 1D-BPP. Most of these works use improvements or variants of classical heuristics that are used to address this problem. As is known, heuristics are generally scalable and very fast in generating solutions, however, optimality is not guaranteed [3].

Fleszar and Hindi [4] developed a high-quality algorithm (Perturbation-MBS') that incorporates a modified version of the MBS heuristic from Gupta and Ho [5], a procedure that is based on a variable neighborhood search and lower bounds, to successively build a new solution by perturbing the current solution.

Alvim et al. [6] presented a hybrid improvement heuristic (HI_BP) in which the dual strategy used by Scholl et al. [7] is reintroduced, in addition to the search space reduction techniques of Martello and Toth [2] and the use of lower bounding strategies. This procedure uses a load redistribution method that is based on dominance, differencing, and unbalancing and the inclusión of an improvement process that utilizes a tabú search.

Singh and Gupta [8] proposed a combined approach (C_BP) that uses two heuristics: a hybrid steady-state grouping genetic algorithm and an improved version of the Perturbation-MBS heuristic from Fleszar and Hindi [4].

Loh et al. [9] developed a weight annealing procedure (WA), making use of the concept of weight annealing to expand and speed up the neighborhood search by creating distortions in different parts of the search space. The proposed algorithm is simple and easy to implement, and the authors reported a high-quality performance.

Fleszar and Charalambous [10] proposed a modification to the Perturbation-MBS method [4] that uses a new sufficient average weight (SAW) principle to control the average weight of the items that are packed in each bin (Perturbation-SAW-MBS).

Cruz-Reyes et al. [11] proposed a hybrid genetic clustering algorithm called HGGA-BP, which is inspired by the Falkenauer group representation scheme that applies evolutionary operators at the container level. This algorithm includes efficient heuristics to generate the initial population and perform group mutation and crossover; as well as hybrid strategies for the accommodation of objects that were left free when applying the group operators.

Sim and Hart [12] presented an island model that uses a form of cooperative coevolution to generate a set of deterministic heuristics through single-node genetic programming, which interacts cooperatively to collectively minimize the number of containers used.

Buljubašić and Vasquez [13] offered a consistent neighborhood search approach for 1D-BPP. They used different techniques such as problem reduction through the first fit heuristic to obtain a partial solution, Taboo Search to add or remove elements from the containers, and the hill climbing/descent procedure to minimize the given objective function (see Table 1).

2.2 *Metaheuristic Techniques*

Metaheuristic algorithms have become dominant for solving challenging optimization problems in various fields because they are widely used tools for finding (near) optimal solutions to solve large problem instances in reasonable execution times. Metaheuristic algorithms can be divided into evolutionary algorithms (EA) like the genetic algorithm (GA) or swarm intelligence algorithms like ant colony optimization

Table 1 Heuristic techniques for 1D-BPP

Author	Algorithm	Techniques	Year	Quotes
Fleszar and Hindi [4]	Perturbation-MBS	Variable neighborhood search and lower bounds	2002	257
Alvim et al. [6]	HI_BP	Reduction techniques, dual and lower bounding strategies	2004	192
Singh and Gupta [8]	C_BP	Hybrid steady-state grouping genetic algorithm and an improved version of the Perturbation-MBS heuristic [4]	2007	51
Loh et al. [9]	WABP	Weight annealing procedure and neighborhood search	2008	95
Fleszar and Charalambous [10]	Perturbation-SAW-MBS'	Modification of the MBS method [4] that uses a new sufficient average weight (SAW)	2011	72
Cruz-Reyes et al. [11]	HGGA-BP	Genetic algorithm, different simple heuristics, and evolutionary operators at the container level	2012	2
Sim and Hart [12]	GSMCH_BP	Single node genetic programming, hyper heuristics, and island models	2013	31
Buljubašić and Vasquez [13]	CNS_BP	Reduction techniques, a variant of first fit heuristic and local searches: taboo search and hill climbing descent	2016	21

(ACO), particle swarm optimization (PSO), and whale optimization, among others. Bhatia and Basu [14] introduced a multi-chromosomal grouping genetic algorithm (MGGA) with grouping genetic operators and a rearrangement procedure based on the better-fit heuristic.

Levine and Ducatelle [15] proposed a hybrid procedure that implements the ant colony optimization metaheuristic (HACO-BP), which includes a strategy for a local search that relies on the dominance criterion from Martello and Toth [2].

Layeb and Boussalia [16] presented an approach based on the quantum-inspired cuckoo search algorithm and define an appropriate quantum representation based on qubit representation to represent bin packing solutions. Also, they proposed a new hybrid quantum measure operation that uses the first fit heuristic to pack no filled objects by the standard measure operation.

Layeb and Chenche [17] presented an approach based on the GRASP procedure. They make use of two phases; the first is based on a new random heuristic based on the hybridization between the First Fit and Best Fit heuristics. while in the second phase the Taboo search algorithm is used to improve the solutions found in phase one.

Dokeroglu and Cosar [18] proposed a set of robust and scalable hybrid parallel algorithms that take advantage of parallel computation techniques, evolutionary grouping genetic metaheuristics, and bin-oriented heuristics to obtain solutions for large-scale one-dimensional BPP instances.

Bayraktar et al. [19] integrated an Artificial Bee Colony (ABC) algorithm and equipped it with a memory component that combines two main search methods for diversification and intensification: neighborhood search and random search. Memory integration is a component of a well-known metaheuristic, called Tabu Search, which steps up to neighborhood search with some awareness, where short-term and long-term memory models are recruited.

Quiroz-Castellanos et al. [20] propound the GGA-CGT algorithm that focuses on promoting the transmission of the best genes on the chromosomes while maintaining the balance between selection pressure and population diversity. The transmission of the genes is generated through a set of genetic clustering operators, while the evolution is balanced by a reproduction technique that controls the exploration of the search space and prevents premature convergence of the algorithm.

Kucukyilmaz and Kiziloz [21] proposed a scalable Island-parallel grouping genetic algorithm (IPGGA), where on a given run, the population is partitioned into semi-isolated subpopulations. With this parallelization approach, they manage to explore and exploit the search space more efficiently.

El-Ashmawi and Abd Elminaam [22] presented a modified version of the optimization algorithm (SSA) squirrel search algorithm. The proposed algorithm is based on an improved best-fit heuristic to generate a feasible initial packing of elements into containers. During solution improvement, operator strategy can take place to obtain an optimized solution for the BPP.

Borgulya [23] proposed a hybrid evolutionary algorithm where an individual is a feasible solution and contains the description of the containers. The algorithm works without recombination; uses two new mutation operators and improves the quality of solutions with local search procedures. The work of the mutation operators is based on a frequency matrix of relative pairs, and based on this matrix, the frequency of each pair of elements is known, that is, how often they are included in the same container in the best solutions. The frequency matrix helps to pack elements into subsets of elements; these subsets are the containers in the problem.

Hartono et al. [24] present a hybrid classical-quantum approach called H-BPP. The algorithm consists of two modules, each one designed to be executed in different computational ecosystems. First, a quantum subroutine searches for a set of feasible container configurations of the problem at hand. Second, a classical computation subroutine builds complete solutions to the problem from the subsets given by the quantum subroutine (see Table 2).

2.3 Exact Techniques

The exact methods, as their name indicates, obtain the optimal solution to a problem, however, in instances that are very difficult to solve, the execution times of this type of algorithm are usually exponential. Nevertheless, in the state-of-the-art, few exact methods are used to address 1D-BPP.

Brandao and Pedroso [25] presented an exact method to solve bin packing and cutting stock problems that include multiple constraint variants. Their proposal is based on a laterally constrained arc-flow formulation equivalent to Gilmore and Gomory, including a graph compression algorithm that builds a graph, where paths from the source to the target node represent every valid packing pattern.

Wei et al. [26] proposed a branch-and-price-and-cut algorithm, this is another exact method based on the classical set-partitioning model for the One-dimensional bin packing problems and the subset row inequalities. They implemented an ad hoc label-setting algorithm as well as dominance and fathoming rules used to speed up its computation (see Table 3).

3 Relevant Optimization Methods

We present a brief description of the three best works of the twenty-one reviewed with respect to the results reported by each author.

3.1 Grouping Genetic Algorithm with Controlled Gene Transmission

The Grouping Genetic Algorithm with Controlled Gene Transmission algorithm (GGA-CGT) developed by Quiroz-Castellanos et al. [20], is an intelligent packing metaheuristic that simplifies and improves the packing of objects, makes use of a rearrangement procedure that allows the exploration and exploitation of the search space and a set of genetic clustering operators that promote the transmission of the best genes on the chromosomes without losing the balance between selection pressure and population diversity. The transmission of the best genes (the fullest containers) is carried out through a set of genetic clustering operators, while the evolutionary process is balanced using a reproduction technique that controls the exploration of the search space and avoids the premature convergence of the algorithm.

Table 2 Metaheuristic techniques for 1D-BPP

Author	Algorithm	Techniques	Year	Quotes
Bhatia and Basu [14]	MGGA	Genetic algorithm with grouping genetic operators and a rearrangement procedure based on the better-fit heuristic	2004	34
Levine and Ducatelle [15]	HACO-BP	Ant colony optimization metaheuristic and a strategy for a local search that relies on the dominance criterion from [2]	2004	289
Layeb and Boussalia [16]	QICSABP	Quantum-inspired cuckoo search algorithm containing a Gubit representation for the search space and quantum operators	2012	72
Layeb and Chenche [17]	GRASP-BP	GRASP procedure and hybridization between the First Fit and Best Fit heuristics and Taboo search algorithm	2012	19
Dokeroglu and Cosar [18]	Parallel Exon-MBS-BFD	Hybrid parallel algorithms, evolutionary grouping genetic metaheuristics, and bin-oriented heuristics	2014	62
Bayraktar et al. [19]	MEABC	Integrated an Artificial Bee Colony (ABC) algorithm equipped it with a memory component that combines a neighborhood search (Taboo) and random search	2014	6
Quiroz-Castellanos et al. [20]	GGA-CGT	A genetic algorithm that focuses on promoting the transmission of the best genes on the chromosomes	2015	103
Kucukyilmaz and Kiziloz [21]	IPGGA	Grouping genetic algorithm with parallel island model	2019	42

(continued)

Table 2 (continued)

Author	Algorithm	Techniques	Year	Quotes
El-Ashmawi and Abd Elminaam [22]	MSBPP	A squirrel search algorithm based on an improved best-fit heuristic	2019	42
Borgulya [23]	HEA	A hybrid evolutionary algorithm that works without recombination. It uses two mutation operators	2021	6
Hartono et al. [24]	H-BPP	A hybrid classical-quantum approach with a quantum subroutine searches for a set of feasible container configurations and a classical computation subroutine builds complete solutions	2022	0

Table 3 Exact techniques for 1D-BPP

Author	Algorithm	Techniques	Year	Quotes
Brandao and Pedroso [25]	General Arc-Flow Formulation	Laterally constrained arc-flow formulation, including a graph compression algorithm that builds a graph	2016	102
Wei et al. [26]	EXM	Branch-and-Price-and-Cut Algorithm formulated with Set-Partitioning and SR Inequalities	2020	31

Algorithm 1: GGA-CGT

Generate the initial population P with FF-$\tilde{n}$
while *generation* < *max_gen* y size(*best_solution*) > L_2
    Select n_c individuals to cross by Controlled_Selection
    Apply Gen_Level_Crossover_FFD to n_c selected individuals
    Apply Controlled_Replacement to introduce offspring
    Select n_m individuals and clone elite solutions by Controlled_Selection
    Apply Adaptative_Mutation_RP to the best n_m individuals
    Apply Controlled_Replacement to introduce clones
    Update *best_solution*
end while

3.2 Consistent Neighborhood Search for Bin Packing Problem

The Consistent Neighborhood Search for one-dimensional Bin Packing (CNS_BP) algorithm is a heuristic proposed by Buljubašić and Vasquez [13] that consists of performing a local search to derive a feasible solution with a given number of bins, m, starting from $m = UB\text{-}1$, where UB is an upper bound obtained by using a variant of the classic First Fit heuristic. It makes use of a reduction to simplify the problem and then applies two local searches: Taboo Search and Hill Climbing Descent. This algorithm shows successful results in its experimentations since it obtains equal or better solutions than the other algorithms in the state of the art, including the Hard28 instances that are considered one of the most difficult within BPP.

Algorithm 2: CNS_BP

Remove item pairs (i,j) such that $w_i + w_j = C$
Compute lower bound LB
Randomly shuffle the set of items
$\boldsymbol{S} \leftarrow$ complete solution obtained by First Fit
$\boldsymbol{m} \leftarrow$ number of bins in S
while $m > LB$ **and** time limit not exceded **do**
    $\boldsymbol{m \leftarrow m - 1}$
    Build partial solution P with m-2 bins ▹ Delete 3 bins from S
    $\boldsymbol{S' \leftarrow CNS(P)}$ ▹ Try to find complete solution with m bins
    If solution $\boldsymbol{S'}$ not complete **then** TERMINATE
    $\boldsymbol{S \leftarrow S'}$
end while
return $\boldsymbol{S}$

3.3 General Arc-Flow Formulation with Graph Compression

This work was proposed by Brandao and Pedroso [25], which is an exact method based on an arc flow formulation with lateral constraints. It is used to solve Bin Packing and Cutting Stock problems. It includes a graph compression algorithm that reduces the size of the underlying graph substantially without weakening the model. The formulation used is equivalent to that of Gilmore and Gomory [27]. However, instead of using column generation in an iterative process, the method builds a graph, where the paths from source to destination node represent each valid packing pattern.

4 Comparison of Results

In the literature, different instances are used to solve the one-dimensional Bin Packing problem. These instances can be classified into easy and difficult [25]. Table 4 presents the instances that were used to evaluate the performance of the three most relevant algorithms of this study and Table 5 presents the comparative results.

For the comparison of results, tests were carried out on the most relevant algorithms of those reviewed in this work; GGA-CGT, General Arc-Flow Formulation, and CNS_BP. For the first two algorithms, the experimentation was carried out in a Cluster with operating system CentOS version 6.7, GCC version 5.1.0, and Intel Xeon E5-2650 2.30 GHz processor, while CNS_BP was executed under a virtual machine with operating system. system Ubuntu 14.04.4, GCC version 4.8.4, and Intel Celeron 2.16 GHz processor, the latter was run in a virtual machine because it needed different conditions than the cluster.

Table 4 Set of test instances used by the most relevant algorithms in the literature

Author	Class name	Number of instances
Falkenauer [1]	Uniform	80
	Triplets	80
Scholl et al. [7]	Data set 1	720
	Data set 2	480
	Data set 3	10
Schwerin & Wäscher [28, 29], Gau [30]	Was 1	100
	Was 2	100
	Gau 1	17
Schoenfield [31]	Hard28	28

Table 5 Comparison of results among the three most relevant works of the state of the art

Class	Inst	CNS_BP		GGA-CGT		General Arc-Flow F	
		Opt	T(s)	Opt	T(s)	Opt	T(s)
Uniform	80	80	0.07	80	0.23	80	0.16
Triplets	80	80	0.02	80	0.41	80	0.17
Data Set 1	720	720	0.07	720	0.35	720	3.33
Data Set 2	480	480	0.03	480	0.12	480	3.33
Data Set 3	10	10	0.00	10	1.99	10	3.33
Was 1	100	100	0.00	100	0.00	100	0.10
Was 2	100	100	0.00	100	1.07	100	0.10
Gau 1	17	17	2.68	16	0.27	17	9.58
Hard28	28	25	7.21	16	2.40	28	0.82
Total	1615	1612	10.08	1602	6.84	1615	20.92

We can see that the exact algorithm manages to solve the set of test instances optimally, however, as expected, it takes longer than the heuristic and metaheuristic algorithms. On the other hand, the CNS_BP heuristic manages to solve 99.81% of the instances in almost half the time that the exact algorithm takes.

5 Conclusion

This work has reviewed the algorithms developed in the last twenty years that have been used to solve the 1D-BPP. We find several works that make use of different classical heuristics as well as nature-inspired and population-based metaheuristics. The Metaheuristics approaches are widely used optimization tools for finding near-optimal solutions to large graph-sized problem instances of the BPP in reasonable execution times. These algorithms continue to be promising tools for numerous NP-Hard optimization problems where the exact solution methods tend to fail because of the exponential search spaces. However, some disadvantages of the metaheuristic algorithms include the issue of premature convergence, which can lead the algorithms into getting trapped in local optimum, and the aspect of parameter fine-tuning, as some of the algorithms would require having to set the control parameter to meet a certain specified threshold. It would be interesting to perform hybridization of metaheuristics with machine learning strategies, this would help to resolve certain limitations such as parameter tuning. It would also be an area of opportunity since, to our knowledge, no approach in the literature uses computational learning to solve the bin packing problem. We hope that with the experimental results provided, the best strategies can be taken to obtain more promising results if these are combined or implemented with some variation or improvement.

References

1. Falkenauer Emanuel. (1996). A Hybrid Grouping Genetic Algorithm for Bin Packing. *Journal of Heuristics*, 5–30.
2. Martello, S., & Toth, P. (1990). Lower bounds and reduction procedures for the bin packing problem. *Discrete Applied Mathematics, 28*(1), 59–70.
3. Ahmad, I., AlFailakawi, M. G., AlMutawa, A., & Alsalman, L. (2021). Container scheduling techniques: A survey and assessment. *Journal of King Saud University-Computer and Information Sciences.*
4. Fleszar, K., & Hindi, K. S. (2002). New heuristics for one-dimensional bin-packing. *Computers & Operations Research, 29*(7), 821–839.
5. Gupta, J. N., & Ho, J. C. (1999). A new heuristic algorithm for the one-dimensional bin-packing problem. *Production Planning & Control, 10*(6), 598–603.
6. Alvim, A. C., Ribeiro, C. C., Glover, F., & Aloise, D. J. (2004). A hybrid improvement heuristic for the one-dimensional bin packing problem. *Journal of Heuristics, 10*(2), 205–229.
7. Scholl, A., Klein, R., Jurgens, C. (1997). Bison: A fast hybrid procedure for exactly solving the one-dimensional bin packing problem. *Computers & Operations Research, 24*(7), 627 _ 645

8. Singh, A., & Gupta, A. K. (2007). Two heuristics for the one-dimensional bin-packing problem. *OR Spectrum, 29*(4), 765–781.
9. Loh, K. H., Golden, B., & Wasil, E. (2008). Solving the one-dimensional bin packing problem with a weight annealing heuristic. *Computers & Operations Research, 35*(7), 2283–2291.
10. Fleszar, K., & Charalambous, C. (2011). Average-weight-controlled bin-oriented heuristics for the one-dimensional bin-packing problem. *European Journal of Operational Research, 210*(2), 176–184.
11. Cruz-Reyes, L., Marcela Quiroz, C., Alvim, A. C. F., FraireHuacuja, H. J., Claudia Gómez, S., Torres-Jiménez, J. (2012). Efficient hybrid grouping heuristics for the bin packing problem. *Computación y Sistemas, 16*(3), 349–360.
12. Sim, K., & Hart, E. (2013, July). Generating single and multiple cooperative heuristics for the one dimensional bin packing problem using a single node genetic programming island model. In *Proceedings of the 15th Annual Conference on Genetic and Evolutionary Computation* (pp. 1549–1556).
13. Buljubašić, M., & Vasquez, M. (2016). Consistent neighborhood search for one-dimensional bin packing and two-dimensional vector packing. *Computers & Operations Research, 76*, 12–21.
14. Bhatia, A. K., & Basu, S. K. (2004, November). Packing bins using multi-chromosomal genetic representation and better-fit heuristic. In *International conference on neural information processing* (pp. 181–186). Springer.
15. Levine, J., & Ducatelle, F. (2004). Ant colony optimization and local search for bin packing and cutting stock problems. *Journal of the Operational Research Society, 55*(7), 705–716.
16. Layeb, A., & Boussalia, S. R. (2012). A novel quantum inspired cuckoo search algorithm for bin packing problem. *International Journal of Information Technology and Computer Science, 4*(5), 58–67.
17. Layeb, A., & Chenche, S. (2012). A novel grasp algorithm for solving the bin packing problem. *International Journal of Information Engineering and Electronic Business, 4*(2), 8.
18. Dokeroglu, T., & Cosar, A. (2014). Optimization of one-dimensional bin packing problem with island parallel grouping genetic algorithms. *Computers & Industrial Engineering, 75*, 176–186.
19. Bayraktar, T., Aydin, M. E., & Dugenci, M. (2014). A memory-integrated artificial bee algorithm for 1-D bin packing problems. In *Proceedings of the CIE IMSS* (pp. 1023–1034).
20. Quiroz-Castellanos, M., Cruz-Reyes, L., Torres-Jimenez, J., Gómez, C., Huacuja, H. J. F., & Alvim, A. C. (2015). A grouping genetic algorithm with controlled gene transmission for the bin packing problem. *Computers & Operations Research, 55*, 52–64.
21. Kucukyilmaz, T., & Kiziloz, H. E. (2018). Cooperative parallel grouping genetic algorithm for the one-dimensional bin packing problem. *Computers & Industrial Engineering, 125*, 157–170.
22. El-Ashmawi, W. H., & Abd Elminaam, D. S. (2019). A modified squirrel search algorithm based on improved best fit heuristic and operator strategy for bin packing problem. *Applied Soft Computing, 82*, 105565.
23. Borgulya, I. (2021). A hybrid evolutionary algorithm for the offline Bin Packing Problem. *Central European Journal of Operations Research, 29*(2), 425–445.
24. Hartono, N., Ismail, A. H., Zeybek, S., Caterino, M., Jiang, K., Sahin, M., & Natalia, C. (2022). The 1-D bin packing problem optimisation using bees algorithm. *Journal Industrial Servicess, 7*(2), 323–326.
25. Brandao, F., & Pedroso, J. P. (2016). Bin packing and related problems: General arc-flow formulation with graph compression. *Computers & Operations Research, 69*, 56–67.
26. Wei, L., Luo, Z., Baldacci, R., & Lim, A. (2020). A new branch-and-price-and-cut algorithm for one-dimensional bin-packing problems. *INFORMS Journal on Computing, 32*(2), 428–443.
27. Gilmore P. C., & Gomory R. E. (1963). A linear programming approach to the cutting stock problem—Part II. *Annals of Operations Research*, 863–88. https://doi.org/10.1287/opre.11.6.863.
28. Schwerin, P., & Wäscher, G. (1997). The bin-packing problem: A problem generator and some numerical experiments with FFD packing and MTP. *International Transactions in Operational Research, 4*, 337–389.

29. Schwerin, P., & Wäscher, G. (1999). A new lower bound for the bin-packing problem and its integration to MTP. *Pesquisa Operacional, 19*, 111–129.
30. Wäscher, G., & Gau, T. (1996). Heuristics for the one-dimensional cutting stock problem: A computational study. *OR Spektrum, 18*, 131–144.
31. Schoenfield, J. E. (2002). Fast, exact solution of open bin packing problems without linear programming. *Draft, US Army Space & Missile Defense Command, Huntsville, Alabama, USA.*

Experimental Evaluation of Adaptive Operators Selection Methods for the Dynamic Multiobjective Evolutionary Algorithm Based on Decomposition (DMOEA/D)

José A. Brambila-Hernández, Miguel Á. García-Morales, Héctor J. Fraire-Huacuja, Armando Becerra del Angel, Eduardo Villegas-Huerta, and Ricardo Carbajal-López

1 Introduction

Currently, both in scientific research and industrial production, dynamic multi-objective problems can be found which do not remain static over time. Due to their dynamic nature, these problems change their objective functions and constraints from time to time [1]. In recent years, these problems have increased their number of objectives, so better and more efficient algorithms are required, which is why new strategies have been applied to improve them.

The work of Deb [2] has been taken as a reference for the implementation of a dynamic multi-objective algorithm. In particular, the decomposition-based multi-objective evolutionary algorithm (MOEA/D) was used. We call this dynamic version of MOEAD/D decomposition-based dynamic multi-objective evolutionary algorithm (DMOEA/D). Three adaptive operator selection (AOS) methods have been applied to this dynamic multi-objective algorithm. Three metrics have been used for the behavior of each of the methods and to determine which one behaves better applied to the dynamic algorithm: hypervolume, generalized dispersion, and inverted generational distance.

J. A. Brambila-Hernández (✉) · M. Á. García-Morales · H. J. Fraire-Huacuja · A. B. del Angel · E. Villegas-Huerta · R. Carbajal-López
Tecnológico Nacional de México/Instituto Tecnológico de Ciudad Madero, Ciudad Madero, Tamaulipas, Mexico
e-mail: alfredo.brambila@outlook.com; automatas2002@yahoo.com.mx

O. Castillo and P. Melin (eds.), *Hybrid Intelligent Systems Based on Extensions of Fuzzy Logic, Neural Networks and Metaheuristics*, Studies in Computational Intelligence 1096,
https://doi.org/10.1007/978-3-031-28999-6_20

2 The Overview of DMOEA/D

The multiobjective evolutionary algorithm based on decomposition (MOEA/D) is an algorithm proposed by Qingfu and Hui [3], which uses an approach to decompose a problem into several subproblems and optimize them simultaneously. Each of the generated subproblems is optimized from the information of its neighboring problems. The population comprises the best solutions found so far for the subproblems, and this is achieved by exploiting the most recent solutions of the neighboring subproblems to be optimized.

For the implementation of the dynamic version of the MOEA/D algorithm, the work of Deb [2] has been taken as a reference, in which he proposes some modifications to the NSGA-II algorithm so that it can solve dynamic multidimensional problems. Objective problems that have been applied to the MOEA/D algorithm. The proposed modifications are two, the first is the detection of the change of the problem, and the second is a mechanism of reaction to the change.

2.1 *Problem Change Detection*

The first thing is to randomly take some solutions from the parent population (in this case, 10%), which we call detectors, as shown in Algorithm 2.1.

Algorithm 2.1 GetDetectorList(Pop)

Inputs: *Pop* – population list
nPop – size of population
Outputs: *L* – List of detectors

```
L = ∅
nDetectors = nPop × 0.1
for i ← 1 to nDetectors do
    index = random number [0,1]
    newDetector = Pop_index
    add newDetector to L
end for
return L
```

The time function is updated in each iteration, and the detectors are reassessed. If there is a change in their objectives, it can be established that there was a change in the problem (see Algorithm 2.2); if this occurs, the entire population is reevaluated.

Algorithm 2.2 ProblemChangeDetection(L)

Input: L – List of detectors
Output: changeFlag – returns true if there was a change, false if there was no change

```
changeFlag = False
foreach element in L do
   i = 0
   for m ← 1 to M do /*  M is the number of objectives   */
      tempValue_{i,m} = element.objective_m
   end for
   i = i + 1
end for
evaluate detector list L
foreach element in L do
   i = 0
   for m ← 1 to M do /*  M is the number of objectives   */
      if element.objective_m != tempValue_{i,m} do
         changeFlag = True, and break for and foreach cycle
      end if
   end for
   i = i + 1
end for
return changeFlag
```

2.2 Mechanism of Reaction to Problem Change

Deb proposes two versions of the algorithm. The first is called DNSGA-II A, in which, after detecting a change, a percentage of the population ($\zeta\%$) is replaced with new random solutions (see Algorithm 2.3). The second version, DNSGA-II B, replaces a percentage of the population ($\zeta\%$) with mutated solutions. In this case, version A will be used.

Algorithm 2.3 ChangeResponseMechanism

Input: Pop – population list
$nPop$ – size of population
$zeta$ – percentage of individuals to replace
Output: Pop – updated Pop list
1: $nReplacements = nPop \times zeta$
2: $\boldsymbol{for}\ i \leftarrow 1\ \boldsymbol{to}\ nReplacements\ \boldsymbol{do}$
3: $index = random\ number\ [1, nPop]$
4: $newIndividual$ = generate random individual based on problem
5: evaluate $newIndividual$
6: replace the current individual in Pop_{index} with $newIndividual$
7: $\boldsymbol{end\ for}$
8: $\boldsymbol{return}\ L$

As part of the modification for MOEA/D, it is considered that, when a change is detected, before reevaluating the population, instead of storing the first front of the population (POP) as the optimal front at time t, the Current Estimated Pareto List (EP).

Figure 1 presents the general framework to convert a static MOEA to a DMOEA. This general framework also serves as a guide for what is applied to DMOEA/D.

Below is the pseudocode of the dynamic implementation of the DMOEA/D algorithm (see Algorithm 2.4).

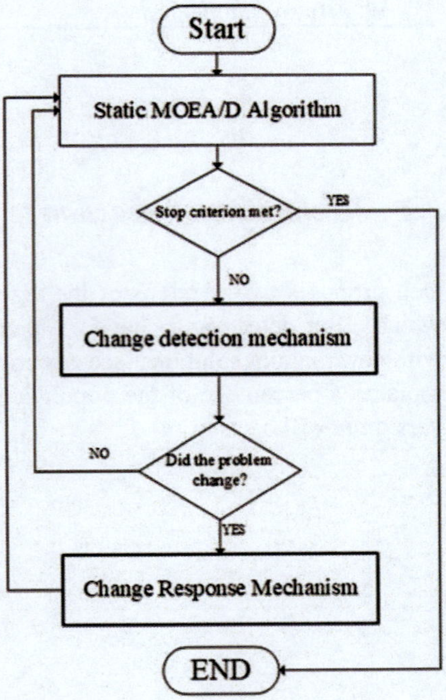

Fig. 1 General framework of a DMOEA applied to DMOEA/D [4]

Algorithm 2.4 DMOEA/D

Input: DMOP – dynamic multiobjective problem
N – number of the subproblems
stopping criterion
T – number of the weight vectors in the neighborhood of each weight vector
uniform spread of N weight vectors: $\lambda^1, \ldots, \lambda^N$
nT – change frequency
$tauT$ – change severity
Output: $Front_f$ – list of EP fronts

```
EP = ∅
Get the Euclidean distances between two weight vectors (any) and then compute the T closest weight vectors to each weight vector.
for i ← 1 to N do
  B(i) = {i_1, ..., i_T}  /* where λ^{i_1}, ..., λ^{i_T} are T closest weight vectors to λ^i */
end for
Generate initial population (Pop) x^1, ..., x^N
FV^i = F(x^i)
initialize z = (z_1, ..., z_m)^N by a problem specific method
time = 0
detectors = GetDetectorList(Pop)
it = 0
f = 1
while stopping criteria is not satisfied do
  for i ← 1 to N do
    Choose two indexes at random k, l from B(i), and generate a new solution y from x^k and x^l by using genetic operators
    Repair y to produce y' using a repair or improvement heuristic
    for j ← 1 to m do
      if z_j < f_j(y') then
        z_j = f_j(y')
      end if
    end for
    foreach index j ∈ B(i) do
      if g^te(y'|λ^j, z) ≤ g^te(x^j|λ^j, z) then
        x^j = y'
        FV^j = F(y')
      end if
    end foreach
  end for
  remove from EP list all solutions dominated by F(y')
  add solution F(y') to EP list if no solutions in EP list dominate solution F(y')
  time = 1/nT × ⌊1×it / tauT⌋
  if ProblemChangeDetection (detectors) then
    add EP list to Front_f
    initialize z = (z_1, ..., z_m)^N
    evaluate Pop
    run ChangeResponseMechanism
    FV^i = F(x^i)
    remove from EP list all solutions dominated by F(y')
    add solution F(y') to EP list if no solutions in EP list dominate solution F(y')
    f = f + 1;
  end if
  it = it + 1
end while
return Front_f
```

3 The AOS Methods

Adaptive operator selection (AOS) is a good technique for increasing the generalization of heuristic optimization methods. This method explores the solution space by choosing an operator at each iteration based on its past performance and is rewarded for solutions.

3.1 Fitness-Rate-Rank-Based Multiarmed Bandit Adaptive Operator Selection (FRRMAB)

Fitness-Rate-Rank-Based Multiarmed Bandit (FRRMAB) is a method for the AOS problem proposed by Li et al. [5], which pays awareness to the dynamic essence of AOS. This method mainly consists of two modules. The first is the allocation of credits, and the second is the selection of operators.

3.1.1 Credit Assignment

Using the raw values of the improvements generated by the recent use of the operator being evaluated is one of the most used approaches; however, the scope of raw fitness improvements can vary from one problem to another and even at different stages of the optimization process. It is common for the gross value of fitness improvement to be more significant in the early stages than in the later stages. As discussed in work by Fialho [6] the direct use of raw fitness enhancement could impair the algorithm's robustness. FRRMAB uses "Fitness Improvement Rates" (FIR) [5] to avoid this problem.

$$FIR_{i,t} = \frac{pf_{i,t} - cf_{i,t}}{pf_{i,t}} \tag{1}$$

where $pf_{i,t}$ is the fitness value of the parent, and $cf_{i,t}$ is the fitness value of the offspring.

The FIR value of the operator that has been most recently used is stored in a structure called a "sliding window" [5]. Which is of type FIFO, of fixed size W, so that the lately used operator is added to the queue and the oldest used is removed from it to maintain the size of the window. Each slot in the sliding window stores the operator index and the operator FIR value (Fig. 2).

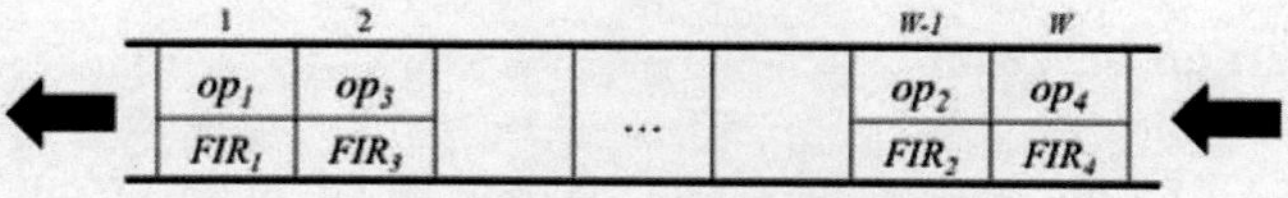

Fig. 2 Sliding window structure (FIFO) [5]

The authors have proposed a sliding window structure because, in dynamic environments, the operator's performance in the primary stages may be irrelevant. Furthermore, the sliding window ensures that the stored FIR information is for the current search situation. The reward of each operator ($Reward_i$) is obtained by adding all the FIR values set to each operator in the present sliding window. $Reward_i$ values are ordered in descending order, where $Rank_i$ is the ranking of operator *i*. Finally, to give more opportunities to the best operators, $D \in [0, 1]$ is a decay factor is introduced to transform $Reward_i$ into:

$$Decay_i = D^{Rank_i} \times Reward_i \tag{2}$$

Next, credit is assigned to operator *i* as follows:

$$FRR_{i,t} = \frac{Decay_i}{\sum_{j=1}^{K} Decay_j} \tag{3}$$

The lower the value of D, the greater the impact for the most promising operator.

Algorithm 3.1.1 Credit Assignment

Initialize each $Reward_i = 0$
Initialize $n_i = 0$;
for $i \leftarrow 1$ ***to*** Sliding_Window.length ***do***
    $op =$ Sliding_Window.GetIndexOp(i)
    $FIR =$ Sliding_Window.GetFIR(i)
    $Reward_{op} = Reward_{op} + FIR$
    n_{op} ++
end

Rank $Reward_i$ in descending order and set $Rank_i$ to be the rank value of operator *i*
for *op* $\leftarrow$ ***to*** *K* ***do***
    $Decay_{op} = D^{Rank_{op}} \times Reward_{op}$
end
    $DecaySum = \sum_{op=1}^{K} Decay_{op}$
for *op* $\leftarrow$ ***to*** *K* ***do***
    $FRR_{op} = Decay_{op}/DecaySum$
end

3.1.2 Operator Selection

The operator selection scheme selects an operator based on the credits received to generate new solutions. This method uses a bandit-based scheme to select an operator, using the *FRR* values as a quality index.

Algorithm 3.1.2 Bandit-based operator selection

if Some operators have not been selected ***then***
    op_t = uniformly randomly select an operator from th operators pool
else
    $op_t = \underset{i=\{1,\dots,K\}}{\text{argmax}} \left(FRR_{i,t} + C \times \sqrt{\frac{2\times\ln\left(\sum_{j=1}^{K} n_{j,t}\right)}{n_{i,t}}} \right)$
end

Fitness-Rate-Rank-Based Multiarmed Bandit Adaptive Operator selection (FRRMAB) has been added to the DMOEA/D algorithm. Below is the pseudocode:

Algorithm 3.1.3 DMOEA/D-FRRMAB

Inputs: DMOP – dynamic multiobjective problem
N – number of the subproblems
stopping criterion
T – number of the weight vectors in the neighborhood of each weight vector
uniform spread of N weight vectors: $\lambda^1, \dots, \lambda^N$
nT – change frequency
$tauT$ – change severity
K – number of genetic operators
Outputs: $Front_f$ – list of EP fronts

$EP = \emptyset$
Get the distances (Euclidean) between two weight vectors (any) and then compute the T closest weight vectors to each weight vector.
for $i \leftarrow 1$ ***to*** N ***do***
    $B(i) = \{i_1, \dots, i_T\}$ /* where $\lambda^{i_1}, \dots, \lambda^{i_T}$ are T closest weight vectors to λ^i */
end for
Generate initial population (*Pop*) $x^1, \dots, x^N$
$FV^i = F(x^i)$
initialize $z = (z_1, \dots, z_m)^N$
$time = 0$
$detectors$ = GetDetectorList(Pop)
$it = 0$
$f = 1$
while stopping criteria is not satisfied ***do***
    for $i \leftarrow 1$ ***to*** N ***do***
        $op = FRRMAB(FRR)$

Generate a new solution y = reproduction (i, op, Pop)
Repair y to produce y' using a repair or improvement heuristic
for $j \leftarrow 1$ ***to*** m ***do***
if $z_j < f_j(y')$ *then*
$z_j = f_j(y')$
end if
end for
foreach index $j \in B(i)$ ***do***
$\eta = \frac{g^{te}(x^j|\lambda^j,z) - g^{te}(y'|\lambda^j,z)}{g^{te}(x^j|\lambda^j,z)}$
if $g^{te}(y'|\lambda^j,z) \le g^{te}(x^j|\lambda^j,z)$ ***then***
$x^j = y'$
$FV^j = F(y')$
$FIR_{op} = FIR_{op} + \eta$
end if
end foreach
SlidingWindows.SetIndex(op)
SlidingWindows.SetFIR(FIR_{op})
CreditAssignement(*Reward, SlidingWindow*)
Decay (*Reward,FRR*)
end for
remove from EP list all solutions dominated by *F*(*y'*)
add solution *F*(*y'*) to EP list if no solutions in EP list dominate solution *F*(*y'*)
$time = \frac{1}{nT} \times \lfloor \frac{1 \times it}{tauT} \rfloor$
if ProblemChangeDetection ($detectors$) ***then***
add EP to $Front_f$
initialize $z = (z_1, \ldots, z_m)^N$
evaluate *Pop*
run ChangeResponseMechanism
$FV^i = F(x^i)$
remove from EP list all solutions dominated by *F*(*y'*)
add solution *F*(*y'*) to EP list if no solutions in EP list dominate solution *F*(*y'*)
$f = f + 1;$
end if
$it = it + 1$
end while
return $Front_f$

3.2 Adaptive Operator Selection Based on Dynamic Thompson Sampling

Dynamic Thompson Sampling (DYTS) is a method Lei Sun and Ke Li proposed in 2020 [7]. Using Thompson sampling, Bayesian estimation is exploited for online decision problems where actions are applied successively, assigning to each arm an independent previous reward distribution of $\theta_i, i \in \{1, \ldots, K\}$. In this method, the Beta distribution is used with the parameters $\alpha = (\alpha_1, \ldots, \alpha_k)$ and $\beta = (\beta_1, \ldots, \beta_k)$ to represent the following prior distribution in Eq. 4:

$$\mathrm{P}^{Beta}(\theta_i) = \frac{\Gamma(\alpha_i + \beta_i)}{\Gamma(\alpha_i)\Gamma(\beta_i)}\theta_i^{\alpha_i - 1}(1 - \theta_i)^{\beta_i - 1} \tag{4}$$

where parameters $\alpha_i$andβ_i are associated with the ith arm $i \in \{1, \ldots, k\}$ and the gamma function is $\Gamma(\cdot)$. After receiving the updated rewards, according to the Bayes, rule the parameters are updated. For a given action $a_i, i \in \{1, \ldots, k\}$ depending on the reward r_i, $\alpha_i$andβ_i are updated as follows in Eq. 5:

If $r_i = 1$

$$\alpha_i = \alpha_i + 1, \beta_i = \beta_i \tag{5}$$

If $r_1 = 0$

$$\alpha_i = \alpha_i + 1, \beta_i = \beta_i$$

The arm with the highest sampling value of the distribution is chosen in the decision-making stage, which will be applied in the next step. It is the underlying process that balances exploitation and exploration. Each component of αandβ is initialized to 1, while $P^{Beta}(\theta_i)$ is set to a uniform distribution between [0, 1]. The Beta distribution has a mean $\alpha_i/(\alpha_i + \beta_i)$ and its distribution becomes more concentrated in its mean as $\alpha_i + \beta_i$ increases.

Since the distribution of rewards used in Thomson's sampling is fixed in advance, it does not fit the non-stationary nature of EAs. This work considers using Dynamic Thomson Sampling (DYTS) to address this problem, which uses a threshold C to control the two different parameter update rules. If $\alpha_i + \beta_i < C$ the parameters of the Beta distribution are updated with the Thomson sample as seen above. On the other hand, if $\alpha_i + \beta_i \geq C$, the rule aims to adjust the parameters so that $\alpha_i + \beta_i = C$ is obtained, and the pseudocode in Algorithm 3.2.1 can be observed.

3.2.1 Credit Assignment

The fitness improvement (FI) is used as a measure to evaluate the credit of the operator chosen by the AOS. FI is used to avoid the problems generated when considering a reward in the Bernoulli distribution, which is a binary value. Each operator is considered a bandit's arm. The operator receives a reward of 1 if his application improves the evolutionary population towards the optimum; otherwise, his reward is 0.

Algorithm 3.2.1 Parameter update

Input: Parameters of $i-$ th operator α_i, β_i, it's up to date reward *reward* and threshold
Output: Updated (α_i, β_i)
1: $\boldsymbol{if} \alpha_i + \beta_i < C$ ***then***
2: $\alpha_i = \alpha_i + r;$
3: $\beta_i = \beta_i + 1 - r;$
4: ***else***
5: $\alpha_i = (\alpha_i + r)\frac{C}{C+1};$
6: $\beta_i = (\beta_i + 1 - r)\frac{C}{C+1};$
7: ***return*** (α_i, β_i)

3.2.2 Operator Selection

The idea proposed by [7] for the AOS method based on the DYTS strategy is simple and intuitive. The operator with the wide sample range of the updated Beta distribution is chosen to be used in the next iteration.

Algorithm 3.2.2 DYTS AOS

Input: Parameters of Beta distribution $\alpha = (\alpha_1, \ldots, \alpha_k)$ *and* $\beta = (\beta_1, \ldots, \beta_k)$
Output: Index of the selected reproduction operator
1: ***for*** $i \leftarrow 1$ ***to*** k ***do***
2: sample the estimated mean reward $\hat{\theta}_i$ from Beta distribution P^{Beta}
3: $op = argmax_{i \in \{1, \ldots k\}} \hat{\theta}_i$
4: ***return*** a

AOS Based on Dynamic Thompson Sampling (DYTS) has been added to the DMOEA/D algorithm. Below is the pseudocode:

Algorithm 3.2.3 DMOEA/D-DYTS

Inputs: DMOP – dynamic multiobjective problem
N – number of the subproblems
stopping criterion
T – number of the weight vectors in the neighborhood of each weight vector
uniform spread of N weight vectors: $\lambda^1, \ldots, \lambda^N$
nT – change frequency
$tauT$ – change severity
C – distribution threshold /* C=100 */
K – number of genetic operators
Outputs: $Front_f$ – list of EP fronts

$EP = \emptyset$
Get the distances (Euclidean) between two weight vectors (any) and then compute the T closest weight vectors to each weight vector.
for $i \leftarrow 1$ ***to*** N ***do***
  $B(i) = \{i_1, \ldots, i_T\}$ /* where $\lambda^{i_1}, \ldots, \lambda^{i_T}$ are T closest weight vectors to λ^i */
end for
Generate initial population (*Pop*) $x^1, \ldots, x^N$
initialize parameters of the Beta distribution α, β
$FV^i = F(x^i)$
initialize $z = (z_1, \ldots, z_m)^N$
$time = 0$
$detectors =$ GetDetectorList(Pop)
$it = 0$
$f = 1$
while stopping criteria is not satisfied ***do***
  for $i \leftarrow 1$ ***to*** N ***do***
    $op = AOS(\alpha, \beta)$
    Generate a new solution y = reproduction (*i*, *op*, *Pop*)
    Repair y to produce y' using a repair or improvement heuristic
    for $j \leftarrow 1$ ***to*** m ***do***
      if $z_j < f_j(y')$ *then*
        $z_j = f_j(y')$
      end if
    end for
    foreach index $j \in B(i)$ ***do***
      $FI = max_{x_j \in Pop}\{ g^{te}(x^j|\lambda^j, z) - g^{te}(y'|\lambda^j, z)\}$
      $Reward = 0$

```
        if g^te(y'|λ^j, z) ≤ g^te(x^j|λ^j, z) then
            x^j = y'
            FV^j = F(y')
            Reward = 1
        end if
        (α_op, β_op) = ParameterUpdate(α_op, β_op, Reward, C)
    end foreach
  end for
  remove from EP list all solutions dominated by F(y')
  add solution F(y') to EP list if no solutions in EP list dominate solution F(y')
  time = 1/nT × ⌊1×it / tauT⌋
  if ProblemChangeDetection (detectors) then
     add EP list to Front_f
     initialize z = (z_1, ..., z_m)^N
     evaluate Pop
     run ChangeResponseMechanism
     FV^i = F(x^i)
     remove from EP list all solutions dominated by F(y')
     add solution F(y') to EP list if no solutions in EP list dominate solution F(y')
     f = f + 1;
  end if
  it = it + 1
end while
return Front_f
```

3.3 *Adaptive Operator Selection with Test-and-Apply Structure (TAOS)*

This method is proposed by Lisha Dong, which is split into various successive segments, and each segment is built with two steps: "the test phase and the application phase" [8].

3.3.1 Test Phase

In the testing phase, each operator must be tested in the same environment, and each operator must be evaluated on the same evaluation function number. Therefore, the test phase is divided into N parts, for example, *test1*, …, *testN* (see Fig. 3). All operators will be evaluated within the sequence described. Therefore, the number of function evaluations for the testing phase can be defined as follows:

$$FE_{test} = K \times N \tag{6}$$

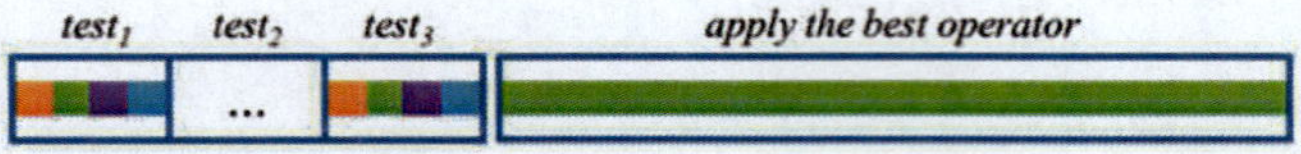

Fig. 3 Test-and-apply structure [8]

where K is the number of operators and N is the size of the population.

After applying an operator in the search process, it is necessary to measure its impact. For this work, a successful update indicator (SUI) is introduced for the allocation of credits since they consider that the existing methods FI and FIR are not fair in assigning credit to each operator. In the case of FI, the reasons that stand out are: that its range varies from different subproblems and changes dynamically at distinct phases of the optimization process. In FIR it lies in the fact that a second solution can substitute a diverse number of solutions in the population. SUI is defined as follows:

$$SUI = \begin{cases} 1, & if\ the\ generated\ solution\ is\ better\ than \\ & at\ least\ one\ solution\ in\ the\ population \\ 0, & otherwise \end{cases} \tag{7}$$

When a child updates the first solution, the value of SUI changes from 0 to 1. However, if the child keeps updating, the SUI solutions will remain unchanged, making comparing different operators more stable and fairer.

3.3.2 Application Phase

At the end of the test phase, the successful update count (SUC) is obtained for each operator. Its formulation is presented below in Eq. 8.

$$SUC_{op} = \sum_{i=1} SUI_{op}^{i} \tag{8}$$

where SUI_{op}^{i} means that the *op* operator is tested in evaluating the i-th function of the test phase, compared to existing approaches, SUC has the advantage that it is easy to calculate without additional parameters and allows comparisons between different operators.

The operator selection is made by comparing them, and the operator with the SUC with the highest value will be the one selected to be applied in the remaining phase of that segment. Therefore, the number of feature evaluations for the application phase in each segment is defined as follows:

$$FE_{apply} = FE_{test} \times k \tag{9}$$

where the variable k controls the resources for the test and application phases. $\frac{1}{k+1}$ is used for the test phase and $\frac{k}{k+1}$ for the application phase. A smaller value of k, i.e., $k < 1$, allocates more resources to the test phase versus the apply phase, a larger value of k makes fewer resources available for the test phase.

Adaptive operator selection with a test-and-apply structure (TAOS) has been added to the DMOEA/D algorithm. Below is the pseudocode:

Algorithm 3.3.2 DMOEA/D-TAOS

Inputs: DMOP – dynamic multiobjective problem
N – number of the subproblems
stopping criterion
T – number of the weight vectors in the neighborhood of each weight vector
uniform spread of N weight vectors: $\lambda^1, \ldots, \lambda^N$
nT – change frequency
$tauT$ – change severity
K – number of genetic operators
$NPop$ – population size
Outputs: $Front_f$ – list of EP fronts

$EP = \emptyset$
Get the distances (Euclidean) between two weight vectors (any) and then compute the T closest weight vectors to each weight vector.
for $i \leftarrow 1$ ***to*** N ***do***
  $B(i) = \{i_1, \ldots, i_T\}$ /* where $\lambda^{i_1}, \ldots, \lambda^{i_T}$ are T closest weight vectors to λ^i */
end for
Generate initial population (Pop) $x^1, \ldots, x^N$
$FV^i = F(x^i)$
initialize $z = (z_1, \ldots, z_m)^N$
$SUC_i = 0, \forall i \in \{1, \ldots K\}$
$k = 1$
$FE_{test} = K \times NPop, FE_{apply} = k \times FE_{test}$
$time = 0$
$detectors$ = GetDetectorList(Pop)
$it = 0$
$f = 1$
while stopping criteria is not satisfied ***do***
  for $i \leftarrow 1$ ***to*** N ***do***
    if $mod(it, FE_{test}, FE_{apply}) < FE_{apply}$ ***then***
      if $mod(it, FE_{test}, FE_{apply}) == 0$ ***then*** /* test phase */
        $SUC_i = 0, \forall i \in \{1, \ldots K\}$
      end if
      $op = mod(it, K) + 1$
    else /* apply phase */
      $op = arg\ max_{i \in \{1, \ldots, K\}}\{SUC\}$
    end if

```
        Generate a new solution y = reproduction (i, op, Pop)
        Repair y to produce y' using a repair or improvement heuristic
        for j ← 1 to m do
            if z_j < f_j(y') then
                z_j = f_j(y')
            end if
        end for
        foreach index j ∈ B(i) do
            if g^te(y'|λ^j, z) ≤ g^te(x^j|λ^j, z) then
                x^j = y'
                FV^j = F(y')
                SUI = 1
            end if
        end foreach
        if it is in test phase then
            SUC_op = SUC_op + SUI
        end if
    end for
    remove from EP list all solutions dominated by F(y')
    add solution F(y') to EP list if no solutions in EP list dominate solution F(y')
    time = 1/nT × ⌊1×it / tauT⌋
    if ProblemChangeDetection (detectors) then
        add EP to Front_f
        initialize z = (z_1, ..., z_m)^N
        evaluate Pop
        run ChangeResponseMechanism
        FV^i = F(x^i)
        remove from EP list all solutions dominated by F(y')
        add solution F(y') to EP list if no solutions in EP list dominate solution F(y')
        f = f + 1;
    end if
    it = it + 1
end while
return Front_f
```

3.4 Operator Pool

In this work, five operators are considered. First, four variants of the differential evolution (DE) crossover operator, while the other is the uniform mutation (UM). Then, in the AOS, each of the four crossover operators is used as a bandit's arm, as follows:

(1) DE/rand/1

$$v^i = x^i + F \times (x^{r1} - x^{r2})$$

(2) DE/rand/2

$$v^i = x^i + F \times \left(x^{r1} - x^{r2}\right) + F \times \left(x^{r3} - x^{r4}\right)$$

(3) DE/current-to-rand/1

$$v^i = x^i + K \times \left(x^i - x^{r1}\right) + F \times \left(x^{r2} - x^{r3}\right) + F \times \left(x^{r4} - x^{r5}\right)$$

(4) DE/current-to-rand/2

$$v^i = x^i + K \times \left(x^i - x^{r1}\right) + F \times \left(x^{r2} - x^{r3}\right)$$

The parameters used for the four crossover operators are $CR = 1.0$ and $F = 0.5$. The mutation operator is the polynomial mutation with a mutation probability of 20% and a distribution index of 20. In Algorithm 3.4 the operation of the pool of operators is shown.

Algorithm 3.4 y=reproduction (*i, op, S*)

Input: $i \in \{1, \ldots, N\}$ – the current solution index
$op \in \{1, \ldots, K\}$ – the selected operator index
S – the scope to select parents
Output: y – the produced child solutions

switch *op* ***do***
  case 1
    $CR = 1.0$ /* currenDE with CR = 1.0, F = 0.5 */
    Randomly select two solutions x^{r1}, x^{r2} from S
    Apply DE/rand/1 *on* x^{r1}, x^{r2} to produce offspring y
  case 2
    $CR = 1.0$ /* currenDE with CR = 1.0, F = 0.5 */
    Randomly select four solutions $x^{r1}, x^{r2}, x^{r3}, x^{r4}$ from S
    Apply DE/rand/2 *on* $x^{r1}, x^{r2}, x^{r3}, x^{r4}$ to produce offspring y
  case 3
    $CR = 1.0$ /* currenDE with CR = 1.0, F = 0.5 */
    Randomly select four solutions $x^{r1}, x^{r2}, x^{r3}, x^{r4}, x^{r5}$ from S
    Apply DE/current − to − rand/1 *on* $x^{r1}, x^{r2}, x^{r3}, x^{r4}, x^{r5}$ to produce offspring y
  case 4
    $CR = 1.0$ /* currenDE with CR = 1.0, F = 0.5 */
    Randomly select four solutions x^{r1}, x^{r2}, x^{r3} from S
    Apply DE/current − to − rand/2 *on* x^{r1}, x^{r2}, x^{r3} to produce offspring y
Apply polynomial mutation on y /* $p_m = 0.2, \eta = 20$ */
return *y*

Table 1 Dynamic multiobjective problems

Problem	Change severity τ_t	Change frequency n_t
dMOP1	20	10
dMOP2	20	10
FDA1	20	10
FDA3	20	10

Table 2 Variables and parameters

Variables/parameters	DMOEA/D-FRRMAB	DMOEA/D-DYTS	DMOEA/D-TAOS
maxIt	200	200	200
nPop	100	100	100
fileSize	100	100	100
zeta	0.2	0.2	0.2
decayFactor	1	–	–
C	–	1	–
k	–	–	1

4 Computational Experiments

The Table 1 shows four dynamic multi-objective problems and their frequency and severity of change values. Each problem produces ten fronts. For each algorithm and multi-objective dynamic problem, 30 independent executions have been carried out (Table 2).

5 Results

The result of the experimentation is presented below in a table by metric. In each table, the first column corresponds to the problem and the front number of said problem. The columns two to four correspond to the results of each of the algorithms. The algorithm in the last column is taken as the reference. The following symbols are included in the results tables: the symbol ▲ means that there is statistical significance in favor of the reference algorithm, ▼ that there is statistical significance in favor of the algorithm that is compared with the reference algorithm, and == means there is no significance statistics.

5.1 Hypervolume

The hypervolume (HV) gives the multidimensional volume of the portion of target space that is weakly dominated by an approximation set (Table 3).

The *P-value* calculated for the DMOEA/D-FRRMAB algorithm by the Friedman test is 8.875455925760889E-11, with a significance level of 5%. Therefore, it can be said that it is significant.

5.2 Generalized Spread

The generalized spread (GS) measures the uniformity of the solutions found and their dispersion (Table 4).

The *P-value* calculated for the DMOEAD-DYTS algorithm by the Friedman test is 5.30191446301842E-11, with a significance level of 5%. Therefore, it can be said that it is significant.

5.3 Inverted Generational Distance

The inverted generational distance (IGD) gives the average distance between any point from the reference set and its closest point from the approximation set (Table 5).

The P-*value* calculated for the DMOEA/D-FRRMAB by the Friedman test is 1.2058720688656877E-10, with a significance level of 5%. Therefore, it can be said that it is significant.

6 Conclusions

In the analysis carried out in this work it can be observed that according to the experiments carried out, the TAOS method applied to the static MOEA/D algorithm outperforms the FRRMAB algorithm, however in dynamic problems the following can be observed: in the evaluation of the metric for hypervolume (HV), the DMOEA/D FRRMAB algorithm obtains the best value in 32 of 40 obtained fronts, and the non-parametric Friedman test shows a significant difference in favor of this algorithm. In the evaluation of the generalized spread metric (GS) the DMOEA/DYTS algorithm obtains the best value in 27 of 40 obtained fronts, and the non-parametric Friedman test shows a significant difference in favor of this algorithm.

In the evaluation of the inverted generational distance (IGD), the DMOEA/D FRRMAB algorithm obtains the best value in 27 of 40 obtained fronts, and the non-parametric Friedman test shows a significant difference in favor of this algorithm.

Table 3 Comparison of the three algorithms (median and IQR values) for HV

HV			
Problem	DMOEA/D-FRRMAB	DMOEA/D-DYTS	DMOEA/D-TAOS
DMOP1-F1	$0.00e+00_{0.00e+00}$ ==	$0.00e+00_{0.00e+00}$ ==	$0.00e+00_{0.00e+00}$
DMOP1-F2	$0.00e+00_{3.98e-01}$ ==	$0.00e+00_{0.00e+00}$ ==	$0.00e+00_{3.97e-01}$
DMOP1-F3	$3.73e-01_{3.89e-01}$ ▼	$0.00e+00_{3.84e-01}$ ▲	$1.94e-01_{3.85e-01}$
DMOP1-F4	$3.71e-01_{3.73e-01}$ ▼	$3.62e-01_{3.72e-01}$ ==	$3.67e-01_{3.71e-01}$
DMOP1-F5	$3.56e-01_{9.40e-02}$ ▼	$3.52e-01_{3.58e-01}$ ▲	$3.56e-01_{1.92e-02}$
DMOP1-F6	$3.46e-01_{7.98e-03}$ ▼	$3.45e-01_{3.47e-01}$ ▼	$3.45e-01_{7.12e-03}$
DMOP1-F7	$3.36e-01_{3.28e-03}$ ==	$3.36e-01_{3.37e-01}$ ▲	$3.36e-01_{3.88e-03}$
DMOP1-F8	$3.30e-01_{1.94e-03}$ ▼	$3.28e-01_{3.30e-01}$ ▲	$3.29e-01_{2.57e-03}$
DMOP1-F9	$3.25e-01_{2.36e-03}$ ▼	$3.23e-01_{3.25e-01}$ ▲	$3.24e-01_{2.26e-03}$
DMOP1-F10	$3.22e-01_{2.73e-03}$ ▼	$3.20e-01_{3.22e-01}$ ▲	$3.21e-01_{3.93e-03}$
DMOP2-F1	$2.29e-01_{3.77e-01}$ ▲	$8.76e-02_{3.22e-01}$ ▲	$2.71e-01_{3.37e-01}$
DMOP2-F2	$3.90e-01_{2.68e-02}$ ▲	$3.77e-01_{8.35e-02}$ ▲	$3.95e-01_{2.82e-02}$
DMOP2-F3	$3.61e-01_{2.73e-02}$ ▼	$3.28e-01_{5.54e-02}$ ▲	$3.58e-01_{1.64e-02}$
DMOP2-F4	$3.30e-01_{2.79e-02}$ ==	$2.78e-01_{6.89e-02}$ ▲	$3.31e-01_{3.62e-02}$
DMOP2-F5	$3.07e-01_{6.35e-02}$ ▼	$2.34e-01_{9.47e-02}$ ▲	$2.83e-01_{6.63e-02}$
DMOP2-F6	$2.92e-01_{4.76e-02}$ ▼	$1.90e-01_{1.14e-01}$ ▲	$2.71e-01_{7.21e-02}$
DMOP2-F7	$3.02e-01_{4.41e-02}$ ▼	$2.11e-01_{1.60e-01}$ ▲	$2.96e-01_{3.57e-02}$
DMOP2-F8	$3.21e-01_{9.91e-03}$ ▼	$2.55e-01_{2.00e-01}$ ▲	$3.17e-01_{1.50e-02}$
DMOP2-F9	$3.21e-01_{3.36e-03}$ ▼	$3.11e-01_{9.94e-02}$ ▲	$3.19e-01_{7.01e-03}$
DMOP2-F10	$3.20e-01_{2.74e-03}$ ▼	$3.15e-01_{1.40e-02}$ ▲	$3.19e-01_{2.82e-03}$
FDA1-F1	$2.91e-01_{7.29e-02}$ ▲	$2.66e-01_{1.07e-01}$ ▲	$3.07e-01_{1.11e-01}$
FDA1-F2	$2.07e-01_{1.91e-01}$ ==	$1.29e-01_{1.71e-01}$ ▲	$2.06e-01_{9.12e-02}$
FDA1-F3	$0.00e+00_{1.05e-01}$ ==	$0.00e+00_{2.18e-02}$ ==	$0.00e+00_{4.59e-02}$
FDA1-F4	$0.00e+00_{0.00e+00}$ ==	$0.00e+00_{0.00e+00}$ ==	$0.00e+00_{0.00e+00}$
FDA1-F5	$0.00e+00_{0.00e+00}$ ==	$0.00e+00_{0.00e+00}$ ==	$0.00e+00_{0.00e+00}$
FDA1-F6	$0.00e+00_{0.00e+00}$ ==	$0.00e+00_{0.00e+00}$ ==	$0.00e+00_{0.00e+00}$
FDA1-F7	$0.00e+00_{0.00e+00}$ ==	$0.00e+00_{0.00e+00}$ ==	$0.00e+00_{0.00e+00}$
FDA1-F8	$0.00e+00_{0.00e+00}$ ==	$0.00e+00_{0.00e+00}$ ==	$0.00e+00_{0.00e+00}$
FDA1-F9	$0.00e+00_{7.86e-02}$ ==	$0.00e+00_{0.00e+00}$ ==	$0.00e+00_{0.00e+00}$
FDA1-F10	$0.00e+00_{5.72e-01}$ ==	$0.00e+00_{0.00e+00}$ ==	$0.00e+00_{0.00e+00}$
FDA3-F1	$8.90e-01_{1.01e-01}$ ▲	$8.86e-01_{7.26e-02}$ ==	$9.21e-01_{8.60e-02}$
FDA3-F2	$8.57e-01_{6.54e-02}$ ▼	$8.51e-01_{9.80e-02}$ ==	$8.49e-01_{6.95e-02}$
FDA3-F3	$7.15e-01_{1.60e-01}$ ▼	$6.83e-01_{1.46e-01}$ ==	$6.05e-01_{1.75e-01}$
FDA3-F4	$4.76e-01_{2.40e-01}$ ▼	$3.87e-01_{2.36e-01}$ ==	$3.90e-01_{4.17e-01}$
FDA3-F5	$3.13e-01_{3.97e-01}$ ▼	$1.90e-01_{3.41e-01}$ ▲	$3.04e-01_{4.17e-01}$

(continued)

Table 3 (continued)

HV			
Problem	DMOEA/D-FRRMAB	DMOEA/D-DYTS	DMOEA/D-TAOS
FDA3-F6	3.44e-01$_{2.68e-01}$ ▼	1.03e-01$_{2.91e-01}$ ▲	2.44e-01$_{3.77e-01}$
FDA3-F7	2.78e-01$_{1.81e-01}$ ▼	1.55e-01$_{2.50e-01}$ ▲	2.72e-01$_{2.86e-01}$
FDA3-F8	2.05e-01$_{4.31e-03}$ ▼	1.94e-01$_{4.46e-02}$ ▲	2.01e-01$_{4.38e-02}$
FDA3-F9	1.46e-01$_{2.40e-03}$ ▼	1.45e-01$_{6.45e-03}$ ▼	1.45e-01$_{5.92e-03}$
FDA3-F10	1.14e-01$_{1.56e-03}$ ▼	1.13e-01$_{1.49e-03}$ ==	1.13e-01$_{2.21e-03}$

Table 4 Comparison of the three algorithms (median and IQR values) for GS

GS			
Problem	DMOEAD-FRRMAB	DMOEAD-DYTS	DMOEAD-TAOS
DMOP1-F1	4.44e-02$_{3.83e-02}$ ▲	4.30e-02$_{1.90e-02}$ ▲	5.27e-02$_{6.55e-02}$
DMOP1-F2	1.27e-01$_{6.40e-01}$ ▼	5.51e-02$_{1.49e-01}$ ▲	1.13e-01$_{5.55e-01}$
DMOP1-F3	5.53e-01$_{6.36e-01}$ ▼	2.69e-01$_{6.23e-01}$ ==	4.38e-01$_{6.23e-01}$
DMOP1-F4	6.60e-01$_{5.14e-01}$ ==	6.43e-01$_{4.34e-01}$ ▲	6.83e-01$_{5.95e-01}$
DMOP1-F5	7.29e-01$_{2.01e-01}$ ▼	6.03e-01$_{1.99e-01}$ ▲	7.27e-01$_{1.60e-01}$
DMOP1-F6	7.08e-01$_{9.04e-02}$ ▲	6.80e-01$_{1.84e-01}$ ▲	7.32e-01$_{9.98e-02}$
DMOP1-F7	7.26e-01$_{6.21e-02}$ ==	6.85e-01$_{1.77e-01}$ ▲	7.16e-01$_{9.21e-02}$
DMOP1-F8	7.28e-01$_{8.85e-02}$ ▲	7.14e-01$_{1.91e-01}$ ▲	7.40e-01$_{8.36e-02}$
DMOP1-F9	7.29e-01$_{7.38e-02}$ ==	7.00e-01$_{1.88e-01}$ ▲	7.31e-01$_{6.97e-02}$
DMOP1-F10	7.33e-01$_{7.63e-02}$ ==	6.97e-01$_{2.16e-01}$ ▲	7.54e-01$_{9.14e-02}$
DMOP2-F1	2.70e-01$_{4.72e-01}$ ▲	1.49e-01$_{5.40e-01}$ ▲	3.53e-01$_{5.64e-01}$
DMOP2-F2	4.08e-01$_{2.76e-01}$ ▲	3.52e-01$_{2.96e-01}$ ▲	5.19e-01$_{2.52e-01}$
DMOP2-F3	2.86e-01$_{2.29e-01}$ ==	2.23e-01$_{1.60e-01}$ ▲	2.84e-01$_{1.41e-01}$
DMOP2-F4	2.10e-01$_{1.42e-01}$ ▲	1.93e-01$_{1.53e-01}$ ▲	2.11e-01$_{1.82e-01}$
DMOP2-F5	1.79e-01$_{1.57e-01}$ ▼	1.54e-01$_{8.55e-02}$ ==	1.45e-01$_{1.48e-01}$
DMOP2-F6	1.77e-01$_{2.31e-01}$ ▼	1.41e-01$_{5.24e-02}$ ▲	1.63e-01$_{1.35e-01}$
DMOP2-F7	3.25e-01$_{2.23e-01}$ ▼	1.73e-01$_{1.18e-01}$ ▲	1.96e-01$_{2.82e-01}$
DMOP2-F8	5.11e-01$_{1.42e-01}$ ▼	2.65e-01$_{3.20e-01}$ ▲	5.00e-01$_{1.79e-01}$
DMOP2-F9	6.81e-01$_{6.56e-02}$ ▼	5.54e-01$_{4.63e-01}$ ▲	6.68e-01$_{1.42e-01}$
DMOP2-F10	6.95e-01$_{9.95e-02}$ ▲	6.74e-01$_{1.81e-01}$ ▲	7.00e-01$_{6.63e-02}$
FDA1-F1	2.79e-01$_{1.36e-01}$ ▼	2.63e-01$_{1.17e-01}$ ▼	2.51e-01$_{1.02e-01}$
FDA1-F2	2.01e-01$_{1.45e-01}$ ▼	1.86e-01$_{8.38e-02}$ ▼	1.50e-01$_{1.36e-01}$

(continued)

Table 4 (continued)

GS			
Problem	DMOEAD-FRRMAB	DMOEAD-DYTS	DMOEAD-TAOS
FDA1-F3	1.13e-01$_{9.83e-02}$ ▼	1.11e-01$_{7.88e-02}$ ▼	1.00e-01$_{8.53e-02}$
FDA1-F4	7.26e-02$_{2.60e-02}$ ==	6.98e-02$_{2.66e-02}$ ==	6.65e-02$_{3.38e-02}$
FDA1-F5	5.69e-02$_{1.11e-02}$ ▼	5.63e-02$_{1.58e-02}$ ▼	5.53e-02$_{1.15e-02}$
FDA1-F6	5.24e-02$_{1.94e-02}$ ▼	4.59e-02$_{1.63e-02}$ ▼	4.53e-02$_{2.21e-02}$
FDA1-F7	4.36e-02$_{1.89e-02}$ ▼	3.58e-02$_{1.07e-02}$ ▲	3.76e-02$_{1.50e-02}$
FDA1-F8	3.96e-02$_{2.30e-02}$ ▼	3.25e-02$_{1.05e-02}$ ▲	3.88e-02$_{2.42e-02}$
FDA1-F9	5.16e-02$_{3.07e-02}$ ▼	3.16e-02$_{1.65e-02}$ ▲	4.19e-02$_{2.89e-02}$
FDA1-F10	4.99e-02$_{2.76e-01}$ ==	3.57e-02$_{3.09e-02}$ ▲	5.02e-02$_{2.38e-02}$
FDA3-F1	1.26e-01$_{4.88e-02}$ ▼	1.26e-01$_{3.99e-02}$ ▼	1.21e-01$_{3.58e-02}$
FDA3-F2	1.59e-01$_{2.83e-02}$ ==	1.55e-01$_{2.49e-02}$ ▲	1.56e-01$_{3.88e-02}$
FDA3-F3	2.46e-01$_{5.61e-02}$ ▲	2.64e-01$_{7.67e-02}$ ▼	2.61e-01$_{8.50e-02}$
FDA3-F4	2.20e-01$_{2.06e-01}$ ▼	2.06e-01$_{1.21e-01}$ ▼	1.86e-01$_{2.06e-01}$
FDA3-F5	1.91e-01$_{1.51e-01}$ ▼	1.63e-01$_{1.05e-01}$ ▲	1.75e-01$_{1.44e-01}$
FDA3-F6	1.70e-01$_{1.12e-01}$ ▲	1.37e-01$_{8.38e-02}$ ▲	1.82e-01$_{1.28e-01}$
FDA3-F7	1.31e-01$_{4.26e-02}$ ▲	1.13e-01$_{6.11e-02}$ ==	1.31e-01$_{5.96e-02}$
FDA3-F8	7.71e-02$_{5.82e-03}$ ▼	7.69e-02$_{1.15e-02}$ ▼	7.36e-02$_{8.84e-03}$
FDA3-F9	4.35e-02$_{1.33e-03}$ ==	4.36e-02$_{2.43e-03}$ ▼	4.32e-02$_{1.51e-03}$
FDA3-F10	3.24e-02$_{5.08e-04}$ ▼	3.25e-02$_{4.97e-04}$ ▼	3.21e-02$_{7.05e-04}$

Table 5 Comparison of the three algorithms (median and IQR values) for IGD

IGD			
Problem	DMOEA/D-FRRMAB	DMOEA/D-DYTS	DMOEA/D-TAOS
DMOP1-F1	1.82e-01$_{1.41e-01}$ ▼	1.86e-01$_{8.86e-02}$ ▼	1.30e-01$_{1.69e-01}$
DMOP1-F2	3.76e-02$_{1.92e-01}$ ==	1.16e-01$_{1.21e-01}$ ▼	3.97e-02$_{1.87e-01}$
DMOP1-F3	6.38e-04$_{1.04e-01}$ ==	2.52e-02$_{1.24e-01}$ ▼	5.93e-03$_{1.02e-01}$
DMOP1-F4	5.52e-04$_{3.88e-02}$ ▲	5.81e-04$_{3.78e-02}$ ==	6.56e-04$_{3.80e-02}$
DMOP1-F5	6.07e-04$_{2.45e-03}$ ==	5.91e-04$_{2.32e-02}$ ▼	5.83e-04$_{5.79e-04}$
DMOP1-F6	5.07e-04$_{2.60e-04}$ ==	5.74e-04$_{2.30e-02}$ ▼	5.33e-04$_{3.88e-04}$
DMOP1-F7	5.20e-04$_{1.74e-04}$ ==	5.21e-04$_{2.29e-02}$ ▲	5.27e-04$_{3.31e-04}$
DMOP1-F8	4.91e-04$_{1.93e-04}$ ▲	5.55e-04$_{2.27e-02}$ ▼	5.05e-04$_{2.13e-04}$
DMOP1-F9	5.11e-04$_{2.34e-04}$ ▼	6.20e-04$_{2.27e-02}$ ▼	4.93e-04$_{1.95e-04}$
DMOP1-F10	5.06e-04$_{2.09e-04}$ ▲	5.86e-04$_{2.26e-02}$ ▼	5.45e-04$_{2.10e-04}$
DMOP2-F1	5.54e-03$_{1.50e-02}$ ▼	1.04e-02$_{1.27e-02}$ ▼	5.03e-03$_{1.28e-02}$

(continued)

Table 5 (continued)

IGD			
Problem	DMOEA/D-FRRMAB	DMOEA/D-DYTS	DMOEA/D-TAOS
DMOP2-F2	6.92e-04$_{6.23e-04}$ ▼	9.54e-04$_{1.97e-03}$ ▼	6.86e-04$_{6.67e-04}$
DMOP2-F3	9.20e-04$_{8.64e-04}$ ▲	1.62e-03$_{1.49e-03}$ ▼	9.32e-04$_{4.12e-04}$
DMOP2-F4	1.08e-03$_{6.80e-04}$ ==	2.30e-03$_{2.63e-03}$ ▼	1.20e-03$_{8.51e-04}$
DMOP2-F5	1.35e-03$_{1.55e-03}$ ▲	3.87e-03$_{3.55e-03}$ ▼	1.88e-03$_{1.72e-03}$
DMOP2-F6	1.42e-03$_{1.17e-03}$ ▲	4.93e-03$_{5.18e-03}$ ▼	2.02e-03$_{2.01e-03}$
DMOP2-F7	1.07e-03$_{9.59e-04}$ ▲	3.93e-03$_{6.45e-03}$ ▼	1.27e-03$_{9.29e-04}$
DMOP2-F8	5.59e-04$_{3.76e-04}$ ▲	2.12e-03$_{7.40e-03}$ ▼	6.89e-04$_{2.42e-04}$
DMOP2-F9	5.01e-04$_{1.05e-04}$ ▲	6.31e-04$_{2.55e-03}$ ▼	5.58e-04$_{1.77e-04}$
DMOP2-F10	4.97e-04$_{1.37e-04}$ ▼	5.68e-04$_{4.36e-04}$ ▼	4.80e-04$_{1.45e-04}$
FDA1-F1	1.09e-02$_{2.44e-03}$ ▼	1.10e-02$_{3.71e-03}$ ▼	9.55e-03$_{3.72e-03}$
FDA1-F2	1.34e-02$_{8.22e-03}$ ==	1.56e-02$_{8.63e-03}$ ▼	1.30e-02$_{4.54e-03}$
FDA1-F3	2.57e-02$_{1.68e-02}$ ==	3.37e-02$_{1.79e-02}$ ▼	2.90e-02$_{1.45e-02}$
FDA1-F4	4.93e-02$_{3.01e-02}$ ▼	5.55e-02$_{2.99e-02}$ ▼	4.90e-02$_{3.15e-02}$
FDA1-F5	7.29e-02$_{5.52e-02}$ ▲	8.85e-02$_{5.95e-02}$ ▼	8.66e-02$_{4.55e-02}$
FDA1-F6	9.01e-02$_{8.65e-02}$ ▲	1.20e-01$_{7.86e-02}$ ▼	1.13e-01$_{7.69e-02}$
FDA1-F7	1.06e-01$_{9.86e-02}$ ▲	1.59e-01$_{6.82e-02}$ ▼	1.50e-01$_{8.67e-02}$
FDA1-F8	1.09e-01$_{1.12e-01}$ ▲	1.80e-01$_{6.63e-02}$ ▼	1.44e-01$_{1.12e-01}$
FDA1-F9	5.57e-02$_{1.19e-01}$ ▲	1.82e-01$_{1.05e-01}$ ▼	1.15e-01$_{1.26e-01}$
FDA1-F10	3.29e-02$_{1.48e-01}$ ▲	1.58e-01$_{1.41e-01}$ ▼	9.54e-02$_{9.53e-02}$
FDA3-F1	2.18e-02$_{1.65e-02}$ ▲	2.29e-02$_{1.79e-02}$ ▲	2.58e-02$_{1.87e-02}$
FDA3-F2	1.70e-02$_{6.85e-03}$ ==	1.92e-02$_{6.83e-03}$ ▼	1.57e-02$_{6.12e-03}$
FDA3-F3	7.21e-03$_{5.20e-03}$ ▼	8.11e-03$_{6.49e-03}$ ▼	7.16e-03$_{6.82e-03}$
FDA3-F4	4.68e-03$_{5.20e-03}$ ▲	4.80e-03$_{4.27e-03}$ ▼	4.69e-03$_{7.89e-03}$
FDA3-F5	6.00e-03$_{1.14e-02}$ ▼	1.00e-02$_{1.34e-02}$ ▼	5.98e-03$_{1.35e-02}$
FDA3-F6	6.22e-03$_{8.39e-03}$ ▲	1.33e-02$_{1.76e-02}$ ▼	6.99e-03$_{1.84e-02}$
FDA3-F7	8.96e-03$_{6.73e-03}$ ▼	1.04e-02$_{1.15e-02}$ ==	8.93e-03$_{2.40e-02}$
FDA3-F8	1.19e-02$_{1.78e-04}$ ▲	1.22e-02$_{1.56e-03}$ ==	1.20e-02$_{1.87e-03}$
FDA3-F9	1.45e-02$_{5.67e-05}$ ▲	1.45e-02$_{2.89e-04}$ ▼	1.45e-02$_{3.79e-04}$
FDA3-F10	1.61e-02$_{2.77e-05}$ ▲	1.61e-02$_{4.62e-05}$ ==	1.61e-02$_{4.20e-05}$

References

1. Azzouz, R., Bechikh, S., & Ben Said, L. (2016). Dynamic multi-objective optimization using evolutionary algorithms: A survey. In: *Recent advances in evolutionary multi-objective optimization* (pp. 31–70). Springer International Publishing.
2. Deb, K., Rao N., U. B., & Karthik, S. (n.d.). Dynamic multi-objective optimization and decision-making using modified NSGA-II: A case study on hydro-thermal power scheduling. In: Lecture notes in computer science (pp. 803–817). Berlin: Springer.

3. Zhang, Q., & Li, H. (2007). MOEA/D: A multiobjective evolutionary algorithm based on decomposition. *IEEE Transactions on Evolutionary Computation, 11*(6), 712–731. Institute of Electrical and Electronics Engineers.
4. Yu, Q., Zhong, S., Liu, Z., Lin, Q., & Huang, P. (2020). Dynamic multiobjective optimization with multiple response strategies based on linear environment detection. In: Q. Wang (Ed.), Complexity (Vol. 2020, pp. 1–26). Hindawi Limited.
5. Li, K., Fialho, A., Kwong, S., & Zhang, Q. (2014). Adaptive operator selection with bandits for a multiobjective evolutionary algorithm based on decomposition. *IEEE Transactions on Evolutionary Computation, 18*(1), pp. 114–130. Institute of Electrical and Electronics Engineers (IEEE).
6. Fialho, Á. (2010). Adaptive operator selection for optimization. Computer Science [cs]. Université Paris Sud - Paris XI, 2010. English. tel-00578431
7. Sun, L., & Li, K. (2020). Adaptive operator selection based on dynamic thompson sampling for MOEA/D (Version 1).
8. Dong, L., Lin, Q., Zhou, Y., & Jiang, J. (2022). Adaptive operator selection with test-and-apply structure for decomposition-based multi-objective optimization. *Swarm and Evolutionary Computation, 68*, 101013. Elsevier BV.

Automated Machine Learning to Improve Stock-Market Forecasting Using PSO and LSTM Networks

Francisco J. Pedroza-Castro, Alfonso Rojas-Domínguez, and Martín Carpio

1 Introduction

Forecasting the time series of financial stocks is a challenge for investors, academics and researchers. Currently, these three have advanced the use of Computational Intelligence (CI) as a tool for financial forecasting from fuzzy logic [1, 2], evolutionary algorithms (EA) [3], different Machine Learning (ML) algorithms such as Deep Learning (DL) [4], to hybridizing algorithms such as DL with EA [5]. In the present work, we hybridize a Long-Short Term Memory (LSTM) network with an EA, particularly the Particle Swarm Optimization (PSO) algorithm as the optimization engine to perform the Hyper-parameter Optimization (HPO) [6–8] and the feature selection tasks in a simplified version of Automated Machine Learning (AutoML).

Regarding the feature selection task, in the literature some authors report that the use of Convolutional Neural Networks (CNNs) to extract or generate new features from raw data provides better forecasting results [9], while other authors state that the use of Technical Indicators (TI) achieve better performance [10]. In the present work during the AutoML process the optimization metaheuristic selects whether to use CNN or TI in a process directed toward improving the model performance in forecasting of Google's stocks.

In Sect. 2 we review works related to HPO and AutoML in financial forecasting. Section 3 describes the CNN, LSTM, and Particle Swarm Optimization. In Sect. 4 we describe our methodology to perform AutoML. The experimental results are reported in Sect. 5. Finally, our conclusions and directions for future work are offered in Sect. 6.

F. J. Pedroza-Castro · A. Rojas-Domínguez (✉) · M. Carpio
National Technological Institute of Mexico / Technological Institute of León, 37290 León, Guanajuato, México
e-mail: alfonso.rojas@gmail.com

O. Castillo and P. Melin (eds.), *Hybrid Intelligent Systems Based on Extensions of Fuzzy Logic, Neural Networks and Metaheuristics*, Studies in Computational Intelligence 1096,
https://doi.org/10.1007/978-3-031-28999-6_21

2 Related Work

According to our review of the literature, the most widely used DL models are the LSTM networks and their variations or hybridization with other techniques. A summary of the review is provided in Table 1. In less recent research the LSTM network is also the most commonly used [10].

The differences between the most recent research and previous works are: (1) That more recently the objective is not limited to perform forecasting only, but also to design automated trading systems. (2) That recent works do not use the LSTM networks by themselves, but hybridize these with other algorithms, for example to perform feature selection and hyper-parameter optimization (in an AutoML approach). (3) That not only is the LSTM network hybridized to perform some part of the AutoML process, but it is also combined with other models to generate forecasts or investment signals; for example, LSTM networks have been used with an ARIMA (AutoRegressive Integrated Moving Average) model to generate financial forecasts [12].

Regarding the Data Preprocessing and Feature Engineering task, there are different approaches to financial forecasting, using as input either technical indicators or the raw data with simple preprocessing, such as normalization. In those cases that use inputs without complex preprocessing, LSTM networks are used in conjunction with other types of networks, such as CNNs which generate the internal representations for forecasting. In one work the reported performance was up to 6 orders of magnitude better than that of simple LSTM networks [15].

Table 1 Models optimized as reported in the literature

Author	Model	Hyper-parameters	Optimization method
Chung and Shin [5]	LSTM	Number of cells LSTM (Cl_s), Hidden strates (H_s) and Window size (W_s)	Genetic Algorithm (GA)
Bhandari et al. [11]	LSTM	Cl_s, H_s, Epochs (E_s), Optimizer, learning rate (α), W_s	GA
Kumar et al. [12]	Bi-LSTM-ARIMA	H_s, α, E_s, dropout, E_s and feature selection	GA
Dang et al. [13]	LSTM-MEMD	H_s, W_s, α, Batch size (B_s) and E_s	Artificial Bee Colony (ABC)
Kumar et al. [7]	LSTM	Cl_s, H_s, W_s, E_s and feature selection	PSO and FPA
Dang et al. [14]	LightGBM-NSGA-II-SW	α, max leaf size, and min leaf size	Elitist Non-Dominated Sorting Genetic Algorithm (NSGA-II)

In another work an LSTM network with Complete Ensemble Empirical Mode Decomposition with Adaptive Noise (LSTM-CEEMDAN) has been used for financial forecasting; this model was compared against a Support Vector Machine (SVM) and a simple LSTM, the results show that LSTM-CEEMDAN outperforms the SVM and the LSTM, with up to 3 and 10 orders of magnitude better respectively [16].

Chen et al. [17] reported that they used Graph CNN (GC-CNN) to analyze stock behavior and market information such as volatility and correlated stocks. The experimental results show that the model has an annual return of up to 33.42%, outperforming models such as CNN-LSTM that obtained an annual return of 6.65%. However, hyper-parameter and feature tuning is not considered, which would be like comparing models that are not in their best conditions.

The AutoML process to perform financial forecasting was used with population metaheuristic algorithms (p-metaheuristics) including GA, ABC, FPA, PSO and NSGA-II [18].

3 Theoretical Background

The two machine learning models available to the AutoML process are briefly described in this section. The Optimization engine (PSO, a bioinspired metaheuristic) is also described.

3.1 Convolutional Neural Networks

A Convolutional Neural Network (CNN) is a type of multilayer network used to transform raw data (i.e. perform feature extraction) by means of convolution filters which constitute the weights or parameters of the network [19]. In ML parlance, the convolution operation is a simple linear combination of the input data where the coefficients of the combination are the weights of the network. Additionally, a CNN includes non-linear activation functions (typically the Rectified Linear Unit, ReLU), and pooling layers that enable dimensionality reduction. After the feature-extraction layers a classification or regression stage can be included, in the form of a Support Vector Machine or of traditional Fully-Connected layers. Figure 1 illustrates the structure of a typical CNN for image classification.

The weights of a CNN, i.e. the convolution filters, are adjusted iteratively through application of gradient descent by means of the Stochastic Backpropagation (BP) algorithm, which computes the gradient of the network weights with respect to a loss function (a measure of the network's error) and performs the required adjustments to minimize the error of the network; this process is called *training* of the network.

To provide stability to the BP algorithm, the training instances (the data over which the network's error for a given iteration is computed) are provided to the network in

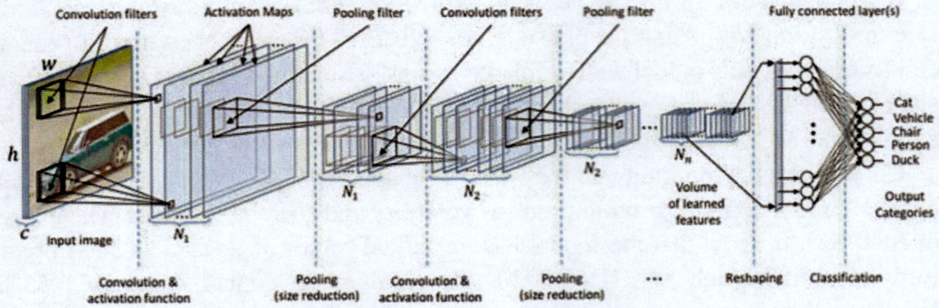

Fig. 1 An schematic diagram of a typical CNN for classification

small groups called mini-batches. The magnitude of the weight adjustments, which affects the training speed, is regulated by a hyper-parameter called *learning rate*.

As can be seen, the performance of a CNN depends on the training process and the architecture of the network. Thus, the hyper-parameters of a CNN are the number of training iterations, called *epochs*; the number of convolutional layers; the filters per layer; the learning rate; and the size of the mini-batches. In this work, all of these hyper-parameters are automatically adjusted by means of an optimization engine, which is described in Sect. 3.2.

Long-Short Term Memory

The second type of ML model that can be selected is the Long-Short Term Memory (LSTM). LSTM networks are designed to process data sequences; it has been shown that an LSTM can process sequences of up to 1000 values at a time, while avoiding the exploding-gradient and the vanishing-gradient problems [20].

An LSTM network works by simulating logic gates that allow deleting or adding information to a Cell State (7). First a Forget gate (4) decides what information is forgotten or kept; then, the Input gate (3) and the Candidate gate (6) decide what information is added; finally the Hidden state (8) is generated using the Output gate (5) and the Cell state (7) [20]. The equations for this process are shown below. Fig. 2 illustrates the structure of an LSMT cell [20].

$$\textbf{Input gate}: \ i_t = sig(x_t U_i + H_{t-1} W_i + b_i), \ i_t \in \mathbb{R}^{d_H} \tag{1}$$

$$\textbf{Forgetgate}: \ F_t = sig\left(x_t U_f + H_{t-1} W_f + b_f\right), \ f_t \in \mathbb{R}^{d_H} \tag{2}$$

$$\textbf{Output gate}: \ O_t = sig(x_t U_o + H_{t-1} W_o + b_o), \ O_t \in \mathbb{R}^{d_H} \tag{3}$$

$$\textbf{Candidate gate}: \ G_t = tanh\left(x_t U_g + H_{t-1} W_g + b_g\right), \ G_t \in \mathbb{R}^{d_H} \tag{4}$$

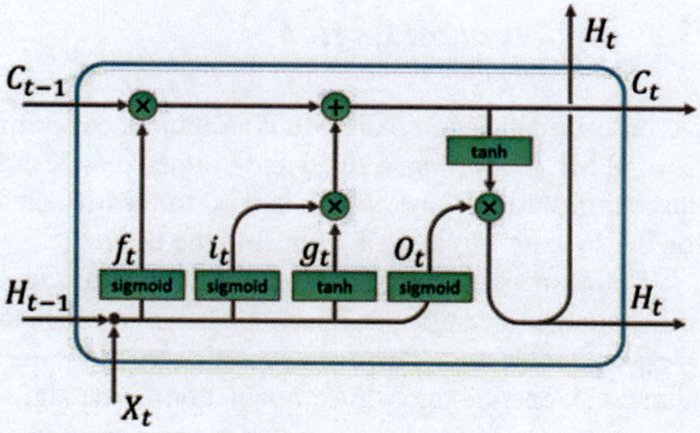

Fig. 2 Schematic diagram of an LSTM Cell

$$\textbf{Cell state}: \; C_t = f_t \otimes c - t - 1 + i_t \otimes G_t, \; C_t \in \mathbb{R}^{d_H} \tag{5}$$

$$\textbf{Hidden state}: H_t = O_t \otimes tanh(C_t) \tag{6}$$

$$\textbf{State}: \; s_t = R_{LSTM}(s_{t-1}, x_t) = [c_t; H_t], \; s_t \in \mathbb{R}^{2d_H} \tag{7}$$

where $U \in \mathbb{R}^{d_x \times d_H}$ and $W \in \mathbb{R}^{d_x \times d_H}$ are the network weights, $x_t \in \mathbb{R}$ is the input, $sig(\bullet)$ is the sigmoid function, $tanh(\bullet)$ is the hyperbolic tangent function, and $\otimes$ represents the element-wise product [20].

An LSTM network can be formed by connecting C_t and H_t in a loop, back into the cell's inputs, making the LSTM a type of Recurrent Neural Network. This is equivalent to having several LSTM cells connected in sequence and forming a layer, as illustrated in Fig. 3. An LSTM network may also include multiple of those layers.

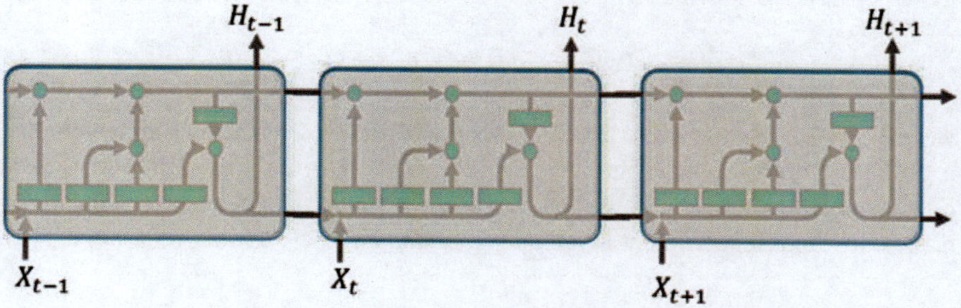

Fig. 3 Schematic diagram of an LSTM layer with several cells

3.2 Optimization Engine

As described in Sect. 1, AutoML is an automated optimization process through which several ML models are trained and evaluated on test data. This process is driven by an optimization engine, which is used to identify the model with the smallest error on the specific ML task of interest to the user.

There are many algorithms, generally known as metaheuristics that can be used as the optimization engine in AutoML, and these are organized in families according to their characteristics. There are evolutionary algorithms, such as Differential Evolution and Genetic Algorithms; population-based algorithms such as Particle Swarm Optimization (PSO, also characterized as swarm intelligence); estimation of distribution algorithms such as the Univariate Marginal Distribution Algorithm; and many bio-inspired metaheuristics, such as the Bat Algorithm, Ant Colony Optimization, etc.

Due to its flexibility and effectiveness, in this work the PSO is employed as the optimization engine for AutoML. PSO is a metaheuristic inspired by the social behavior of insects, animals and humans. This algorithm starts with a random population of so called *particles*; each particle constitutes a potential solution to the optimization problem at hand. Then, while a termination criterion is not met, PSO will generate new candidate solutions by moving the existing particles in the search space, with the goal of finding better solutions than the currently known candidate solutions. The population of particles is thus iteratively updated until the termination criterion (e.g. a maximum number of iterations) is met and PSO returns the best solution that was found. The pseudocode of PSO is given in Algorithm 1 and the equations employed to update the particles are described below.

According to PSO, a particle $\overrightarrow{x}$ that moves with velocity $\overrightarrow{v}$ will be updated using the best position $\overrightarrow{p}$ that it has previously occupied and the best solution $\overrightarrow{p}_b$ in the current population. This idea is shown in the update velocity (8),

$$\vec{v}_i \leftarrow \gamma(w\vec{v}_i + \varphi_1(\vec{p}_i - \vec{x}_i) + \varphi_2(\vec{p}_b - \vec{x}_i)) \tag{8}$$

were $\overrightarrow{v}_l$ is the velocity of particle i, $\overrightarrow{x}_i$ is the particle i, w is the inertial weight, $\overrightarrow{p}_b$ is the best particle found so far, $\varphi_1 = R_1c_1, \varphi_2 = R_2c_2$, c_1 is the cognitive coefficient, c_2 is the social coefficient, R_1 and R_2 are uniform random values between 0 and 1, γ is the constriction coefficient given by (9), and the particle update equation is (10).

$$\gamma = \frac{2}{\left|2 - \phi^2 - 4\phi\right|}, \quad \phi = c_1 + c_2 > 4 \tag{9}$$

$$\vec{x}_i \leftarrow \vec{x}_i + \vec{v}_i \tag{10}$$

Algorithm 1 PSO, assuming minitization

Generate random population $X = \{\vec{x}_1, \vec{x}_2, ..., \vec{x}_n\}$.
Finding the best solution $\vec{p}_b$ in X.
while *Stop criterion* is not met, **do**
  for particle $\vec{x}_i$ **do**
    if $f(\vec{x}_i) < f(\vec{p}_i)$ **then**
      $\vec{p}_i \leftarrow \vec{x}_i$; update the best position of particle i.
    end if
    Identify the best position of the iteration $\vec{p}_b$.
    Update velocity $\vec{v}_i$ via (10).
    Update position $\vec{x}_i$ via (12).
  end for
end while
return The best solution found $\vec{p}_b$.

4 Methodology

This section describes the experiments performed to obtain the solutions via the AutoML process using PSO as the optimization engine. Provided that PSO, as any other metaheuristic is a semi-stochastic method which depends on initial solutions and other factors, the AutoML solutions found may vary between different executions of this process. Because of this, the AutoML process is executed 10 times, to generate 10 different "optimal" solutions. After the AutoML process, the 10 solutions returned are compared with each other by performing 33 trials, each consisting in re-training and evaluating their performance in forecasting of the stock prices. This multi-trial evaluation is performed to provide statistical support to our final conclusions.

To perform the AutoML process the PSO algorithm and the rest of the system was implemented using several Python libraries including NumPy, Pandas, Scikit-Learn and TensorFlow. The metaheuristic iteratively performs feature selection and design of the LSTM network with the objective of finding a configuration that minimizes the error of the network for the forecasting of the closing price of Google's stocks. The fitness function used is the Mean Squared Error (MSE). In each iteration, the metaheuristic creates a set of candidate solutions; each solution is a vector containing the hyper-parameters and features that will be given as inputs to the network. The ultimate objective is to iteratively improve the solution, so that the metaheuristic finally returns the network with the lowest possible MSE. This process is shown in Fig. 4.

In each fitness-function call, an LSTM network is designed, the features that will be the input to this network are selected, and the training of the LSTM network is performed. For the training, the closing price of Google's stock from 2004-11-08 to 2022-01-21 was used, with 64% of the data used as the training set; 21% of the data used for validation; and 15% employed for testing of the network. To reduce

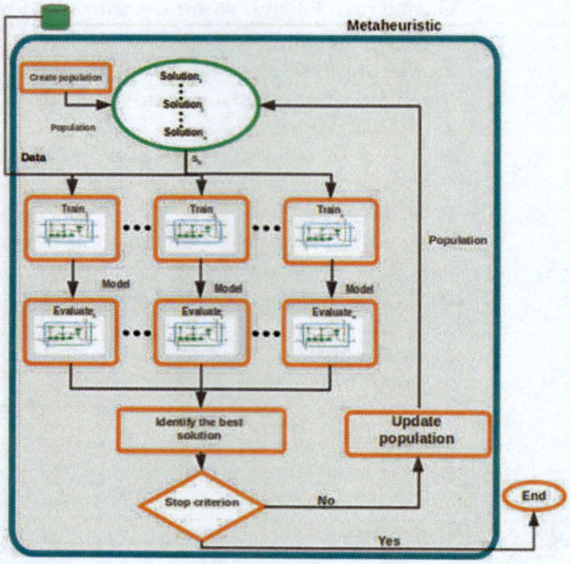

Fig. 4 AutoML Methodology based on PSO

the execution time and avoid overfitting, early stopping was employed. The Adam optimizer was used to train the LSTM network.

In this work, the selection of characteristics is based on a set of technical indicators (TIs, shown in Table 2) which were selected according to [5, 7, 17, 20]. The selection of hyper-parameters (Table 4) was carried out according to [7, 8]. The parameters of the PSO algorithm (Table 3) where set according to [21].

Table 2 Technical Indicators

	Name	Symbol	Search space
1	Price rate of change	ROC	{0, 1}
2	Simple moving average	SMA	{0, 1}
3	Exponential moving average	EMA	{0, 1}
4	Moving average convergence divergence	MACD	{0, 1}
5	Relative strength index	RSI	{0, 1}
6	Commodity channel index	CCI	{0, 1}
7	William R	WR	{0, 1}
8	Money flow index	MFR	{0, 1}
9	Force index	RFI	{0, 1}
10	Average true range	TR	{0, 1}
11	Ease of movement indicator	EMI	{0, 1}
12	Close	Close	{0, 1}

Table 3 Parameters of the PSO metaheuristic

Parameter	Value
Constriction coefficient	$c_1 = c_2 = 2.05$
Inertial weight	0.8
Population size	86
Iterations	16

Table 4 Hyper-parameters and model

	Hyper-parameters and model	Symbol	Search space
1	CNN layer	CNN	{0, 1}
2	No. Filters CNN layer	FC_s	{1, 20, 50, 80, 96, 110, 150}
3	Kernel size	KC_s	{1, 20, 50, 80, 96}
4	LSTM cell	Cl_s	{1, 2, 3, 4, 5, 6, 7, 8}
5	Hidden state	H_s	{1, 20, 50, 80, 96, 110, 150}
6	Windows size	W_s	{20, 40, 60, 80, 90, 100}
7	Learning rate	α	{1E-1, 1E-2, 1E-3, 1E-4, 1E-5}
8	Batch size	B_s	{8, 16, 32, 64}
9	Epochs	E_s	{12, 25, 100, 150}

5 Experimental Results

The execution time for each experiment was between 3 and 4 h. Each solution achieves an MSE in the order of 1E-5. The MSE on test data are in the range of 1E-1 to 1E-5. The 10 AutoML solutions (S-1 to S-10) obtained are shown as the columns in Table 5. The last three rows in Table 5 report the statistics of the MSE obtained by each solution in the multi-trial evaluation. The whole set of results are presented as boxplots in Fig. 5. Additionally, the forecasting results of the top four solutions are presented in Fig. 6.

The results show that even though a network may be deeper, it does not necessarily achieve a better performance. This may be due to the fact that a deeper network can be more prone to overfitting. The best performing solutions have 110 and 150 hidden states, while the solutions with 96 hidden states have the lowest performance. However, the best solutions (S-2, S-3 and S-4) contain very different amount of parameters. The best solution (S-2) contains approximately 53 k parameters, while the second and third best solutions (S-3 and S-4, respectively according to Fig. 3) contain about 94 k parameters each (see Table 5).

In all but one solution (S-3), the learning rate was set to α=1E-3, indicating a consistency in the optimal value of this parameter for this problem. The batch size

Table 5 The ten AutoML-generated Solutions and their performance metrics

	Symbol	S-1	S-2	S-3	S-4	S-5	S-6	S-7	S-8	S-9	S-10
Features	Close	0	1	1	1	1	1	1	1	1	1
	ROC	0	1	1	1	0	1	0	0	0	1
	SMA	0	1	1	1	1	1	1	1	1	0
	EMA	1	1	0	1	0	0	0	0	0	1
	MACD	1	0	0	0	0	1	0	0	0	0
	RSI	0	0	0	0	0	1	0	0	0	1
	CCI	1	1	0	0	1	1	1	1	1	0
	WR	1	1	1	0	0	0	0	1	0	0
	MFR	1	0	0	1	1	1	1	1	1	1
	RFI	1	0	1	1	1	0	0	0	0	0
	TR	1	0	0	0	1	1	0	0	0	0
	EMI	1	1	0	1	0	0	0	0	0	0
Sum of features		8	7	5	7	6	8	4	5	4	5
Hyper-parameters	CNN	0	1	0	0	0	0	0	0	0	0
	FC_s	0	110	0	0	0	0	0	0	0	0
	KC_s	0	2	0	0	0	0	0	0	0	0
	Cl_s	6	1	1	1	1	1	1	1	1	1
	H_s	96	110	150	150	150	96	150	110	110	150
	W_s	100	40	60	60	60	60	60	90	80	90
	α	1E-3	1E-3	1E-4	1E-3	1E-3	1E-3	1E-3	1E-3	1E-3	1E-3
	B_s	16	32	16	16	16	32	32	16	16	32

(continued)

Table 5 (continued)

	Symbol	S-1	S-2	S-3	S-4	S-5	S-6	S-7	S-8	S-9	S-10
	E_s	100	50	100	25	25	100	100	50	100	50
Amount of parameters		410,977	55,333	93,751	94,951	39,649	40,801	93,151	39,265	38,881	93,751
Performance	Fitness	3E-5	4E-5	5E-5	3E-5	5E-5	4E-5	4E-5	3E-5	5E-5	6E-5
	mean	1E-2	9E-4	1E-3	1E-3	2E-3	5E-3	2E-3	2E-3	3E-3	2E-3
	Std. Dev	5E-3	5E-4	5E-4	1E-3	1E-3	1E-3	9E-4	1E-3	1E-3	9E-4

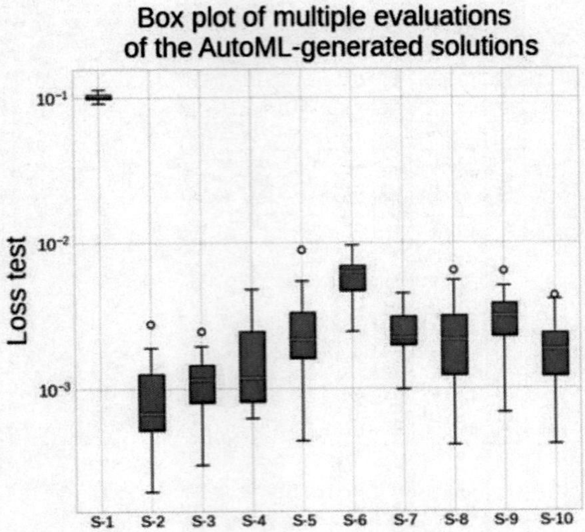

Fig. 5 Loss test of 33 experiments of train solutions

values of the solutions are either 16 or 32, thus giving indications that these two batch sizes promote the satisfactory performance of the model.

Only one solution (S-2) has a CNN layer, and this solution is the one with the best performance, as shown in Fig. 5. This is consistent with previous work that has found that the use of CNNs promotes better forecasting. However, we have no explanation as to why the AutoML process only generated one solution with CNN layers.

Regarding the technical indicators, SMA is selected in solutions S-2 to S-9. SMA is a technical indicator for smoothing the time series, and its repeated selection in most solutions indicates that smoothing the time series promotes a better forecasting.

On the other hand, RSI is only found in two solutions (S-6 and S-10), indicating that knowing whether or not the financial stock is overbought has little influence on the performance of the model (in other words, the trend has no significant effect on the performance of the solution obtained by the metaheuristic).

Similarly, EMA also serves to determine the trend and is found in solutions S-1, S-2, S-4 and S-10. S-2, obtained the best performance and S-1 the worst performance. However, S-2 processes the features through a CNN layer to generate new features, and this may have a significant impact on the performance of the network well beyond the particular selection of features that constituted the input to the network.

6 Conclusions

PSO manages to carry out the AutoML process, although being a non-deterministic algorithm, the solutions it finds are not always the best. It was found that a CNN-LSTM network has better performance than an LSTM network with technical indicators, but this solution was found only in one out of the 10 experiments performed.

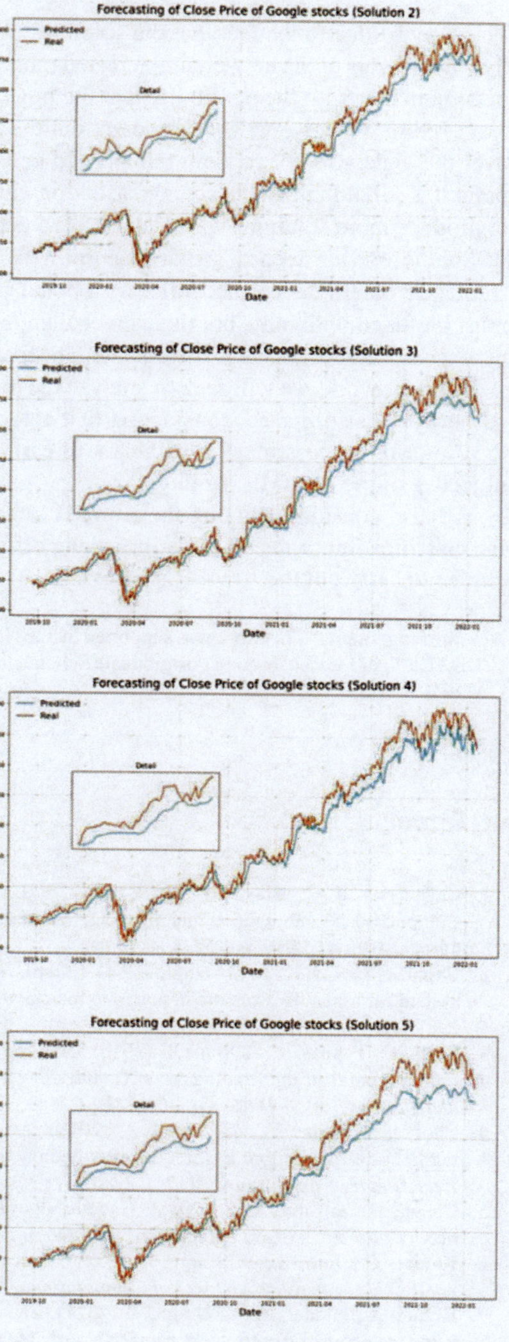

Fig. 6 Stock forecasting of AutoML Solutions 2, 3, 4 and 5 (top to bottom, respectively)

This may be due to the fact that the solutions obtained are in the order of 1E-4 and that the weights of the networks are created randomly with a different seed each time, in conjunction with the possibility that the fitness function is not bijective.

In future work, we will seek to carry out the AutoML process with more control over the generation of random numbers to better understand this problem. At this point it is still unknown the actual reason for which the metaheuristic did not choose to produce more solutions with CNN-LSTM networks.

Solutions with deeper artificial neural networks do not always generate better results, to obtain deeper networks with better performance their hyper-parameters must be tuned optimally, but this is a challenge when the networks contain a large number of parameters, as is the case with deeper networks.

In future work, we will seek to analyze the influence of independent data sources on network performance, considering that a good amount of solutions obtained via the AutoML process may be sufficient to carry out this analysis, which is also an objective of the AutoML paradigm.

We also consider applying the AutoML process on different problems, such as the classification problem, as well as using different algorithms as the optimization engine to carry out the AutoML process.

Acknowledgements This work was supported by the National Council of Science and Technology (CONACYT) of Mexico through Postgraduate Scholarship 774627 (F. Pedroza) and Research Grant CÁTEDRAS-2598 (A. Rojas).

References

1. Cavalcante, R. C., Brasileiro, R. C., Souza, V. L., Nobrega, J. P., & Oliveira, A. L. (2016). Computational intelligence and financial markets: A survey and future directions. *Expert Systems with Applications, 55*, 194–211.
2. Chourmouziadis, K., & Chatzoglou, P. D. (2016). An intelligent short term stock trading fuzzy system for assisting investors in portfolio management. *Expert Systems with Applications, 43*, 298–311.
3. Evans, C., Pappas, K., & Xhafa, F. (2013). Utilizing artificial neural networks and genetic algorithms to build an algo-trading model for intra-day foreign exchange speculation. *Mathematical and Computer Modelling, 58*(5–6), 1249–1266.
4. Chopra, S., Yadav, D., & Chopra, A. (2019). Artificial neural networks based Indian stock market price prediction: Before and after demonetization. *Journal of Swarm Intelligence and Evolutionary Computation, 8*(174), 2.
5. Chung, H., & Shin, K. S. (2018). Genetic algorithm-optimized Long Short-Term Memory network for stock market prediction. *Sustainability, 10*(10), 3765.
6. Kumar, G., Jain, S., & Singh, U. (2021). Stock market forecasting using computational intelligence: A survey. *Archives of Computational Methods in Engineering, 28*(3), 1069–1101.
7. Kumar, K., Haider, M., & Uddin, T. (2021). Enhanced prediction of intra-day stock market using metaheuristic optimization on RNN–LSTM network. *New Generation Computing, 39*(1), 231–272.
8. Silva, T. R., Li, A. W., & Pamplona, E. O. (2020). Automated trading system for stock index using LSTM neural networks and risk management. In *2020 International Joint Conference on Neural Networks (IJCNN)* (pp. 1–8). IEEE.

9. Tsantekidis, A., Passalis, N., Tefas, A., Kanniainen, J., Gabbouj, M., & Iosifidis, A. (2020). Using deep learning for price prediction by exploiting stationary limit order book features. *Applied Soft Computing*, *93*, 106401.
10. Sezer, O. B., Gudelek, M. U., & Ozbayoglu, A. M. (2020). Financial time series forecasting with deep learning: A systematic literature review: 2005–2019. *Applied soft computing*, *90*, 106181.
11. Bhandari, H. N., Rimal, B., Pokhrel, N. R., Rimal, R., Dahal, K. R., & Khatri, R. K. (2022). Predicting stock market index using LSTM. *Machine Learning with Applications*, 100320.
12. Kumar, R., Kumar, P., & Kumar, Y. (2022). Three stage fusion for effective time series forecasting using Bi-LSTM-ARIMA and improved DE-ABC algorithm. *Neural Computing and Applications*, 1–17.
13. Deng, C., Huang, Y., Hasan, N., & Bao, Y. (2022). Multi-step-ahead stock price index forecasting using long short-term memory model with multivariate empirical mode decomposition. *Information Sciences*.
14. Deng, S., Xiao, C., Zhu, Y., Tian, Y., Liu, Z., & Yang, T. (2022). Dynamic forecasting of the Shanghai Stock Exchange index movement using multiple types of investor sentiment. *Applied Soft Computing*, 109132.
15. Kanwal, K., Lau, M. F., Ng, S. P., Sim, K. Y., & Chandrasekaran, S. (2022). BiCuDNNLSTM1dCNN—A hybrid deep learning-based predictive model for stock price prediction. *Expert Systems with Applications*, *202*, 117123.
16. Lin, Y., Yan, Y., Xu, J., Liao, Y., Ma, F. (2021). Forecasting stock index price using the CEEMDAN-LSTM model. *The North American Journal of Economics and Finance*, *57*, 101421.
17. Chen, W., Jiang, M., Zhang, W. G., & Chen, Z. (2021). A novel graph convolutional feature based convolutional neural network for stock trend prediction. *Information Sciences*, *556*, 67–94.
18. Gu, J., Wang, Z., Kuen, J., Ma, L., Shahroudy, A., Shuai, B., Liu, T., Wang, X., Wang, G., Cai, J., et al. (2018). Recent advances in convolutional neural networks. *Pattern Recognition*, *77*, 354–377.
19. Schmidhuber, J., et al. (1997). Long short-term memory. *Neural Computing*, *9*(8), 1735–1780.
20. Bustos, O., Pomares-Quimbaya, A. (2020). Stock market movement forecast: A systematic review. *Expert Systems with Applications*, *156*, 113464.
21. Gendreau, M., Potvin, J. Y., et al. (2019). *Handbook of metaheuristics* (3rd ed., Vol. 272). Springer.

Evolutionary Gaussian-Gradient: A New Optimization Algorithm for the Electromechanical Design of Gravitational Batteries

Juan de Anda-Suárez, Felipe J. Flores-Calva, Daniel Jiménez-Mendoza, and Germán Pérez-Zúñiga

1 Introduction

Historically, electromechanical design has been a crucial element in the development of technology advance; Leonardo Da Vinci was one of the founders of this concept, who combined his engineering knowledge and drawing to create prototypes without the need for physical experimentation [1]. Subsequently, the industrial revolution demanded the creation of multiple technologies to satisfy their needs, thus creating the line of research in electromechanical design. Currently, computing has generated a new approach based on computer-aided design and simulation; from this perspective—simulation has laid the foundations for optimizing electromechanical systems [2, 3].

In the sense of optimization, a problem to be optimized must have an optimum x^* in a solution domain $a \leq x \leq b$ under a cost function $f(x)$ and its constraints $r : \{r_1, \ldots, r_n\}$; analytically, x^* can be calculated from the derivative of the cost function [4]. However, in electromechanical design, having an explicit derivative function only occurs in some cases since electrical and mechanical design generally connects different mathematical models and their constraints [5]. Therefore—the problem of proposing a technique to optimize a cost function without having an explicit derivative arises [6, 7].

Metaheuristic optimization algorithms have been a solution in optimal electromechanical design: in mechanics, research has proposed models of materials that distribute loads or have specific geometry for formation resistance [8, 9]; in fluid, Metaheuristics have generated solutions to problems of flow resistance, wave mitigation in tsunamis and modulation of simulation parameters [10–12]; in

J. de Anda-Suárez (✉) · F. J. Flores-Calva · D. Jiménez-Mendoza · G. Pérez-Zúñiga
Tecnológico Nacional de México-Instituto Tecnológico Superior de Purísima del Rincón, Purísima del Rincón, Gto, México
e-mail: Juan.ds@purisima.tecnm.mx

O. Castillo and P. Melin (eds.), *Hybrid Intelligent Systems Based on Extensions of Fuzzy Logic, Neural Networks and Metaheuristics*, Studies in Computational Intelligence 1096,
https://doi.org/10.1007/978-3-031-28999-6_22

electrical, Metaheuristics have modeled communication antennas and telecommunication systems [13–15]; finally, in renewable energies, evolutionary algorithms have been employed for the control of clean energy systems and optimization of hybrid power scheduling elements [16–18]. However, optimizing electromechanical systems applied to renewable energies is an opportunity area with an impact on the environment.

Currently, clean energy production systems for lighting need to store the produced energy by the contact of a generating source; the predominant one in the market the lead-acid batteries, which both in their production and disposal generate soil pollutants and affections to the ecosystem; where the risk falls on people who have contact with lead [19–21]. Therefore, the problem of the double moral of renewable energies arises.

Researchers have proposed to use gravitational potential energy as an accumulator, as an agent of change to lead-acid batteries; this method is known as the gravitational battery, which consists of having a transmission that lifts a weight using an electric motor powered by a clean energy source, when the weight is at its maximum point it transfers the potential energy to electrical energy [22–25]. However, a problem arises since a dual-purpose transmission has to be designed: firstly, to lift the weight and accumulate the energy; secondly, to discharge the potential energy as slowly as necessary since the free fall will be proportional to the gravitational acceleration. Consequently, our proposal's crucial point lies in designing an optimization algorithm to build gravitational batteries.

In the population-based metaheuristic sense, evolutionary algorithms are characterized by containing a population, each individual in the population constituting a solution to the optimization problem, which is improved under a set of decision-making operators in a given number of iterations [26]. Mathematically, evolutionary algorithms are guaranteed to converge to an optimal solution in infinite states [27, 28]. Compared to the problem of the optimal design of a gravity battery, the computational needs of simulation do not allow an infinite time of solution refinements. Therefore—the need arises to design an evolutionary optimization algorithm that converges in a shorter amount of time.

A possible improvement to the convergence of evolutionary algorithms lies in using the information available from the individuals to know the optimal direction. However, this direction implies estimating the fitness function's gradient; as already mentioned, the fitness function does not necessarily have a direct derivative. Consequently, to capture the direction of the global optimum, the literature has used the local vector distance to infer a possible direction using an elite sample [29, 30]. The elitist vector difference, nevertheless, is only limited to local changes, which are susceptible to misdirection, thus compromising convergence.

We propose a Gaussian gradient as an approximation to the gradient direction of the fitness function and its application to gravitational battery transmission design. This fact results in the following contributions:

- Analytical deduction of a fitness function gradient from the analytical Gaussian gradient.

- Fitness function gradient fitting in the population-based metaheuristic approach.
- Proposal of a new Gaussian-Gradient-Differential Evolution (GG-DE) optimization algorithm.
- Implementation of GG-DE in the optimization of gravitational batteries.

2 Methodology

In this section, we will deal with an analytical estimation of the gradient of a fitness function under the population-based Metaheuristic approach. In Sect. 2.1, we start from the interpolation of the fitness function using the Dirac delta and estimate its approximation to a theoretical kernel. In Sect. 2.2, we compute the gradient of the fitness function in terms of the generalized analytic gradient in the hypervolume. Finally, in Sect. 2.3, we derive a theoretical population gradient of the fitness function in terms of the Gaussian kernel in the solution hypervolume.

2.1 Interpolation of the Fitness Function in the Hypervolume

The impact of our proposal lies in estimating a gradient of the fitness function without having a direct derivative. In this subsection, we present a new scheme for approximating the change direction of a fitness function in the population-based Metaheuristic scheme. Accordingly, we start by defining a fitness function $f(\vec{r})$ in the hypervolume [a, b]^n, which we interpolate by Dirac delta, giving the expression of Eq. (1):

$$f(\vec{r}) = \int\int \ldots \int_{\Omega} f(\vec{r}\,')\delta(\vec{r} - \vec{r}\,')d^n\vec{r} \tag{1}$$

where $\delta(\vec{r} - \vec{r})$ is the Dirac function, and is denoted by Eq. (2):

$$\delta(\vec{r} - \vec{r}) = \begin{cases} 1\ \vec{r} = \vec{r}\,' \\ 0\ \vec{r} \neq \vec{r}\,' \end{cases} \tag{2}$$

In practice, the Dirac delta interpolation is replaced by an approximation, called a smoothed interpolation kernel, which is defined in Eq. (3):

$$\langle f(\vec{r})\rangle = \int\int \cdots \int_{\Omega} f(\vec{r}\,')W(||\vec{r} - \vec{r}\,'||, h)d^n\vec{r} \tag{3}$$

where $\boldsymbol{h}$ is the kernel smoothing parameter; consequently, the smoothed kernel must fulfill a set of properties, which we define in Eq. (4): first, the kernel, when integrated into the hypervolume of the Metaheuristic search space, has the value of 1; second, in the limit when the smoothing parameter tends to zero the kernel is equal to the Dirac Delta function; third, the kernel must fulfill the compactness criterion.

$$1.\quad \int_{\Omega} W\big(||\vec{r}-\vec{r}\,'||, h\big) d\vec{r} = 1$$

$$2.\quad \lim_{h\to 0} W\big(||\vec{r}-\vec{r}\,'||, h\big) = \delta\big(\vec{r}-\vec{r}\,'\big)$$

$$3.\quad W\big(||\vec{r}-\vec{r}\,'||, h\big)\ when\ |\vec{r}-\vec{r}| > \kappa h \tag{4}$$

Up to this point, we have derived an expression of the fitness function in a smoothed kernel. However, we are interested in generating a gradient that interpolates the fitness function. In Sect. 2.2, we present an analytical deduction of the gradient of the fitness function in terms of the smoothed gradient.

2.2 *Gradient Interpolation of the Fitness Function*

To approximate a gradient of the fitness function, we start from the expression of Eq. (3); and apply the gradient operator $\overrightarrow{\nabla} = (\frac{\partial}{\partial r})\hat{r}$. Then, we obtain the result of Eq. (5):

$$\langle \vec{\nabla} f(\vec{r}) \rangle = \int\int \cdots \int_{\Omega} \vec{\nabla} f(\vec{r}\,') W\big(||\vec{r}-\vec{r}\,'||, h\big) d^n\vec{r} \tag{5}$$

As a result of the expression of Eq. (5), we observe that estimating the interpolated gradient of the fitness function implies knowing the gradient of the remaining points of the solution hypervolume, which goes against our primary objective. However, following property $\nabla[a \cdot b] = \nabla a \cdot b + a \cdot \nabla b$, we find the result of Eq. (6):

$$\vec{\nabla} f(\vec{r}\,') W\big(||\vec{r}-\vec{r}\,'||, h\big) = \int_{\Omega} \vec{\nabla} \cdot \big[f(\vec{r}\,') W\big(||\vec{r}-\vec{r}\,'||, h\big)\big] d^n\vec{r} - \int_{\Omega} f(\vec{r}\,') \vec{\nabla} W\big(||\vec{r}-\vec{r}\,', h||\big) d^n\vec{r} \tag{6}$$

To simplify the result of Eq. (6), we rewrite the $\int_{\Omega} \overrightarrow{\nabla} \cdot \big[f(\vec{r}\,') W\big(\|\vec{r}-\vec{r}\,'\|, h\big)\big] d^n\vec{r} = \int_S f(\vec{r}\,') W\big(\|\vec{r}-\vec{r}\,'\|, h\big) \cdot \hat{n}\, ds$ term and consider that the kernel is a compact function and the surface integral equals zero. Therefore, we find the expression of Eq. (7):

$$\langle \vec{\nabla} f(\vec{r}) \rangle = -\int\int \cdots \int_{\Omega} f(\vec{r}\,')\vec{\nabla} W(\|\vec{r} - \vec{r}\,'\|, h) d^n\vec{r} \tag{7}$$

In summary of the section, we have presented an analytical derivation of the gradient of the fitness function involving only the evaluations of the said function in the search space. Consequently, Eq. (7) allows us to know the optimal direction of the problem to be optimized—without the need for an explicit derivative. Nevertheless, Eq. (7) needs to define a kernel whose derivative is known. In Sect. 2.3, we will discuss the approach of a kernel that meets the constraints of Eq. (4) and has a direct derivative.

2.3 Gaussian-Gradient Approximation

From the result of Eq. (7), in the present section, we will design a kernel based on the Gaussian distribution due to its direct approximation to the Dirac delta. In Fig. 1, we show a comparison between the Dirac delta function and the proposed Gaussian kernel model; we observe that the parameter is the one that modulates the degree of approximation to the Dirac delta; this means that the smaller h, the better the approximation. In relation to the behavior presented in Fig. 1, we propose the Gaussian model of Eq. (8):

$$G(\|\vec{r} - \vec{r}\,'\|, h) = A_n \cdot \exp\left(-\frac{\|\vec{r} - \vec{r}\,'\|^2}{h^2}\right) \tag{8}$$

where h is the smoothing parameter of the Gaussian kernel, and A_n is the normalization constant described in Eq. (9):

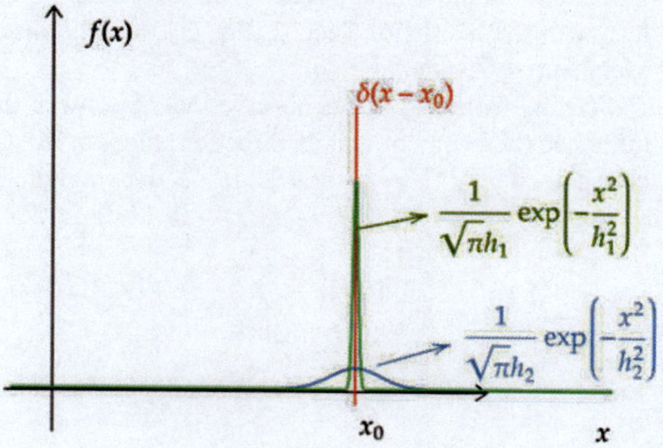

Fig. 1 Comparison between Dirac delta function and Gaussian kernel: In red, it represents the Dirac delta function; In green, a Gaussian kernel with $h_1 = 0.01$; In blue, a kernel with $h_2 = 1$

$$A_n = \frac{1}{\int\int\cdots\int_{-\infty}^{\infty} f(\vec{r}\,')W(||\vec{r}-\vec{r}\,'||, h)d^n\vec{r}} = \frac{1}{\pi^{\frac{n}{2}}h^n} \tag{9}$$

At this point, we have the elements to express the interpolation of the gradient of the fitness function from the results Eqs. (7), (8), and (9). Therefore, Eq. (10) expresses the function's gradient to be optimized in terms of the Gaussian gradient; hereafter, we will call it Smoothed Gradient (Gs).

$$\langle\vec{\nabla} f(\vec{r})\rangle = -\int\int\cdots\int_{\Omega} f(\vec{r}\,')\vec{\nabla}\left[\frac{1}{\pi^{\frac{n}{2}}h^n}\exp\left(-\frac{||\vec{r}-\vec{r}\,'||^2}{h^2}\right)\right]d^n\vec{r} \tag{10}$$

As a summary of Sect. 2, in Sect. 2.1, we discuss the analytical interpolation of the fitness function in an optimization problem; in Sect. 2.2, we calculate the gradient of the function using the gradient of a generalized kernel; finally, in Sect. 2.3, we present the smoothed kernel of the fitness function interpolated by the direct derivative of the Gaussian kernel. In conclusion, we have shown that it is possible to calculate the gradient of the fitness function without an analytical expression of its derivative.

In Sect. 3, we will describe the uses of the results of Eq. (10) in the context of population-based Metaheuristic algorithms, as well as the modification of Differential Evolution with the smoothed-gradient.

3 Smoothed Gradient in Population-Based Metaheuristics

In Metaheuristic theory, a population contains a number N of individuals, which are solutions to the optimization problem. For example, in Fig. 2, we show a two-dimensional optimization problem where the blue dots represent the individuals of the population, which cover the search space, where the neighborhood (dashed circle) influences the individual's state change decision-making (r_i). Connecting the above fact with the results of Sect. 2, Eq. (10) can be discretized in a population-based Metaheuristic context.

Starting from Fig. 2, we observe that at point ri, there is a set of individuals that influence the behavior and its direction; then, in Eq. (10), $d^n\overrightarrow{r}$ and the integral can be replaced by $d^n\overrightarrow{r} = \frac{1}{\rho}$ and Σ, respectively, giving; as a result of Eq. (5):

$$\langle f(\overrightarrow{r_i})\rangle = \sum_{i=1}^{n} \frac{1}{\rho_j} \cdot f(\overrightarrow{r_j})G(||\overrightarrow{r_i} - \overrightarrow{r_j}||, h) \tag{11}$$

where ρ represents the population density and is calculated by Eq. (12):

$$\rho_j = \sum_{i=1}^{n} f(\overrightarrow{r_j})G(||\overrightarrow{r_i} - \overrightarrow{r_j}||, h) \tag{12}$$

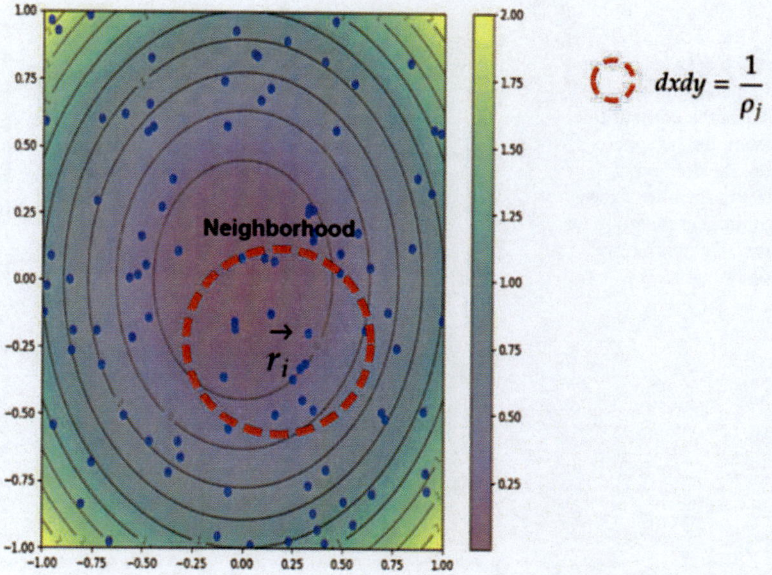

Fig. 2 Illustration of an optimization problem in the Smoothed gradient approach: on contour lines, we present a 2D optimization problem; the blue dots represent the individuals of the population of a Metaheuristic; the dotted circle group the neighborhood of influence of the individuals

On the other hand, following the same methodology as Eqs. (11) and (13) expresses the gradient approximation of the fitness function

$$\langle \nabla f(\vec{r_i}) \rangle = \sum_{i=1}^{n} \frac{1}{\rho_j} \cdot f(\vec{r_j}) \nabla w(||\vec{r_i} - \vec{r_j}||, h) \tag{13}$$

As an example of comparison of Eq. (13), in Fig. 3, we show the contour lines of the Sphere function in 2D and present in the blue arrow the analytical gradient, while the red arrow the gradient of approximation by Eq. (13). We observe that the blue arrow in the upper left corner has a higher approximation error than the blue arrow in the center; the difference between results is because the interpolation point is close to the boundary of the search domain, while the one in the center is in a higher density region.

Up to this point, in Sect. 2, we have shown an analytical derivation of the gradient interpolation of the fitness function; in Sect. 3.1, we have generated a numerical approximation of Eq. (10) in a population-based metaheuristic context. Consequently, in Sect. 3.1, we show the use of the Smoothed Gradient in the mutation scheme of the differential evolution algorithm.

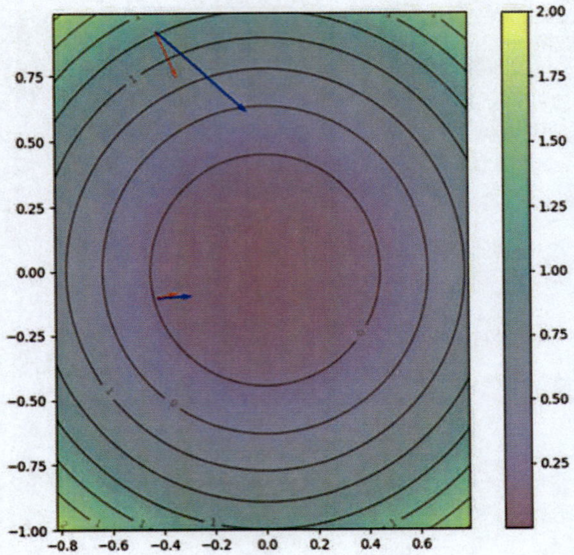

Fig. 3 Graphical comparison between analytical and approximate gradient: In the contour line, we present the 2D sphere function, the domain of $[-1, 1]^2$; in the blue arrow, the analytical gradient; in the red arrow, the approximate interpolation gradient

3.1 Diferential Evolution and Smoothed Gradient

In metaheuristic theory, Differential Evolution (DE) is an evolutionary algorithm that uses mutation, crossover, and selection operators in its primary state [31]. DE has been designed to solve continuous optimization problems, so its applications are mainly in designing and tuning parameters in different areas of electrical and mechanical engineering. The fundamental elements of DE are crossover and mutation, which are defined in Eqs. (14) and (15):

$$u_i = \vec{r}_i + F \cdot (\vec{r}_a - \vec{r}_b) \tag{14}$$

$$t_i = \begin{cases} u_i \; si \; rand(0, 1) < CR \\ \vec{r}_i \qquad otherwise \end{cases} \tag{15}$$

where F is a parameter in the domain [0, 2], and CR represents the combination criterion of the individuals in [0, 1].

As a contribution of information from Smoothed Gradient to Differential Evolution, we propose substituting $\vec{\nabla} f(\vec{r}_i)$ instead of $\vec{r}_a - \vec{r}_b$ of Eq. (14) since the vector difference of DE is the one that provides the direction in an elitist context. Therefore, Eq. (16) calculates the mutation with the direction of the gradient of the fitness function:

$$u_i = \vec{r}_i + F \cdot \langle \vec{\nabla} f(\vec{r_i}) \rangle \tag{16}$$

In retrospect, In Sect. 3, we present a new approach to estimating the gradient of a fitness function using only the information provided by the population evaluation of a Metaheuristic. In addition, in Sect. 3.1, we modify the mutation scheme of Differential Evolution, which demonstrates the application of the result of Eq. (13). However, the fundamental goal of our proposal is the optimal design of gravitational batteries; accordingly, in Sect. 4, we present a general model of a gravitational battery, while in Sect. 5 we demonstrate the performance of GG-DE versus DE and Particle Swarm Optimization (PSO, [32]); finally, in Sect. 5.1, we optimize a gravitational battery model.

4 Analytical Gravity Battery Model

To define a gravity battery optimization model, we will address in Sect. 4.1 the general system of a triple reducer, which aims to transmit the electrical energy from a solar panel to gravitational potential energy that will later be discharged for consumption needs. In adittion, In Sect. 4.2, we will model the physics of weight drop in the gravitational battery connected to the triple reducer transmission.

4.1 Gravitational Energy Transmission

The main element of a gravity battery is the gear transmission system in charge of transmitting the electrical energy to potential. In our model, we use a triple gear system, illustrated in Fig. 4, where the input angular velocity (ω_{in}) to a gear train determines the angular velocity of the output train (ω_{out}) and its direction of rotation. Therefore, Eq. (17) calculates the ratio of the input angular velocity to the output angular velocity:

$$\frac{\omega_{in}}{\omega_{out}} < 1 \tag{17}$$

From Eq. (17), in Sect. 4.2, we will derive a physical model of battery gravitation where the triple reducer will be a function of the number of teeth and the radius of each gear in Fig. 4. Consequently, we will demonstrate a differential equation of different variables to be optimized.

4.2 Mathematical Model of the Gravitational Battery

In order to define the gravitational battery model, we start with Newton's second law and relate the torque $\vec{\tau}$ in the flywheel, presented in Fig. 3, obtaining Eq. (18):

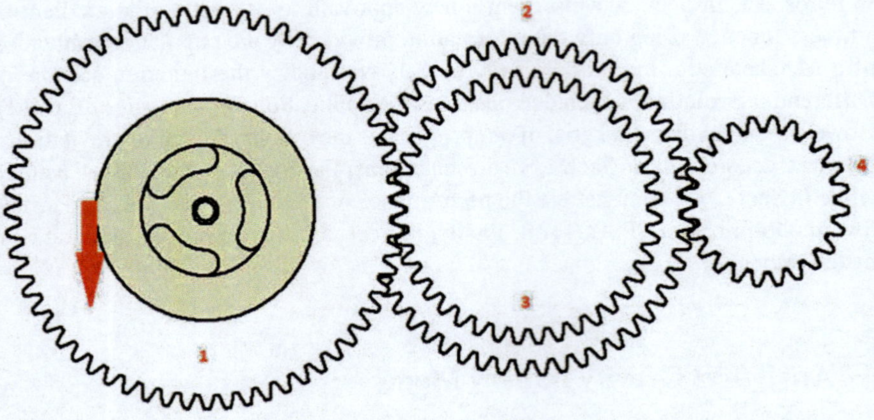

Fig. 4 Triple reduction system with pulley for the transmission of potential to electrical energy

$$\vec{F} = \frac{\vec{\tau}}{R_p} - m\vec{g} \tag{18}$$

where R_p represents the radius of the kite, $\vec{g}$ the gravitational acceleration. Consequently, we write Eq. (18) in its differential form, obtaining Eq. (19):

$$\frac{d^2\vec{y}}{dt^2} = \frac{\vec{\tau}}{R_p \cdot m} - \vec{g} \tag{19}$$

According to the ratio of gear teeth in Fig. 4, we write $\vec{\tau} = \left(\frac{N_1 \cdot N_3}{N_2 \cdot N_4}\right)\vec{\tau}_s$ in Eq. (19), and obtain Eq. (20):

$$\frac{d^2\vec{y}}{dt^2} = \frac{\left(\frac{N_1 \cdot N_3}{N_2 \cdot N_4}\right)}{R_p \cdot m}\vec{\tau}_s - \vec{g} \tag{20}$$

where N_1, N_2, N_3, and N_4 represent the number of teeth in the transmission of Fig. 4, and $\vec{\tau}_s$ is the output torque required to rotate the electric generator.

As a conclusion of Sects. 4.1 and 4.2, we have described in Subsection 4.1 a triple gear drive system as a proof of concept for storing gravitational energy by lifting a weight by the drive of clean energy, either solar or wind. Subsequently, in Sect. 4.2, we analyzed the relationship between the triple-reducer drive and the physics of gravitational energy storage, resulting in the differential discharge acceleration equation. However, our research aims to generate an optimal design for power generation. Therefore, Sect. 4.3 presents a fitness function suitable for evolutionary optimization algorithms, notably the Gradient Smooth algorithm.

4.3 Gravitational Energy Transmission

In the present section, we define the fitness function in terms of the required acceleration of gravitational to dynamic potential energy transfer of the electric generator in a time [0, t_f]. Mathematically, the fitness function is defined in Eq. (21):

$$Fitness = \left[\left\| \vec{a}_c - \left(\frac{\left(\frac{N_A \cdot N_c}{N_B \cdot N_D}\right) \vec{\tau}_s - \vec{g}}{R_p \cdot m} \right) \right\| \right]^2 \tag{21}$$

where $\vec{a}_c$ is the desired drop acceleration of the gravity battery design; specifically, Eq. (20) relates the desired acceleration of energy transfer to the theoretical acceleration of transmission derived from lifting the weight by a renewable energy system, whether solar, wind, or dynamic flow.

5 Results

In the present section, we will address the experimental comparisons that validate the theoretical results of Sect. 2 and the algorithmic implementation of the Smoothed Gradient presented in Sect. 3 using a statistical contrast with the control algorithms Particle Swarm Optimization and Differential Evolution (Sects. 4.1 and 4.2). Finally, in Sect. 4.3, we generate a case study of optimal gravity battery design using Smoothed Gradient.

5.1 Sperimental Setup

To generate a statistical comparison of Smoothed Gradient performance, we designed a set of 31 independent experiments using 10 test fitness functions (Table 1). In statistical tests, we aligned the experiments to those required by the methodology of Derrac et al. [33], which uses non-parametric statistical tests [33].

The software used to implement PSO, DE and GG-DE are Python: PSO was implemented in its basic form with parameters c1 = 2.5 and c2 = 0.9; DE from the library implemented in Scipy with F = 0.5 and CR = 05; finally, SG was implemented by us. As population-based Metaheuristcos parameters, we used a set of 100 individuals for each compared algorithm.

Table 1 Test functions for algorithm comparison

Function	Model	Domain
Ackley	$20 - 20e^{-0.2\sqrt{\sum \frac{x_i^2}{n}}} + e^{-0.2\sqrt{\sum \frac{\cos(2\pi x_i)}{n}}} + e$	[−32.76, 32.76]
Griewangk	$1 + \sum \frac{x_i^2}{4000} - \prod_{i=1}^{n} \cos\left(\frac{x_i}{\sqrt{i}}\right)$	[−600, 600]
Sphere	$\sum_{i=1}^{n} x_i^2$	[−10, 5]
Ellipsoid	$\sum_{i=1}^{n} 10^{6\frac{i-1}{n-1}} x_i^2$	[−10, 5]
Cigar	$x_1^2 + 10^6 \sum_{i=2}^{n} x_i^2$	[−10, 5]
Tablet	$10^6 x_1^2 + \sum_{i=2}^{n} x_i^2$	[−10, 5]
Cigar Tablet	$x_1^2 + 10^4 \sum_{i=2}^{n-1} x_i^2 + 10^8 x_n^2$	[−10, 5]
Two Axes	$\sum_{i=1}^{n/2} 10^6 x_i^2 + \sum_{i=\frac{n}{2}+1}^{n} x_i^2$	[−10, 5]
Rastregin	$10n + \sum_i^n \left[x_i^2 - 10\cos(2\pi x_i)\right]$	[−5.12, 5.12]
Rosenbrock	$\sum_{i=1}^{n-1} \left[100(x_i - x_{i-1}^2)^2 + (1 - x_i)^2\right]$	[−10, 5]

5.2 *Comparison of Smoothed Gradient with Control Algorithms*

In an experimental comparison, we evaluated 31 parallel runs of each function in Table 1, in 10 and 20 dimensions, to generate sufficient statistical evidence. Consequently, we ran the DE, PSO, and GG-DE algorithms using the same Hardware features: i7 eighth generation processor; 16 of RAM; Ubuntu 20 operating system. Finally, the statistical comparison results are reported in Table 2; where we present the median and standard deviation of the 31 experiments.

In statistical performance comparison, we highlight that SGA converges to the desired error point for the first six functions in Table 2, both in 10 and 20 dimensions. In another perspective, GG-DE has a better median approach to the optimum than DE and PSO in the Two Axes, Rastregin, and Rosenbrock functions for 20 dimensions.

5.3 *Statistical Tests of Algorithmic Performance*

To validate the convergence results in Table 2, we use the Friedman's Ranking, Aligned Friedman, and Quade tests, which guarantee that they are independent of the distribution profile of the experiment, having as a particular case the Normal distribution. Consequently, Table 3 shows the convergence rankings of the experiments in Table 2, for a value of alpha = 0.05; we highlight that SGa has the lowest ranking value for all three tests. Therefore, GG-DE indicates sufficient evidence of statistical differences compared to DE and PSO.

Table 2 Experimental results of comparison between DE, GG-DE, and PSO; reporting median and standard deviation of 31 parallel runs in 10 and 20 dimensions

Function		DE	GG-DE	PSO
Ackley	10D	$2.1 \times 10^{-1} \pm 0.09$	$8.8 \times 10^{-7} \pm 1.3 \times 10^{-7}$	$9.2 \times 10^{-7} \pm 6.6 \times 10^{-1}$
	20D	5.4 ± 0.53	$1.5 \times 10^{-5} \pm 6.9 \times 10^{-7}$	2.6 ± 1.04
Griewangk	10D	0.61 ± 0.10	$1.5 \times 10^{-1} \pm 3.8 \times 10^{-2}$	$9.5 \times 10^{-2} \pm 8.5 \times 10^{-2}$
	20D	2.21 ± 0.52	$3.2 \times 10^{-1} \pm 7.2 \times 10^{-2}$	0.04 ± 0.11
Sphere	10D	$4.8 \times 10^{-5} \pm 3.2 \times 10^{-4}$	$7.8 \times 10^{-7} \pm 2.1 \times 10^{-7}$	$8.4 \times 10^{-7} \pm 1.5 \times 10^{-7}$
	20D	0.72 ± 0.21	$8.1 \times 10^{-7} \pm 1.8 \times 10^{-7}$	$1.2 \times 10^{-4} \pm 8.8 \times 10^{-2}$
Ellipsoid	10D	0.47 ± 0.25	$2.1 \times 10^{-1} \pm 0.09$	$1.1 \times 10^{2} \pm 4.2 \times 10^{2}$
	20D	650.8 ± 163.4	$8.1 \times 10^{-7} \pm 1.7 \times 10^{-7}$	650.8 ± 163.4
Cigar	10D	113.28 ± 69.87	$6.5 \times 10^{-7} \pm 1.9 \times 10^{-7}$	$3.6 \times 10^{1} \pm 3.6 \times 10^{1}$
	20D	$4.9 \times 10^{5} \pm 1.2 \times 10^{5}$	$7.8 \times 10^{-7} \pm 4.7 \times 10^{-7}$	$2.9 \times 10^{3} \pm 4.0 \times 10^{3}$
Tablet	10D	$2.5 \times 10^{-7} \pm 7.9 \times 10^{-7}$	$6.7 \times 10^{-7} \pm 2.0 \times 10^{-7}$	$1.6 \times 10^{1} \pm 2.3 \times 10^{1}$
	20D	1.5 ± 0.59	$5.5 \times 10^{-7} \pm 1.5 \times 10^{-7}$	68.8 ± 49.7
Cigar Tablet	10D	$2.5 \times 10^{-3} \pm 1.1 \times 10^{-3}$	$7.6 \times 10^{-7} \pm 2.3 \times 10^{-7}$	9.3 ± 2.1
	20D	$7.1 \times 10^{3} \pm 3.1 \times 10^{3}$	$7.4 \times 10^{-7} \pm 1.9 \times 10^{-7}$	$6.1 \times 10^{4} \pm 6.8 \times 10^{4}$
Two Axes	10D	$4.1 \times 10^{-1} \pm 1.7 \times 10^{-1}$	$7.1 \times 10^{-7} \pm 2.1 \times 10^{-7}$	$6.8 \times 10^{1} \pm 6.2 \times 10^{1}$
	20D	$6.91 \times 10^{2} \pm 2.7 \times 10^{2}$	$7.9 \times 10^{-7} \pm 1.9 \times 10^{-7}$	$2.1 \times 10^{2} \pm 1.1 \times 10^{2}$
Rastregin	10D	$3.1 \times 10^{1} \pm 5.6$	2.6 ± 1.1	$1.5 \times 10^{1} \pm 8.2$
	20D	$1.2 \times 10^{2} \pm 9.1$	$1.1 \times 10^{1} \pm 2.3 \times 10^{1}$	42.7 ± 15.2
Rosenbrock	10D	$1.1 \times 10^{1} \pm 3.41$	$9.7 \times 10^{-1} \pm 4.3 \times 10^{-1}$	6.7 ± 1.8
	20D	$7.8 \times 10^{2} \pm 2.5 \times 10^{2}$	3.2 ± 5.9	$5.3 \times 10^{1} \pm 5.6 \times 10^{1}$

Table 3 Raking comparison results based on Friedman, Aligned Friedman, and Quade tests

Algorithms	Friedman	Aligned Friedman	Quade
DE	2.6	39.5	2.6
GG-DE	**1.1**	**18.3**	**1.0**
PSO	2.1	33.6	2.2

Based on the ranking results in Table 4, we proceed to validate by pairwise comparison the statistical differences with a P-value = 0.05. Table 5 shows the test values, where we highlight that GG-DE demonstrates sufficient statistical evidence that indicates better performance than DE and PSO under our experimental scheme; furthermore, we observe from Table 4 that there is not enough difference between the performance of PSO and DE. In conclusion, we demonstrate that the mathematics deduced in Sect. 2 and the adaptation to population-based Meta-heuristics from Sect. 3 improve convergence performance.

To illustrate the convergence comparisons, Fig. 4 presents in logarithmic scale the performance of DE, GG-DE, and PSO by several functions calls: for Ackley, it

Table 4 Pairwise statistical comparison between DE, SGa, and PSO under Holm's and Shaffer's nonparametric tests; for p = 0.05

Algorithms	Holm	Shaffer
DE versus GG-DE	**0.016**	**0.016**
GG-DE versus PSO	**0.025**	0.05
DE versus PSO	0.05	0.05

Table 5 Optimized gravity battery parameters

m	1Kg
R_p	10 cm
N_1	75
N_2	55
N_3	45
N_4	25

is observed that in 10 dimensions, SGA competes in convergence with PSO, while in 20 dimensions, GG-DE performs better than DE and PSO; for Tablet, GG-DE presents the best performance in both 10 and 20 dimensions. In general conclusion, the results in Tables 3 and 4, as well as the graphical comparisons, demonstrate that the GG-DE Algorithm performs better in higher problem dimensions, besides being competitive in dimensions less than 10.

In conclusion, we have experimentally validated the importance of using a gradient direction, deduced in Sects. 2 and 3. In statistical comparison, we found under the results of Tables 3 and 4 that GG-DE has better statistical performance than PSO and DE; furthermore, in Sect. 4, the Smoothed Gradient showed that the population-based Metaheuristic requirement does not exceed the needs of other members of the same family. Based on the results of Sects. 5.1 and 5.2. In Sect. 5.3, we will present the results of using SGA in the electromechanical design of a gravity battery.

5.4 *Example of an Optimized Gravity Battery*

To show an example of gravity battery design, we used the fitness function described in Sect. 4.3, then we proceeded to generate optimization proposals of triple-reducer system based on SGA; the solution domain is defined as: mass [1, 2] kg, Rp: [5, 29] cm, N1 [100,50], N2 [60,30], N3 [50,20], N4 [10, 29]. Consequently, Table 5. shows the optimization results of the system, where we emphasize that the optimized results coincide with the set of gears presented in Fig. 4.

Based on the results in Table 5, we elaborated a coupling design of the gravitational battery to the electric generator, which we present in Fig. 6. On the left, we show an isometric view of the gravitational battery configured with the number of teeth corresponding to each gear in Table 5, gear 1 contains the weight-loading pulley and has the corresponding dimensions for Rp from the Table 5, on the other hand,

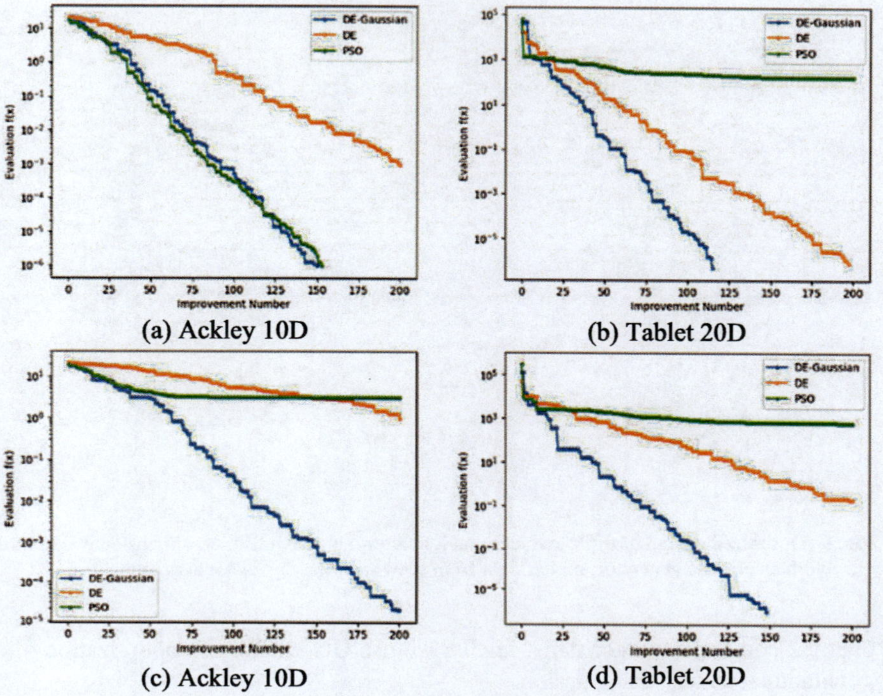

Fig. 5 Comparison of performance in logarithmic scale versus number of improvements: in green color, GG-DE results are shown; in Blue PSO; in orange DE

gear 4 is in charge of generating the rotational motion in the electric generator; on the right, we represent the two-dimensional drawing of the optimized transmission design eliminating the pulley and the generator.

Here, we demonstrate the electromechanical design feasibility of a triple reduction gear drive for gravity battery using a new optimization scheme called Smoothed Gradient Algorithm. The results are in Table 5 shows that GG-DE is an algorithm that can find suitable designs for the constraints imposed on the battery discharge time. Finally, in Fig. 5. We show a feasible coupling design of the optimized results.

6 Conclusions

We present a new single-objective optimization algorithm for electromechanical design applied to gravity batteries. Consequently, our contribution derives from the mathematical modeling of an interpolation of the gradient of the fitness function using a Gaussian kernel, in addition to its adaptation in the Differential Evolution methodology (GG-DE); showing that GG-DE has a higher equal performance

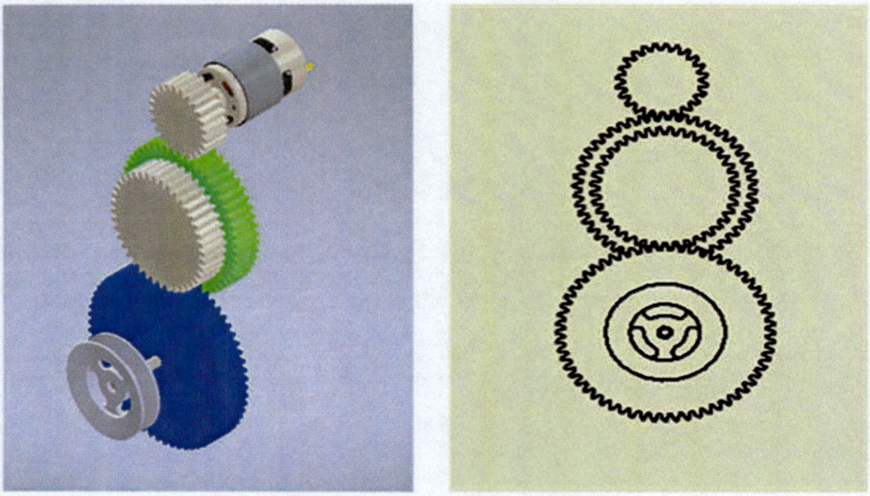

Fig. 6 Optimized design of triple-reducer gravity battery: on the left three-dimensional isometric view with an electric generator; on the right front view in a two-dimensional drawing

than the comparison algorithms; finally, we use GG-DE for the optimization of a gravitational battery transmission.

In the context of computational optimization, GG-DE was compared with Particle Swarm Optimization and Differential Evolution in a 10- and 20-dimensional experimental scheme on ten optimization functions (Table 2); result that DE-SG has enough statistical evidence to distinguish better convergence performance (Tables 3 and 4). Therefore, we show that the analytical results in Sects. 2 and 3 are sufficient to improve the convergence of a population-based optimization algorithm.

From the point of view of electromechanical design, GG-DE generated a model of a triple-reducer gravitational battery limited to the needs of electric discharge time. Accordingly, Table 5 presents the optimum design results, and in Fig. 6, we represent a possible arrangement of the triple reducer coupled to the pulley and the electric generator. Therefore, GG-DE is a reliable method of convergence in electromechanical design.

References

1. Dixit, U. S., Hazarika, M., & Davim, J. P. (2017). *A brief history of mechanical engineering*. Springer International Publishing.
2. Scherer, P. O. J. (2010). *Computational physics*. Springer.
3. Scherer, P.O. (2017). *Computational physics: Simulation of classical and quantum systems*. Springer.
4. Corriou, J.-P. (2021). Analytical Methods for Optimization. In J.-P. Corriou (Ed.), *Numerical methods and optimization: Theory and practice for engineers* (pp. 455–503). Springer International Publishing.

5. Potuzak, T., & Lipka, R. (2018). Analysis and optimization of fitness function of genetic algorithm for road traffic network division. In: S. Hippe Zdzisław, J. L. Kulikowski, & T. Mroczek (Eds.), *Human-computer systems interaction: Backgrounds and applications* (Vol. 4, pp. 91–102). Springer International Publishing
6. Corradi, G. (2011). A method for non-differentiable optimization problems. *International Journal of Computer Mathematics, 88*, 3750–3761. https://doi.org/10.1080/00207160.2011.620095
7. Bertsekas, D. P. (1973). Stochastic optimization problems with nondifferentiable cost functionals. *Journal of Optimization Theory and Applications, 12*, 218–231. https://doi.org/10.1007/BF00934819
8. Treanţă, S. (2022). Robust optimality in constrained optimization problems with application in mechanics. *Journal of Mathematical Analysis and Applications, 515*, 126440. https://doi.org/10.1016/j.jmaa.2022.126440.
9. Tanskanen, P. (2002). The evolutionary structural optimization method: Theoretical aspects. *Computer Methods in Applied Mechanics and Engineering, 191*, 5485–5498. https://doi.org/10.1016/S0045-7825(02)00464-4
10. Poloni, C., Giurgevich, A., Onesti, L., & Pediroda, V. (2000). Hybridization of a multi-objective genetic algorithm, a neural network and a classical optimizer for a complex design problem in fluid dynamics. *Computer Methods in Applied Mechanics and Engineering, 186*, 403–420. https://doi.org/10.1016/S0045-7825(99)00394-1
11. Yoshimura, M., Shimoyama, K., Misaka, T., & Obayashi, S. (2017). Topology optimization of fluid problems using genetic algorithm assisted by the Kriging model. *International Journal for Numerical Methods in Engineering, 109*, 514–532. https://doi.org/10.1002/nme.5295
12. de Anda-Suárez, J., Jeyakumar, S., Carpio, M., et al. (2019). Parameter optimization for the smoothed-particle hydrodynamics method by means of evolutionary metaheuristics. *Computer Physics Communications*. https://doi.org/10.1016/j.cpc.2019.05.008
13. Janamala, V., & Sreenivasulu Reddy, D. (2021). Coyote optimization algorithm for optimal allocation of interline–Photovoltaic battery storage system in islanded electrical distribution network considering EV load penetration. *Journal of Energy Storage, 41*. https://doi.org/10.1016/j.est.2021.102981.
14. Sahay, K.B., Balachander, B., Jagadeesh, B., et al. (2022). A real time crime scene intelligent video surveillance systems in violence detection framework using deep learning techniques. *Computers and Electrical Engineering, 103*, 108319. https://doi.org/10.1016/j.compeleceng.2022.108319.
15. Cheng, J., & Shi, T. (2022). Structural optimization of transmission line tower based on improved fruit fly optimization algorithm. *Computers and Electrical Engineering, 103*, 108320. https://doi.org/10.1016/j.compeleceng.2022.108320.
16. Dujardin, J., Kahl, A., & Lehning, M. (2021). Synergistic optimization of renewable energy installations through evolution strategy. *Environmental Research Letters, 16*. https://doi.org/10.1088/1748-9326/abfc75.
17. Khosla, A., & Aggarwal, M. (2022). *Renewable energy optimization, planning and control*. Springer
18. Baños, R., Manzano-Agugliaro, F., Montoya, F. G., et al. (2011). Optimization methods applied to renewable and sustainable energy: A review. *Renewable and Sustainable Energy Reviews, 15*, 1753–1766.
19. Zhang, J., Chen, C., Zhang, X., & Liu, S. (2016). Study on the environmental risk assessment of lead-acid batteries. *Procedia Environmental Sciences, 31*, 873–879. https://doi.org/10.1016/j.proenv.2016.02.103
20. Gottesfeld, P., Were, F. H., Adogame, L., et al. (2018). Soil contamination from lead battery manufacturing and recycling in seven African countries. *Environmental Research, 161*, 609–614. https://doi.org/10.1016/j.envres.2017.11.055
21. Jera, A., Ncube, F., & Kanda, A. (2017). Contamination of soil with Pb and Sb at a lead-acid battery dumpsite and their potential early uptake by phragmites Australis. *Applied and Environmental Soil Science*. https://doi.org/10.1155/2017/8010453.

22. Loudiyi, K., & Berrada, A. (2017). Experimental validation of gravity energy storage hydraulic modeling. In *Energy Procedia* (pp. 845–854). Elsevier Ltd.
23. Berrada, A., & Loudiyi, K. (2019). Gravity energy storage applications. In *Gravity energy storage* (pp. 75–103). Elsevier
24. Berrada, A., & Loudiyi, K. (2019). Technical design of gravity energy storage. In *Gravity energy storage* (pp. 25–49). Elsevier
25. Berrada, A., & Loudiyi, K. (2019). Energy storage. In *Gravity energy storage* (pp 1–23). Elsevier
26. Oliva, D., Elaziz, M. A., & Hinojosa, S. (2019). Studies in computational intelligence. In *Metaheuristic algorithms for image segmentation: Theory and applications*
27. He, J., & Yu, X. (2001) Conditions for the convergence of evolutionary algorithms q.
28. Hu, Z., Xiong, S., Su, Q., & Zhang, X. (2013). Sufficient conditions for global convergence of differential evolution algorithm. *Journal of Applied Mathematics*. https://doi.org/10.1155/2013/193196.
29. Zhang, J., Bi, S., & Zhang, G. (2021). A directional Gaussian smoothing optimization method for computational inverse design in nanophotonics. *Materials and Design, 197*. https://doi.org/10.1016/j.matdes.2020.109213.
30. Bosman, P. A. N. (2012). On gradients and hybrid evolutionary algorithms for real-valued multiobjective optimization. *IEEE Transactions on Evolutionary Computation, 16*, 51–69. https://doi.org/10.1109/TEVC.2010.2051445
31. Ahmad, M. F., Isa, N. A. M., Lim, W. H., & Ang, K. M. (2022). Differential evolution: A recent review based on state-of-the-art works. *Alexandria Engineering Journal, 61*, 3831–3872. https://doi.org/10.1016/j.aej.2021.09.013.
32. Marini, F., & Walczak, B. (2015). Particle swarm optimization (PSO). A tutorial. *Chemometrics and Intelligent Laboratory Systems, 149*, 153–165. https://doi.org/10.1016/j.chemolab.2015.08.020
33. Derrac, J., García, S., Molina, D., & Herrera, F. (2011). A practical tutorial on the use of nonparametric statistical tests as a methodology for comparing evolutionary and swarm intelligence algorithms. *Swarm and Evolutionary Computation, 1*, 3–18. https://doi.org/10.1016/j.swevo.2011.02.002

A Comparison Between Selection Operators Heuristics of Perturbation in CSP

Lucero Ortiz-Aguilar, Hernández-Aguirre Yeovanna, M. Benitez, Sergio Rodriguez-Miranda, and Fernando Mendoza-Vazquez

1 Introduction

In the context of combinatorial optimization problems, it is essential to give a solution to a set of problems or instances which belongs to different families [22]. Usually, for each set of instances, an algorithm, metaheuristic, or another is designed ad-hoc. Nevertheless, if a new variant of each problem is designed, the algorithm must be redesigned or adjusted to solve the problem. However, the algorithm design for a given problem set requires expert knowledge and a deep analysis of each variable and restriction.

Likewise, real problems have different characteristics, restrictions, and variables, making applying the same algorithms challenging to get a good solution [2]. Various methods have been used in the literature to graph Coloring Problems, Vertex Cover, and Vehicle routing Problems, among others [17, 26]. Therefore, these problems are well-known and studied, and it is possible to propose a solution. It is important to remember that real problems need to be solved in a short period before the problem expires. Some Combinatorial Problems can be solved easily by heuristic and metaheuristic ad-hoc, but their performances depend strongly on their design and parameter tuning.

By another hand, Hyperheuristics have been defined recently as "automated methodologies for selecting or generation heuristics to solve hard computational

L. Ortiz-Aguilar (✉) · M. Benitez · S. Rodriguez-Miranda · F. Mendoza-Vazquez
División de Sistemas Automotrices, Tecnológico Nacional de México, Instituto Tecnológico Superior de Purísima del Rincón, Purísima del Rincón, Guanajuato, México
e-mail: lucero.oa@purisima.tecnm.mx; ldm_oa@hotmail.com

H.-A. Yeovanna
División de Estudios de Posgrado e Investigación, Tecnológico Nacional de México, Instituto Tecnológico de León, León, Guanajuato, México

O. Castillo and P. Melin (eds.), *Hybrid Intelligent Systems Based on Extensions of Fuzzy Logic, Neural Networks and Metaheuristics*, Studies in Computational Intelligence 1096,
https://doi.org/10.1007/978-3-031-28999-6_23

search problems" [22], which are demonstrated a good performance in different optimization areas [12, 19, 25]. This chapter aims to provide a methodology to design new heuristics considering the various components. These components have been studied before in a wide variety of algorithms [1, 12, 20].

The remaining content of this article contains a description of the Background in Sect. 2, which covers a review of heuristics, Hyperheuristics, and the constraint satisfaction problem used. The methodology used is shown in Sect. 3. The found results and findings, including performance comparison, are described in Sect. 4. Finally, concluding remarks are presented in Sect. 5.

2 Background

In this Section, we focus on concepts about the Constraint satisfaction problem, the Acceptance criteria method, and heuristic information that we will use in this work.

2.1 *Constraint Satisfaction Problems*

The formal definition of Constraint Satisfaction Problems (CSP) was given by Montanari in [16] with $V = \{v_1, v_2, \ldots, v_n\}$ a set of variables, $W = \{w_1, w_2, \ldots, w_m\}$ a set of values of each variable and $R = \{r_{v1}, r_{v2}, \ldots r_{vn}\}$ a set of restrictions for each variable denoted. Montanari [16] applied the CSP to image processing and some years later Jeavons in [13] defined this problem as an instance of a General Combinatorial problem. The CSP problem has been studied by other researchers and its applications in Sequencing Mixed-model Assembly Lines [10], Knapsack [6], Location [7], Cutting Stock [15], Scheduling [21], and among others.

In this work, we use The Graph Coloring Problem to test our methodology because this problem is well-known studied, and defined in state of the art, therefore for several instances, it is the best know solution or optimal fitness.

Graph Coloring Problem (GC-P)

The Graph Coloring Problem (GC-P) consists of given a graph G, which must be colored with q colors, and it was defined as NP by [14]. The vertexes of a graph can not be colored with the same color if these are adjacent. The set of instances used in this work can be found at mat.gsia.cmu.edu..[1]

Formally the GC-P is defined as:

Given a graph G with V a set of nodes $v_1, \ldots, v_n$, Q a set of colors $q_1, \ldots, q_l$, and $E = e_1, \ldots, e_m$ edges. For each solution, the fitness is the sum of the nodes connected with the same color as:

[1] http://mat.gsia.cmu.edu/COLOR/instances.htmlmat.gsia.cmu.edu.

$$H(G) = \sum_{i=0}^{n} \sum_{i=0}^{n} h(m) \tag{1}$$

$$h(m) = \begin{cases} if(v_i, v_j \epsilon E) \cap (q_{vi} == q_{vj})) & 1 \\ othercase & 0 \end{cases} \tag{2}$$

2.2 Heuristics

In this work, the aim is to identify the different components of low-level heuristics. A simple definition of a heuristic is "the art of discovering new rules to solve problems," and its origin is Greek [23].

The pool of low-level heuristics is the main component of hyper-heuristics; therefore, according to the problem, this pool changes. According to the classification proposed by Burke et al. [5] about Hyperheuristics, these use perturbative and constructive heuristics. Perturbative heuristics are low-level, as they operate directly on the candidate solutions of the underlying optimization problem. On the other hand, constructive heuristics build a solution starting with an empty or partial solution.

2.2.1 Components

In this work, we used the perturbative heuristics for Graph Coloring Problem. Previous research conducted by Ortiz-Aguilar et al. [20] classified the heuristics into three groups according to the information used in the heuristic process: problem information, no problem information, and other solutions information (see Fig. 1 taken from [20]).

Notwithstanding the above, the classification does not consider other characteristics such as acceptance criteria, selection, solutions generated, and memory, among others. Nevertheless, no study focuses on the effects of each component and how it

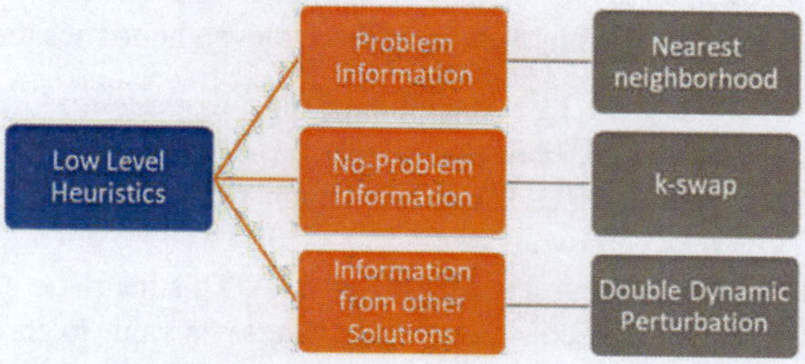

Fig. 1 Low level heuristic aggrupation

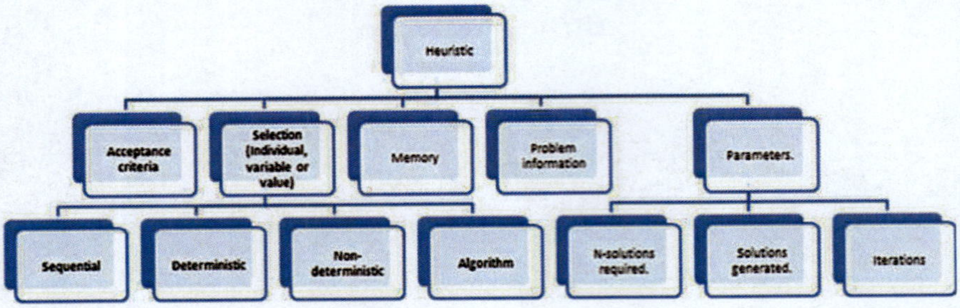

Fig. 2 Perturbation heuristics components

affects the process search. Consequently, after deep research on perturbative heuristics, we proposed the diagram presented in Fig. 2, which contains different components and their derivatives. In the present work, it was possible to identify some parts that influenced the performance of heuristics, such as selection or acceptance criteria.

In a more detailed way, the components for these heuristics are:

- **Acceptance Criteria**. The Acceptance criteria were studied by Jackson et al. [12], and they propose a classification based on the Nature of accepting/rejecting decisions divided into Stochastic and Non-stochastic (Basic and threshold); Nature of how the algorithmic parameters with Memoryless, Static, and Dynamic Memory.
- **Selection (variable or value)**. Each heuristic can select a variable or variable value of a given solution. This selection can be sequential, deterministic, non-deterministic, or guided by algorithms.
- **Memory**. Some heuristics need a memory or historical records to guide their movements to perturb the solution.
- **Problem information**. This information lets us search for feasible solutions.
- **Parameters**. Some heuristics need parameters to make or guide their search process. Some examples are the number of solutions required and solutions generated, among others.

2.2.2 Algorithms

In this work, we selected 11 heuristics to prove our methodology in Graph Coloring Problem. The algorithms of these eleven heuristics are:

1. **k-flip (H_1)** For the Graph Coloring, problem *k*-flip changes the color of a node to another feasible [8].
2. **K-Swap (H_2)**. For the Graph Coloring Problem, it interchanges the colors between two nodes.
3. **Best Single Perturbation (H_3)**. This heuristic chooses a variable according to the list of constraints and changes its value to another feasible one. The selected variable is modified, it is determined according to the next position, respecting

the last variable that was chosen (sequentially). As a result, the solution with the best fitness is returned.

4. **Statical Dynamic Perturbation (H_4).** This heuristic chooses a variable and changes its value randomly [24], these are selected according to a probability distribution of the frequency in the k iterations.
5. **Two Points Perturbation (H_5).** This is a particular case of the K-Swap, with a value of $k = 2$.
6. **Double dynamic Perturbation (H_6).** This heuristic returns the best solution between an initial solution and a solution modify by Statical Dynamic Perturbation [24].
7. **Move to less conflict (H_7).** This heuristic selects a random variable and assigns it to another feasible value that generates the least cost for the solution [25].
8. **Min-Conflicts (H_8).** This heuristic chooses a random variable and assigns a value that generates the lowest cost [25].
9. **First-Fit (H_9).** This heuristic changes the value of a variable to other feasible, which is the least repeated in other variables [25].
10. **Worse-fit (H_{10}).** This heuristic assigns the most repeated value to a randomly selected variable [11].
11. **Burke-Abdullah (BA) (H_{11}).** This heuristic selects a variable applying Fail-First or Brelaz Heuristic [9] and its value is changing to another according to the application of the algorithms such as Minimum conflict, Randomly, Sequential selection, and least Constrained and the solution with the best fitness is selected.

2.3 Acceptance Criteria

Warren et al. [12] proposed a classification for different acceptance criteria methods, and each criterion was applied in single-objective local search metaheuristics. These methods have a generic algorithm scheme as Initial solution or solutions, choice of variable to modify, change the value, evaluation of the new solution, apply acceptance criteria method, and return solution. Warren G. et al. propose the next classification [12]:

- Nature of accept/reject decision.
  - Stochastic
  - Non-stochastic: Basic and threshold
- Parameter settings:
  - Memoryless. Static and dynamic
  - Memory. Adaptive.

2.4 Selection Method

Each perturbation heuristic studied in this work has a component called "*selection.*" The "*selection*" type components were identified in three different contexts through the solution search process:

- **Variable**. Sometimes the perturbation heuristics need to select one or more variables to make or improve the value of these variables. Therefore, the way to choose these variables is meaningful because if the same variable is selected or not, this affects the solution fitness.
- **Value**. After selecting the variable, it is essential to choose the value which can improve the solution according to the problem restriction.
- **Individual**. If the heuristic needs two or more solutions to generate a new one, only the first solution is given as a parameter; therefore, the additional solution must be selected or generated by another alternative.

The selection process, according to its nature, can be:

1. **Sequential**. In this case, the heuristic stores a memory that indicates the next element that should be selected.
2. **Deterministic**. The heuristic always selects the same element (value, solution, or Variable). This can be the first or the last element; it depends on the heuristic steps. An example of a heuristic with these components is Best Single perturbation.
3. **Non-deterministic**. This selection is probabilistic or random. An example of this selection in heuristic can be found in *k*-flip, *k*-swap, and double dynamic perturbation, among others.
4. **Algorithms**. In genetic algorithms, the selection of individuals is, for example, a roulette wheel, Vasconcelos or another algorithm.

2.5 Hyperheuristics

Hyperheuristics have been defined as "*heuristics for choosing heuristics*"; this definition agrees with the traditional concept proposed by [4]. Unlike conventional metaheuristic algorithms, Hyperheuristics aim to solve different combinatorial optimization problems by operating in the space of low-level heuristics. The hyperheuristic approach can be divided into selection and generation heuristics. According to low-level heuristics, there are two groups of construction and perturbation heuristics.

The main components of Hyperheuristics are: (a) The heuristic selection method, which aims to choose a heuristic from a heuristic's set. (b) The low-level-heuristic pool can be perturbation or construction heuristics (only one kind per group). (c) Move acceptance criteria that decide whether to reject or not the new solution. This investigation is focused on five of the best pool of heuristics for Hyperheuristics.

3 Methodology

The low-level heuristics has three stages: inputs, process, and outputs, and this is a simple schema that lets to identify components in each step (see Fig. 3).

In detail, the basic steps in the heuristic are shown in Fig. 4. The inputs are step 1, where the information of the problem is provided, such as the initial solutions or other parameters. The variable to be modified for each solution is selected in step 2. The next step is to select the new value or calculate it (step 3). If the problem has restrictions, a list of feasible values is given from which one is chosen, or if a value can be calculated for the variable, it is done in this way (step 3). After making all the corresponding changes to the solution(s), it is necessary to update the fitness value of each solution. In step 6, according to the acceptance criteria, the new solution was chosen, and finally, in the output stage, the solution was returned.

Taking into account the information shown in Fig. 4, in this research, the aim is to select the best selection method for each heuristic. Our methodology for choosing the best *select method* is described as:

(a) **Step 1**. To choose the set of instances to solve by partitions, due to the processing instances, is thought the methodologies of design proposed by [3, 18, 24]. The methodology of design lets consider restrictions in a simple way which facilitates the search process.
(b) **Step 2**. In experimentation, the acceptance criteria are always set as the original parameter of each algorithm. In each experiment, two possibilities can be used in the selection method.
(c) **Step 3**. Each heuristic with the different selection method and acceptance criteria is applied to instances set.
(d) **Step 4**. With the results of applying the different heuristics, the non-parametric statistical tests were used to determine which of them had better performance.

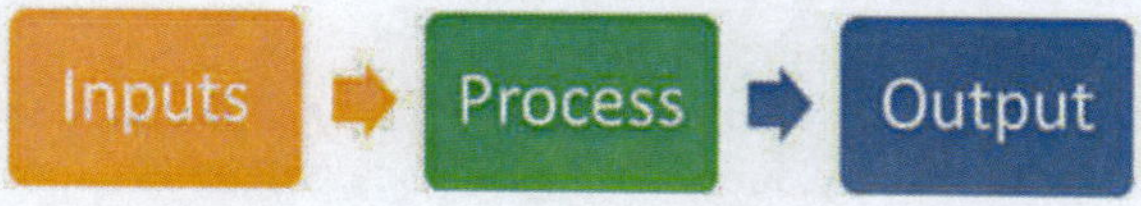

Fig. 3 Low-level heuristic structure

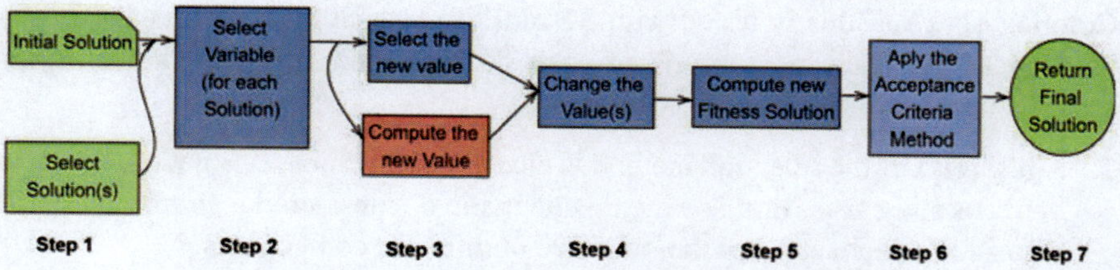

Fig. 4 Low-level heuristic detailed process

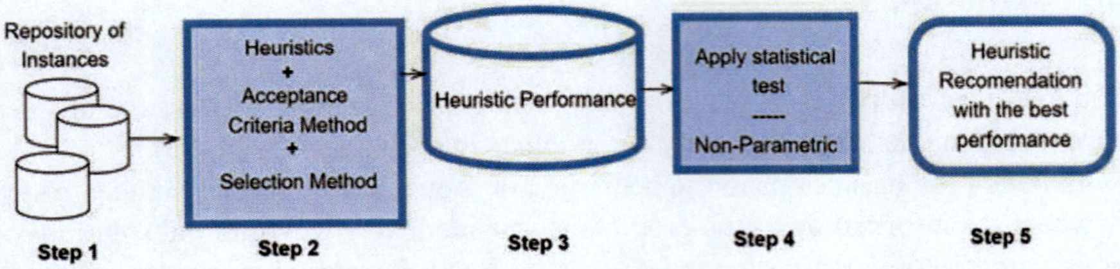

Fig. 5 Methodology of heuristic selection with different selection and acceptance criteria

(e) **Step 5**. The heuristics with the best performance are returned according to the results of the statistical tests.

The previous steps are summarized in the diagram in Fig. 5.

4 Results

This section describes our experiments for the Graph Coloring (GC-P) benchmark used in this paper and the configuration for implementing the 11 heuristics tested. Our approach was implemented in JAVA language with JDK 1.8 using the IDE NetBeans IDE 8.2. The experiments were executed on a computer with processor Intel i3-1005G1, 1.19 GHz, 8 GB DDR3 RAM, and operational system Windows 11 Home.

Finally, the statistical tests were described to compare our results of experimental methodology. Each heuristic was applied to the instance until 100,00 function calls were waisted in each run. We used the Shapiro-Wilks test to check whether the data results were normal, choosing a better representative (average or Medium). If the data behavior is according to a normal distribution, the average was taken as representative, and otherwise, the median.

The description of all the heuristics used with their different selection method according to their variables, values, and acceptance criteria is shown in Table 1. In the columns Variable, Value, and Solution, in this case, all the heuristics do not need to select the initial solution.

We used the benchmark proposed for the second DIMACS challenge on Graph Coloring [14] and this is tested with 31 runs. To reduce the experimental experimentation, we made an analysis of each heuristic and conclude the following propositions:

1. **k-flip (H_1)**. If the selection method is changing with another, a new variation of the heuristic exists. In this way, this heuristic is considered a simple and basic algorithm which let proof the influence of different components.
2. **K-Swap (H_2)**. In this heuristic, the behavior is similar to k-flip.

Table 1 Heuristic description. Acceptance criteria (AC)

Heuristic	Variable	Value	Acceptance criteria	Solutions required
k-flip	Stochastic	Stochastic	Replace	1
k-swap/2 pp	Stochastic	Stochastic	Replace	2
Best single perturbation	Sequential	Stochastic	Improving or equal	1
Double dynamic perturbation	Probability	Stochastic	Replace	2
Move less conflict	Non-Stochastic	Non-stochastic	Improving or equal	1
Min-Conflicts	Stochastic	Memory	Improving or equal	1
First-Fit	Non-Stochastic	Strategy	Improving or equal	1
Worse-Fit	Strategy	Strategy	The best solution	1
Statical dynamic perturbation (SDP)	Probability	Stochastic	Improving or equal	1
Burke-Abdullah (BA)	Strategy	Strategy	The best solution	1

3. **$\mathbf{H_3}$, $\mathbf{H_5}$, $\mathbf{H_6}$, $\mathbf{H_7}$, $\mathbf{H_8}$, $\mathbf{H_9}$,$\mathbf{H_{10}}$ and $\mathbf{H_{11}}$**. If we fix the position as first or last, the result is an algorithm similar to k-flip. Therefore, for this work, we decided to let this set heuristic out of experimentation.
4. **Statical Dynamic Perturbation ($\mathbf{H_4}$)**. This heuristic also presents a different performance.

In Tables 2 and 3 we showed our results. We denote the best results with a bold face.

We applied the same procedure with the statistical tests of Friedman (FT), Alienated Friedman (AFT), and Quade (Qt) to distinguish the behavior of the heuristics set. We established $\alpha = 0.05$ and $h_0 =$ as there are no differences between the performance of the heuristics and h_a as there are differences between the performance of the heuristics. Table 4 shows the ranks obtained in the three statistical tests.

We summarize according to the criteria, which were the heuristics that had the best performance:

- The heuristic with best performance is $\boldsymbol{H}_5$ replace first
- The heuristic with worst performance is $\boldsymbol{H}_1$ replace first.

5 Discussion and Future Work

A good algorithm design to solve a wide range of problems is still a research area with different challenges. To generate a more complex algorithm that can solve any kind of problem, it is necessary to get more information and mine the knowledge of each family problem. Therefore, this problem has two parts the problem definition and the algorithm solution. We focus on the algorithms in this investigation and clarify their

Table 2. Heuristics Results for Graph Coloring Instances H_1, H_2 and H_5.. Where Median (M), Average (Avg), Minimum (Min), Maximum (Max) and Standard Deviation (DE)

		H_1					H_2					H_5				
Name	No	M	Avg	Min	Max	DE	M	Avg	Min	Max	DE	M	Avg	Min	Max	DE
2-Insertions_3	37	18	17.8	13	24	3.1	18	18	10	26	4	0	0.1	0	1	0.3
3-Insertions_3	56	26	27	15	39	5.7	28	28	20	40	4	0	0.1	0	1	0.3
Myciel3	11	4	4.6	2	10	1.8	4	4.2	1	6	1	0	0.0	0	0	0.0
Myciel4	23	14	14.1	9	22	2.8	14	14	9	19	3	0	0.1	0	1	0.3
Myciel5	47	40	40.1	30	50	4.6	39	38	29	47	5	1	0.9	0	2	0.7
Queen10_10	100	129	133.6	117	163	12.1	137	136	114	152	10	21	22.1	16	31	3.9
Queen11_11	121	181	181	157	203	11.3	179	178	151	207	10	38	38.1	30	44	2.7
Queen12_12	144	201	199.5	176	225	12.2	195	197	172	225	13	35	34.2	26	41	3.8
Queen13_13	169	255	255.1	221	300	17.6	260	261	239	289	14	51	52.2	46	59	3.9
Queen14_14	196	261	264	228	301	17.8	262	263	234	288	11	33	32.7	22	43	4.4
Queen15_15	225	327	325.2	298	351	14.7	324	324	289	352	15	52	51.0	42	58	4.3
Queen16_16	256	375	375.5	342	417	19	371	371	342	415	16	56	58.2	49	70	5.2
Queen5_5	25	31	30.5	26	38	3.3	32	32	21	41	5	7	6.5	0	13	3.1
Queen6_6	36	42	40.9	29	50	5.7	40	40	30	54	6	6	6.2	3	9	1.6
Queen7_7	49	69	69.1	49	91	8.9	68	69	53	89	7	16	16.4	11	23	2.7
Queen8_8	64	82	81.8	67	102	8.5	79	80	61	109	11	13	12.8	7	20	3.1
Zeroin.i.1	211	81	81.9	66	105	9.7	83	85	67	107	9	9	9.0	6	13	1.7
Zeroin.i.2	211	117	117.5	101	132	7.1	118	116	91	128	9	8	8.3	6	10	1.0
Zeroin.i.3	206	115	116.3	95	133	9.6	123	121	103	141	10	8	8.2	5	11	1.5

Table 3 Heuristics Results with selection first for Graph Coloring Instances H_1, H_2 and H_5. . Where Median (M), Average (Avg), Minimum (Min), Maximum (Max) and Standard Deviation(DE)

		H_1					H_2					H_5				
Name	Nodes	M	Avg	Min	Max	DE	M	Avg	Min	Max	DE	M	Avg	Min	Max	DE
2-Insertions_3	37	18	18.3	12	25	3.5	18	18.0	10	26	3.4	19	19.1	9	28	4.0
3-Insertions_3	56	27	27.2	20	35	3.7	26	26.7	16	36	4.3	27	27.2	17	37	4.4
Myciel3	11	5	4.8	0	10	2.4	5	4.8	1	14	2.4	5	5.1	2	8	1.6
Myciel4	23	14	13.8	7	20	3.6	14	14.2	7	21	3.5	15	14.6	5	22	3.6
Myciel5	47	42	41.9	31	57	6.3	41	40.2	27	54	6.6	39	39.6	30	49	4.9
Queen5_5	25	31	31.1	23	41	4.3	31	31.0	23	37	3.6	29	30.0	21	38	4.4
Queen6_6	36	39	39.2	31	62	5.8	42	41.7	33	51	4.8	41	41.5	33	50	4.3
Queen7_7	49	70	69.8	57	86	6.5	69	67.9	50	83	7.3	65	65.5	52	80	7.2
Queen8_8	64	82	81.9	66	100	7.5	82	83.3	63	103	10.5	81	79.5	59	100	8.6
Queen10_10	100	134	135.1	118	163	11	134	135.3	120	153	8.3	138	137.9	113	164	11.7
Queen11_11	121	178	180.4	162	203	11.1	180	181.7	162	206	13.8	177	179.0	152	207	13.1
Queen12_12	144	202	200.7	179	232	13	201	201.2	174	235	13.7	199	200.4	185	223	10.3
Queen13_13	169	258	259	232	286	15.8	255	255.6	222	287	13.1	251	253.5	223	282	16.1
Queen14_14	196	258	259.8	233	298	14	264	263.2	231	308	17.3	265	265.5	233	306	16.1
Queen15_15	225	322	323.3	300	344	10.5	317	316.8	289	350	16.0	326	326.7	286	367	20.1
Queen16_16	256	364	367.7	338	430	21.3	369	371.4	330	416	23.8	371	369.4	330	405	16.7
Zeroin.i.1	211	85	84.8	68	108	9.9	85	84.9	66	106	7.2	84	84.7	74	106	7.5
Zeroin.i.2	211	115	116.2	100	147	10.2	117	119.0	104	143	9.4	115	118.3	93	145	12.7
Zeroin.i.3	206	115	115.1	98	145	10.3	122	120.5	99	136	9.0	115	116.1	91	143	10.5

Table 4 Heuristics ranks per Classes, Friedman (Fr), Aligned Friedman (A-Fr) and Quade (Qu). Where normal (N) and simple replace first (SF)

	H_1		H_2		H_5	
	N	SF	N	SF	N	SF
Friedman ranks	3.9	4.0	3.8	3.9	4.2	1
Quade ranks	4	4.0	3.9	3.7	4.2	1
Friedman aligned	1201	1206	1200.5	1274.5	1279	394

main components. First, we describe all the different heuristics components and give a definition and examples.

In previous work, we identify that the acceptance criteria were an essential component on the heuristics, and now we complement with the different selection method. As we see on experimental results, the was significantly differences between the select the initial value or variable and the default algorithm settings.

Nevertheless, we recommended proofing our methodology with a different instance problem and other heuristics. The identification of components in heuristics to find the solution to complex problems is an optimization strategy that can improve the hyper-heuristic design. The methodology presented allows for generating a better heuristic according to a specific problem. Finally, in future work, we propose to experiment with different issues and components.

References

1. Abdulaziz, H., Elnahas, A., Daffalla, A., Noureldien, Y., Kheiri, A., & Özcan, E. (2018). Late acceptance selection hyper-heuristic for wind farm layout optimisation problem. In *2018 International Conference on Computer, Control, Electrical, and Electronics Engineering* (ICCCEEE) (pp. 1–5). IEEE.
2. Adam, S. P., Alexandropoulos, S. A. N., Pardalos, P. M., & Vrahatis, M. N. (2019). No free lunch theorem: A review. *Approximation and Optimization, 57–82.*
3. Alba, E., & Dorronsoro, B. (2006). Computing nine new best-so-far solutions for capaci-tated VRP with a cellular genetic algorithm. *Information Processing Letters, 98*(6), 225–230.
4. Burke, E. K., Gendreau, M., Hyde, M., Kendall, G., Ochoa, G., Özcan, E., & Qu, R. (2013). Hyper-heuristics: A survey of the state of the art. *Journal of the Operational Research Society, 64*(12), 1695–1724.
5. Burke, E. K., Hyde, M. R., Kendall, G., Ochoa, G., Özcan, E., & Woodward, J. R. (2019). A classification of hyper-heuristic approaches: revisited. In *Handbook of metaheuristics* (pp. 453–477). Cham: Springer.
6. Cacchiani, V., Iori, M., Locatelli, A., & Martello, S. (2022). Knapsack problems-An overview of recent advances. Part II: Multiple, multidimensional, and quadratic knapsack problems. *Computers and Operations Research, 105693.*
7. Chan, H., Filos-Ratsikas, A., Li, B., Li, M., & Wang, C. (2021). Mechanism design for facility location problems: a survey. arXiv preprint arXiv:2106.03457.
8. de Werra, D. (1990). Heuristics for graph coloring. In: *Computational graph theory* (pp. 191–208). Springer.
9. Gent, I. P., MacIntyre, E., Presser, P., Smith, B. M., & Walsh, T. (1996). An empirical study of dynamic variable ordering heuristics for the constraint satisfaction problem. In *International*

Conference on Principles and Practice of Constraint Programming (pp. 179–193). Berlin, Heidelberg: Springer.
10. Guo, G., & Ryan, S. M. (2022). Sequencing mixed-model assembly lines with risk-averse stochastic mixed-integer programming. *International Journal of Production Research, 60*(12), 3774–3791.
11. Habashi, S. S., Salama, C., Yousef, A. H. et al. (2018). Adaptive diversifying hyper-heuristic based approach for timetabling problems. In: *2018 IEEE 9th Annual Information Technology, Electronics and Mobile Communication Conference* (IEMCON) (pp. 259–266). IEEE.
12. Jackson, W. G., Özcan, E., & John, R. I. (2018). Move acceptance in local search metaheuristics for cross-domain search. *Expert Systems with Applications, 109*, 131–151.
13. Jeavons, P. (1998). On the algebraic structure of combinatorial problems. *Theoretical Computer Science, 200*(1–2), 185–204.
14. Karp, R. (1972). *Complexity of Computer Computations*. Springer USA.
15. Melega, G. M., de Araujo, S. A., & Jans, R. (2018). Classification and literature review of integrated lot-sizing and cutting stock problems. *European Journal of Operational Research, 271*(1), 1–19.
16. Montanari, U. (1974). Networks of constraints: Fundamental properties and applications to picture processing. *Information Sciences, 7*, 95–132.
17. Mostafaie, T., Khiyabani, F. M., & Navimipour, N. J. (2020). A systematic study on meta-heuristic approaches for solving the graph coloring problem. *Computers & Operations Research, 120*, 104850.
18. Ortiz-Aguilar, L. D. M., Carpio, M., Soria-Alcaraz, J. A., Puga, H., Díaz, C., Lino, C., & Tapia, V. (2016). Training OFF-Line Hyperheuristics For Course Timetabling Using K-Folds Cross Validation. la revista programación matemática y software.
19. Ortiz-Aguilar, L., Carpio, M., Rojas-Domínguez, A., Ornelas-Rodriguez, M., Puga-Soberanes, H. J., & Soria-Alcaraz, J. A. (2021). A methodology to determine the subset of heuristics for hyperheuristics through metalearning for solving graph coloring and capacitated vehicle routing problems. *Complexity*.
20. Ortiz-Aguilar Lucero, C.-L. V., Anda-Suárez Juan, B.-S. R., & Zapata-Gonzales, N. (2022). A comparison of replacement operators in heuristics for CSP problems. In *New Perspectives on Hybrid Intelligent System Design based on Fuzzy Logic, Neural Networks and Metaheuristics*. Cham: Springer.
21. Pellerin, R., Perrier, N., & Berthaut, F. (2020). A survey of hybrid metaheuristics for the resource-constrained project scheduling problem. *European Journal of Operational Research, 280*(2), 395–416.
22. Pillay, N. (2016). A review of hyper-heuristics for educational timetabling. *Annals of Operations Research, 239*(1), 3–38.
23. Talbi, E. G. (2009). *Metaheuristics: From design to implementation*. Wiley.
24. Soria-Alcaraz, J. A., Ochoa, G., Swan, J., Carpio, M., Puga, H., & Burke, E. K. (2014). Effective learning hyper-heuristics for the course timetabling problem. *European Journal of Operational Research, 238*(1), 77–86.
25. Soria-Alcaraz, J. A., Özcan, E., Swan, J., et al. (2016). Iterated local search using an add and delete hyper-heuristic for university course timetabling. *Applied Soft Computing, 40*, 581–593.
26. Vidal, T., Laporte, G., & Matl, P. (2020). A concise guide to existing and emerging vehicle routing problem variants. *European Journal of Operational Research, 286*(2), 401–416.

Hybrid Intelligent Systems

Trajectory Tracking Control of Wheeled Mobile Robots Using Neural Networks and Feedback Control Techniques

Victor D. Cruz, Jesus A. Rodriguez, Luis T. Aguilar, and Roger Miranda Colorado

1 Introduction

Wheeled Mobile Robots (WMRs) are appealing research and industry application systems. Their use includes tasks such as space exploration, autonomous driving, and surveillance, among others [1]. The usual functions demanded by WMRs include regulation, path following, and trajectory tracking. In the literature, applications such as coordination among different vehicles, generation of the trajectory to reach the desired place, and autonomous parking, among others, show the importance of the trajectory tracking problem. Then, in this work, we focus on controlling a WMR and generating the desired reference signal for the trajectory tracking problem.

When designing a reference signal, it is essential to consider the nonholonomic restrictions owned by a WMR so that the vehicle can follow it properly. Besides, this stage must be robust to environmental conditions (e.g., road conditions, illumination variations) to generate a reliable and proper reference trajectory for the WMR. Furthermore, if the vehicle operates on a road or track, a computer algorithm can use lane lines as a reference to obtain the desired trajectory.

V. D. Cruz · J. A. Rodriguez · L. T. Aguilar
Instituto Politécnico Nacional, Avenida Instituto Politécnico Nacional 1310 Col. Nueva Tijuana, 22435 Tijuana, BC, México
e-mail: vcruz@citedi.mx

J. A. Rodriguez
e-mail: jrodriguez@citedi.mx

L. T. Aguilar
e-mail: laguilarb@ipn.mx

R. M. Colorado (✉)
Avenida Instituto Politécnico Nacional 1310 Col. Nueva Tijuana, 22435 Tijuana, BC, México
e-mail: rmirandaco@gmail.com; rmirandaco@conacyt.mx

O. Castillo and P. Melin (eds.), *Hybrid Intelligent Systems Based on Extensions of Fuzzy Logic, Neural Networks and Metaheuristics*, Studies in Computational Intelligence 1096,
https://doi.org/10.1007/978-3-031-28999-6_24

We can find several approaches to solving the trajectory generation problem in the literature. Some schemes use conventional vision techniques, while others utilize intelligent procedures such as neural networks (NNs). For instance, reference [2] develops an algorithm for line detection based on the edge detection method through the Hough transform.

Regarding NNS methodologies, Chen et al. [3] propose a technique for lane line identification. They use an adaptive region of interest extraction approach as the foundation for the RANSAC algorithm to generate the detection of lane lines. Similarly, Haixia et al. [4] provide flexible lane detection algorithms using convolutional NNs. This approach resizes the input pictures to produce lane line detection using the *TuSimple* dataset. Also, Khan et al. [5] provide a vision-based line identification method employing a convolutional NNs architecture, presenting real-world vehicle navigation with deep learning algorithm assistance and producing a robust detection. Reference [6] develops a framework utilizing deep neural networks for visual detection and lane line identification in traffic situations.

After designing the reference signal, the control stage for solving the trajectory tracking problem is in order. Several schemes in the literature provide good results even in the presence of disturbances. A model-based Proportional Integral Derivative controller combined with a disturbance observer is developed in [1] for solving the trajectory tracking problem in car-like robots. The author obtains good tracking results despite the presence of kinematic disturbances. In [7], a trajectory tracking controller, combined with a disturbance observer, is developed and implemented in a car-like robot. Reference [8] produces a predictive-model controller, exhibiting smooth trajectories in dynamic environments. A technique of deep reinforcement learning and intelligent control for solving the trajectory tracking problem is developed in [9]. Through the Gazebo simulator, the author assesses the proposed scheme and demonstrates that the control objective is achieved. Furthermore, reference [10] employs dynamic decoupling and sliding mode control to follow specific trajectories.

Contribution

The previous literature review unveils that, although the earlier works show promising results, some issues deserve further study. For instance, most existing methodologies are applied to WMRs with a simplified kinematic model. Besides, many control methodologies do not consider the disturbance effect in the controller design stage. Also, the combination of intelligent techniques for trajectory generation and robust controllers requires further development. Hence, in this paper, we give a step ahead by implementing a feedback controller taken from [10] and combining it with two intelligent NNs-based algorithms to solve a trajectory tracking problem. Also, numerical simulations demonstrate that the proposed methodology satisfactorily fulfills the trajectory tracking objective.

Specifically, two different neural network architectures for trajectory design are presented along with a controller that solves the trajectory tracking problem. Both techniques are assessed numerically through the Robotic Operative System (ROS) and the Gazebo simulator [12]. Then the main contributions of this work are:

- With the Autominy V3 simulator [12], using Gazebo and ROS, two NNs algorithms are fully described and implemented to obtain lane detection.
- Implementing a feedback control technique for a car-like robot in a trajectory tracking problem, using the Autominy V3 Simulator for simulating the vehicle and Matlab-Simulink for developing the controller.
- Show how to solve the trajectory tracking problem by combining NNs, generating the desired trajectory, and a feedback controller.

Organization

The remainder of this paper is organized as follows. Section 2 fully describes the two intelligent methods used to generate the desired reference. Section 3 presents the kinematic model of the car-like robot. Also, we show the control goal and the structure of the controller used to track the desired reference, as well as the procedure for combining the reference generation and control design stages to accomplish the trajectory tracking task. Section 4 presents the numerical simulations, combining the intelligent NNs algorithms and the controller. Finally, Sect. 5 wraps up this paper with some concluding remarks.

2 Intelligent Methods for Trajectory Generation

As indicated above, trajectory generation is essential for a trajectory tracking task. In this Section, we provide a complete description of two neural network techniques employed for detecting lane lines. Then, based on this procedure, a reference trajectory for the WMR is generated. Specifically, HybridNets [13] and Ultra Fast-Structure Aware Deep Lane Detection [14] schemes are utilized. The structure of each NN is described in the following.

2.1 *HybridNets*

HybridNets is an end-to-end NN architecture built on PyTorch [13]. It serves for lane detection, lane segmentation, and vehicle identification. Also, it utilizes the BDD100K dataset to train the current network [15]. The HybridNets general structure is depicted in Fig. 1. The whole architecture possesses an encoder and a decoder, which accomplish the NN objective. Through the GitHub page given in [16], we may access the code of this NN. Using the Mean Intersection Over Union ($mIoU$) approach, the algorithm achieves 90.5% accuracy in lane segmentation and 85.4% in lane detection. Besides, the model is exported to an ONNX (*Open Neural Network Exchange*) format to make it an available framework.

The HybridNets network uses a picture with a resolution of 1280 $\times$ 720 pixels, which is received and scaled to 640 $\times$ 384 pixels to optimize efficiency and precision.

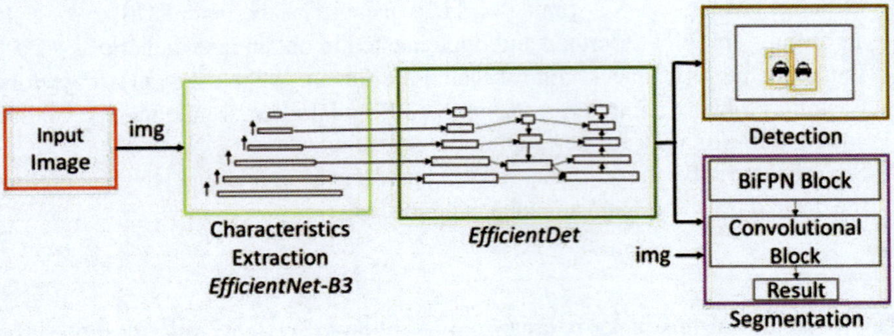

Fig. 1 HybridNets architecture

This scaling stage preserves the image's original size and prevents a general loss in image quality. Then, the input image is sent to a network called EfficientNet-B3 to perform characteristics extraction. A highly efficient convolutional NN, called EfficientNet-B3, is employed to generate this feature extraction. Also, this NN has a scaling function that uniformly increases the image depth, width, and resolution image through a composite coefficient.

Subsequently, a second NN called EfficientDet uses the features mixed at a different resolution based on the scaling connection. This second network uses a BiFPN module, where each node for each path bi-directionally adds the value of the weight of the face characteristic to learn the significance of each level. Then, by using the attributes of the preceding layer, the NN delimits the item to be identified (decoding) and provides the prediction probability and reliability.

2.2 *Ultra-Fast Structure Aware Deep Lane Detection Algorithm*

The second algorithm employed for trajectory generation is the *Ultra Fast Structure-aware Deep Lane Detection (UFSDLD)* algorithm [14], whose code can be accessed at the GitHub repository in [17]. Its general structure is depicted in Fig. 2. This NN uses a PyTorch-based architecture; its goal is to recognize lanes to address the lack of visual cues. In contrast to other segmentation methodologies, this technique scans the picture to detect lanes rather than segmenting every pixel. The NN displays a 96.04% accuracy rate, exhibits quick inference while running the model, and utilizes the *TuSimple* dataset for training purposes [18].

By detecting lanes in a set of rows, known as anchor rows, and using the global characteristics in the picture, this network aims to offer a rapid solution to the lane identification problem. By separating each anchor row's position into many cells, the NN segments these anchor rows, choosing specific cells over the specified anchor vectors.

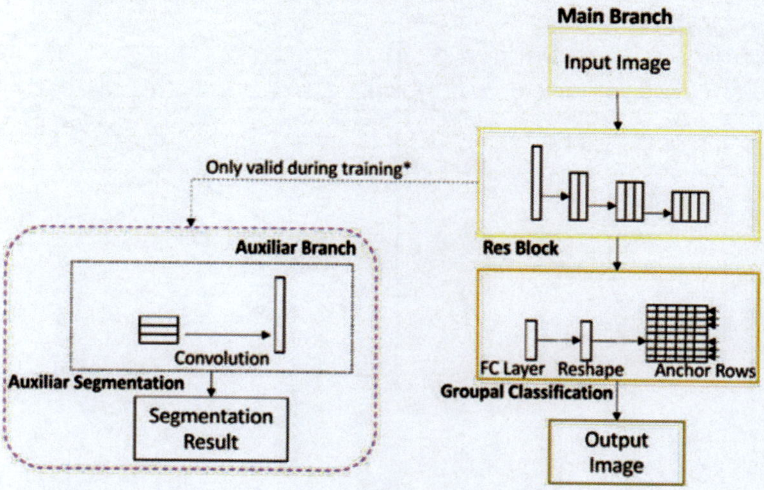

Fig. 2 Ultra fast network architecture

The NN has one main branch and an auxiliary one. In the main branch, the NN uses the input image to create a block and perform characteristics extraction through a residual block. The characteristics are sent to a classification block, which puts them together into a Fully Connected (FC) layer. Then, a re-scaling procedure is performed. Each row in the picture is examined, and a set of anchor rows is chosen. The collection of anchor rows aids in determining if a lane line is visible in the segmented picture. Once those vectors are found, they are used to create the line.

The auxiliary branch is only utilized during training and is discarded during testing. Hence, it does not influence the model's speed. This branch uses a convolutional layer for segmentation. Then, it provides the data gathered from the attributes extraction stage.

3 Kinematic Model and Control Design

The car-like robot can be schematically described as indicated in Fig. 3. For control purposes, it is expressed through the kinematic model given by [1, 10]

$$\begin{bmatrix} \dot{x} \\ \dot{y} \\ \dot{\theta} \\ \dot{\phi} \end{bmatrix} = \begin{bmatrix} \cos(\theta) \\ \sin(\theta) \\ \frac{\tan(\phi)}{l} \\ 0 \end{bmatrix} v_1 + \begin{bmatrix} 0 \\ 0 \\ 0 \\ 1 \end{bmatrix} v_2, \tag{1}$$

where $v_1(t) \in \mathbb{R}$ is the linear velocity, $v_2(t) \in \mathbb{R}$ is the angular velocity, $x(t) \in \mathbb{R}$, $y(t) \in \mathbb{R}$ are the cartesian coordinates located in the middle of the rear axle (point

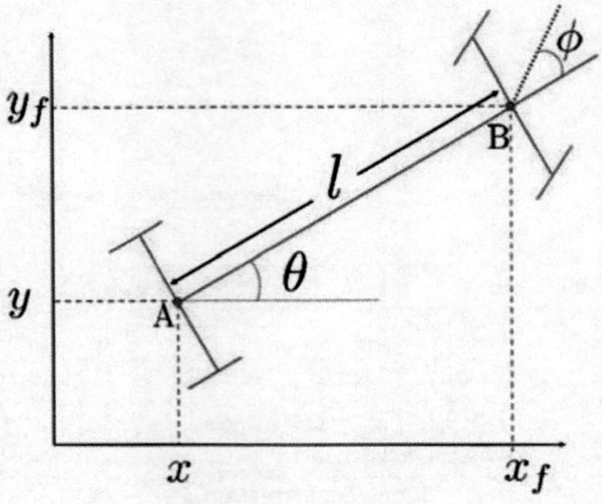

Fig. 3 Diagram of a Car-Like robot

A shown in Fig. 3), $\theta(t) \in \mathbb{R}$ is the vehicle's orientation with respect to the x axis, and $\phi(t) \in \mathbb{R}$ is the steering angle of the front wheels, while l is the distance between the front and rear axle.

In practical situations, the WMR has mechanical restrictions and actuator limitations. In the simulator used in this document, the WMR model also possesses these mechanical restrictions. Hence, we assume that the vehicle has mechanical restrictions in the steering angle rotation given by $|\phi| < 0.35$ [rad], and the linear velocity is limited by $|v_1| < 2.5$ m/s. Moreover, we assume that the vehicle's kinematic model fulfills the following nonholonomic constraints [1]

$$\begin{aligned} \dot{x}\sin(\theta) + \dot{y}\cos(\theta) &= 0, \\ \dot{x}_f\sin(\theta+\phi) - \dot{y}_f\cos(\theta+\phi) &= 0, \end{aligned} \tag{2}$$

which limit the vehicle's motion.

The kinematic model (1) is first transformed into the chained form to handle the trajectory tracking problem. To this end, the following coordinate transformation is used

$$\begin{aligned} x_1 &= x, \\ x_2 &= \tfrac{\tan(\phi)}{l cos^3(\theta)}, \\ x_3 &= \tan(\theta), \\ x_4 &= y, \end{aligned} \tag{3}$$

together with the input transformations

$$\begin{aligned} v_1 &= \tfrac{u_1}{\cos(\theta)}, \\ v_2 &= u_2 l\cos^2(\phi)\cos^3(\theta) - \tfrac{3\sin^2(\phi)\sin(\theta)}{l\cos^2(\theta)}u_1. \end{aligned} \tag{4}$$

Transformation (3)–(4) excludes the case $\theta = \pm\pi/2$, and permits obtaining the following so called chained model

$$\begin{aligned} \dot{x}_1 &= u_1, \\ \dot{x}_2 &= u_2, \\ \dot{x}_3 &= x_2 u_1, \\ \dot{x}_4 &= x_3 u_1, \end{aligned} \tag{5}$$

where u_1 and u_2 are the inputs and x_j for $j = 1, ..., 4$ are the states.

This paper aims to follow a trajectory in terms of the Cartesian position given by

$$x_d = x_d(t),\ y_d = y_d(t),\ \forall t > 0. \tag{6}$$

The reference trajectory is generated through the following equations

$$\begin{aligned} x_1 &= x_d, \\ x_2 &= \frac{\ddot{y}_d \dot{x} - \ddot{x}_d \dot{y}_d}{\dot{x}_d^3} \\ x_3 &= \frac{\dot{y}_d}{\dot{x}_d} \\ x_4 &= y_d, \end{aligned} \tag{7}$$

The previous development showed the generation of the desired reference trajectory and the structure used to describe the WMR kinematics. Then, we may state the control goal.

Control goal. *Given the chained model (5), find a set of control inputs $u_1(t), u_2(t)$ such that the limit*

$$\lim_{t \to \infty} |x_{di}(t) - x_i(t)| = 0, \tag{8}$$

with $i = 1, ..., 4$, holds.

3.1 Controller Design

To track the desired trajectory, we implement a controller with full-state linearization via dynamic feedback [10], whose structure uses the chained form (5) and adds two auxiliary variables defined as

$$\begin{aligned} \dot{\psi}_1 &= \psi_2, \\ \dot{\psi}_2 &= r_1, \end{aligned} \tag{9}$$

where

$$
\begin{aligned}
u_1 &= \psi_1, \\
u_1 &= \frac{r_2 - x_3 r_1 - x_2 \psi_1 \psi_2}{\psi_1^2}, \\
r_1 &= \dddot{x}_{d1} + k_{a1}(\ddot{x}_{d1} - \ddot{x}_1) + k_{v1}(\dot{x}_{d1} - \dot{x}_1) + k_{a1}(x_{d1} - x_1), \\
r_2 &= \dddot{x}_{d4} + k_{a2}(\ddot{x}_{d4} - \ddot{x}_4) + k_{v2}(\dot{x}_{d4} - \dot{x}_4) + k_{p2}(x_{d4} - x_4),
\end{aligned}
\tag{10}
$$

With $r_1 = \dddot{x}_1$ and $r_2 = \dddot{x}_4$.

Let us define a new error variable $\tilde{z}_i = x_{dj} - x_j$ for $j = 1, 4$ and $i = 1, 2$ correspondingly. Hence, the last two expressions of (10) can be written as follows

$$
\dddot{\tilde{z}}_i + k_{ai}\ddot{\tilde{z}}_i + k_{vi}\dot{\tilde{z}}_i + k_{pi}\tilde{z}_i = 0, \text{ i} = 1, 2, \tag{11}
$$

which corresponds to a linear representation of the system (5) in closed-loop with (9)–(10).

The characteristic polynomial corresponding to the systems given in (11) is.

$$
\rho(\lambda) = \lambda^3 + k_{ai}\lambda^2 + k_{vi}\lambda + k_{pi}, i = 1, 2 \tag{12}
$$

Hence, to guarantee asymptotic stability of the system (5), the roots of the polynomial (12) are placed in the left half-complex plane, which makes (12) Hurwitz. Hence, under this pole assignment condition, we ensure the asymptotic stability of the closed-loop system, i.e., the control objective (8) is achieved.

3.2 Control Architecture

The two NNs described in Sect. 2, the Hybridnets and *UFSDLD, will be used to generate* the vehicle's reference trajectory. This procedure will be accomplished in the Autominy Simulator [12]. Then, the controller performance will be assessed using the reference signal produced by the neural networks.

Figure 4 shows how the trajectory generation and control design stages are combined. The whole procedure is accomplished through the ROS environment and Matlab. First, an image is captured from the camera via a ROS topic. This data is sent to the NN, which uses the vehicle's real-time vision to generate lane recognition and segmentation. Then, the trajectory is created using computer vision techniques following the information gathered by the neural networks. Also, the sensor's data is used to implement the non-linear controller. As indicated in Sect. 3, this controller can achieve asymptotic stability. Hence, the trajectory tracking objective is accomplished.

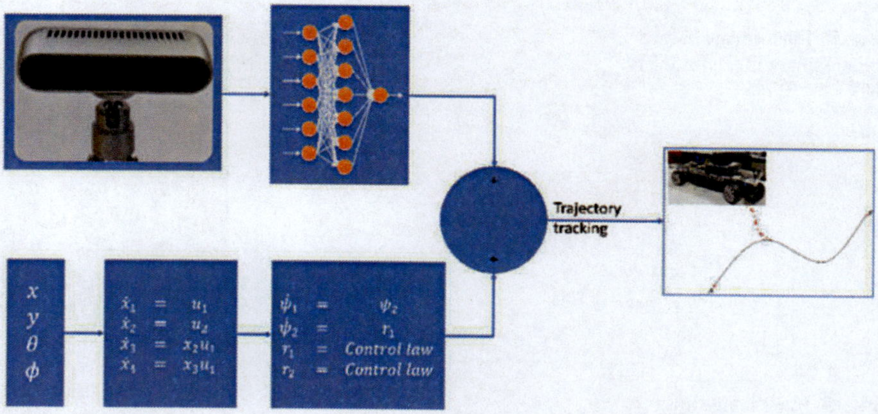

Fig. 4 Combination of the reference trajectory and control design stages

4 Numerical Results

The following numerical simulations were performed using the Gazebo Simulator for implementing the NNs. This simulator employs a model of the WMR Autominy vehicle. The distance between the rear and front wheels is 25.5 [cm]. The nonlinear controller was implemented on a different computer running Matlab-Simulink, with an ode4 Runge–Kutta solver. Both computers were connected via WiFi. The sampling period on each computer was 0.01 [s].

The results obtained by utilizing both NNs are as follows:

- **HybridNets network**. The NN was tested in the Gazebo simulator. A Python node was created to subscribe to the camera's topic and obtain the car's image. These results can be visualized in Figs. 5 and 6. The NN uses the image data from Fig. 5, which has been scaled to an image with 384 × 512 pixels. This procedure creates a vertical field of vision to extend the lanes' field of vision instead of a perspective with horizontal dimensions. The results after processing the image with the NN are depicted in Fig. 6. We observe that an effective straight-line lane recognition is accomplished. Besides, lane segmentation detects not only the lane in front of the vehicle but also other places that are in between two lines.

- **Ultra Fast-Structure Aware Deep Lane Detection NN**. With this methodology, the road picture is re-scaled with dimensions of 384 × 512 pixels, showing improved performance in recognizing straight rail lines with this model. From Fig. 7a, we visualize that, after introducing the image to the NN, fast and accurate identification of the straight line is obtained, revealing the edges of the lane or anchor rows and creating a segmentation of the region that is available for driving. The segmented area is considered for the trajectory's development. To this end, a region of interest is defined in the vehicle's front area, and the image's characteristics are modified to black and white, as illustrated in Fig. 7b. The points created in

Fig. 5 Input image before processing with HybridNets NN

Fig. 6 Input image after processing with HybridNets NN

both lane lines with the anchor rows are used to generate a polygon. Then, we can use the OpenCV skeleton function to construct the polygon's skeleton while delimiting its center to make it into a straight line. Figure 7c depicts the aforementioned straight line, which will be used as the vehicle's reference trajectory.

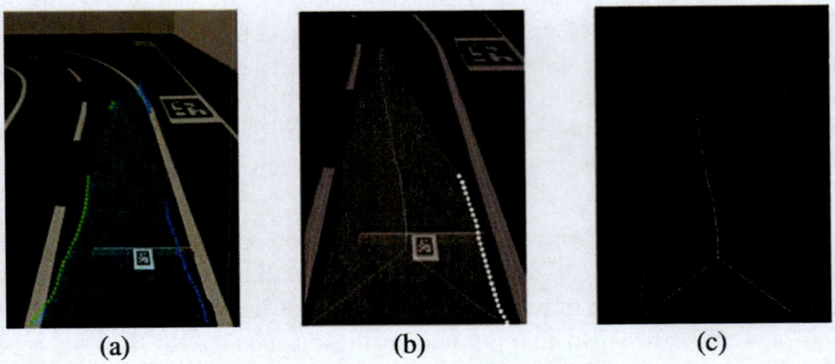

Fig. 7 **a** represents the image after processing with Ultra Fast NN. **b** represents the region of interest converted to black and white. Finally, Fig. 3 shows the skeleton of the polygon created by the lane segmentation

Once the reference trajectory is generated, the controller implementation is in order. For the following numerical simulations, the controller gains were selected by placing the eigenvalues of the characteristic polynomial (12) in $\lambda = -4$. This generates the values $k_{ai} = 12$, $k_{vi} = 48$, and $k_{pi} = 64$, $i = 1, 2$.

The desired trajectory is

$$x(t) = t,\ y(t) = \mathrm{A}\sin(\omega t), \tag{13}$$

with $\mathrm{A} = 1$ and $\omega = \pi/3$. The initial conditions were $x_1(0) = 0$, $x_2(0) = 0$, $x_3(0) = 0$, $x_4(0) = 1$.

Figure 8 shows the behavior of the dynamic controller. It can be visualized that the WMR follows the desired trajectory along the $\boldsymbol{x}$ and $\boldsymbol{y}$ axes. Besides, we observe that the WMR properly follows the selected vehicle's orientation ($\theta(t)$) and steering angle ($\phi(t)$). Hence, we verify that the control objective is accomplished satisfactorily.

The corresponding tracking errors for the chained model (7) are depicted in Fig. 9. We observe that all the errors keep on around a vicinity of the origin.

The previous results show that the proposed methodology for controlling WMRs by combining intelligent schemes for generating the reference trajectory and a feedback controller is efficient. It is important to note that, despite the proven asymptotic stability of the closed-loop signals, Figs. 8 and 9 show the existence of small tracking errors. This behavior is because we have many disturbances that were not considered in the controller design stage. Specifically, we estimated the velocity signals and had quantization errors in the position values obtained from the simulator's sensors. Furthermore, we employed two computers to accomplish the trajectory

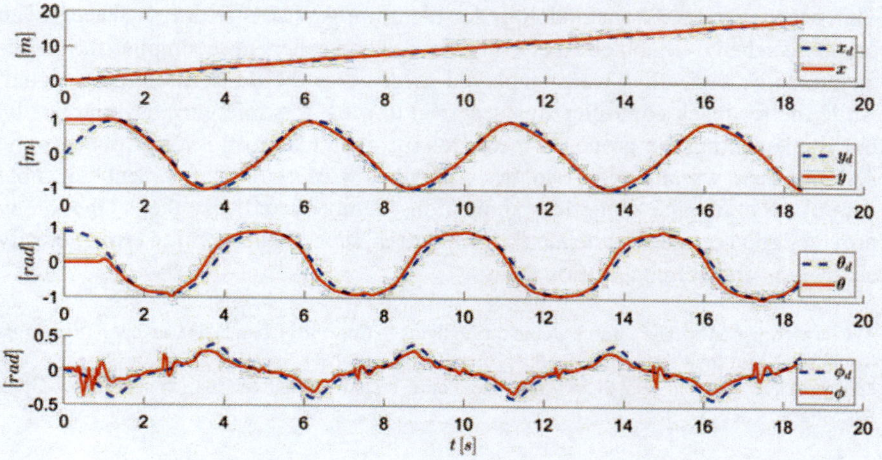

Fig. 8 Time evolution of the vehicle's coordinates x, y, θ, ϕ (solid red lines) and the corresponding reference signals (dashed blue lines)

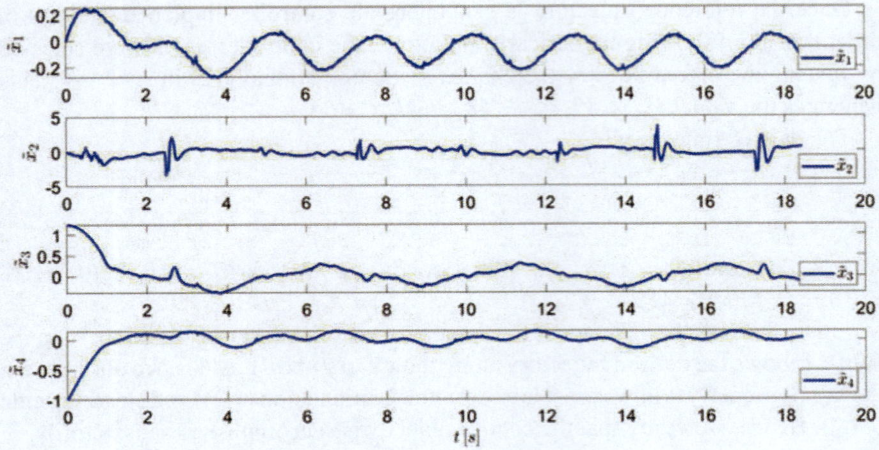

Fig. 9 State errors in chained model transformation

tracking task. Hence, communication delays were also affecting the whole closed-loop system. However, the results demonstrate that the tracking errors remain close to zero. Accordingly, the proposed methodology performs satisfactorily.

5 Conclusion

This paper presented a methodology for controlling WMRs with four wheels. The proposed scheme combined NNs with feedback controllers to accomplish the trajectory tracking task. NNs were employed to generate the desired reference signal, while the feedback controller was designed to track this reference asymptotically. For implementing the proposed methodology, a simulator of a WMR, namely the Autominy car, was utilized. Also, two computers were used to implement the whole control scheme. The numerical simulations demonstrated that the methodology provides good results despite many disturbances, such as quantization error, velocity estimation, and communication delays.

Acknowledgements The authors would like to thank to Consejo Nacional de Ciencia y Tecnología (CONACyT) for their support under the project Investigadoras e Investigadores por México, Cátedras 1537. L. Aguilar also thanks Instituto Politecnico Nacional for his support under Grant 2023–1268.

References

1. Miranda-Colorado, R. (2022). Observer-based proportional integral derivative control for trajectory tracking of wheeled mobile robots with kinematic disturbances. *Applied Mathematics and Computation, 432.*
2. Deng, G., & Wu,Y. (2018). Double lane line edge detection method based on constraint conditions Hough transform. In *2018 17th International Symposium on Distributed Computing and Applications for Business Engineering and Science* (DCABES) (pp. 107–110).
3. Chen, Y., Wong, P. K., & Yang, Z.-X. (2021). A new adaptive region of interest extraction method for two-lane detection. *International Journal of Automotive Technology*, *22*(6), 1631–1649. https://doi.org/10.1007/s12239-021-0141-0.
4. Haixia, L., & Xizhou, L. (2021). Flexible lane detection using CNNs. In *2021 International Conference on Computer Technology and Media Convergence Design* (CTMCD) (pp. 235–238).
5. Khan, M. A.-M., Kee, S.-H., Sikder, N., Al Mamun, M. A., Zohora, F. T., Hasan, M. T., Bairagi, A. K., & Nahid, A.-A. (2021). A vision-based lane detection approach for autonomous vehicles using a convolutional neural network architecture. In *2021 Joint 10th International Conference on Informatics, Electronics Vision* (ICIEV) and *2021 5th International Conference on Imaging, Vision Pattern Recognition* (icIVPR) (pp. 1–10).
6. Li, J., Mei, X., Prokhorov, D., & Tao, D. (2017). Deep neural network for structural prediction and lane detection in traffic scene. *IEEE Transactions on Neural Networks and Learning Systems, 28*(3), 690–703.
7. Rosas-Vilchis, A., Loza, A. F. D., Aguilar, L. T., Cieslak, J., Henry, D., & Montiel-Ross, O. (2020). Trajectory tracking control for an autonomo vehicle using a decoupling approach. In *2020 28th Mediterranean Conference on Control and Automation*, MED 2020 (pp. 375–380), 9 2020.
8. Wang, C., Chen, X., Li, C., Song, R., Li, Y., Meng, M. Q.-H., & Are, Y. L. (2023). Chase and track: Toward safe and smooth trajectory planning for robotic navigation in dynamic environments; chase and track: Toward safe and smooth trajectory planning for robotic navigation in dynamic environments. *IEEE Transactions on Industrial Electronics, 70*, 2023. https://doi.org/10.1109/TIE.2022.3148753.
9. Alomari, K. (2020). Trajectory following with deep reinforcement learning for autonomous cars. Master thesis, Freie Universitat of Berlin (2020).
10. Vilchis, A. J. R. (2018). Algoritmos de observación y control robustos para el vehículo autónomo. Master thesis, CITEDI-IPN.
11. Luca, A. D., Oriolo, G., Samson, C., & Laumond, J.-P. (1998). Feedback control of a nonholonomic car-like robot robot motion planning and control feedback control of a nonholonomic car-like robot. *Lectures Notes in Control and Information Sciences, 229*, 343.
12. Autominy, Quick-start guide–autominy. https://autominy.github.io/AutoMiny/docs/quick-start-guide/.
13. Vu, D., Ngo, B., & Phan, H. (2022). Hybridnets: End-to-end perception network. *Computer Science.*
14. Qin, Z., Wang, H., & Li, X. (2020). Ultra fast structure-aware deep lane detection. *European Conference on Computer Vision* (ECCV) (pp. 276–291).
15. Yu, F., Chen, H., Wang, X., Xian, W., Chen, Y., Liu, F., Madhavan, V., & Darrell, T. (2020). Bdd100k: A diverse driving dataset for heterogeneous multitask learning. In *2020 IEEE/CVF Conference on Computer Vision and Pattern Recognition* (CVPR), 13–19 June, Seattle, WA, USA.
16. Vu, D., Ngo, B., & Phan, H. Github–datvuthanh/hybridnets: Hybridnets: End-to-end perception network. https://bit.ly/395hxV6. https://github.com/datvuthanh/HybridNets.
17. Qin, Z., Wang, H., & Li, X. (2020). Ultra fast structure-aware deep lane detection (eccv 2020). https://github.com/cfzd/Ultra-Fast-Lane-Detection.
18. TuSimple, Tusimple-benchmark. https://github.com/TuSimple/tusimple-benchmark.

An Evolutionary Bilevel Optimization Approach for Neuroevolution

Rocío Salinas-Guerra, Jesús-Adolfo Mejía-Dios, Efrén Mezura-Montes, and Aldo Márquez-Grajales

1 Introduction

Neural Architecture Search (NAS) has attracted the attention of various researchers and has been illustrated as providing benefits to many applications for computing vision and machine learning [33]. Architecture topology and architecture size have been considered two of the most critical aspects of the Deep learning model performance [22]. The community has generated many search algorithms for both aspects of neural architectures [16]. However, the performance gain of these search algorithms is achieved in different search spaces and training settings.

We can find different investigations and algorithms proposed for the neural architecture search for Convolutional Neural Networks (CNNs) in the literature. Recently, different searches for manually designed neural architectures have been proposed, such as ResNet, AlexNet, DenseNet, and VGGNet [22]. However, given that these designs are a complex task and could require a great deal of experience on the part of the designer, recent work suggests reinforcement learning methods to sample the search space with possible architectures and hopefully find a suitable CNN.

Different approaches have been developed to address this search problem regarding the automated search of convolutional neural networks. For instance, the

R. Salinas-Guerra (✉) · E. Mezura-Montes · A. Márquez-Grajales
Artificial Intelligence Research Institute, University of Veracruz, Xalapa, Mexico
e-mail: sague9503@gmail.com

E. Mezura-Montes
e-mail: emezura@uv.mx

A. Márquez-Grajales
e-mail: li.aldomg@gmail.com

J.-A. Mejía-Dios
Applied Mathematics Research Center, Autonomous University of Coahuila, Coahuila, Mexico
e-mail: adolfomejia@uadec.edu.mx

O. Castillo and P. Melin (eds.), *Hybrid Intelligent Systems Based on Extensions of Fuzzy Logic, Neural Networks and Metaheuristics*, Studies in Computational Intelligence 1096,
https://doi.org/10.1007/978-3-031-28999-6_25

most used algorithms in this area are NASRL, BlockQNN, and MetaQNN from meta-learning approaches [5]. On the other hand, the algorithms used in the evolutionary computation, to mention a few, are Large-Scale Evo, Genetic-CNN, CNN-GA, AE-CNN, NSGA-3, and NSGA-4 [31], which have an approach of a single-level Currently, if we look for a hierarchical optimization approach, we realize that few works with a bilevel approach are found in the literature. One of them is the BLOP-CNN algorithm [19].

As a result, we are encouraged to conduct a preliminary investigation of novel bilevel modeling for CNN optimization and to test our proposed technique on NAS benchmarks known as NAS-Bench-101 [39].

The rest of the paper is organized as follows: Sect. 2 gives the theoretical frame, including concepts on convolutional neural networks, neuroevolution, and bilevel optimization. After that, Sect. 3 details the proposed bilevel model for the neural architecture. The resulting multi-objective bilevel problem is solved by the proposed bilevel algorithm given in Sect. 4. The experiments and discussion are reported in Sect. 5. Finally, the final comments are in the conclusions (Sect. 6).

2 Theoretical Foundations and Background

This section gives the essential neuroevolution concepts and multi-objective bilevel optimization problems. Firstly, we describe the convolutional neural networks. Then, the neural architecture search problem is detailed before giving an overview of bilevel optimization problems.

2.1 Convolutional Neural Networks

Convolutional neural networks are models of multilayer artificial neural networks. According to the structure of the animal visual cortex described by Hubel and Weisel [14], the model can be interpreted visually, as shown in Fig. 1.

Each block represents a different layer of the convolutional neural network, which has height, width, and depth. From left to right, we can see the input, hidden (convolutional, pooling, and dropout layers), and fully connected layers. After the last layer, the model generates a classification [8].

Fully connected layers enforce local connectivity between neurons and adjacent layers [15]. As such, the hidden layer inputs are a subset of neurons from the layer preceding that hidden layer. This situation ensures that the neurons in the learning subset produce the best response [20]. Furthermore, the units share the same weight and bias in the activation map so that all neurons in each layer analyze the same feature.

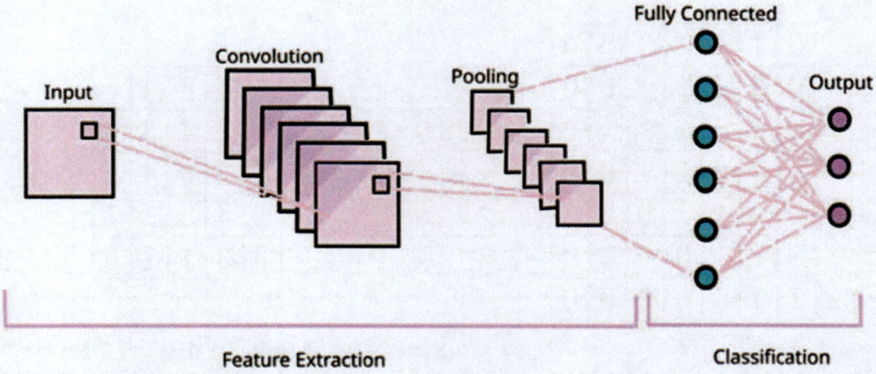

Fig. 1 Architecture or a CNN

As its name indicates, convolutional neural networks come from applying convolution's mathematical operation [8]. Given two functions, convolution is defined as the product's integral, where one is displaced and reflected.

Applying this concept to matrices, we can use the definition of a discrete convolution. Given two matrices (which can be considered discrete functions of two variables $f : N \times N \rightarrow \mathcal{R}$), we can define the operation by matrix multiplication of the submatrices of the input and a kernel or filter (fixed matrix with which all operations are performed).

When performing the translation of the kernel matrix, it is necessary to decide how many jumps to make. In terms of matrix values, this number of jumps is called stride. Finally, we can notice how the dimension of the resulting matrix is reduced. This condition is because the kernel matrix has a dimension greater than 1×1. If the resulting matrix with dimensions equal to the original one is required, extending the matrix through padding or filling in the matrix limits is necessary. For two dimensions, the result of a convolution g between a matrix f and a kernel h is formally defined as:

$$(f \cdot g)(x, y) = \sum_{i=1}^{N-1} \sum_{j=1}^{M-1} f(x - i, y - j)g(i, j)$$

These convolutional layers where filters are optimized to extract information in the best way to solve the problem are shown in Fig. 2.

There are different neural network layers, each with a set of neurons. The Table 1, some of the most common neural network layer types will be presented.

2.1.1 Batch Normalization

Batch normalization consists of adding an extra step, usually between the neurons and the activation function, with the idea of normalizing the output activations [4].

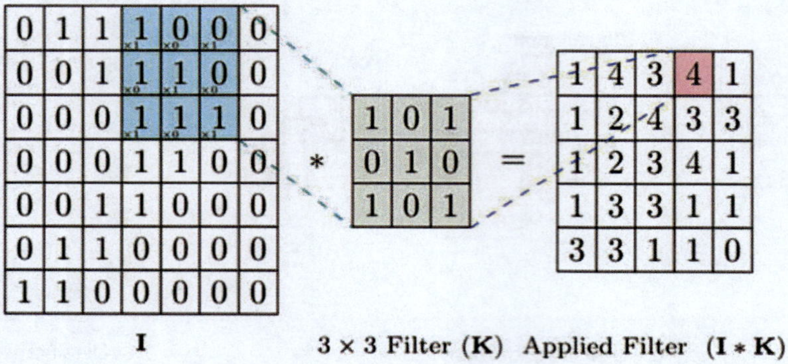

Fig. 2 Convolutional applied to matrix I with filter K

Fig. 3 Fully-Connected Layer, each circle represents a neuron, while the edges correspond to the weighted connections

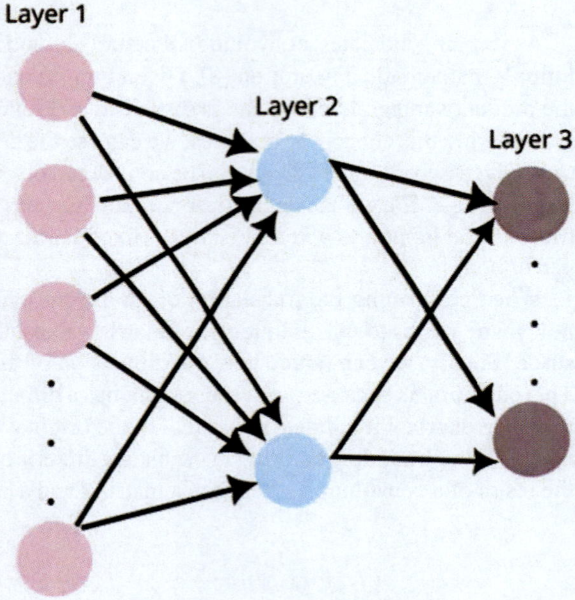

Table 1 Activation functions

Activation function	Expression
Linear	$f(x) = x$
Sigmoid	$f(x) = \frac{1}{1+e^{-x}}$
Hyperbolic tangent (tanh)	$f(x) = \frac{e^x - e^{-x}}{e^x + e^{-x}}$
Rectified linear unit (ReLu)	$f(x) = \begin{cases} x\, if\, x \geq 0 \\ 0\, otherwise \end{cases}$
Softmax	$f(x) = \frac{e^{x_i}}{\sum_{j=1}^{N} e^{x_j}}, i = 1, 2, \ldots, N$

Ideally, the normalization would be done using the mean and variance of the entire training set, but if we apply stochastic gradient descent to train the network, the mean and variance of each input mini-batch will be used. The output of each neuron is normalized independently. The current mini-batch can be analyzed in detail. [36].

2.1.2 Pooling Layers

A pooling layer is usually applied between two convolution layers. This layer receives as input the feature maps formed in the convolution layer output, and its role is to reduce the image size and, at the same time, preserve their most essential characteristics. Among the most used, we find the Max-pooling, which works as follows, if the image is divided into regions of equal size and the maximum value of each region corresponds to a pixel in the resulting extracted image. For each part, the maximum value corresponding to a pixel in the resulting image is extracted, or even the Average pooling, which consists of keeping in each step the average value of the filter window [11]. Finally, the same number of feature maps as in the output is obtained in the output of that Pooling layer but considerably compressed.

2.1.3 Fully Connected Layers

This layer is equal to the classical artificial neural network layers, but its architecture is fully connected. All neurons in this layer are connected to all neurons in the previous layer, and more specifically, they are connected to all feature maps in the last layer [18].

Unlike what has been seen so far in convolutional neural networks, this type of layer does not use the property of local connectivity since a fully connected layer is connected to all input volumes, so it has numerous connections [25]. In contrast, the convolutional layers are connected to a single local region in the input and many of the neurons in the convolutional layer share parameters. The only parameter configured in this type of layer is the number of neurons that make it up. In these layers, its K neurons are connected with all the input volumes of each of its K neurons. Its output will be a single $1 \times 1 \times K$ vector containing the calculated activations. Going from a 3-dimensional input volume to a single $1-$ dimensional output vector suggests there is no convolution layer after this layer. The primary function of the fully connected layers is to carry out a kind of grouping of the information that has been obtained up to that moment, which will be used in subsequent calculations for the final classification.

In general, in convolutional neural networks, more than one layer completely connected in series is usually used, and the last one will have the parameter K, which is the number of classes present in the data set. Then, the final values of K will be fed to the output layer, which will carry out the classification through a specific probabilistic function. Finally, more parameters can be configured, such as the values of the weights and the bias, but it is not strictly necessary, and default values can be used.

2.2 Neural Architecture Search

Choosing the exemplary neural network architecture for a given task remains a non-trivial task, often requiring a time-consuming trial-and-error process, which is why there are recent efforts to automate this process [12]. Automating neural network architecture design is known as neural architecture search. Neural architecture search is a field in machine learning where it seeks to discover the optimal neural network architecture for a task. This approach has led to the discovery of modern architectures that can multitask better than manually designed architectures and consume minimal computing resources.

The search space is determined by a collection of neural networks allowing the algorithm to choose. The choice of search space is often based on previous experience. However, this can be misleading because the architectures considered in the search process may be similar to those previously used. Therefore, solving problems that the architecture is designed to deal with, excluding different architectures, might work better. The optimization method provides the means to explore the search space and discover high-performance neural network architectures. A popular selection of optimization methods in the search for neural architecture includes evolutionary, deep learning, Bayesian optimization algorithms, and others [41].

Finally, the candidate evaluation method is the process used to evaluate the candidate architecture of a neural network that occurs during the execution of an algorithm. Candidate evaluation guides research toward discovering high-performance neural network architectures [24]. A typical candidate evaluation process is to perform a formal training and validation process for each candidate architecture on a specific data set and evaluate the architecture based on the performance statistics calculated in the validation process.

This process is slow because it requires training and evaluating many architectures. Consequently, alternative methods have been proposed to speed up this stage of the neural architecture discovery process. Some approaches for reducing the required time to test applicants include reducing the amount of training time, running the training and validation process on less computationally demanding datasets, which then transfer the results to computer cells or filters used, and use of predictive models to evaluate the candidate architecture and search management performance based on these predictions.

2.2.1 Nas-Bench-101

The Neural Architecture Search Reference Dataset has been developed to provide a quick and easy way to test new algorithms in practice and compare different approaches [39]. This situation is because the candidate evaluation process generally involves training and testing steps that require much time and computational effort. As a result, it has become challenging when we have limited computational resources for experimentation. That is why recent research has opted for using benchmark data, which contains precomputed performance statistics for the entire search space, allowing fast and efficient evaluation and comparison of neural architecture search algorithms, for example, the NAS-Bench-101 dataset [39]. In particular, all networks share the same "skeleton" seen in Fig. 4, differing just the "module" part in all models. It is a collection of neural network operations linked in an arbitrary graph-like structure.

NAS-Bench-101 is the first publicly published neural architecture search benchmark dataset. It consists of 423 K convolutional neural network architectures trained on the CIFAR-10 dataset and intended to be used for the image classification task. Each architecture was trained with three different initialization, and performance statistics were calculated after training each architecture for 4, 12, 36, and 108 epochs. Therefore, the final data set contains more than 5 million architectures. NAS-Bench-101 architectures are built by defining a cell and using a predefined structure to

Fig. 4 Skeleton model

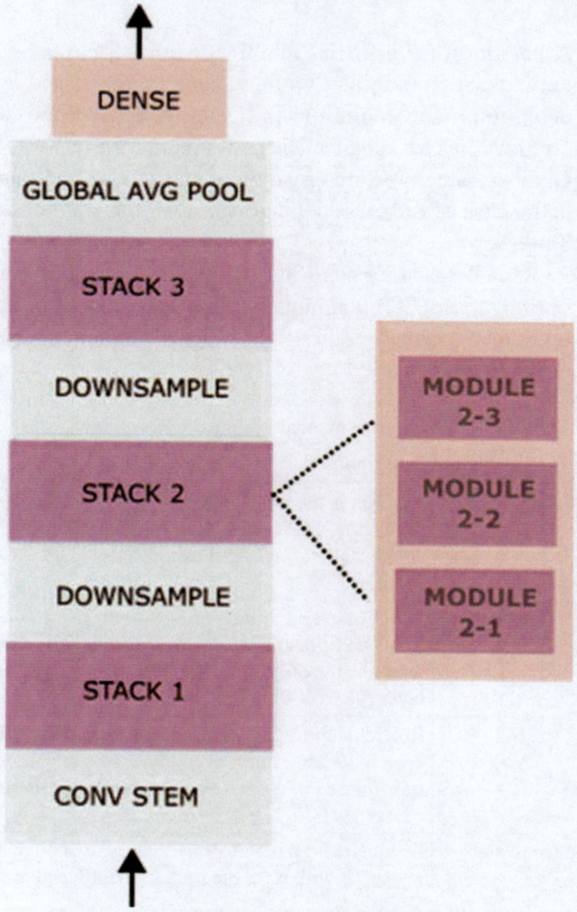

build the network architecture by repeatedly stacking identical cells. It is important to mention that the Nas-Bench-101 database contains a fixed number of available filters:

- conv3 × 3-bn-relu: Convolution of size 3 × 3 with RELU activation function.
- conv1 × 1-bn-relu: Convolution of size 1 × 1 with RELU activation function.
- maxpool3 × 3: Maxpool filter size 3 × 3.

Furthermore, the network nodes are limited to 7, including the input and output layer, and nine connections among layers. As a result, the best architecture in NAS-Bench-101 achieves an average test accuracy of 94.32% on the CIFAR-10 dataset [39].

2.3 Neuroevolution

A paradigm for artificial intelligence that focuses on designing an artificial neural network, including its topology, ideal weights, and learning rules, using simulated evolution "neuroevolution" [32]. Automated methods may be used to find better CNN architectures for specific tasks, saving time and effort. However, the most well-known CNN architectures are created manually based on knowledge and experience, such is the case of the classical algorithms of the state of the art that we can represent in Table 2.

Recent research suggests using reinforcement learning (RL) and evolutionary optimization (EO) techniques to design CNN architectures. This is due to the high level of expertise required for manual design, which can be time-consuming and

Table 2 Representative manual designed CNN architectures

Architecture	Description
AlexNet	AlexNet is the first CNN that employs ReLU as an activation function in place of the Sigmoid function. Moreover, deep learning network training times were significantly accelerated using ReLU as an activation function. Finally, AlexNet contains [21]
VGGNet	It was based on a study of how to make these networks deeper. The network uses 3 × 3 tiny filters and is distinguished by its simplicity in addition to these other features; the only additional elements are pooling layers and a fully connected layer [9]
ResNet	ResNet is the VGG-19-inspired design in which the shortcut connections or skip connections are introduced to a 34-layer plain network. The design is then transformed into the residual network via these skip connections or residual blocks [34]
DenseNet	DenseNet is an architecture that focuses on making deep learning networks even deeper. It will make the train more efficient and shorten the connections between layers

complex. RL-based methodologies provide a superior architecture compared to the manual approach; The most used algorithms in this area are NASRI, Block-ANN [40] and MetaQNN [2]. Unfortunately, their computational costs are high for what was started, which led researchers to seek strategies to reduce computational complexity. For this reason, Evolutionary Algorithms (EAs) are available for overcoming limitations. Although EAs do not guarantee convergence to the global optimum of the objective space, different measures, such as maintaining diversity [32], are taken by reducing precision error and complexity. The work in neuroevolution for designing a CNN has offered a diversity of methods in evolutionary and swarm intelligence. It is essential to mention the impact and growth this area has had; from 2010 to date, around 200 articles have been published using metaheuristics [35].

Figure 5 shows the methods used in the field of neuroevolution as they are: Genetic Algorithms (GA), Particle swarm Optimization (PSO), and Differential Evolution (DE), among others [35]. Although this area of study is a novelty, especially given that many publications have been produced recently, Fig. 5 Neuroevolution methods for CNNs. shows the methods used for this field of study where mono-objective and multi-objective optimization algorithms stand out, it also shows that there are works that have opted for bi-level optimization algorithms.

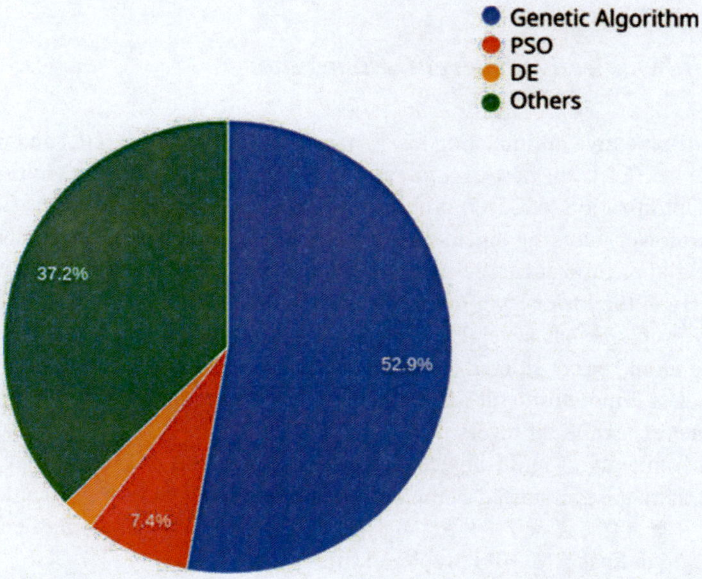

Fig. 5 Neuroevolution methods for CNNs

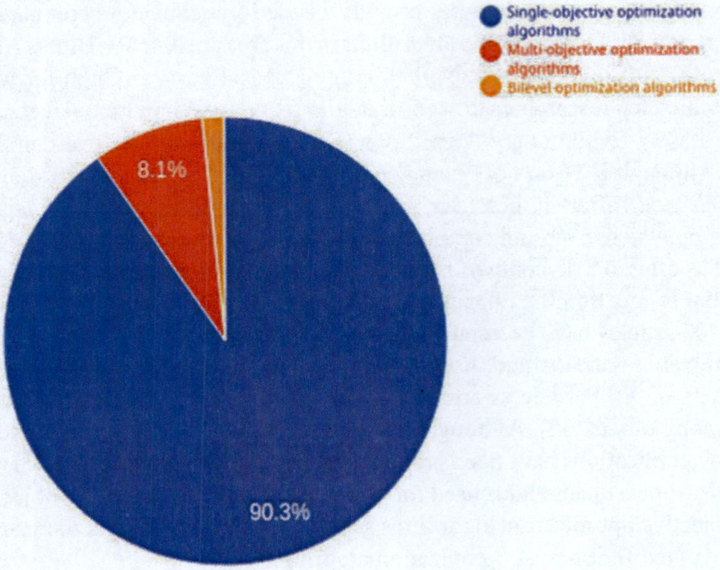

Fig. 6 Optimization algorithms for CNN

2.4 Multi-objective Bilevel Optimization

Two multi-objective optimization tasks, one at the Upper Level (UL) and one at the Lower Level (LL), are part of a hierarchical structure known as a Multi-objective Bilevel Optimization (MOBO) problem [10, 28]. Two Decision-Makers (the leader and the follower) working together often ensure the optimization process. Numerous real-world applications can be modeled using this hierarchical optimization architecture. Here, the lower-level optimization problem in multi-objective bilevel optimization problems has several trade-offs for optimal solutions. All these trade-off solutions should be considered optimum reactions to evaluate one upper answer. This factor makes it more difficult for the leader to complete their search for answers set as parameters for the followers' ideal trade-off solutions.

The complexity of multi-objective bilevel optimization problems come from the fact that there are two nested optimization problems. For example, assume that $F : X \times Y \rightarrow \mathbb{R}^M$, $f : X \times Y \rightarrow \mathbb{R}^m$ are two objective functions. Then, the following formulation defines a MOBO problem [10].

Minimize (upper level):

$$F(x, y) = (F_1(x, y), F_2(x, y), \ldots, F_M(x, y)),\ x \in X$$

subject to (lower level):

$$y \in \underset{y \in Y}{\operatorname{argmin}}\{f(x, y) = (f_1(x, y), f_2(x, y), \ldots, f_m(x, y))\}$$

Here, a multi-objective optimization problem is present on both levels. Because the leader expects the follower to choose from a set of Pareto-optimal options, the feasibility is different. If the LL solutions are the best, evolutionary MOBO seeks to approximate the UL Pareto-optimal front [10]. Figure 7 shows how the upper-level Pareto-optimal front is obtained.

Numerous conventional and evolutionary techniques have been used in the study of bi-level optimization. Bi-level problems are being studied in-depth in classical optimization using approximate solution strategies. Despite their effectiveness, these strategies only apply to minor BLOP situations. Additionally, most of these studies are constrained by regularity qualities, such as smoothness, linearity, convexity, or differentiability. However, they cannot be used for complex, non-convex, non-linear, or high-dimensional bilevel problems because of this [1]. On the other hand, numerous problem-solving techniques for EAs have proven successful in resolving similar issues [29].

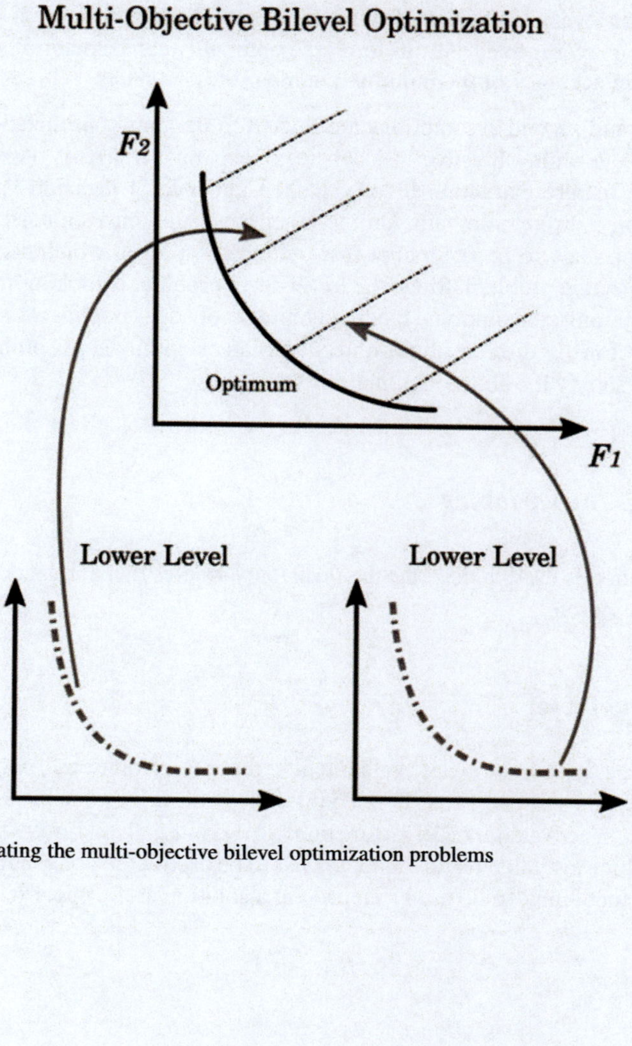

Fig. 7 Illustrating the multi-objective bilevel optimization problems

3 Neural Architecture Search as a Bilevel Optimization Problem

As aforementioned, different optimization models have been proposed for neuro-evolution [17, 38]. This work proposes a hierarchical optimization structure for the neural architecture search. The main idea is to find convolutional neural networks with the minimum number of trainable parameters and maximum accuracy, such that the neural topology is optimal and accurate. The desired bilevel optimization model must satisfy the following assumptions:

1. The upper-level search space contains the arrangements of available filters (convolution, pooling, among others) for each feature extraction layer.
2. The upper-level objective function has to minimize the number of trainable parameters and maximize the validation accuracy (simultaneously) subject to the neural architecture is optimal.
3. The lower-level problem is to find the best neural architecture when each feature extraction layer is fixed. At this level, the optimality can be determined by either maximum accuracy or the minimum number of parameters.

The first and second assumptions are related to the main optimization problem (upper-level), a multi-objective task where a decision variable is associated with an available feature extraction block, i.e., the upper-level decision variables can be defined on a discrete domain. On the other hand, the third supposition restricts the neural topology to be optimal at that optimization level, which can be defined as single-objective problem. Since the lower-level problem is looking for a suitable connection among convolutional blocks, a finite set of edges implies a discrete search space. Therefore, a discrete multi-objective bilevel optimization problem can be proposed to satisfy the above assumptions.

3.1 Decision Variables

This subsection is used to describe the decision variables that are encoding a neural architecture search.

3.1.1 Upper-Level

Let us assume that a finite set of available filter maps (convolutional, pooling, among others) are given a priori. Then, $1 \to \text{conv}1 \times 1 - \text{relu}$, $2 \to \text{conv}3 \times 3 - \text{relu}$, $3 \to$ Pooling, 3×3 activated by RELU function. $x = [x_1, x_2, x_3, x_4]$, $x_i \in X = \{1, 2, 3\}$ indicates which available feature mapping has to be used at the corresponding hidden layer. This encoding can be used to represent a solution at the upper level.

3.1.2 Lower-Level

Once the upper level provides the arrangement of feature mappings (blocks), the lower-level optimizer has to find the optimal architecture for the upper-level blocks. We use the adjacency matrix for a graph representation to represent a neural topology.

Assume that y contains the graph topology information associated with the adjacency matrix, i.e., $y = (a_1, a_2, a_3, b_1, b_2, c_1)$ represents the following adjacency matrix:

$$A = \begin{pmatrix} 0 & a_1 & a_2 & a_3 \\ 0 & 0 & b_1 & b_2 \\ 0 & 0 & 0 & c_1 \\ 0 & 0 & 0 & 0 \end{pmatrix} \begin{matrix} \text{input} \\ x_1 \rightarrow \text{filter 1} \\ x_2 \rightarrow \text{filter 2} \\ \text{output} \end{matrix}$$

Note that binary arrays represent lower-level decision variables, i.e., $y \in Y = \{0, 1\}^m$ where m is the number of elements in the upper triangular matrix. The number of lower-level solutions is based on the upper-level dimension. That is, if the upper-level search space is with dimension n, then the adjacency matrix represented by an upper-triangular matrix with m available entries, where:

$$m = \sum_{i=1}^{n-1}(n - i) = (n - 1)n - \frac{(n - 1)n}{2} = \frac{(n - 1)n}{2}$$

Thus, the number of solutions for the lower level is growing exponentially at a rate of $2^m = 2^{(n-1)n/2}$ (as showed in Fig. 8), implying that a heuristics search can be considered to reduce the computational cost related to finding lower-level optimal solutions. At this point, the upper-and lower-level solutions representation has been proposed. The next part is on defining the upper- and lower-level objective functions.

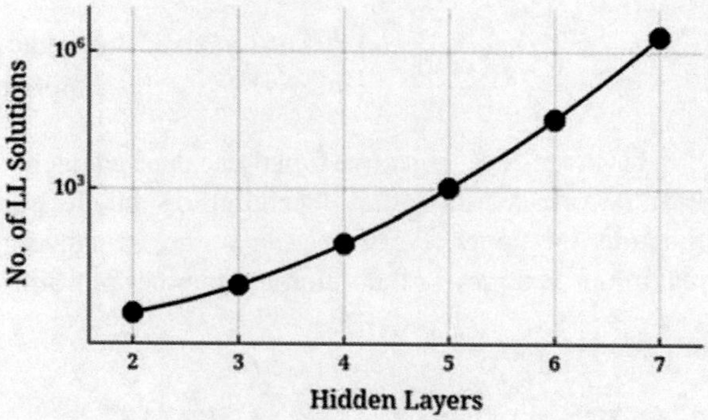

Fig. 8 The number of lower-level solutions. Log scale is used for visualization purposes

3.2 Upper-Level Objective Function

This part aims to define the upper-level objective function, which is in charge of minimizing the number of network parameters while accuracy is simultaneously maximized. Firstly, assume that $x = [x_1, x_2, \ldots, x_n]$ is a vector representing a convolutional neural network with at most n hidden layers, and y contains the optimal network topology (regarding x). Then, if $\rho(x, y)$ is the number of trainable parameters for the convolutional, and Accuracy(x, y) the accuracy after training the neural network defined by *(x,y)*, we found the following objective function:

$$F(x, y) = \begin{pmatrix} \rho(x, y) \\ -\text{Accuracy}(x, y) \end{pmatrix}$$

Note that the second objective is the negative of the accuracy considered to define a minimization problem further. Moreover, it is well known that the accuracy increases when the number of neural network parameters increases, inducing a conflicting behavior implying that F has to be minimized within multi-objective optimization.

3.3 Lower-Level Objective Function

The lower-level optimization problem is, generally, the most expensive component in a bilevel optimization problem due to the parametrized optimization task having to be solved to generate a feasible solution for the bilevel problem. As mentioned above, the lower-level optimization problem is to find the optimal network topology for given feature mappings (filters) that either maximize the convolutional neural network accuracy or minimize the number of trainable parameters. Therefore, the lower-level objective function is defined as follows:

$$f(x, y) = \begin{cases} \text{Accuracy}(x, y) & \text{if accuracy is required.} \\ \rho(x, y) & \text{if simplicity is required.} \end{cases}$$

This lower-level objective function can be adapted to the application's necessities. In this work, we show that depending on the election of the lower-level objective function, the upper-level front can suggest a convolution neural network with the maximum accuracy or the minimum number of trainable parameters.

3.4 Bilevel Optimization Model

The upper- and lower-level objective functions have been defined. Thus, we can define the bilevel optimization problem to find optimal compromises between accuracy and network complexity (number of trainable parameters). Therefore, the bilevel optimization model is given as follows:

Minimize:

$$\min_{x \in X} \left\{ F(x, y) = \begin{pmatrix} \rho(x, y) \\ -\text{Accuracy}(x, y) \end{pmatrix} \right\}$$

subject to:

$$y \in \underset{y \in Y}{\text{argmin}} f(x, y).$$

Note that this problem defines a multi-objective bilevel task which is solved when the upper-level Pareto-optimal front is approximated, subject to the lower-level's mono-objective function is minimized according to y for a fixed upper-level decision vector.

Since the upper- and lower-level objective functions require training a convolutional neural network to compute the accuracy and network complexity for a given network configuration (x, y), exact methods nor single-level reduction cannot be applied. Therefore, an evolutionary approach will be used to approximate solutions to the proposed bilevel model. Figure 9 represents the bilevel solution for the bilevel problem defined in this part.

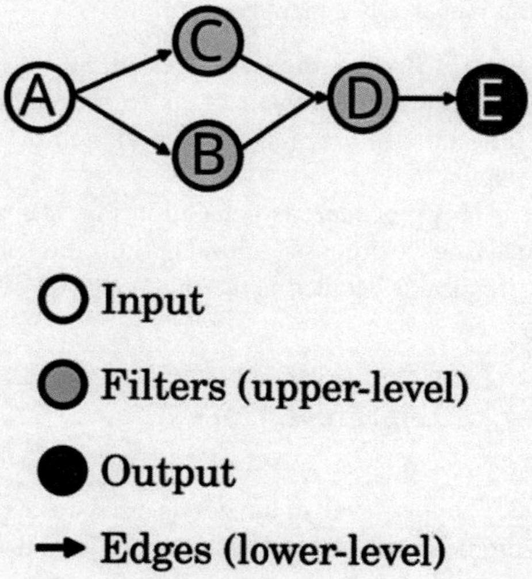

Fig. 9 Representation of the upper- and lower-level variables decoding. Note that A and E are the input and output, respectively. Also, it has to exist an acyclic path from A to E

4 Solution Proposal

This section aims to describe the bilevel approach for the neural architecture search. Firstly, we describe the solution representation at both levels, and the variation operators are described. It is worth mentioning that each bilevel solution encodes a neural architecture, where the upper-level solution encodes the available filters, while the corresponding lower-level solution represents the neural architecture (topology), which represents the adjacency matrix. Thus, the upper-level optimizer must select the optimal filters minimizing the network complexity (number of trainable parameters) subject to the lower-level find the best neural architecture to maximize the validation accuracy of the model.

The following steps summarize the solution proposal:

Step 1: Initialization. Initialize upper-level solutions in the upper-level search space. Then, use the lower-level optimizer to find the optimal network design.

Step 2: Offspring Generation. Generate new upper-level decision vectors using variation operators.

Step 3: Lower-level Optimizer. Approximate optimal lower-level solutions for each newly generated upper-level solution.

Step 4: Evaluation. Once the lower-level solutions are computed in the previous step, evaluate the upper-level objective function. Finally, combine the current population and the newly generated solutions.

Step 5: Environmental selection. Sort the combined population lexicographically using the non-dominated ranking procedure and the crowding distance. If the stopping criteria are not met, go to step 2. Otherwise, stop and report the current population.

This procedure is detailed in Fig. 10, which illustrates the bilevel structure for solving the proposed bilevel optimization problem. Each component of this solutions proposal is detailed in the following sub-section.

4.1 Upper-Level Optimizer

The upper-level optimizer is used to approximate solutions for the upper-level problem. In this work, each upper-level decision vector has a fixed length and only contains integer numbers. Therefore, we can use any common genetic operators to produce new candidate solutions. Moreover, a binary tournament (regarding), simulated binary crossover, and polynomial mutation with the round repair strategy [13, 37] are considered here.

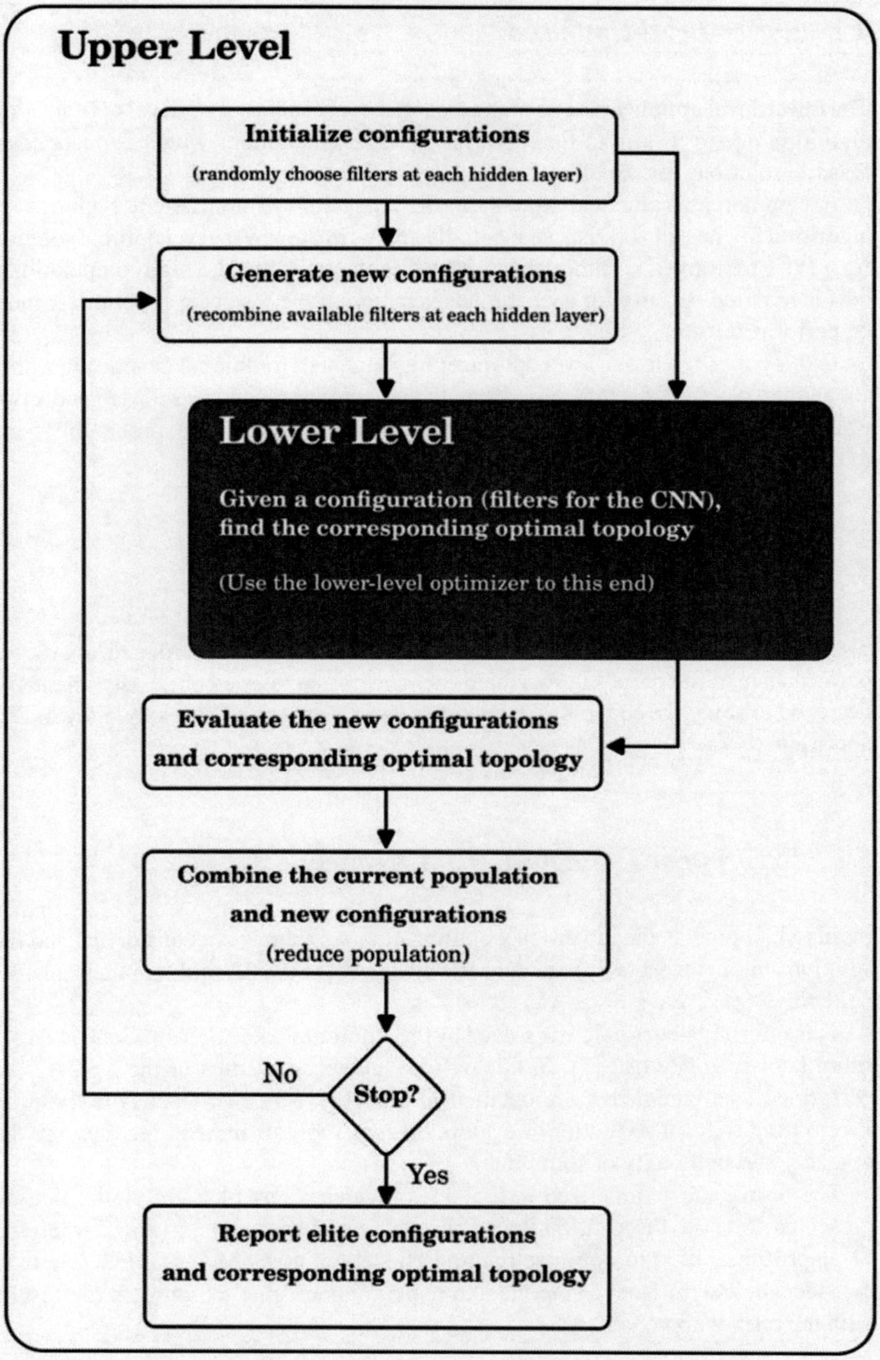

Fig. 10 Diagram containing the bilevel proposal algorithm

4.2 Lower-Level Optimizer

The lower-level optimizer is one of the essential components in a bilevel optimization algorithm due to its aim to find optimal lower-level solutions which correspond to feasible solutions for the bilevel problem [27].

A common and practical strategy is the adaptation of an existent evolutionary algorithm in a nested way, i.e., sequentially approximate lower-level optimal solution for $f(x, \cdot)$ for a given x. Although this nested strategy is helpful, a high computational cost is required because for each x, one execution of the lower-level optimizer must be performed [6].

In this paper, the lower-level optimizer implements a traditional genetic algorithm for a binary vector representation. Here, the parents are randomly selected and a two offspring are generated using the half-uniform crossover (crossover probability fixed to 0.9) with bit-flip mutation (probability fixed to 0.1) [30].

5 Experiments and Discussion

This section describes the experiment settings and results. In order to assess the performance of the proposal, two main experiments are carried out. Experiment 1 is designed to study the convolutional neural network suggested by a single-level (SL) approach.

5.1 Experiment 1: Single-Level Approach

In this SL approach, the lower-level optimization problem is not considered, and the solution at both levels (x, y) are concatenated to solve the SL problem traditionally, i.e., $\min_z F(z)$ where $z = (x, y) \in X \times Y$.

This reformulated problem is solved by the Non-dominated Sorting Genetic Algorithm known as NSGA2 [7]. In this part, 31 independent runs of the NSGA2 are performed (the parameters are detailed in Table 3). Since NSGA2 is reporting a Pareto front for each execution, we adopt the hypervolume indicator to measure the diversity and optimality of solutions.

The results are summarized in Fig. 11, in which a box plot and violin plot are presented to show the distribution of the obtained hypervolume values when the SL approach is used to optimize the upper-level optimization task. Here, we only consider the Pareto front at the median hypervolume value to compare the results with the bilevel approach.

Table 3 Parameters adopted by NSGA2 during the experimentation

	Operator	Parameters
Crossover	Two Point Crossover	0.9 (crossover probability)
Mutation	Polynomial Mutation	0.1, and 20 (mutation probability, and η index)
Population	Random sampling initialization in bounds	100 (population size)
Stopping criteria	Maximum number of generations	500

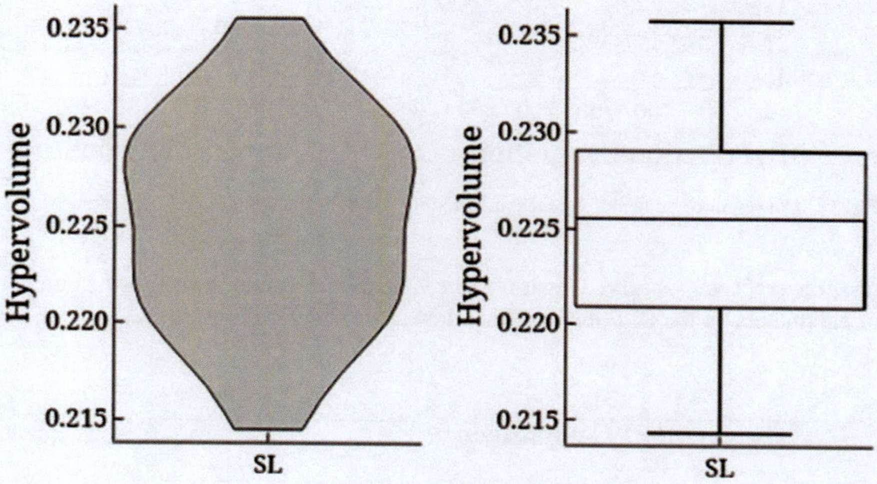

Fig. 11 Violin and box plot from results (31 independent runs of the SL approach) regarding hypervolume values

5.2 *Experiment 2: Bilevel Approach*

The proposed bilevel approach is used to approximate solutions for the proposed bilevel model. Since the upper-level search space is limited by the NAS-bench-101 pre-trained convolutional models [39], we adopt a grid search at the upper level, while the lower-level optimizer is used to approximate solutions for each upper-level decision vector. Since the upper-level optimizer is a deterministic procedure, each independent run will lead to similar outcomes. It is worth mentioning that the grid search (at the upper level) is a free parameter technique, while the parameters for the lower-level optimizer are described in Sect. 4.2.

Moreover, when the lower-level objective function can search either maximum accuracy or the minimum number of parameters, we solve both bilevel problems indicating each variant by BLA (when the lower-level aims to maximize the accuracy) and BLP (when the lower-level aims to minimize the number of trainable parameters). Figure 12 shows the convergence plot for both BLA and BLP approaches. Note that

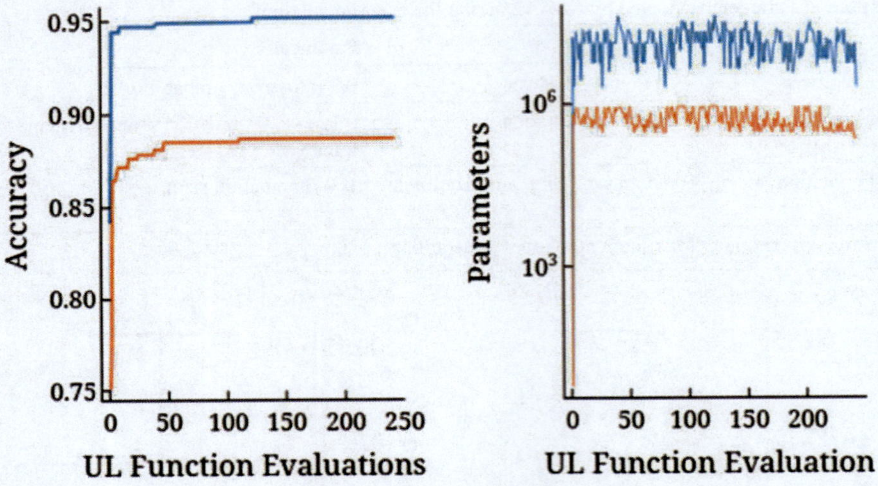

Fig. 12 Convergence graphs of the median execution for both BLA (blue) and BLP (orange)

fast convergence is observed regarding the validation accuracy, although the number of parameters on the BLP increases at a low rated compared with BLA.

5.3 *Experiment 3: Comparison*

This part of the experimentation compares the bilevel approaches against SL, RN [39] a random algorithm for neural architecture search (a free parameter technique that sample 100,000 random solutions and save the network design with the minimum accuracy), and state-of-the-art convolutional neural networks designed via trail-and-error strategy by an expert human [23]. Here, we perform 31 independent runs of each approach (BLA and BLP), and from the obtained Pareto front, a decision-making strategy is performed to select one solution for each execution (the corresponding parameters are given in Sects. 5.1 and 5.2). Here, a Compromise Programming [3] method is used. The main idea is to apply the Tchebysheff scalarization mapping with $w = [0, 1]$ as the reference direction (this w is adopted to prefer accuracy). Once a solution is obtained from each algorithm execution, the results are divided into two tables to analyze the accuracy and the number of trainable parameters.

Table 4 shows the statistical results regarding the accuracy values from the approach BLA and BLP. It is worth mentioning that BLA obtained the best accuracy value; we were expecting this behavior due to the lower-lower objective function for this approach is optimizing the network topology based on the accuracy after training the convolutional neural networks. Again, our approach (BLP) obtains the best configuration with the minimum number of parameters regarding the BLA, SL, and RN approaches, as shown in.

Table 4 Statistical results (accuracy values) from 31 independent BLA, BLP, SL, and RN runs. The best value is in bold font in the corresponding row (based on the Kruskal–Wallis rank sum test)

Approach	Minimum	Median	Mean	Maximum	Std
BLA	**9.518E-01**	**9.518E-01**	**9.518E-01**	**9.518E-01**	**0.000E + 00**
BLP	8.869E-01	8.869E-01	8.869E-01	8.869E-01	**0.000E + 00**
SL	9.472E-01	9.489E-01	9.489E-01	9.518E-01	1.007E-03
RN	9.472E-01	9.474E-01	9.479E-01	9.504E-01	8.116E-04

Table 5 moreover, overall comparing results are given in Table 6 contains median accuracy and the number of parameter values of our bilevel approaches (BLA/BLP) and SL, RN, RESNET, and DenseNet.

Figure 13 presents the Pareto front obtained by BLA, BLP, and SL at the median hypervolume values. It can be observed (Figs. 14, 15, 16, 17, 18, 19, 20, 21, 22, and 23) that BLA suggests the compromise between accuracy and complexity, prioritizing the neural network accuracy, but the number of trainable parameters increases in this approach. On the other hand, BLP suggests low-cost neural networks with limitations in the accuracy rate. Finally, note that SL suggests a good compromise between accuracy and the number of parameters.

Therefore, based on the statistical results, BLA is suitable for providing convolutional neural models with the best accuracy due to exploiting the region where that model lies. BLP can be used when network simplicity is required with acceptable accuracy values. The SL approach can be used to find compromise solutions.

Table 5 Statistical results (number of trainable parameters) from 31 independent BLA, BLP, SL, and RN runs. Note that values are presented in the log10 scale

Approach	Minimum	Median	Mean	Maximum	Std
BLA	7.619E + 00	7.619E + 00	7.619E + 00	7.619E + 00	4.514E-15
BLP	**5.916E + 00**	**5.916E + 00**	**5.916E + 00**	**5.916E + 00**	**2.709E-15**
SL	6.696E + 00	7.364E + 00	7.258E + 00	7.619E + 00	2.352E-01
RN	7.060E + 00	7.347E + 00	7.264E + 00	7.522E + 00	1.681E-01

Table 6 Comparison among the results obtained by our approach (median accuracy value) and manually generated convolutional network design

Approach	Accuracy	Parameters
BLA	**9.518E-01**	7.619E + 00
BLP	8.869E-01	**5.916E + 00**
SL	9.489E-01	7.364E + 00
RN	9.474E-01	7.347E + 00
RESNET101	9.357E-01	6.230E + 00
DenseNet	9.417E-01	6.792E + 00

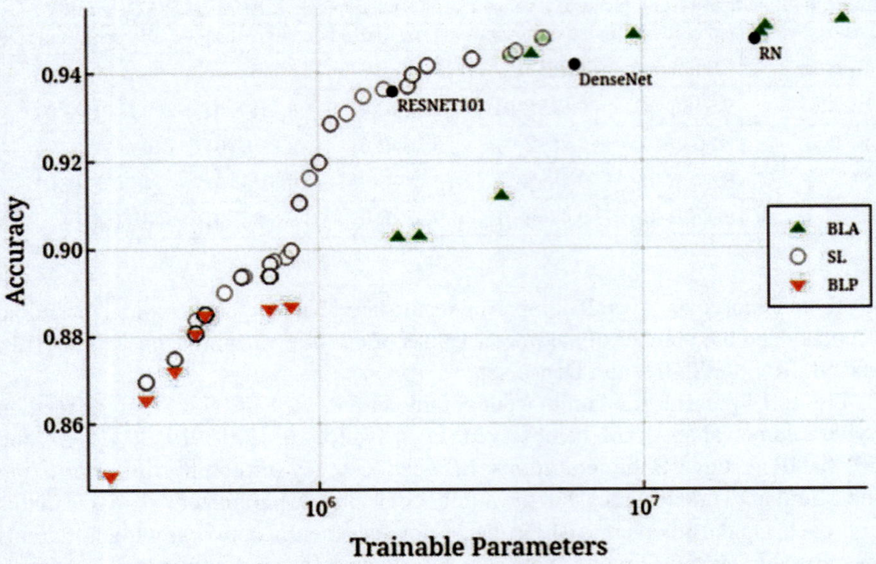

Fig. 13 BLA, BLP, and SL results at median hypervolume value

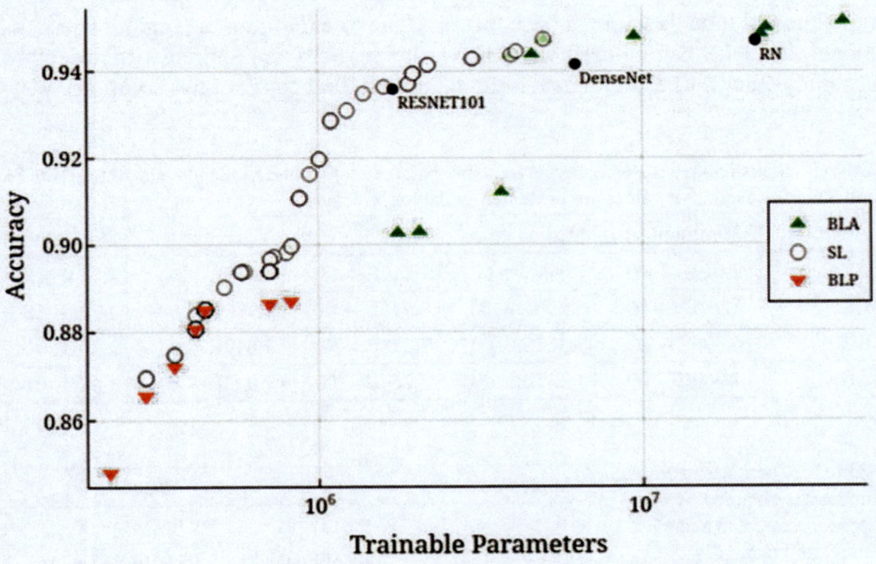

Fig. 14 BLA, BLP, and SL results from 1st execution

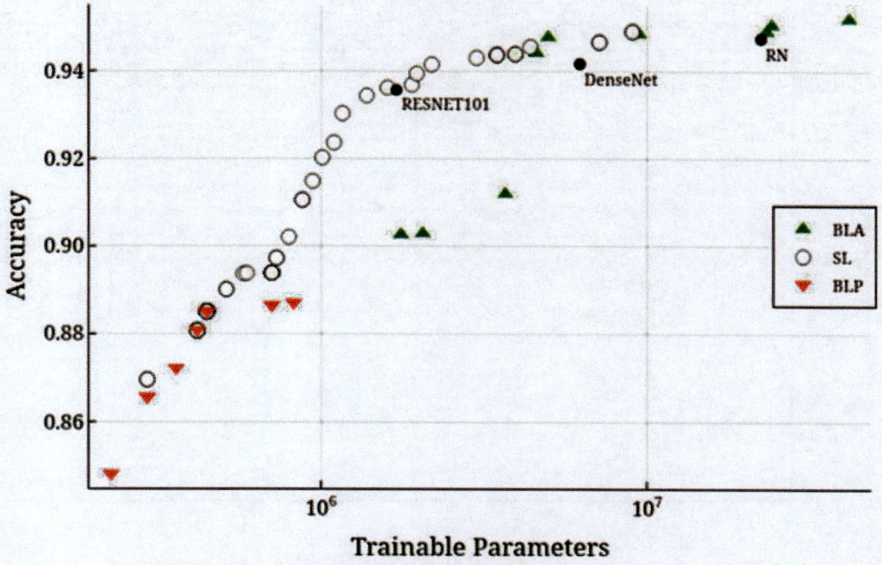

Fig. 15 Results from obtained by BLA, BLP, and SL from 2nd execution

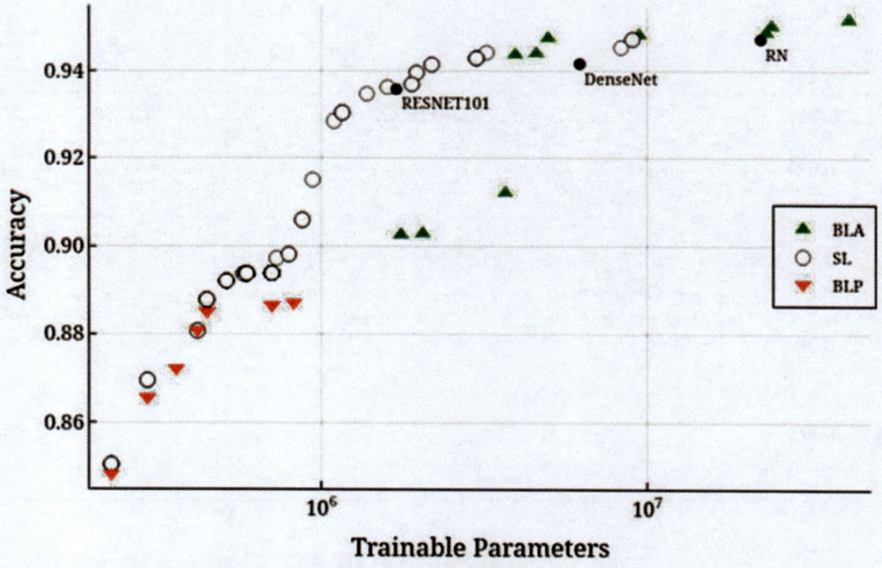

Fig. 16 Results from obtained by BLA, BLP, and SL from 3rd execution

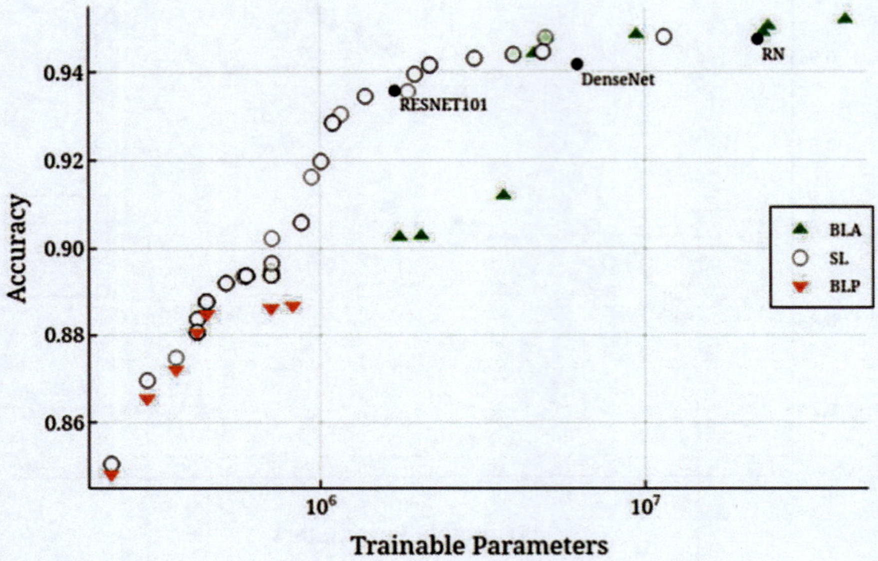

Fig. 17 Results from obtained by BLA, BLP, and SL from 4th execution

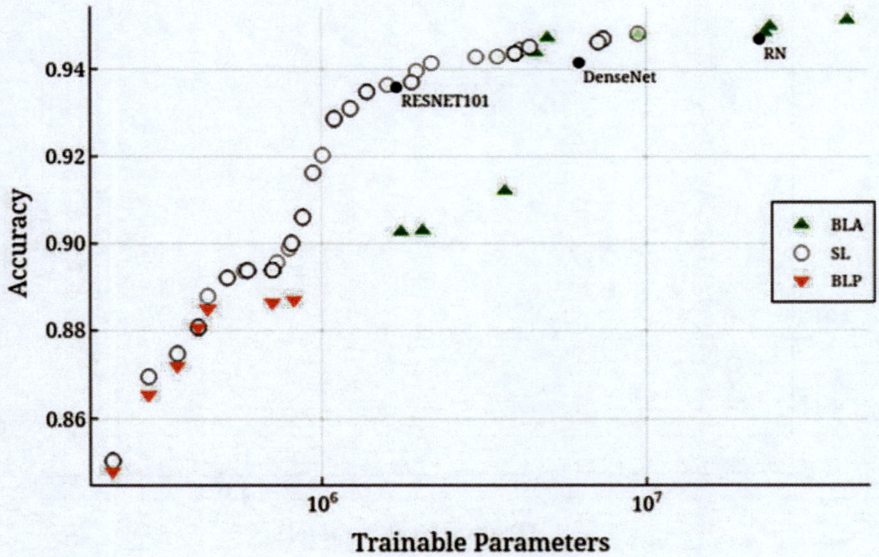

Fig. 18 Results from obtained by BLA, BLP, and SL from 5th execution

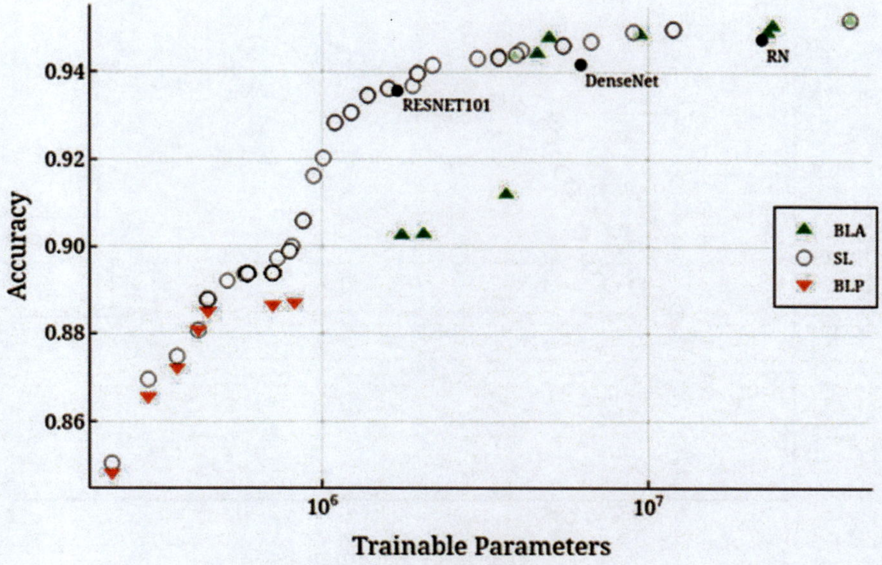

Fig. 19 Results from obtained by BLA, BLP, and SL from 6th execution

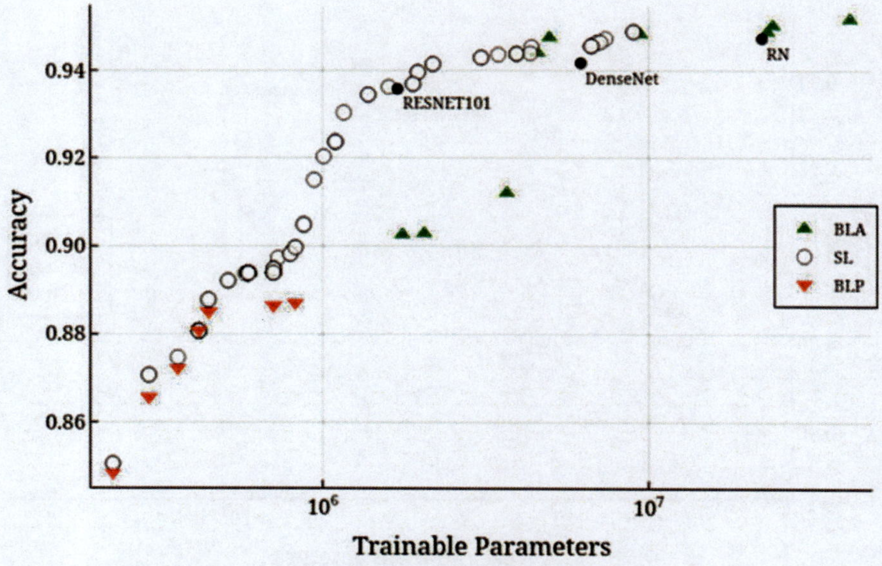

Fig. 20 Results from obtained by BLA, BLP, and SL from 7th execution

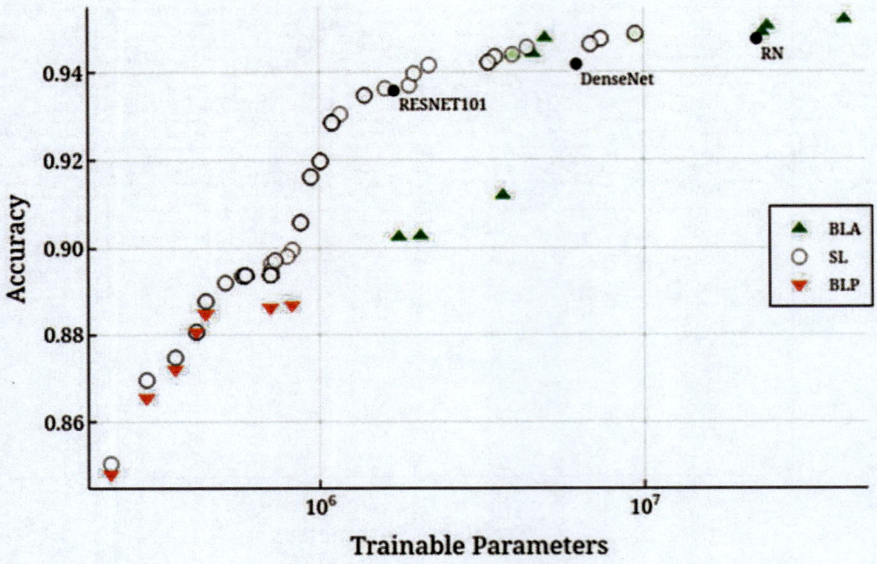

Fig. 21 Results from obtained by BLA, BLP, and SL from 8th execution

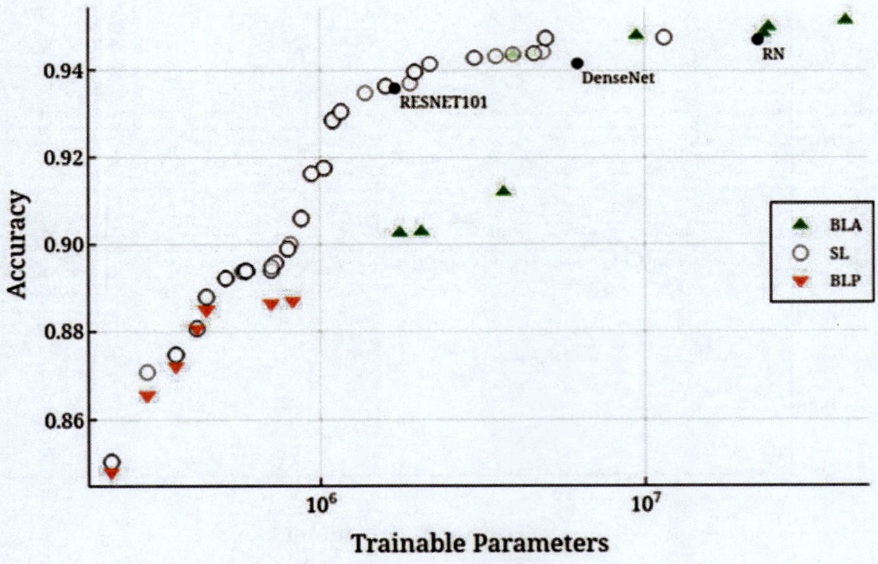

Fig. 22 Results from obtained by BLA, BLP, and SL from 9th execution

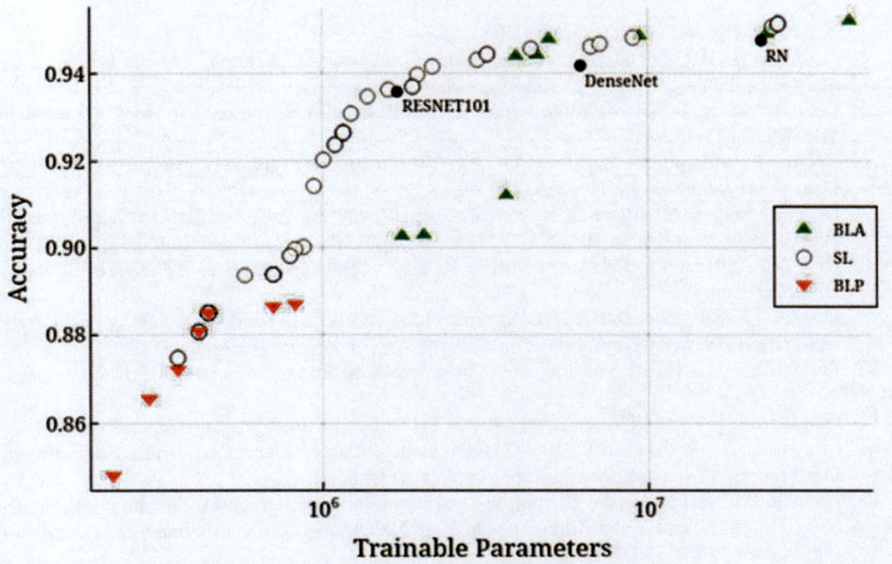

Fig. 23 Results from obtained by BLA, BLP, and SL from 10th execution

6 Conclusion

This work presented a model for the neural architecture search by a multi-objective bilevel optimization problem. The proposed model properties differ from existing approaches because our model was used to minimize the complexity of the network (described by the number of its parameters) at the upper level. At the same time, the lower level optimizes the topology of the network structure for maximum accuracy or network complexity (the user can determine this preference at the lower level). We also solved the two instances of the bilevel model using a bilevel algorithm proposed in this work (BLA and BLP approaches). The statistical results suggest that BLA was suitable for convolutional neural models with the best accuracy due to exploiting the region where those models lie. On the other hand, BLP reported simpler networks with acceptable accuracy values.

References

1. Abbassi, M., Chaabani, A., Said, L. B., & Absi, N. (2020). Bi-level multi-objective combinatorial optimization using reference approximation of the lower level reaction. *Procedia Computer Science, 176*, 2098–2107.
2. Baker, B., Gupta, O., Naik, N., & Raskar, R. (2016). Designing neural network architectures using reinforcement learning. *MedRxiv,* Issue. https://doi.org/10.1101/2020.02.14.20023028.
3. Ballestero, E. (2007). Compromise programming: A utility-based linear-quadratic composite metric from the trade-off between achievement and balanced (non-corner) solutions. *European*

Journal of Operational Research, 182(3), 1369–1382.
4. Beysolow, T., II. (2017). *Introduction to deep learning using R: A step-by-step guide to learning and implementing deep learning models using R.* Apress.
5. Chen, B., & Lu, W. (2018). Meta-Learning with Hessian Free Approach in Deep Neural Nets Training.. *arXiv: Learning.*
6. Colson, B., Marcotte, P., & Savard, G. (2007). An overview of bilevel optimization. *Annals of Operations Research, 153*(1), 235–256.
7. Deb, K., Pratap, A., Agarwal, S., & Meyarivan, T. (2002). A fast and elitist multiobjective genetic algorithm: NSGA-II. *IEEE Transactions on Evolutionary Computation, 6*(2), 182.
8. Dertat, A. (2017). *Applied Deep Learning—Part 4: Convolutional Neural Networks.* Towards Data Science Inc.
9. Dong, C.-Y., Shi, Y., & Tao, R., (2018). Convolutional neural networks for clothing image style recognition. *DEStech Transactions on Computer Science and Engineering.*
10. Eichfelder, G. (2010). Multiobjective bilevel optimization. *Mathematical Programming, 123*(2), 419–449.
11. Goodfellow, I., Bengio, Y., & Courville, A. (2017). Deep learning. *Nature Methods, 13*(1), 35.
12. Liu, H., K. S., & Yang, Y. (2019). DARTS: Differentiable architecture search. *International Conference on Learning Representations,* Volume Online.
13. Hamdan, M. (2012). On the Disruption-level of polynomial mutation for evolutionary multi-objective optimisation algorithms. *Computing and Informatics V Computers and Artificial Intelligence, 29*(5), 783–800.
14. Hubel, D. H., & Wiesel, T. N. (1963). Receptive fields of cells in striate cortex of very young, visually inexperienced kittens. *Journal of Neurophysiology, 26*(6), 994–1002.
15. Jain, G. (2018). *Convolutional Neural Networks : More Dogs, Cats, and Frogs and Cars,* s.l.: linkedin.
16. Kohl, N., & Miikkulainen, R. (2009). Special Issue: Evolving neural networks for strategic decision-making problems. *Neural Networks, 22*(3), 326–337.
17. Koppejan, R., & Whiteson, S. (2011). Neuroevolutionary reinforcement learning for generalized control of simulated helicopters. *Evolutionary Intelligence, 4*(4), 219–241.
18. Lin, C., & Yan (2013). *Network in network,* S.l. arXiv:1312.4400.
19. Louati, H., Bechikh, S., Louati, A., Hung, C. C., & Said, L. B. (2021). Deep convolutional neural network architecture design as a bi-level optimization problem. *Neurocomputing, 439,* 44–62.
20. Montesinos Lopez, O. A., Montesinos Lopez, A., & Crossa, J. (2022). Fundamentals of artificial neural networks and deep learning. *Multivariate Statistical Machine Learning Methods for Genomic Prediction* (pp. 379–425). Springer International Publishing.
21. Muhammad, N. A., Nasir, A. A., Ibrahim, Z., & Sabri, N. (2018). Evaluation of CNN, alexnet and GoogleNet for fruit recognition. *Indonesian Journal of Electrical Engineering and Computer Science, 12*(2), 468–475.
22. Nguyen, G., et al. (2019). Machine Learning and Deep Learning frameworks and libraries for large-scale data mining: A survey. *Artificial Intelligence Review, 52*(1), 77–124.
23. Nielsen, M. A., (2015). *Neural Networks and Deep Learning.* Determination Press.
24. Pham, H. et al. (2028). Efficient neural architecture search via parameters sharing. In: *Proceedings of the 35th International Conference on Machine Learning, ser. Proceedings of Machine Learning Research, J. Dy* (Vol. 80, pp. 4095–4104).
25. Rawat & Wang. (2016). Deep convolutional neural networks for image classification: A comprehensive review. *Neural Computation, 16,* 1120–1132.
26. Schaul, T., & Schmidhuber, J. (2010). Metalearning. *Scholarpedia, 5*(6), 4650.
27. Sinha, A., Malo, P., & Deb, K. (2013). Efficient evolutionary algorithm for single-objective bilevel optimization. *arXiv: Neural and Evolutionary Computing.*
28. Sinha, A., Malo, P., & Deb, K. (2017). Approximated set-valued mapping approach for handling multiobjective bilevel problems. *Computers & Operations Research, 77,* 194–209.
29. Sinha, A., Malo, P., & Deb, K. (2018). A review on bilevel optimization: From classical to evolutionary approaches and applications. *IEEE Transactions on Evolutionary Computation, 22*(2), 276–295.

30. Srinivas, M., & Patnaik, L. M. (1994). Genetic algorithms: A survey. *IEEE Computer, 27*(6), 17–26.
31. Stanley, K. O., Clune, J., Lehman, J., & Miikkulainen, R. (2019). Designing neural networks through neuroevolution. *Nature Machine Intelligence, 1*(1), 24–35.
32. Stanley, K. O., & Miikkulainen, R. (2002). Evolving neural networks through augmenting topologies. Evolutionary Computation, 10(2), 99–127.
33. Szeliski, R. (2010). *Computer Vision: Algorithms and Applications.* ed. Springer Publishing.
34. Targ, S., Almeida, D., & Lyman, K. (2016) Resnet in Resnet: Generalizing Residual Architectures. *arXiv: Learning.*
35. Vargas-Hákim, G.-A., Mezura-Montes, E., & Acosta-Mesa, H.-G. (2022). A review on convolutional neural network encodings for neuroevolution. *IEEE Transactions on Evolutionary Computation, 26*, 17–26. https://doi.org/10.1109/TEVC.2021.3088631.
36. Wang, G. et al. (2018). *Batch Kalman Normalization: Towards Training Deep Neural Networks with Micro-Batches.* ArXiv:abs/1802.03133.
37. Xuân, V. M., & Thủy, N. T. (2012). Real-coded genetic algorithms with simulated binary crossover operator. *Journal of Computer Science and Cybernetics, 22*(2), 134–140.
38. Yao, X., & Liu, Y. (1997). A new evolutionary system for evolving artificial neural networks. *IEEE Transactions on Neural Networks, 8*(3), 694–713.
39. Ying, C. et al. (2019). NAS-Bench-101: Towards reproducible neural architecture search. *arXiv: Learning.*
40. Zhao, Z., Junjie, Y., & Cheng-Lin, L. (2017) Practical network blocks design with Q-learning. *IEEE conference on computer* (pp. 2423–2432).
41. Zoph, B., & Le, Q. V. (2017) Neural architecture search with reinforcement learning. In *5th International Conference on Learning Representations,* Issue. https://openreview.net/forum?id=r1Ue8Hcxg.

Recovering from Population Extinction in the Animal Life Cycle Algorithm (ALCA)

J. C. Felix-Saul and Mario Garcia Valdez

1 Introduction

Biologically inspired algorithms have proven to be very effective when solving complex optimization problems [1–3], but the more challenging the problem, the more computing power it will require to solve them [4]. We consider the main reason it requires more computing power is related to the evaluation of the fitness function, which means calculating the solution to the real problem our algorithm is searching for. One technique to manage this processing power demand is to use distributed computing [5] or the cloud resources [6, 7]. This strategy allows adaptation of the required computing need according to the problem's complexity.

Traditionally, bio-inspired algorithms are developed with a sequential (synchronous) perspective [8, 9], where a process must pause for the previous task to finish before continuing. Some architectures address this issue [10–12] by working on the cloud and finding solutions on distributed technologies. In this research, we present a distributed algorithm totally built as a native cloud solution, where its processes execute asynchronously and in parallel, managing the processing workload among several computers. This strategy allows to elastically increase (or reduce) the computing power according to the nature of the challenge [13].

In previous work, we introduced an algorithm inspired by the biological life-cycle of animal species [14]. As in nature, population individuals grow older and age. In this algorithm's reproduction process, couples must match by mutual attraction, where sometimes individuals won't procreate offspring because of the mate's low appeal.

J. C. Felix-Saul (✉) · M. Garcia Valdez
TecNM, Tijuana Institute of Technology, Tijuana, Mexico
e-mail: jose.felix201@tectijuana.edu.mx

M. Garcia Valdez
e-mail: mario@tectijuana.edu.mx

O. Castillo and P. Melin (eds.), *Hybrid Intelligent Systems Based on Extensions of Fuzzy Logic, Neural Networks and Metaheuristics*, Studies in Computational Intelligence 1096,
https://doi.org/10.1007/978-3-031-28999-6_26

As time passes, the environment kills its individuals either because of low fitness or age, causing occasional population extinction.

We define extinction as a situation in which something no longer exists; in our case scenario, when we refer to extinction in our algorithm, it means the population of individuals no longer exists. When this condition presents itself, it is required to create a new population to continue evolution (restart).

Population restart is the process in which we create a new set of individuals to replace the others that have already died (or were discarded) due to their aptitude. In a previous work [14] we used to generate a new random set of solutions, with the cost of losing all evolution knowledge. In the state of the art, we have found other methodologies that have successfully sought to solve the problem of population extinction with restarts [15–19].

One contribution of this paper is to prove that it is possible to evolve a population of individuals, similar to a Genetic Algorithm (GA), using a distributed, parallel, and asynchronous methodology by algorithm implementation and testing.

The main contribution of this research is to demonstrate how to improve the algorithm's behavior, by solving population extinction caused by nature's pressure in the Animal Life Cycle Algorithm with one of the most promising alternatives: the projection of the Elite towards the Champion. To validate our findings, we compare the results obtained with the latter and the random use of the more traditional alternatives (mutation and crossover), using classic benchmark functions for optimization for comparison.

This paper is organized as follows. First, we illustrate our algorithm model, the encountered problem with the occasional population extinction, and our proposed solution in Sect. 2, followed by our experimental configuration and results in Sect. 3, where we continue to analyze and describe some of our research findings in Sect. 4. We finalize by presenting some inferences based on the results of our experiments in Sect. 5.

2 Proposal

In previous work, we presented an algorithm inspired by nature [14], where we model our algorithm based on the generalization of the life cycle of animal species. Our algorithm, as in nature, consists of four stages [20]: birth, growth, reproduction, and death. One key concept of our idea is that combining those processes evolves the population. We propose to execute all these stages in parallel and asynchronously on a continuously evolving population.

2.1 Algorithm Model

This algorithm was inspired by the traditional Genetic Algorithm, meaning that all individuals have a genotype (chromosome) that is a list of values. We calculate the individual's fitness with the evaluation function and do crossover and mutation to the population. What makes our strategy different is that we don't use the concept of evolutionary generations. We manage our set of solutions as a whole that continuously evolve over time. Allowing individuals of different ages to breed and generate offspring, as it happens in nature. The crossover and mutation execute as independent processes that randomly affect the individuals.

Our algorithm's goal is to mimic the animal life cycle, where at any given moment, new individuals are born to be part of the population and participate in the collective evolution. As time passes, they grow older and mature, suffering changes throughout their lives that we chose to represent as mutations. In our analysis, we considered a couple's attraction a fundamental factor in reproduction. We thought of death's work to maintain balance in the population by enforcing the survival of the fittest. As in life, death can happen to everyone: from a newborn to the elderly, where fitness will determine the individual's longevity. We display the general model concept in Fig. 1.

2.1.1 Birth

Birth is the first step of the population's evolution. For the algorithm representation, we begin with the generation of a randomly set of individuals.

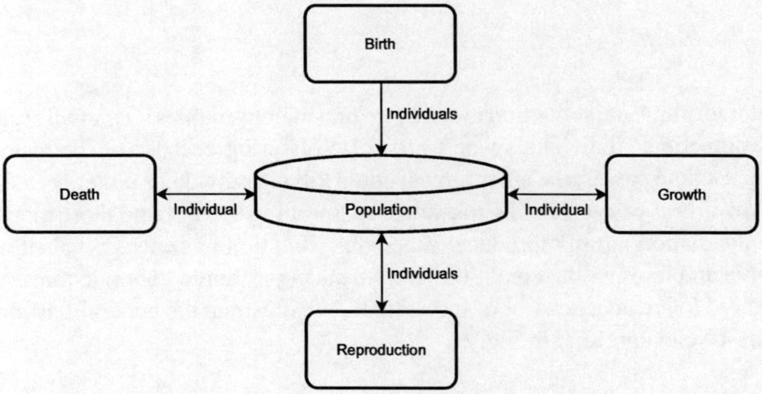

Fig. 1 Animal Life Cycle Algorithm model

2.1.2 Growth

In our algorithm representation, there is a direct correlation between elapsed time and the population's evolution. For this to be possible in our proposal, all individuals must grow older and change as time progresses. With each increment of age, an individual may improve or deteriorate. We chose to manifest this idea by performing a small mutation in each change.

2.1.3 Reproduction

In this step, we select a random pair of individuals and evaluate their couple's attraction, where fitness will be the decisive factor that impacts its value. The higher the fitness of the individual, the more attractive it will appear to fellow individuals. As a consequence of the previous concept, we can anticipate that not all couples will produce offspring, meaning reproduction will not always be successful.

2.1.4 Death

Death is the last step of the animal life cycle, as in our algorithm. We chose death as a representation of all the environmental challenges for survival, where its execution enforces the survival of the fittest. As time progresses, so does increase nature's level of danger. The better the individual fitness will increase its chances of survival.

2.2 Problem

In this algorithm's reproduction process, we find mating couples by mutual attraction, where sometimes individuals won't procreate offspring because of the mate's low appeal. As time passes, the environment could kill its individuals either because they have low fitness or age, leading to extinction, meaning there are no more individuals in the population left to reproduce. When this condition presents, if evaluations are available and termination conditions enable the algorithm to continue the search, it is required to create a new set of individuals. We illustrate the general flowchart for the Life-Cycle algorithm in Fig. 2.

2.3 Proposed Solution

In this article, we test various strategies to generate a new population to replace individuals that have already died (or were discarded) due to their aptitude, as happens

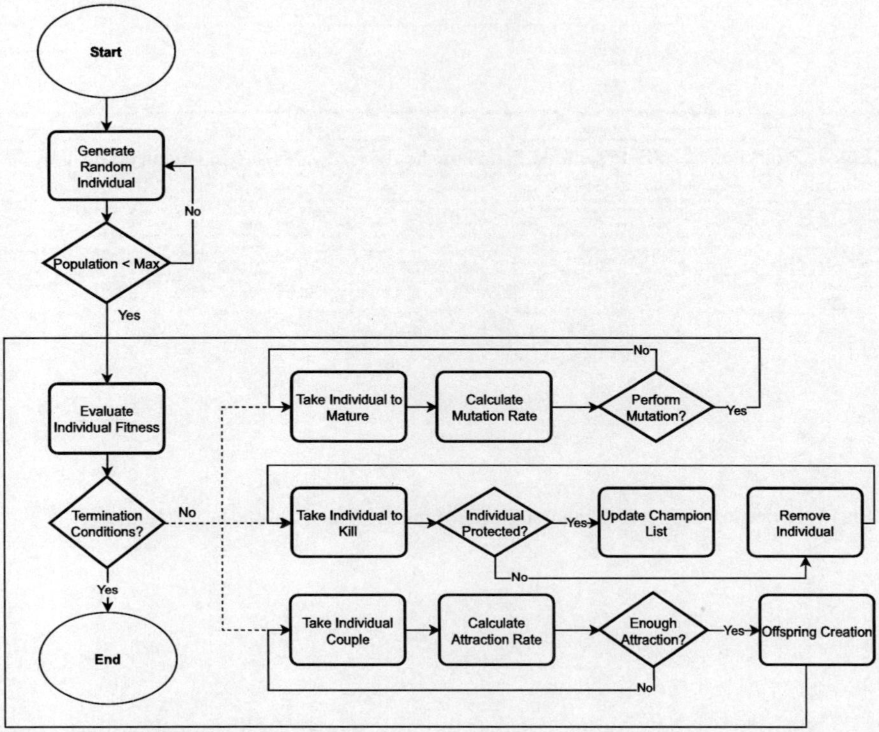

Fig. 2 Animal Life-Cycle Algorithm general flowchart

in the life cycle. We previously used to generate a new random set of solutions, with the cost of losing all evolution knowledge.

This article proposes some strategies to carry out this operation, where the presented distribution strategy obtained the best results. We propose one promising alternative to restart the population, to solve extinction caused by nature's pressure in the Life Cycle algorithm: the projection of the historically best-found candidate solutions (Elite) towards the best-found candidate solution (Champion).

The strategy we describe in this paper use an elite set, that includes the historically best-found candidate solutions to take advantage of the historical evolution. Our goal is for the offspring to be somewhat close to the best individuals of the Elite. We use the golden ratio to compute the distance between new individuals in the population, making an improved distribution. We can consider our strategy an improvement over other alternatives because of the findings of our latter experiments.

Based on the historically best candidate solution and the elite set, we use the Golden Ratio (shown in Eq. 1) to compute the distance between new individuals in the population, making an improved distribution. It randomly selects a solution from the elite set and uses the Golden Ratio [21–25] to compute multiple solutions against the best-found candidate.

Fig. 3 Projection of an Elite solution (found on the right) towards the Champion (at the left)

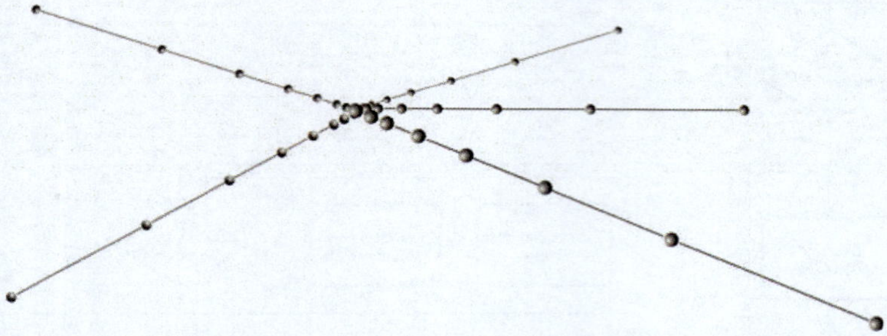

Fig. 4 Projection of five Elite solutions towards the Champion, results in a mini-swarm

$$\alpha = \frac{1 + \sqrt{5}}{2} \tag{1}$$

This strategy is how we can calculate the projection of the historically Elite towards the Champion solution. To illustrate this concept, we include Fig. 3, placing the Elite solution on the far right and the Champion solution on the left corner. We can observe the higher amount of new solutions progressively closer to the Champion.

This process generates new solutions that will be the new population set. It uses the Golden Ratio to progressively approximate from an Elite solution to the Champion. This strategy facilitates to find apparently hidden solutions, having an exploitation emphasis. Figure 4 displays the projection of five Elite solutions (on the corners) towards the Champion (in the center).

3 Experiments

As a proof-of-concept, we implemented the algorithm with Docker containers, where we compare the results obtained with the projection of the Elite towards the Champion strategy and the random use of the more traditional alternatives (such as mutation and crossover), using classic benchmark functions for optimization as implemented by the DEAP (Distributed Evolutionary Algorithms in Python) library [26].

3.1 Experimental Setup

The following tables show the configuration used to execute the experiments for the analysis and comparison in this paper. In Table 1, we can find the General Configuration for the Animal Life Cycle Algorithm. All values from this configuration remained constant during all phases of the experiment runs. The following table, Table 2, shows the values used for the Classic Benchmark Functions for Optimization, where these are adjusted accordingly for each function.

3.2 Experiment Results

For each experiment, we ran 30 independent executions per algorithm and dimensions specified. We recorded the following results: best-found error average, standard deviation, total number of evaluations, and total elapsed time (in seconds). The labels used on the results of our summarized experiments tables are the following:

Table 1 Animal Life Cycle Algorithm (ALCA) general configuration

ALCA general configuration	
Function name	ALL
Population	500
Target error	1.00E-08
Crossover rate	100
Mutation rate	7
Max age	40
Tournament rep.	100
Sample size	20
Base approval	80
Goal approval	200

Table 2 Classic benchmark functions for optimization configuration

Benchmark configuration setup							
Function name	Ackley	Bohachevsky	Griewank	Rastrigin	Sphere	Rosenbrock	Rosenbrock
Dimensions	5, 10	5, 10	5, 10	5, 10	5, 10	5	10
Max evaluations	200,000	200,000	200,000	200,000	200,000	500,000	1,000,000
Max stagnation	50,000	50,000	50,000	50,000	50,000	125,000	250,000
Target fitness	−20.0	−12.0	−150.0	−80.0	−50.0	−3500.0	−3500.0
Bound matrix	[−32, 32]	[−2, 2]	[−500, 500]	[−5, 5]	[−5, 5]	[−2, 2]	[−2, 2]

- Error: is the best-found error average.
- St-Dev: is the standard deviation calculated from the error.
- Evals.: is the total number of evaluations the algorithm executed.
- Time: is the total elapsed or wall clock time, in seconds.

The alternatives to restart this new population are the creation of a new set of candidate solutions based on either: 1. Crossing the Champion with the Elite; 2. The mutation with uniform modification from the Elite; 3. The projection of the Elite towards the Champion; 4. Based on random use of the previous alternatives. We compared three strategies by grouping the previous alternatives in the following:

- Xover Mutation: Crossing the Champion with the Elite, and the mutation with uniform modification from the Elite.
- Elite Projection: The projection of the Elite towards the Champion.
- All Random: Based on random use of the previous alternatives.

3.2.1 Ackley Benchmark Function Experiment

The goal of the first experiment was to confirm that this paper's proposed strategy obtained improved results over our basic technique, which created a new population with a random generation of a new set of candidate solutions. We chose Ackley as the first Classic Benchmark Function for Optimization to test our algorithm behavior, where Fig. 5 shows its corresponding Box and Whisker chart, and we summarized the results in Tables 3 and 4 (for five and ten dimensions, respectively).

3.2.2 Bohachevsky Benchmark Function Experiment

As our second experiment to further validate and prove our algorithm behavior, we chose Bohachevsky Classic Benchmark Function for Optimization. Figure 6 shows its corresponding Box and Whisker chart, where we summarized the results in Tables 5 and 6 (for five and ten dimensions, respectively). Due to the observation

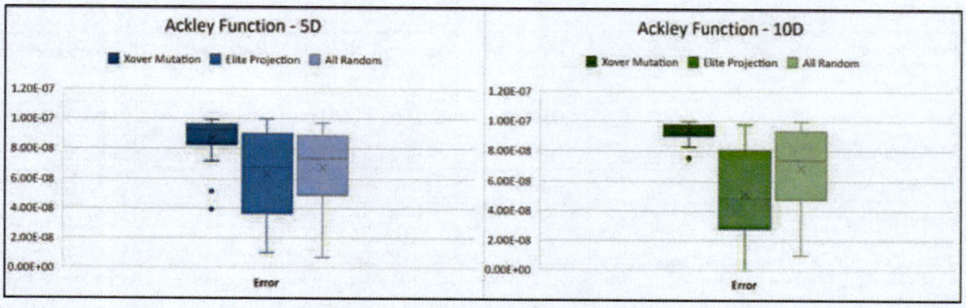

Fig. 5 Box and whisker chart for the Ackley Benchmark Function, 5 verses 10 Dimensions

Table 3 Ackley Benchmark Function summarized experiments table for 5 Dimensions

Ackley function—5 dimensions				
	Error	St-Dev	Evals.	Time
Xover mutation	8.69E-08	1.36E-08	46,331	49.9
Elite projection	6.33E-08	2.70E-08	13,811	18.1
All random	6.78E-08	2.55E-08	30,236	32.8

Table 4 Ackley Benchmark Function summarized experiments table for 10 Dimensions

Ackley function—10 dimensions				
	Error	St-Dev	Evals.	Time
Xover mutation	9.35E-08	5.93E-09	104,842	111.7
Elite projection	5.11E-08	3.13E-08	16,019	20.1
All random	6.89E-08	2.75E-08	43,631	46.3

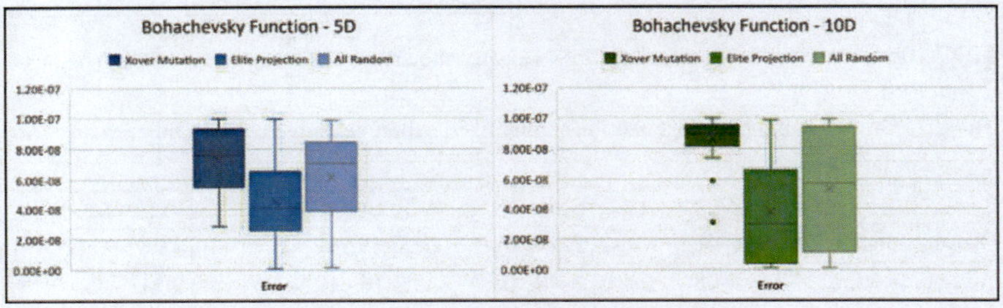

Fig. 6 Box and whisker chart for the Bohachevsky Benchmark Function, 5 verses 10 Dimensions

Table 5 Bohachevsky Benchmark Function summarized experiments table for 5 Dimensions

Bohachevsky function—5 dimensions				
	Error	St-Dev	Evals.	Time
Xover mutation	7.28E-08	2.20E-08	15,010	24.2
Elite projection	4.51E-08	2.94E-08	8,266	12.6
All random	6.16E-08	2.85E-08	11,998	21.7

and analysis of these results, we identify our proposed strategy obtained solutions in the same order as the traditional Xover Mutation, in a significantly reduced number of evaluations.

Table 6 Bohachevsky Benchmark Function summarized experiments table for 10 Dimensions

Bohachevsky function—10 dimensions

	Error	St-Dev	Evals.	Time
Xover mutation	8.66E-08	1.38E-08	40,166	44.4
Elite projection	3.80E-08	3.34E-08	8,506	12.9
All random	5.34E-08	3.76E-08	19,537	26.8

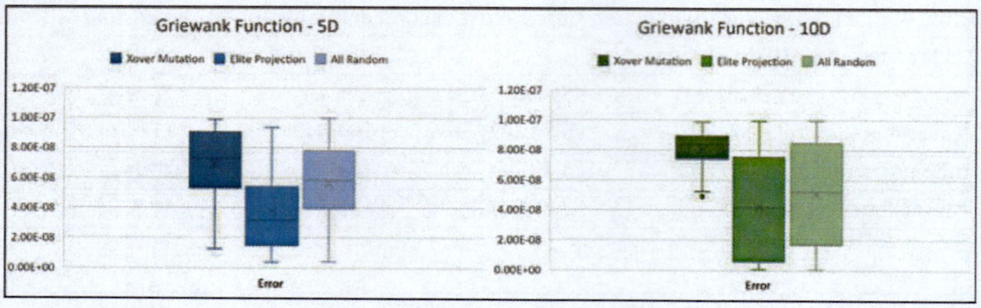

Fig. 7 Box and whisker chart for the Griewank Benchmark Function, 5 verses 10 Dimensions

Table 7 Griewank Benchmark Function summarized experiments table for 5 Dimensions

Griewank function—5 dimensions

	Error	St-Dev	Evals.	Time
Xover mutation	6.92E-08	2.40E-08	44,566	48.3
Elite projection	3.77E-08	2.74E-08	16,412	20.1
All random	5.63E-08	2.99E-08	35,761	38.5

3.2.3 Griewank Benchmark Function Experiment

The third experiment objective was to confirm Animal Life Cycle Algorithm continued finding solutions in a reduced number of evaluations. For this test, we selected the Griewank Benchmark Function for Optimization. Figure 7 shows its corresponding Box and Whisker chart, where we summarized the results in Tables 7 and 8 (for five and ten dimensions, respectively). With this analysis, we confirmed our strategy continued to obtain solutions in the same order as expected, in a reduced number of evaluations and less time.

3.2.4 Rastrigin Benchmark Function Experiment

This experiment's goal was to follow and study the behavior of our proposed strategy, to repeat the results in the same order as expected, in a reduced number of both time and evaluations. We selected Rastrigin Classic Benchmark Function for Optimiza-

Table 8 Griewank Benchmark Function summarized experiments table for 10 Dimensions

Griewank function—10 dimensions				
	Error	St-Dev	Evals.	Time
Xover mutation	8.07E-08	1.39E-08	92,885	100.0
Elite projection	4.21E-08	3.40E-08	13,534	17.3
All random	5.08E-08	3.54E-08	40,544	52.5

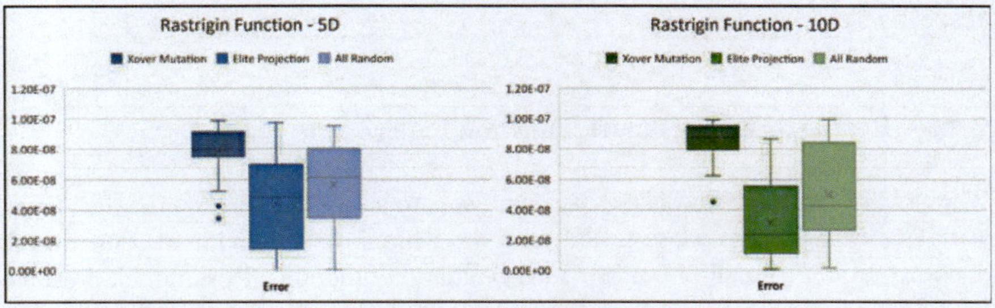

Fig. 8 Box and whisker chart for the Rastrigin Benchmark Function, 5 verses 10 Dimensions

Table 9 Rastrigin Benchmark Function summarized experiments table for 5 Dimensions

Rastrigin function—5 dimensions				
	Error	St-Dev	Evals.	Time
Xover mutation	7.85E-08	1.59E-08	22,322	34.1
Elite projection	4.42E-08	3.26E-08	10,860	14.8
All random	5.66E-08	2.81E-08	21,122	35.0

Table 10 Rastrigin Benchmark Function summarized experiments table for 10 Dimensions

Rastrigin function—10 dimensions				
	Error	St-Dev	Evals.	Time
Xover mutation	8.56E-08	1.23E-08	53,579	58.7
Elite projection	3.17E-08	2.57E-08	8,704	13.0
All random	5.00E-08	3.16E-08	24,376	27.4

tion for this experiment. Figure 8 shows its corresponding Box and Whisker chart, where we summarized the results in Tables 9 and 10 (for five and ten dimensions, respectively).

3.2.5 Sphere Benchmark Function Experiment

For our fifth experiment we chose the Sphere Classic Benchmark Function for Optimization. Even though the sphere function is considered one of the simplest to solve, the same behavior was consistent for the algorithm, where it showed a reduced number of evaluations and execution time. We can confirm this affirmation with the results shown in this experiment. Figure 9 displays its corresponding Box and Whisker chart, where we summarized the results in Tables 11 and 12 (for five and ten dimensions, respectively).

3.2.6 Rosenbrock Benchmark Function Experiment

The goal of our sixth and last experiment was to study the behavior of the Animal Life-Cycle Algorithm when tested with the Rosenbrock Benchmark Function for Optimization. From all of our previous evaluation functions, Rosenbrock presented

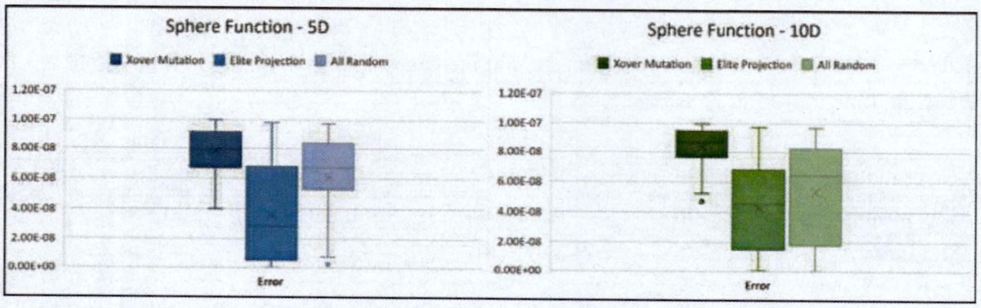

Fig. 9 Box and whisker chart for the Sphere Benchmark Function, 5 verses 10 Dimensions

Table 11 Sphere Benchmark Function summarized experiments table for 5 Dimensions

Sphere function—5 dimensions

	Error	St-Dev	Evals.	Time
Xover mutation	7.82E-08	1.65E-08	11,312	14.5
Elite projection	3.60E-08	3.27E-08	7,056	10.9
All random	6.14E-08	2.79E-08	9,158	12.2

Table 12 Sphere Benchmark Function summarized experiments table for 10 Dimensions

Sphere function—10 dimensions

	Error	St-Dev	Evals.	Time
Xover mutation	8.39E-08	1.45E-08	30,769	34.4
Elite projection	4.28E-08	3.06E-08	9,236	13.3
All random	5.32E-08	3.38E-08	21,465	26.3

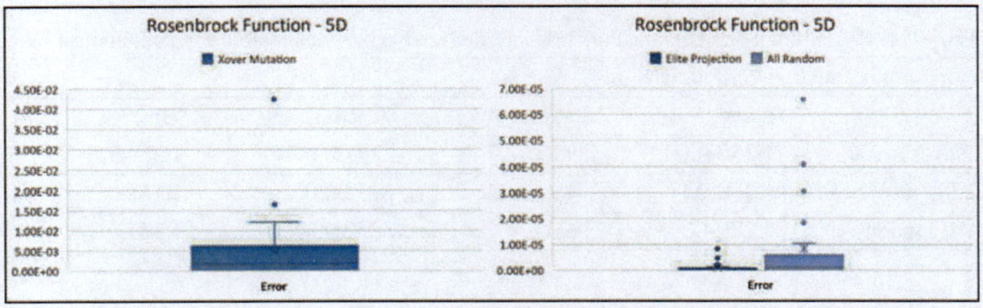

Fig. 10 Box and whisker chart for the Rosenbrock Benchmark Function, 5 Dimensions

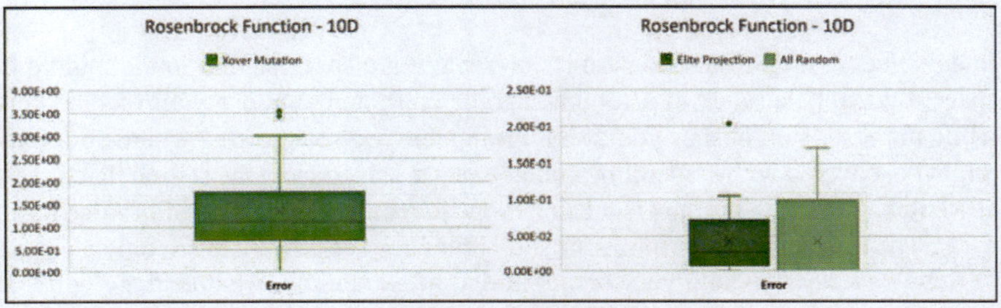

Fig. 11 Box and whisker chart for the Rosenbrock Benchmark Function, 10 Dimensions

Table 13 Rosenbrock Benchmark Function summarized experiments table for 5 Dimensions

Rosenbrock function—5 dimensions				
	Error	St-Dev	Evals.	Time
Xover mutation	4.65E-03	8.29E-03	500,000	646.4
Elite projection	9.62E-07	1.86E-06	307,878	427.0
All random	8.30E-06	1.58E-05	491,400	697.8

the most difficult challenge for our algorithm. This test facilitated the differentiation between the three compared strategies because the solutions found were in different numeric order. After all tests, the Elite Projection proved to be the best alternative. We can corroborate this by analyzing Figs. 10 and 11, which shows its corresponding Box and Whisker chart. We present the summarized results in Tables 13 and 14 (for five and ten dimensions, respectively).

Table 14 Rosenbrock Benchmark Function summarized experiments table for 10 Dimensions

Rosenbrock function—10 dimensions				
	Error	St-Dev	Evals.	Time
Xover mutation	1.35E+00	9.83E-01	1,000,000	1,384.3
Elite projection	4.26E-02	4.34E-02	1,000,000	1,146.0
All random	4.27E-02	6.02E-02	952,038	1,147.8

4 Discussion

Imagine being an astronaut on a mission repairing a space station in the middle of space. Suddenly, a block of asteroids hits the craft, and you lose your oxygen tank while the explosion pushes you away. From afar, you see the area where your tank might be, covered with something similar to a gas leaked from the station that blocks all visibility. Finding the tank is the only hope to return to your main ship some miles away. The oxygen reserve from your suit is running out, and there are only a couple of minutes left. Now hold your breath in real life and think, what strategy will you choose to search for your tank? We repeated this mental exercise several times until we found the best solution (or blacked out).

If your life depended on it, would you choose a Genetic Algorithm, a variant of a Particle Swarm Optimization Algorithm, or another? We searched for a way to mix both, at least in some way, take the main idea or concepts from them and make them work together. We strongly believe it is possible to combine the features, as previous work has proven to obtain outstanding results [2, 10, 11, 27]. Our goal was to create a swarm out of the best individuals found in evolution. We probably have a long way to go, but at least we are getting promising results. From our experiments, all six Classic Benchmark Functions for Optimization (Ackley, Bohachevsky, Griewank, Rastrigin, Sphere, and Rosenbrock) proved the same behavior for the Animal Life Cycle Algorithm: finding the expected solution in a reduced number of evaluations and less execution time.

5 Conclusions

The Animal Life Cycle Algorithm allows for multiple parameters' fine-tuning, facilitating the freedom to experiment with different alternatives to restart the population and solve extinction caused by nature's pressure, which will impact how quickly, and the quality of its found solution. We compared the results obtained with the use of historical elite projection versus other alternatives, using classic benchmark functions for optimization for comparison, where it showed favorable and promising results.

As the nature of the challenge increases, it is indispensable to have an elastic, scalable, and fault-tolerant model designed with cloud collaboration processes, and asynchronous communication. We have proven that it is possible to evolve a population, using a distributed, parallel, and asynchronous strategy, inspired by the Genetic Algorithm (GA). To further validate this work, we could use some more demanding (or control) problem that requires calculating real numbers [28, 29]. We implemented the algorithm using Docker containers.

Acknowledgements This paper has been supported in part by TecNM Project 15340.22-P.

References

1. Castillo, O., Valdez, F., Soria, J., Amador-Angulo, L., Ochoa, P., & Peraza, C. (2019). Comparative study in fuzzy controller optimization using bee colony, differential evolution, and harmony search algorithms. *Algorithms, 12*(1), 9.
2. Valdez, F. (2021). Swarm intelligence: A review of optimization algorithms based on animal behavior. Recent Advances of Hybrid Intelligent Systems Based on Soft Computing (pp. 273–298).
3. Acherjee, B., Maity, D., & Kuar, A. S. (2020). Ultrasonic machining process optimization by cuckoo search and chicken swarm optimization algorithms. *International Journal of Applied Metaheuristic Computing (IJAMC), 11*(2), 1–26.
4. Ontiveros, E., Melin, P., & Castillo, O. (2018). High order α-planes integration: A new approach to computational cost reduction of general type-2 fuzzy systems. *Engineering Applications of Artificial Intelligence, 74,* 186–197.
5. Thain, D., Tannenbaum, T., & Livny, M. (2005). Distributed computing in practice: the condor experience. *Concurrency and Computation: Practice and Experience, 17*(2–4), 323–356.
6. García-Valdez, M., Mancilla, A., Trujillo, L., Merelo, J. J., & Fernández-de Vega, F. (2013). Is there a free lunch for cloud-based evolutionary algorithms? In *2013 IEEE Congress on Evolutionary Computation* (pp. 1255–1262). IEEE.
7. Eshratifar, A. E., Esmaili, A., & Pedram, M. (2019). Bottlenet: A deep learning architecture for intelligent mobile cloud computing services. In *2019 IEEE/ACM International Symposium on Low Power Electronics and Design (ISLPED)* (pp. 1–6). IEEE.
8. Porto, V. W. (2018). Evolutionary programming. *Evolutionary Computation* (vol. 1, pp. 127–140). CRC Press.
9. Back, T. (1996). *Evolutionary algorithms in theory and practice: Evolution strategies, evolutionary programming, genetic algorithms*. Oxford University Press.
10. Valdez, M. G., & Guervós, J. J. M. (2021). A container-based cloud-native architecture for the reproducible execution of multi-population optimization algorithms. *Future Generation Computer Systems, 116*, 234–252.
11. García-Valdez, M., Trujillo, L., Merelo, J. J., de Vega, F. F., & Olague, G. (2015). The evospace model for pool-based evolutionary algorithms. *Journal of Grid Computing, 13*(3), 329–349.
12. Merelo, J. J., García-Valdez, M., Castillo, P. A., García-Sánchez, P., Cuevas, P., & Rico, N. (2016). Nodio, a javascript framework for volunteer-based evolutionary algorithms: First results. arXiv:1601.01607
13. Armbrust, M., Fox, A., Griffith, R., Joseph, A. D., Katz, R., Konwinski, A., Lee, G., Patterson, D., Rabkin, A., Stoica, I., et al. (2010). A view of cloud computing. *Communications of the ACM, 53*(4), 50–58.
14. Felix-Saul, J. C., Valdez, M. G., Guervós, & J. J. M. (2022) *A novel distributed nature-inspired algorithm for solving optimization problems* (pp. 107–119). Cham: Springer International Publishing.
15. Hellwig, M., & Beyer, H. G. (2020). A modified matrix adaptation evolution strategy with restarts for constrained real-world problems. In *2020 IEEE Congress on Evolutionary Computation (CEC)* (pp. 1–8). IEEE.

16. Loshchilov, I. (2013). CMA-ES with restarts for solving CEC 2013 benchmark problems. In *2013 IEEE Congress on Evolutionary Computation* (pp. 369–376). IEEE.
17. Jansen, T., & Zarges, C.: Aging beyond restarts. In *Proceedings of the 12th Annual Conference on Genetic and Evolutionary Computation* (pp. 705–712).
18. Tendresse, I. l., Gottlieb, J., & Kao, O. (2001). The effects of partial restarts in evolutionary search. In *International Conference on Artificial Evolution (Evolution Artificielle)* (pp. 117–127). Springer.
19. Mathias, K. E., Schaffer, J. D., Eshelman, L. J., & Mani, M. (1998). The effects of control parameters and restarts on search stagnation in evolutionary programming. In *International Conference on Parallel Problem Solving from Nature* (pp. 398–407). Springer.
20. Read, K., & Ashford, J. (1968). A system of models for the life cycle of a biological organism. *Biometrika, 55*(1), 211–221.
21. Nematollahi, A. F., Rahiminejad, A., & Vahidi, B. (2020). A novel meta-heuristic optimization method based on golden ratio in nature. *Soft Computing, 24*(2), 1117–1151.
22. Gaikwad, P. S., & Kulkarni, V. B. (2021). Face recognition using golden ratio for door access control system. In *Advances in Signal and Data Processing* (pp. 209–231). Springer.
23. Khesin, B., & Wang, H. (2022). The golden ratio and hydrodynamics. *The Mathematical Intelligencer, 44*(1), 22–27.
24. Battaloglu, R., & Simsek, Y. (2021). On new formulas of fibonacci and lucas numbers involving golden ratio associated with atomic structure in chemistry. *Symmetry, 13*(8). https://www.mdpi.com/2073-8994/13/8/1334
25. Narushin, V. G., Griffin, A. W., Romanov, M. N., & Griffin, D. K. (2022). Measurement of the neutral axis in avian eggshells reveals which species conform to the golden ratio. *Annals of the New York Academy of Sciences, n/a*(n/a). https://nyaspubs.onlinelibrary.wiley.com/doi/abs/10.1111/nyas.14895
26. Fortin, F. A., De Rainville, F. M., Gardner, M. A. G., Parizeau, M., & Gagné, C. (2012). Deap: Evolutionary algorithms made easy. *The Journal of Machine Learning Research, 13*(1), 2171–2175.
27. García-Valdez, M., & Merelo, J. J. (2021). Event-driven multi-algorithm optimization: Mixing swarm and evolutionary strategies. In *International Conference on the Applications of Evolutionary Computation (Part of EvoStar)* (pp. 747–762). Springer.
28. Stanley, K. O., & Miikkulainen, R. (2002). Evolving neural networks through augmenting topologies. *Evolutionary computation, 10*(2), 99–127.
29. Miikkulainen, R., Liang, J., Meyerson, E., Rawal, A., Fink, D., Francon, O., et al. (2019). Evolving deep neural networks. In *Artificial Intelligence in the Age of Neural Networks And Brain Computing* (pp. 293–312). Elsevier.

Multi-objective Optimization Through Coevolution and Outranking Methods with Uncertainty Management

Lorena Rosas-Solórzano, Claudia Gomez-Santillan, Nelson Rangel-Valdez, Eduardo Fernández, Laura Cruz-Reyes, Lucila Morales-Rodriguez, and Hector Fraire-Huacuja

1 Introduction

Multi-objective optimization began in the 80s and has evolved significantly to the present day. However, although it has had a remarkable evolution, important challenges need to be addressed, for example, Algorithmic design, scalability, and handling of expensive objective functions, among others [1].

To solve some of the challenges, coevolutionary algorithms are being applied [2, 3], since these could attack problems, for example, the scalability, which is defined as proposing problems with several decision variables and/or objective functions, which can be increased while maintaining the difficulty of the problem [4].

In addition to the algorithms that provide a solution to the multi-objective problem, other authors such as Fernández [5] and Balderas [6] have applied outranking methods and intervals, that is, they compare the alternatives in pairs based on two measures: agreement and disagreement [7], where the parameters of the preferences model that include the outranking methods make use of uncertainty in some of their

L. Rosas-Solórzano (✉) · C. Gomez-Santillan · N. Rangel-Valdez · L. Cruz-Reyes · L. Morales-Rodriguez · H. Fraire-Huacuja
Tecnológico Nacional de México, Instituto Tecnológico de Ciudad Madero, Ciudad Madero, Mexico
e-mail: lorenarocio52@gmail.com

L. Morales-Rodriguez
e-mail: lucila.mr@cdmadero.tecnm.mx

H. Fraire-Huacuja
e-mail: hector.fh@cdmadero.tecnm.mx

E. Fernández
Facultad de Contaduría Y Administración, Universidad Autonoma de Coahuila, 27000 Torreon, COA, Mexico
e-mail: eduardo.fernadez@uadec.edu.mx

O. Castillo and P. Melin (eds.), *Hybrid Intelligent Systems Based on Extensions of Fuzzy Logic, Neural Networks and Metaheuristics*, Studies in Computational Intelligence 1096,
https://doi.org/10.1007/978-3-031-28999-6_27

parameters, being represented with interval numbers which represent a range of values; this is done to help decision-makers (DM) to find solutions according to the preferences that are consistent with the objectives of the problem and direct the search towards the region of interest (RoI).

To date, some authors have attempted to solve multi-objective optimization problems with and without uncertainty [5, 8–10]. They have also been combined with outranking methods in evolutionary algorithms such as NSGA-II and MOEA/D but have not been used together or with outranking methods. For this reason, it is proposed to integrate them, since there is no coevolutionary algorithm that makes use of outranking methods to be able to solve multi-objective optimization problems with uncertainty, where scalability is not only addressed in the number of decision variables, but also the number of objectives.

This document contains the following sections: Sect. 2 Backgroung, which includes the description of Multi-Objective Optimization, the representation of uncertainty with interval mathematics, Interval outranking model, performance indicators for RoI and Coevolution, Sect. 3 solution methodology, Sect. 4 Experimentation and results. The Sect. 5 is a brief conclusion.

2 Background

2.1 Multi-objective Optimization

Optimization can be defined as a search problem in which you try to find the best solution to that problem (maximization or minimization). Typically, multi-objective optimization (MOP) problems are defined as follows (Eqs. 1 and 2):

$$\textbf{Minimize} \min_{s.t.x\,\varepsilon\Omega} F(x) = \left(f_1(x), f_2(x), \ldots f_m(x)^T\right) \quad (1)$$

$$\textbf{Maximize} \max_{s.t.x\,\varepsilon\Omega} F(x) = \left(f_1(x), f_2(x), \ldots f_m(x)^T\right) \quad (2)$$

where x = $(x_1, x_2, \ldots, x_n)$ is the decision vector, n is the dimension of the decision space, Ω is the feasible region in the decision space, and $F(x)$ is a set of objective functions consisting of m objective functions [11].

MoPs can present conflicts between objectives, DM can express each objective, as an interval, taking into account the imprecision in the information, that is, $f_j(x) = \left[\underline{f_j}(x), \overline{f_j}(x)\right]$. Every element of the set X is treated as an intervals vector $\vec{f}(x)$.

$$\begin{aligned} &max\, F(x) = (f_1, f_2, \ldots, f_m) \\ &Subject\ to: x \in \Omega \end{aligned} \quad (3)$$

2.2 Representing Uncertainty with Interval Mathematics

Uncertainty comes from incomplete knowledge. One way to represent uncertainty is with interval math. Moore and other scholars made additional studies of interval numbers [12, 13]. Moore [12] defines an interval number as a range, where $X = [\underline{X}, \overline{X}]$, being $\underline{X}$ the lower limit and $\overline{X}$ the upper limit. Interval numbers allow you to perform arithmetic operations such as addition, subtraction, multiplication, and division, as shown in Eqs. 4 through 7, respectively.

$$X + Y = [\underline{X} + \underline{Y}, \overline{X} + \overline{Y}] \tag{4}$$

$$X - Y = [\underline{X} - \underline{Y}, \overline{X} - \overline{Y}] \tag{5}$$

$$X \cdot Y = [\min\{\underline{XY}, \underline{X}, \overline{Y}, \overline{X}, \underline{Y}, \overline{XY}\}, \max\{\underline{XY}, \underline{X}, \overline{Y}, \overline{X}, \underline{Y}, \overline{XY}\}] \tag{6}$$

$$X/Y = [\underline{X}, \overline{Y}] \cdot [1/\underline{Y}, 1/\overline{Y}] \tag{7}$$

To calculate the relationship between X and Y with interval numbers Yao [14] defined rules that are based on a probability measure that $Y \geq X$, which is calculated with Eqs. 8–10:

$$P(X \geq Y)\begin{cases} 1 & if p_{XY} > 1 \\ p_{XY} & if 0 \leq p_{XY} \leq 1 \\ 0 & if p_{XY} \leq 0 \end{cases} \tag{8}$$

where:

$$p_{XY} = \frac{\overline{X} - \underline{Y}}{(\overline{X} - \underline{X}) + (\overline{Y} - \underline{Y})} \tag{9}$$

$$P(X \leq Y)\begin{cases} 0.5\ (X = Y) \boldsymbol{Si}\ \underline{X} = \underline{Y} \& \overline{X} = \overline{Y} \\ 1\ (X > Y) \boldsymbol{Si}\ \underline{X} > \overline{Y} \\ 0\ (X < Y) \boldsymbol{Si}\ \overline{X} < \underline{Y} \\ 0.5\ (Y > X) \boldsymbol{Si}\ \underline{X} \leq \underline{Y} \leq \overline{X} \leq \overline{Y} || \underline{X} \leq \underline{Y} \leq \overline{Y} \leq \overline{X} \\ 0.5\ (Y < X) \boldsymbol{Si}\ \underline{X} \leq \underline{Y} \leq \overline{X} \leq \overline{Y} || \underline{X} \leq \underline{Y} \leq \overline{Y} \leq \overline{X} \end{cases} \tag{10}$$

In Roy [15] $P(X \leq Y) = \alpha$ is the degree of probability that given two relations X and Y, the realization x is less than or equal to the realization y. In the model of exceeding intervals of Fernández [9], use these possibility rules to calculate the credibility index. $\sigma(x,y)$, which helps to approximate the Region of Interest of an optimization problem, taking advantage of the uncertainty in the preferences of the DM.

2.3 Interval Outranking Model

The search in a region of the Pareto front preferred by the DM is called Region of Interest (RoI). Preferences are parameters that the DM must provide to define the RoI.

For this work, information is taken from Fernández's preference model [9], where they form the RoI combining the best compromises, with the solutions that are closely related, such that the value system of the DM must be considered, DM = (w, v, λ, β), the strict preference relation defined by R_5 (see Table 1), the net flow of outranking (Eq. 12) and the credibility index $\sigma(y,x)$ (Eq. 13) see in [9] for more details on the calculation of $\sigma(y,x)$.

The multi-objective optimization problem that is the RoI, is based on obtaining the best compromise BC (Eq. 11). Therefore, the RoI is formed by the solutions in BC^{DM} and the solutions that satisfy the condition $\sigma(y,x) > \beta$, such that x belongs to BC^{DM}.

$$BC^{DM} = \left\{ xPF | \{ yPF | yR_5^{DM} x \} = \phi, \varphi(x) = \max_{yPF} \{ \varphi^{DM}(y) \} \right\} \tag{11}$$

$$\varphi(x) = \varphi^+(x) - \varphi^- = \sum_{yPF} \sigma(x, y) - \sum_{yPF} \sigma(y, x) \tag{12}$$

$$\sigma(x, y) = \max_{\gamma \in \Omega} \sigma_\gamma \tag{13}$$

Fernández [9] describes in his work and mentions that from the calculation of the degree of credibility (Eq. 13) that x is at least as good as y, different binary

Table 1 Binary preference relationships [9]

Relation	Definition	Description
R_1	$\sigma(y, x) > \sigma(x, y)$	Indicates some preference in favor of y, although its credibility may be low
R_2	$\sigma(y, x) \geq \beta$	y is at least as good as x with a credibility threshold $\beta > 0.5$
R_3	$\sigma(y, x) \geq \beta and \sigma(x, y) \leq \beta$	Indicates an asymmetric preference in favor of y with a credibility threshold $\beta > 0$
R_4	$\sigma(y, x) > \sigma(x, y) and \sigma(y, x) > 0.5$	Indicates a certain preference in favor of y with a credibility threshold of 0.5
R_5	$\sigma(y, x) \geq \beta and \sigma(x, y) < 0.5$	Indicates strict preference in favor of y

relationships of higher rank arise from the value system of the $DM = \{w, v, \lambda, \beta\}$ that go from R_1 until R_5 (See Table 1).

2.4 Performance Indicators for RoI

To measure the performance of the algorithms, there are indicators such as spacing, generational distance, or hypervolume. Still, these indicators are used when the objective is to find the complete Pareto front. On the other hand, if the objective is to find only a fragment of the front according to the preferences of the DM, that is, the RoI, these indicators would not be suitable.

There are other indicators, considering the solutions that maximize the preferences of the DM, such as those based on distance, which measure the quality in terms of similarity between X* and the approximation to the RoI (A-RoI). When talking about X*, it refers to the last set of solutions of an algorithm and the following indicators can be used [16]:

- **Minimum Euclidean.** This indicator is the Euclidean distance between the closest solution from X* to the A-RoI.
- **Minimum Tchebycheff**. This indicator is the Tchebycheff distance between the closest solution from X* to the A-RoI.

2.5 Coevolution

Coevolution is a reciprocal evolutionary change between interacting species driven by natural selection [17]. Within the computational area, coevolution consists of processes or algorithms that have a mixing relationship with other algorithms, leading them to have reciprocal evolutionary changes between the participants.

Much of the work on coevolutionary algorithms has focused on two types: cooperative systems [18, 19] and competitive systems [20, 21]. These two types of approaches can be subclassified as follows (Fig. 1).

In cooperative coevolutionary algorithms, individuals are rewarded when they perform well. Hence, each subpopulation represents a part of the problem, and the subpopulations gradually evaluate other competing parts of the overall problem and eventually collaborate with other species [5] (Fig. 2).

- **Based on variable decomposition**: Each decision variable is assigned to a population of species, and each population of species optimizes one or more decision variables in tandem.
- **Based on objective function decomposition**: Each objective function of the problem is assigned to a species population, and all populations cooperate to approximate the entire Pareto front.

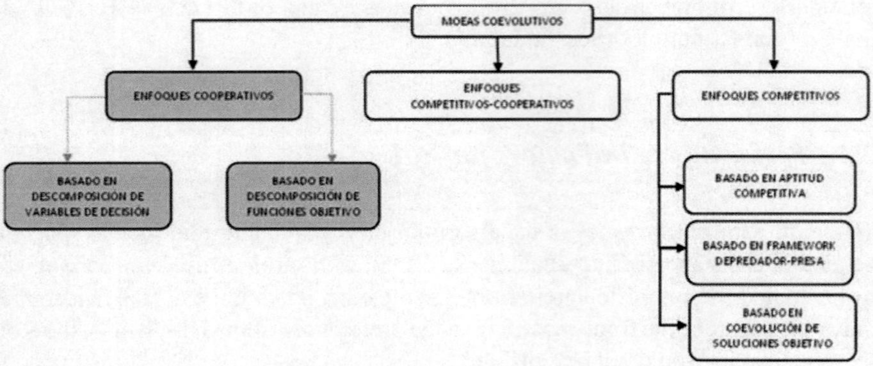

Fig. 1 Taxonomy of coevolutionary MOEAS [9]

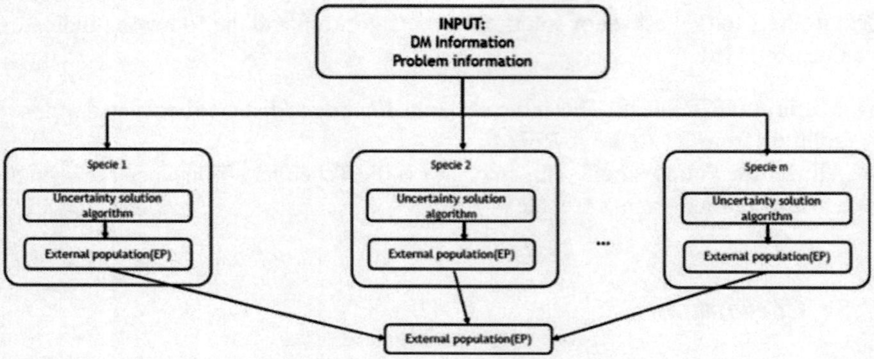

Fig. 2 Cooperative coevolutionary solution proposal with outranking model and uncertainty

3 Solution Methodology

In this work, a cooperative coevolutionary algorithm is proposed with an outranking method with uncertainty, which enters the information of the DM and the problem. This algorithm works with a division of objectives and decision variables in each species, which handles scalability and many objectives. In such a way, if the algorithm makes each subpopulation evolve fragments of the problem, it will allow the problem to be scalable or manageable.

The introduction of coevolution in MOEA/D occurs naturally when the neighborhood is updated, as this step updates the general population based on the results that have been achieved so far.

The MOEA/D update has a small modification, initially, to update the neighborhood it was with the Tchebycheff function, in this work Tchebycheff was added to the neighborhood update combined with the interval bridging model, where the DM preferences are values with uncertainty, in addition to the weights w.

Having a population with N individuals, the first thing the CO-MOEA/D does is divide the population into subspecies (N_1, N_2, ..., N_k). Then, each subspecies N defines a criterion (C_i), such that evolution is considering a subset of decision variables. If Solution S has V decision variables, each subspecies Si considers a subset. Each subspecies works the same as the original MOEA/D, but instead of working with the entire population, CO-MOEA-D works with V' variables. The weight vectors are divided equally among the subspecies S, and the subspecies are fixedly associated with the weight vectors. Thus, the genetic crossover and mutation operators know which subspecies they are working with and thus create the new solutions in each subspecies, which will form part of the general External Population (EP), which is where the EP of each subspecies joins.

4 Experimentation and Results

This section presents the experimental validation of the proposed algorithm, which includes uncertainty and an outranking model within a coevolutionary algorithm. The experimentation consisted of comparing the MOEA/D/O and CO-MOEA/D algorithms, with the DTLZ1 problem being the test instance.

Twenty experiments were carried out for each algorithm, 100,000 evaluations, with a population of 100 individuals, the size of the neighborhood T = 10, the selection of parents was random, the cross was Simulated Binary Crossover (SBX) with a probability of 1.0, and the mutation used was the polynomial with a probability of 0.14. The DM preferences for both algorithms were also manually configured. The credibility threshold: [0.510, 0.750], majority threshold [0.51, 0.67] and dominance threshold [0.51, 0.51].

On the other hand, each algorithm was tested with seven variants that vary the updating of the neighborhood; the variants are VAR1 to VAR6 (97,98,100,104,112 and 127), variant seven is the MOEAD (original 96). The configurations described above were run for each of the mentioned variants. In addition, another parameter that is configured is the number of objectives in 5, 7, and 10 objectives, together with 9 and 100 variables combined with each number of objectives, that is, 5 objectives with 9 variables, 5 objectives with 100 variables, and so on. Successively for the minimum Euclidean indicator (See example in Table 2) and for the minimum Tchebycheff.

Table 2 represents the results of the minimum Euclidean of the MOEA/D/O algorithm with 5 objectives and 9 variables. Each column represents a variant, and the results shown are the minimum Euclidean by experimentation. At the end of each column, the smallest solution of the minimum Euclidean of the variant is shown (indicating that it is the closest to the RoI, that is, the best solution per variant).

This process in Table 2 was carried out for each combination of algorithms with objectives and variables, both for the minimum Euclidean indicator and the minimum Tchebycheff.

Then the solutions for each variant were attached, comparing the solutions by objective, variable, indicator, and algorithm (*Table 3*). In variant 3, which are the

Table 2 Minimum Euclidean for each variant of 5 objectives- 9 variables of MOEA/D/O

MOEA/D/O							
5 objectives-9 variables							
Minimum euclidean							
EXP	MOEAD	VAR1	VAR2	VAR3	VAR4	VAR5	VAR6
1	9.71733907	0.559778974	7.633350514	8.901144875	0.372402592	0.030761461	0.324716027
2	10.46302338	0.0298646	1.667021342	0.031729727	0.097904536	8.588222231	0.279958427
3	9.487296949	0.132609717	0.042503177	1.198404466	0.405252963	2.320579996	0.382775002
4	0.03383936	0.127156483	0.118871792	0.580385061	0.405252963	0.330826161	8.883115491
5	0.689916624	0.260255366	8.845697169	0.035478683	0.104889655	0.682521927	0.111669104
6	0.278997087	0.304979037	0.116299091	0.683532599	0.079284466	0.688016988	0.139960904
7	0.363073456	0.350513486	0.037708218	0.021429037	0.030355266	0.133178026	8.848821565
8	0.031401641	0.266703417	0.113459603	2.461872168	0.109874249	0.08854686	0.379279271
9	7.782979278	0.038984838	8.845901563	0.077289455	0.229450325	0.085099828	0.088353614
10	0.397184232	7.397658435	0.090663486	0.690077717	0.179118023	0.099880318	0.136384457
11	6.702406874	8.857473974	0.68259124	1.114707299	6.633230113	0.103945583	0.093898726
12	0.422458156	0.682338073	0.132064944	0.106496987	0.141941741	3.298667186	0.111229528
13	0.355132964	0.218171303	0.144079067	0.029745016	0.814284118	0.118618105	0.251778875
14	2.892679425	0.031536794	0.15840155	3.265199116	0.029944744	0.599847155	0.253045399
15	0.389491764	0.273510647	0.123661839	0.055434305	0.333208305	0.102520601	0.131136315
16	7.436975226	0.682523315	0.101906642	2.568474311	0.100972495	1.06175883	0.689077277
17	1.462955921	7.177996061	0.188003407	9.494063591	0.087476409	0.307073405	0.163407979
18	6.786962259	0.382536263	0.119760962	15.79742118	0.701099249	0.30195285	0.050008362

(continued)

Table 2 (continued)

MOEA/D/O							
5 objectives-9 variables							
Minimum euclidean							
EXP	MOEAD	VAR1	VAR2	VAR3	VAR4	VAR5	VAR6
19	0.685055409	0.218181653	8.849505775	0.154658073	0.081646807	0.122002362	0.093210956
20	0.559778974	0.178336793	8.867277111	0.052730131	0.409425939	0.030463183	0.029761574
MIN	0.031401641	0.0298646	0.037708218	0.021429037	0.029944744	0.030463183	0.029761574

results that favored our algorithm, it can be observed that, with a more significant number of objectives and variables, the CO-MOEA/D algorithm is closer to the RoI than the MOEA/D/O algorithm.

On the contrary, in VAR1, only in 7 objectives with 9 variables, the CO-MOEA-D algorithm is closer to the RoI than MOEA/D/O. Still, in other situations, the results are not very significant between both algorithms, as in the case of 10 objectives with 9 variables, for both indicators, there is little difference between the distances (Table 4).

Table 3 Experimentation of minimum Euclidean and minimum Tchebycheff of the VAR3

VAR3				
m	Variables	Indicator	Algorithm	DTLZ1
5	9	Min. Euclid	MOEA-D/O	0.02142904
			CO-MOEA-D	17.2527625
		Min. Tchev	MOEA-D/O	0.01723951
			CO-MOEA-D	15.7584624
	100	Min. Euclid	MOEA-D/O	838.301972
			CO-MOEA-D	2054.09583
		Min.Tchev	MOEA-D/O	541.200731
			CO-MOEA-D	1198.53171
7	9	Min. Euclid	MOEA-D/O	0.1458483
			CO-MOEA-D	0.14697418
		Min. Tchev	MOEA-D/O	0.08546923
			CO-MOEA-D	0.08746853
	100	Min. Euclid	MOEA-D/O	672.489006
			CO-MOEA-D	861.743207
		Min.Tchev	MOEA-D/O	451.857122
			CO-MOEA-D	861.743199
10	9	Min. Euclid	MOEA-D/O	0.08486119
			CO-MOEA-D	0.09500747
		Min. Tchev	MOEA-D/O	0.06141803
			CO-MOEA-D	0.05709327
	100	Min. Euclid	MOEA-D/O	654.239235
			CO-MOEA-D	421.450308
		Min.Tchev	MOEA-D/O	1436.10623
			CO-MOEA-D	737.328056

Table 4 Experimentation of the minimum Euclidean and minimum Tchebycheff of the VAR1

VAR1				
m	Variables	Indicator	Algorithm	DTLZ1
5	9	Min. Euclid	MOEA-D/O	0.0298646
			CO-MOEA-D	29.07666133
		Min. Tchev	MOEA-D/O	0.025183938
			CO-MOEA-D	175.4240624
	100	Min. Euclid	MOEA-D/O	657.3979558
			CO-MOEA-D	1231.056145
		Min.Tchev	MOEA-D/O	449.571058
			CO-MOEA-D	729.219219
7	9	Min. Euclid	MOEA-D/O	0.122366405
			CO-MOEA-D	0.08048422
		Min. Tchev	MOEA-D/O	0.078209674
			CO-MOEA-D	0.0573
	100	Min. Euclid	MOEA-D/O	683.6030009
			CO-MOEA-D	975.4348012
		Min.Tchev	MOEA-D/O	415.065164
			CO-MOEA-D	975.4347937
10	9	Min. Euclid	MOEA-D/O	0.092027702
			CO-MOEA-D	0.125661125
		Min. Tchev	MOEA-D/O	0.059789676
			CO-MOEA-D	0.085009836
	100	Min. Euclid	MOEA-D/O	609.6414475
			CO-MOEA-D	823.693297
		Min.Tchev	MOEA-D/O	405.3842319
			CO-MOEA-D	1011.769605

5 Conclusions

- We worked with the CO-MOEA/D algorithm, which provides a new combination of coevolution, uncertainty, and an outranking model.
- The cooperative coevolutionary algorithm was used to handle scalability since solving the problem uses more than five decision variables and more than ten objective functions.
- Uncertainty is present in the outranking model in the credibility, majority, and dominance thresholds; this information is used as an aid for the decision maker to select the best decision and is represented as interval mathematics.

- The results obtained up to this moment give evidence that when the number of objective functions and variables increases, the CO-MOEA/D algorithm begins to show an improvement in the quality of the solutions as it approaches the RoI.

Acknowledgements The authors thanks CONACyT for the support granted through the Scholarship for Postgraduate Studies with CVU 960719, and the projects no. 3058, A1-S-11012. Moreover, we thank the Laboratorio Nacional de Tecnologías de la Información (LANTI), and acknowledge the support of ITCM, and the TECNM project 14612.22-P.

References

1. Zapotecas, S., Coello, C., Aguirre, H., & Tanaka, K. (2019). A review of features and limitations of existing scalable multiobjective test suites. *IEEE Transactions on Evolutionary Computation, 23*(1), 130–142. https://doi.org/10.1109/TEVC.2018.2836912
2. Antonio, L. M., & Coello, C. A. C. (2013). Use of cooperative coevolution for solving large scale multiobjective optimization problems. In *2013 IEEE Congress on Evolutionary Computation* (pp. 2758–2765). IEEE.
3. Antonio, L. M., Coello, C. A. C., Morales, M. A. R., Brambila, S. G., González, J. F., & Tapia, G. C. (2020). Coevolutionary operations for large scale multi-objective optimization. In *2020 IEEE Congress on Evolutionary Computation (CEC)* (pp. 1–8). IEEE.
4. Tse, G. T., Teo, J. T. W., & Hui, K. L. (2007). Performance scalability of a cooperative coevolution Multiobjective Evolutionary Algorithm. In: *2007 International Conference on Computational Intelligence and Security* (CIS'07), 15–19 December 2007, Harbin, Heilongjiang, China.
5. Fernández, E., Figueira, J. R., & Navarro, J. (2019). An indirect elicitation method for the parameters of the ELECTRE TRInB model using genetic algorithms. *Applied Soft Computing, 77*, 723–733. https://doi.org/10.1016/J.ASOC
6. Balderas, F., Fernández, E., Gómez, C., Cruz, L. Y., & Rangel, N. (2016). *International Journal of Combinatorial Optimization Problems and Informatics, 7*(3), 101–118). ISSN: 2007–1558.
7. Ruiz, A. B., Saborido, R., & Luque, M. (2015). A preferencE-based evolutionary algorithm for multiobjective optimization: The weighting achievement scalarizing function genetic algorithm. *Journal of Global Optimization, 62*(1), 101–129.
8. Balderas, F., Fernandez, E., Gomez-Santillan, C., Rangel-Valdez, N., & Cruz, L. (2019). An interval-based approach for evolutionary multi-objective optimization of project porfolios. *International Journal of Information Technology & Decision Making, 18*(04), 1317–1358.
9. Fernandez, E., Rangel-Valdez, N., Cruz-Reyes, L., Gomez-Santillan, C., & Coello Coello, C. A. (2021). Preference incorporation into MOEA/D using an outranking approach with imprecise model parameters. *Social Science Research Network, 2021*, 1–24. https://doi.org/10.2139/ssrn.3960041
10. Tseng, M. L., Wang, R., Chiu, A. S., Geng, Y., & Lin, Y. H. (2013). Improving performance of green innovation practices under uncertainty. *Journal of cleaner production, 40*, 71–82.
11. Liu, R., Wang, R., Feng, W., Huang, J., & Jiao, L. (2016). Interactive reference region based multiobjective evolutionary algorithm through decomposition. *IEEE Access, 4*, 7331–7346.
12. Moore, R. E. (1979). Methods and applications of interval analysis. *Society for Industrial and Applied Mathematics.*
13. Ishihuchi, H., & Tanaka, M. (1990). Multiobjective programming in optimization of the Interval objective function. *European Journal of Operation Research, 48*, 219–25. Holland, J. H. (1975). *Adaptation in Natural and Artificial Systems*, University of Michigan Press.

14. Yao, S., Jiang, Z., Li, N., Zhang, H., & Geng, N. (2011). A multi-objective dynamic scheduling approach using multiple attribute decision making in semiconductor manufacturing. *International Journal of Production Economics, 130*(1), 125–133.
15. Roy, B. (1991). The outranking approach and the foundations of ELECTRE methods. *Theory and Decision, 31*(1), 49–73. https://doi.org/10.1007/978-3-642-75935-28
16. Castellanos, A., Cruz-Reyes, L., Fernández, E., Rivera, G., Gomez-Santillan, C., & Rangel-Valdez, N. (2022). Hybridisation of swarm intelligence algorithms with multi-criteria ordinal classification: A strategy to address many-objective optimisation. *Mathematics, 10*(3), 322. https://doi.org/10.3390/math10030322
17. Thompson, J. (2005). *The Geographic Mosaic of Coevolution.* University of Chicago Press. https://doi.org/10.7208/9780226118697
18. Qi, Y., Li, X., Yu, J., & Miao, Q. (2019). User-preference based decomposition in MOEA/D without using an ideal point. *Swarm and Evolutionary Computation, 44*, 597–611. https://doi.org/10.1016/j.swevo.2018.08.002.
19. Tan, B., Ma, H., Mei, Y., & Zhang, M. (2020). A cooperative coevolution genetic programming hyper-heuristic approach for on-line resource allocation in container-based clouds. *IEEE Transactions on Cloud Computing*.
20. Vu, V. T., Bui, L. T., & Nguyen, T. T. (2020). A competitive co-evolutionary approach for the multi-objective evolutionary algorithms. *IEEE Access, 11*. https://doi.org/10.1109/access.2020.2982251.
21. Harris, S. N., & Tauritz, D. R. (2021). Competitive coevolution for defense and security: Elo-based similar-strength opponent sampling. In *Proceedings of the Genetic and Evolutionary Computation Conference Companion* (pp. 1898–1906).

Experimental Proposal with Mallows Distribution Applied to the Mixed No-Idle Permutation Flowshop Scheduling Problem

E. M. Sánchez Márquez, M. Ornelas-Rodríguez, H. J. Puga-Soberanes, Pérez-Rodríguez, Ricardo, and Martin Carpio

1 Introduction

One of the main conflicts in a continuous production plant is the timely and appropriate orders received daily, which are not uniform and involve a large amount of human and material resources. The No-Idle Permutation Flowshop Scheduling Problem (NPFSP) has as its main characteristic that the machines do not allow idle time between two jobs during production, from this it searches for a sequence that respects restrictions and minimizes the total completion time known as makespan. This type of problem appears in production environments where the times and operating costs of the machines are high and turning off the machines after the process has started is not profitable, therefore a machine must process all its assigned jobs without interruptions. However, in real-life production, shops there is a need to contemplate both active and inactive machines, which gives rise to the Mixed No-Idle Permutation Flowshop Scheduling Problem (MNPFSP) that allows both types of machines. This problem searches a sequence that minimizes the makespan (maximum completion time) and was raised by [1], and was classified as NP-Hard.

This problem considers two types of machines: no-idle, which does not allow idle time between two jobs, and idle, which does allow it. Figure 1 shows an example of how the problem MNPFSP works, it consists of 2 jobs and 5 machines, of which two are classified as no-idle, (they do not allow idle time between jobs) and 3 machines as idle. The generated sequence, shown as the permutation, represents the order of jobs that get into the production line and each machine. The size of each icon of a job

E. M. Sánchez Márquez · M. Ornelas-Rodríguez · H. J. Puga-Soberanes · M. Carpio (✉)
Graduate Program Division, National Technology of México/León Institute of Technology, León, Guanajuato, Mexico
e-mail: juanmartin.carpio@leon.tecnm.mx

Pérez-Rodríguez · Ricardo
National Council of Science and Tecnology of Mexico (CONACYT), Mexico City, Mexico

O. Castillo and P. Melin (eds.), *Hybrid Intelligent Systems Based on Extensions of Fuzzy Logic, Neural Networks and Metaheuristics*, Studies in Computational Intelligence 1096,
https://doi.org/10.1007/978-3-031-28999-6_28

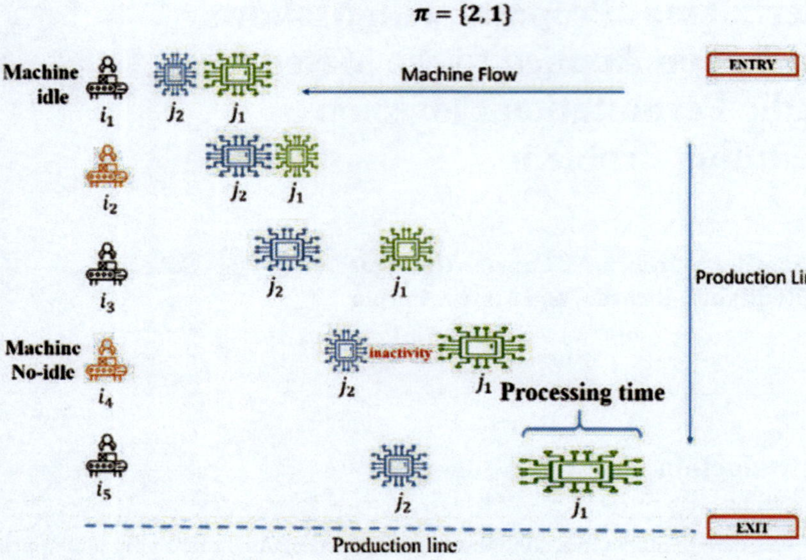

Fig. 1 MNPFSP

j, indicates the processing time on each machine. The space between icons in a job illustrates the magnitude of idle time between two consecutive jobs, for example, on machine 4 it is shown idle time between j is idle greater than on machine 3 with the same jobs. In the case of a no-idle machine i_1, i_4, the idle times seek to be reduced or eliminated. This is not necessary in the case of the remaining machines classified as idle.

Some examples of industries where this type of problem is applied are integrated circuit production, ceramic production, fiberglass processing, and the steel industry, among others.

This work aimed to explore the possibility of reducing the execution time and computational cost in an Estimation Distribution Algorithm (EDA) coupled with the Generalized Mallows Model (GMM) by proposing intuitive and practical changes in the calculation of the EDA parameters [2].

In the classical learning model [3], the Borda algorithm has been used to obtain the central permutation as a position parameter and the Maximum Likelihood Estimation (MLE) method to estimate the spread parameter. Considering the objectives of the problem MNPFSP, it was proposed an EDA-GMMp algorithm was proposed, unlike the classic EDA-GMM that we will denote as EDA-GMMc, it was proposed the estimation of the position and spread parameters in an intuitively and practically search solutions. Select the position parameter, the performance function of the EDA is used and for the spread parameter, a simple function is proposed from the observation of the relationship between the mean sampling distance and the spread parameter found by the MLE.

The rest of the article is organized as follows: Sect. 2 describes the Generalized Mallows Model (GMM), Sect. 3 contains the EDA-GMMp proposal in general; Sect. 4 presents the proposed learning model; Sect. 5 the differences between the proposal and the GMM model are briefly shown; Sect. 6 presents the experimental development and Sect. 7 shows the comparison and statistical analysis of the results obtained by the algorithms; Finally, conclusions and future work are given.

2 Generalized Mallows Model (GMM)

Originally the EDAs were designed for problems with the domain of real or integer values, however, over time and due to the existence of problems based on permutations as is the case of MNPFSP, models that could be adapted were generated.

The Generalized Mallows Model (GMM) is an improved version of the Mallows Model (MM) by Fligner and Verducci in 1986 [4]. In this model the probability value for each permutation depends on two parameters: the central permutation σ_0, and the spread parameter θ, it also, calculates the distance $D(\sigma, \sigma_0)$ between the central permutation σ_0 and any other permutation σ. The GMM model requires a metric that decomposes into multiple terms, in this way the distance is defined as: $D(\sigma, \sigma_0) = \sum_{j=1}^{n-1} S_j(\sigma, \sigma_0)$, where $S_j(\sigma, \sigma_0)$ represents the jth position or component of the permutation. Alike, the model uses $n-1$ spread parameters $\theta = (\theta_1, \theta_2, \ldots, \theta_{n-1})$, where each θ_j is associated with a particular position of the permutation. Figure 2 shows the probability distribution of the GMM model for the permutations with different values of the spread parameter [5, 6].

The expression of the GMM model is:

$$P_\sigma = \frac{e^{\left(\sum_{j=1}^{n-1} -\theta_j S_j(\sigma, \sigma_0)\right)}}{\psi(\theta)} \tag{1}$$

where $\psi(\theta)$ is a normalization term.

$$\psi(\theta) = \prod_{j=1}^{n-1} \psi_j(\theta_j) = \prod_{j=1}^{n-1} \frac{1 - e^{(-\theta_j(n-j+1))}}{1 - e^{(-\theta_j)}} \tag{2}$$

According to [3], the best performing metric in Flow Shop Scheduling Problems is the Kendall distance $D_\tau(\sigma, \pi)$, which counts the minimum number of adjacent transpositions needed to bring the permutation σ to the permutation π [7, 8].

This distance is decomposed $S_j(\sigma, \sigma_0)$ terms, applying properties of invariance and simplification is denoted as $V_j\left(\sigma\sigma_0^{-1}\right)$ [1, 9]. If π represents an arbitrary permutation σ_0^{-1}, the distance calculation can be performed according to:

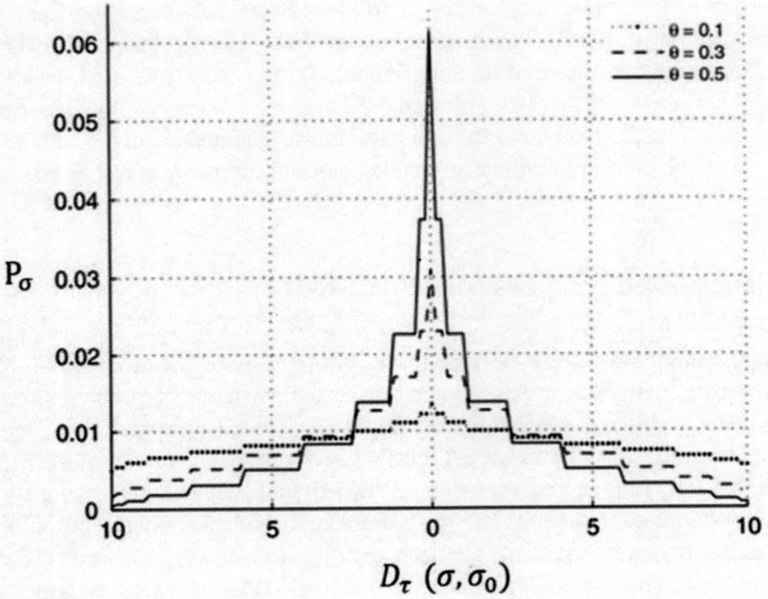

Fig. 2 Probability value for each permutation for different values of the spread parameter θ, [1]

$$V_j(\pi) = \sum_{i=j+1}^{n} I[\pi(j) > \pi(i)] \tag{3}$$

where

$$I = \begin{cases} 1\, if\ \pi(j) > \pi(i)\, is\, true \\ 0\ other\ case \end{cases}$$

$V_j(\pi)$ has a range from 0 to $n - j$ for $1 \leq j < n$ and represents the number of positions to the right with values smaller than $\pi(j)$ in a given permutation. For the above is obtained $D_\tau(\pi) = \sum_{j=1}^{n-1} V_j(\pi)$. Thus (1) is written as:

$$P_\sigma = \frac{e^{\left(\sum_{j=1}^{n-1} -\theta_j V_j\left(\sigma\sigma_0^{-1}\right)\right)}}{\psi(\theta)} \tag{4}$$

The GMM model with the Kendall distance is a “Multistage Ranking Model” or process to build a permutation with $n - 1$ independent components, $V_j\left(\sigma\sigma_0^{-1}\right)$ [10].

This property allows the distribution associated with the permutation σ with distance $V_j\left(\sigma\sigma_0^{-1}\right) = (r_1, \ldots, r_{n-1})$ can be expressed as [3, 9, 11]:

$$P_\sigma = \prod_{j=1}^{n-1} P\left(V_j\left(\sigma\sigma_0^{-1}\right) = r_j\right) \tag{5}$$

where r_j represents a possible value for the position j in V_j.

2.1 Parameter Estimation for GMM

The GMM model depends on the parameters which are the central permutation and the spread parameter respectively.

The central permutation can be obtained exhaustively or heuristically, however, according to the state of the art, the Borda algorithm offers a balance between precision and time for its estimation. In [12] a comparison of various methods for Kemeny's rank rule (central permutation calculus) was performed and it was observed that when the data is generated from the Mallows Model, this algorithm is asymptotically optimal. The Borda algorithm is 10 times faster and cheaper than other algorithms such as Copeland, and CSS, B&B [12].

Borda Algorithm constructs the central permutation σ_0, calculating the average $\tilde{\sigma}(i) = \sum_{s=0}^{m} \sigma_s(i)/m$ per position i of the permutationσ_s, $1 \le s \le m$ in the sample, ordering the elements according to $\tilde{\sigma}(i)$ for to obtain the permutation $\widehat{\sigma}_0$ (Fig. 3) [3, 9, 11, 13].

On the other hand, in the case of the spread parameter, the GMM model uses the principle of Maximum Likelihood Estimation (MLE), where the parameter is obtained from a sample of N permutations $\{\sigma_1, \ldots, \sigma_N\}$ seeking to maximize the likelihood function. The likelihood function for the model is:

$$L(\sigma_1, \ldots, \sigma_N|\theta, \sigma_0) = P(\sigma_1, \ldots, \sigma_N|\theta, \sigma_0) = \prod_{i=1}^{N} \frac{1}{\psi(\theta)} e^{\sum_{j=1}^{n-1} -(\theta_j \overline{V}_j)} \tag{6}$$

where

Fig. 3 Borda algorithm [1]

Algorithm 1 Borda Algorithm for Calculating $\hat{\sigma}_0$

Input: A sample of N permutations $\{\sigma_1, \ldots, \sigma_N\}$
For j 1 **to n**
  $\pi(j) = \sum_{i=1}^{N} \sigma_i(j)/N$;
end For
visited = {};
For j 1 **to n**
  $min\_i \leftarrow \arg\min_i \{\pi(i) | i \notin visited\}$;
  $\hat{\sigma}_0(min\_i) = j$;
  $visited \leftarrow visited \cup min\_i$;
end For
Output: $\hat{\sigma}_0$

$$\overline{V}_j = \frac{1}{N}\sum_{i=1}^{N} V_j(\sigma_i\sigma_0^{-1}) \tag{7}$$

In this case, If $F(\theta) = \ln(L(\sigma_1, \ldots, \sigma_N|\theta, \sigma_0))$ then:

$$F(\theta) = ln(L(\sigma_1, \ldots, \sigma_N|\theta, \sigma_0)) = ln\left(\prod_{i=1}^{N} \frac{e^{\sum_{j=1}^{n-1} -(\theta_j \overline{V}_j(\sigma\sigma_0^{-1}))}}{\psi(\theta)}\right)$$

$$= \sum_{i=1}^{N} (\frac{e^{\sum_{j=1}^{n-1} -(\theta_j \overline{V}_j(\sigma\sigma_0^{-1}))}}{\psi(\theta)}) = N\ln(\frac{e^{\sum_{j=1}^{n-1} -(\theta_j \overline{V}_j(\sigma\sigma_0^{-1}))}}{\prod_{j=1}^{n-1}\psi_j(\theta_j)})$$

$$= N\Big[\ln(e^{\sum_{j=1}^{n-1} -\theta_j \overline{V}_j}) - \ln(\prod_{j=1}^{n-1}\psi_j(\theta_j)\Big]$$

$$= N\left[\sum_{j=1}^{n-1} -\theta_j\overline{V}_j - \sum_{j=1}^{n-1} ln\psi_j\theta_j\right] = -N\left[\sum_{j=1}^{n-1}[\theta_j\overline{V}_j + ln(\psi_j(\theta_j))\right]$$

By differentiating and equating zero

$$F'(\theta) = -N\sum_{j=1}^{n-1}\left[\overline{V}_j + \frac{\psi'_j(\theta_j)}{\psi_j(\theta_j)}\right] = 0$$

$$\overline{V}_j + \frac{\psi'_j(\theta_j)}{\psi_j(\theta_j)} = 0$$

where

$$\frac{\psi'_j(\theta_j)}{\psi_j(\theta_j)} = \frac{n-j+1}{e^{\theta_j(n-j+1)}-1} - \frac{1}{e^{\theta_j}-1}$$

So that

$$\overline{V}_j = \frac{1}{e^{\theta_j}-1} - \frac{n-j+1}{e^{(n-j+1)\theta_j}-1} \tag{8}$$

From (8), $\boldsymbol{\theta_j}$ is estimated by applying strategies based on the Newton–Raphson method [3, 9, 14].

With the estimated parameters for the distribution, the new individuals are generated in the EDA application (9).

$$P\big(V_j\big(\sigma\sigma_0^{-1}\big) = r_j\big) = \frac{e^{(-\theta_j r_j)}}{\psi_j(\theta_j)}; r_j \epsilon \{0, \ldots, n-j\} \tag{9}$$

3 Algorithm Proposal EDA-GMMp

Distribution Estimation Algorithms perform stochastic optimization by exploring the solution space using probabilistic models. To make exploration more efficient, two aspects are taken into account: Aptitude evaluation and the construction and improvement of models. For this, there are several techniques proposed in [15]. In this work, the construction learning models have constructed knowledge problems and learning from experience.

Figure 4 shows the flowchart for the EDA-GMMp Algorithm. In which the construction process of the learning model is highlighted in red, emphasizing that this work focuses mainly on the estimation of the position parameter and the spread parameter. Its characteristics are as follows:

- Population size: $10n$,
- Number of generations: 100,

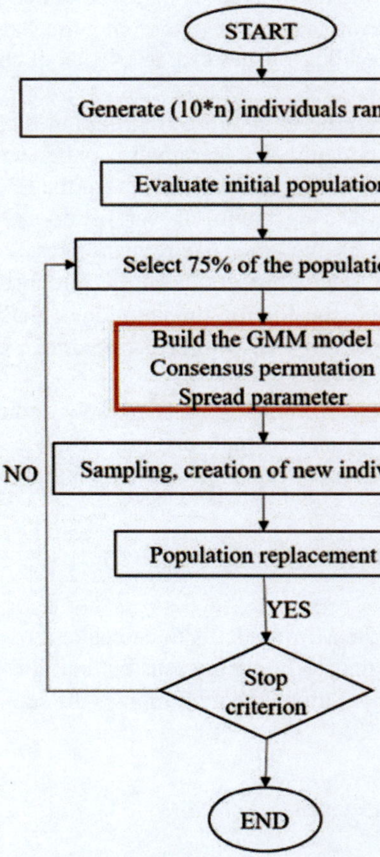

Fig. 4 Flowchart for algorithm EDA-GMMp

– Population selection percentage: 75%,
– Sampling: $populationsize - 1$.

The EDA-GMMp algorithm with the position and spread parameter estimation proposal seeks to reduce the computational cost of using the Borda Algorithm and the MLE method.

The following section presents the details for the estimation of parameters in the proposal applied to the GMM model.

4 Proposed Learning Model

The main difference between an EDA and an evolutionary algorithm EA is the way they generate new individuals. EDAs look for correlations between variables explicitly; this is where the concept of the learning model comes in. This is a formal framework that allows estimating the probability distribution of the set of individuals in a relatively easy way, however, the process of defining the learning model can be complicated and tedious depending on the characteristics of the variables of the problem to solve [16].

The construction of the learning model in a Distribution Estimation Algorithm has two aspects: learning the structure and learning the parameters for the structure [15]. In our case, we focus on learning the parameters for the GMM model.

In the case of the EDA-GMMc algorithm, the parameters are estimated looking for the best fit to the sample, on the other hand, in the present proposal we seek to estimate parameters for the search process of the EDA algorithm; for which the central permutation is assigned considering the aptitude of the individual and the spread parameter from a simple function with behavior similar to that obtained by the MLE method in the usual method.

The proposal seeks to reduce the number of calculations and the execution time, in addition to estimating the parameter simple and practical way.

4.1 Position Parameter

The first step for learning in the MM model is to calculate the central permutation σ_0 from a sample of permutations. The theory tells us that in the MLE the central permutation is given by that permutation that minimizes the sum of the distances to the sample of permutations, (10).

$$\sigma_0 = min\frac{1}{N}\sum_{i=1}^{N} d(\sigma_i, \sigma^*) \tag{10}$$

In the above, $d(\sigma, \sigma^*)$ represents the Kendall distance between two permutations σ_i, σ^*. Estimating the central permutation is a problem, known as the Kemeny Rank Aggregation Problem, which is classified as NP-hard [10, 17]. Because the estimation of the central parameter is expensive, it has been decided to obtain an approximation, with a recursive approach [18]. In [12] it is concluded that the Borda algorithm offers a good balance between time and effectiveness, this algorithm has been applied in the GMM in the most recent works [3, 11].

In this proposal, the permutation with the best fitness in the sample has been selected as the central permutation. This idea has been proposed in the theory [19], but it has not been implemented but his work, it has been taken up again because it seems intuitively more reasonable and practical in the operation of the EDA; because it reuses the calculation to evaluate individuals within the EDA process.

4.2 Spread Parameter

In the EDA-GMMc algorithm, as indicated in Sect. 2.1, when applying the MLE, the expression of (8) is obtained where the spread parameter is implicitly expressed, so to obtain its value, has applied the numerical method of Newton Raphson [3, 9, 13].

Looking to estimate the spread parameter without recourse to numerical calculation as has been done in previous works, the simplest relation is proposed that has a similar behavior between the mean of the Kendall distances [9, 19] (11) and the spread parameter obtained by applying the MLE. Figure 4 shows the behavior of the relationship between the parameters mentioned both when applying the MLE according to (8) and the one proposed in this work (12).

$$\overline{V}_j = \sum_{i=1}^{N} V_j(\sigma\sigma_0^{-1})/N \tag{11}$$

$$\overline{V}_j = \frac{1}{\theta_j} \tag{12}$$

The proposal establishes a practical and simple way to estimate the spread parameter that will be applied in the GMM in the EDA algorithm scheme for the search for better solutions to the MNPFSP problem. An important reason that allowed us to make this proposal is that, even though the spread parameter estimated with the MLE allows us to establish the most appropriate probability distribution for the selected permutation sample, nevertheless it cannot guarantee that it is the best parameter for the search process in the EDA scheme. In the graphs of Fig. 5 it can be seen that both relationships have an asymptotic inverse behavior between their parameters, but that, for the same value of average Kendall distance, the spread parameter of the proposal is greater, which would favor a broader search around the central permutation.

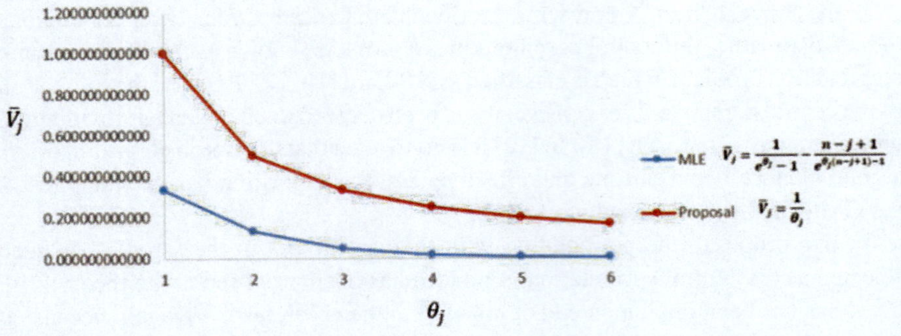

Fig. 5 Behavior of the MLE function compared to the proposal to estimate the spread parameter

The graph of (8) provides the behavior pattern of the variables of the mean distance and the spread parameter as a result of applying MLE, this allows proposing a simple function to directly estimate values of the spread parameter for each value of average distance without recourse to further calculations.

The proposal does not seek to adjust to the values provided by the MLE application, but rather takes them to propose a function that generates a similar behavior pattern to obtain an advantage in the operation of the EDA. In the following sections, the differences, the experiments, and the results obtained are pointed out.

5 Differences Between EDA-GMMc and EDA-GMMp

In this section, the main differences between EDA-GMMc and EDA-GMMp are presented, considering the way of estimating the parameters. Figure 6 shows the differences to estimate the central permutation.

Figure 7 shows the differences to estimate the spread parameter.

Based on the above and to test the proposal, the following experiments were designed.

Central permutation	
Borda Algorithm	**Permutation with better fitness.**
1. Calculates the central permutation from an average permutation obtained from the sample. 2. It is an additional algorithm that must be used in the EDA algorithm, adding overall execution time. 3. It does not consider the fitness of individuals. 4. The central permutation may not be found in the sample.	1. Calculates the central permutation using a fitness function. 2. Reduces the number of calculations, by using the fitness of the EDA algorithm process. 3. Ensures that the selected central permutation is found in the sample. 4. It helps to speed up the search, since from the beginning the search is focused on the best individual in the sample.

Fig. 6 Differences to estimate the position parameter

Spread parameter	
MLE	**Proposal $\theta = \frac{1}{V_j}$**
1. Estimates the spread parameter that best fits the sample. 2. It can be applied for any distance metric of permutations. 3. It requires numerical methods.	1. Estimates the spread parameter as the inverse of the sample mean distance 2. It is suitable for using the Kendall distance. 3. Does not require numerical methods 4. Extends the exploration range.

Fig. 7 Differences for to estimate the spread parameter

6 Experiments

Table 1 shows the cases where the combinations of position and spread parameters were applied in the GMM. In the case of the EDA-GMMp algorithm, the effect is analyzed when applying both parameters of the proposal simultaneously, in Case 3, only the proposed position parameter is applied and in Case 4, only the proposed spread parameter is applied. These last two serve to contrast the effect when applying the parameters of the proposal separately.

For the experiments, a sample of small and large instances was generated taken from place http://soa.iti.es, where specific instances for the problem MNPFSSP were found it used in recent works [1, 13]. The sample was generated from the instance I_3_500_50_1 which contains 500 jobs and 50 machines.

Small and large instances were designed consisting of n jobs and m machines. In the case of small instances, 27 were generated with $n \in \{10, 15, 20, 25, 30, 35, 40, 45, 50\}$ and $m \in \{3, 6, 9\}$. In the case of large instances, 30 instances were generated choosing n from a range of 50 to 500 and $m \in \{5, 10, 15\}$.

Each instance contains the following information:

- Number of jobs and machines.
- Machine processing times for each job.
- Idle machines and no-idle machines are identified with 1 or 0 respectively.

The structure of the instances is shown in Fig. 8.

An example of a complete instance is shown in Fig. 9.

All algorithms were implemented in JAVA and compiled with Intellij IDEA Community Edition 2018. The experiments were run on an Asus with an Intel i7-6700HQ processor and 8 GB of RAM running Windows 10.

Table 1 Different combinations of parameters to the GMM model

Algorithm	EDA-GMMc	EDA-GMMp	Case 3	Case 4
Position parameter	Borda algorithm	Best fitness	Best fitness	Borda algorithm
Spread parameter	MLE	$\theta_j = \frac{1}{V_j}$	MLE	$\theta_j = \frac{1}{V_j}$

#Jobs	#Machines										
20	6										
i_0	Processing Time	i_1	Processing Time	i_2	Processing Time	i_3	Processing Time	i_4	Processing Time	i_5	Processing Time
IDLE 1		NO IDLE 0		IDLE 1		NO IDLE 0		IDLE 1		NO IDLE 0	
0	66	1	18	2	6	3	57	4	85	5	22
0	63	1	79	2	75	3	99	4	26	5	46
0	99	1	74	2	20	3	77	4	63	5	49
. . .	. . .	. . .	. . .	. . .	. . .	. . .	. . .	. . .	. . .	. . .	. . .

Fig. 8 Information contained in an instance

```
20        6
          0    66    1    18    2    6     3    57    4    85    5    22
          0    63    1    79    2    75    3    99    4    26    5    46
          0    99    1    74    2    20    3    77    4    63    5    49
          0    59    1    31    2    31    3    8     4    77    5    89
          0    74    1    19    2    23    3    79    4    21    5    14
          0    8     1    72    2    77    3    79    4    4     5    65
          0    25    1    34    2    45    3    78    4    3     5    66
          0    26    1    67    2    33    3    58    4    49    5    15
          0    62    1    47    2    39    3    83    4    99    5    44
          0    10    1    67    2    48    3    27    4    53    5    18
          0    95    1    63    2    65    3    43    4    40    5    79
          0    92    1    39    2    5     3    39    4    23    5    61
          0    2     1    95    2    53    3    30    4    67    5    18
          0    62    1    28    2    24    3    22    4    47    5    77
          0    53    1    13    2    73    3    94    4    12    5    9
          0    75    1    15    2    2     3    88    4    10    5    32
          0    92    1    70    2    2     3    77    4    78    5    33
          0    37    1    23    2    9     3    64    4    19    5    28
          0    98    1    94    2    64    3    47    4    46    5    82
          0    51    1    62    2    1     3    12    4    53    5    75
NOIDLE
0
1 0 1 0 1 0
```

Fig. 9 Instance example

From the executions for each instance, the fitness value of each execution was obtained, and the time in seconds of the total average. This information will be presented in the next section for each experiment.

The calculation of the fitness function (makespan) was programmed in Java according to the model proposed in [20].

7 Results

For the tests of the algorithms, 30 executions were carried out for each instance. The average execution time was obtained from the executions, as well as the mean and best case of the results when applying the fitness function. Subsequently, a statistical analysis of the results was performed. The results obtained by EDA-GMMp were compared with the EDA-GMMc algorithm and two other versions of the EDA-GMM algorithm (Case 3 and Case 4).

Tables 2 and 3 show the fitness for small and large instances respectively, by executing the algorithm 30 times, each execution with 100 generations. In both, all the experimental cases appear.

Figure 10 shows the execution times (et) for the small instances. The algorithm was executed in 100 generations, in this graph it can be seen that the version of the algorithm in case 3 has longer execution times and the other algorithms have values close to others.

Table 2 Fitness of small instances, *n* represents the number of jobs and *m* the number of machines. With *the only result not improved by EDA-GMMp is identified

Small instances						
Instance	*N*	*m*	EDA-GMMc	EDA-GMMp	Case 3	Case 4
1	10	3	*526.3	527	536.46	527.63
2	10	6	849.9	**828.3**	853.73	843.56
3	10	9	1266.46	**1226.83**	1251.53	1264.83
4	15	3	831.46	**824.1**	829.4	837.06
5	15	6	1149.66	**1108.53**	1126.7	1153.46
6	15	9	1618.73	**1588.5**	1611.83	1634.66
7	20	3	1169.53	**1166.7**	1170	1171.6
8	20	6	1524.03	**1453.9**	1479	1511.23
9	20	9	2106.06	**2015.1**	2054.7	2109
10	25	3	1453.86	**1447.23**	1452.8	1451.2
11	25	6	1799.8	**1740.3**	1752.86	1787.66
12	25	9	2497.06	**2410.26**	2434	2505.36
13	30	3	1758.4	**1748.23**	1753.6	1758
14	30	6	2133.33	**2080.06**	2088	2114.16
15	30	9	2866.1	**2772.3**	2799.13	2856.16
16	35	3	1973.23	**1949.46**	1950.43	1969.26
17	35	6	2340.3	**2294.26**	2295.43	2316.5
18	35	9	3117.66	**3030.56**	3036.2	3089.7
19	40	3	2183.53	**2165.26**	2170.76	2172.23
20	40	6	2584.56	**2534.26**	2542.53	2571.06
21	40	9	3298.43	**3197.26**	3215.5	3276.3
22	45	3	2515.6	**2500**	2507.83	2506.23
23	45	6	2945.2	**2878.53**	2895	2938.96
24	45	9	3705.9	**3595.06**	3624.76	3685.26
25	50	3	2749.8	**2731.46**	2741.73	2737.83
26	50	6	3205.2	**3139.6**	3147.53	3182.06
27	50	9	3959.7	**3859.86**	3859.7	3936.43

Table 3 Fitness of large instances, *n* represents the number of jobs and *m* the number of machines. EDA-GMMp outperforms all other cases

Large instances

Instance	*n*	*m*	EDA-GMMc	EDA-GMMp	Case 3	Case 4
1	50	5	3964.6	**2915.46**	2930.03	2992.26
2	50	10	4065.66	**3941.8**	3966	4024.96
3	50	15	5546.46	**5378.23**	5409.4	5527.96
4	100	5	5418.23	**5314.7**	5332.6	5366.96
5	100	10	6721.5	**6540.8**	6589.83	6644.96
6	100	15	8266.63	**8036.53**	8075.06	8215.1
7	150	5	7995.83	**7891.46**	7904.76	7963.4
8	150	10	9597.9	**9421.63**	9449.66	9553
9	150	15	11,225.13	**10,955.96**	10,972.43	11,194.2333
10	200	5	10,554.03	**10,448.1**	10,463.86	10,517.4
11	200	10	12,517.83	**12,303.83**	12,315.16	12,457.26
12	200	15	14,003.76	**13,670.53**	13,726.8	13,973.66
13	250	5	12,991.56	**12,874.96**	12,878.36	12,925.76
14	250	10	14,904.93	**14,662.43**	14,682.53	14,859.63
15	250	15	16,864.7	**16,540.86**	16,566.26	16,704.06
16	300	5	15,568	**15,404.33**	15,441.33	15,534.93
17	300	10	17,580.13	**17,286.76**	17,347.03	17,548.86
18	300	15	19,530.26	**19,158.53**	19,147.8	19,516
19	350	5	18,294.13	**18,153.8**	18,157.03	19,488.7
20	350	10	20,151.13	**19,865.16**	19,887.13	20,116.06
21	350	15	22,636.2	**22,275.5**	22,300.83	22,625.5
22	400	5	21,103.1	**20,959.6**	20,987.3	21,071.5
23	400	10	21,112.56	**22,542.63**	22,570.96	22,795.93
24	400	15	25,332.5	**24,946.63**	24,968.1	25,269.9
25	450	5	25,325.73	**23,589.53**	23,604.63	23,684.9
26	450	10	25,504.23	**25,230.6**	25,258.93	25,496.43
27	450	15	27,895.6	**25,214.4**	27,520.06	27,852.2
28	500	5	25,886.46	**25,693.3**	25,731.36	25,861.03
29	500	10	28,191.46	**27,859.13**	27,911.8	28,101.86
30	500	15	30,585.23	**30,152.03**	30,180.8	30,560.73

Figure 11 shows the execution times (et) for large instances, executing the algorithm in a total of 100 generations. Where it is observed that from instances 1 to 18, the cases have similar execution times, but from instance 19 the EDA-GMMc has times below the rest of the cases.

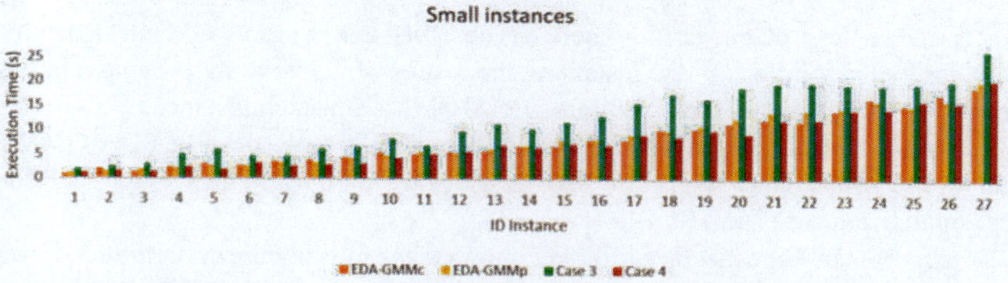

Fig. 10 Execution times for small instances for 100 generations

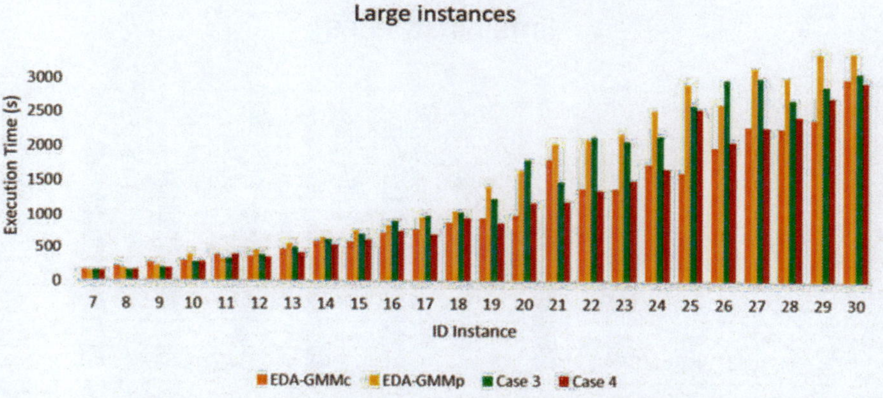

Fig. 11 Execution times for large instances for 100 generations

In Figs. 10 and 11, when running the algorithms 100 generations, contrary to what was expected due to the reduction of calculations in the proposal, the times of EDA-GMMp are shown to be greater than those of EDA-GMMc. This may be because the proposed algorithm performs a more exhaustive search by proposing values of the dispersion parameter greater than those obtained by MLE for the same mean distances (Fig. 12).

To objectively observe the effect of the reduction of time used by any of the algorithms with concerning the best result of EDA-GMMc in 100 generations, the stoppage criterion (Fitness EDA-X) $\leq$ (Fitness EDA-GMMc) was established, where EDA-X represents the algorithm to compare with the reference.

The Time Reduction Percentage (TRP) of the EDA-X algorithm, with concerning the reference, is established in (13), where etEDA-X represents the execution time of EDA-X.

$$TRP = \frac{(etEDA - GMMc) - (etEDA - X)}{etEDA - GMMc} * 100 \tag{13}$$

Based on the previous results where it is observed that the EDA-GMMp algorithm obtains better solutions in the instances, the results of the TPR are presented in the case (Fitness EDA-GMMp) ≤ (Fitness EDA-GMMc), in each instance.

Table 4 shows that EDA-GMMp obtains better TRP results than the EDA-GMMc algorithm with a Time Reduction Percentage of up to 76.8%. The average TRP for all small instances is 55.1%.

In the case of large instances, Table 5 shows the results obtained. Figure 13 shows the Time Reduction Percentage of up to 87.62% and an Average TRP of 60.2% for all large instances, compared to the execution times of the EDA-GMMc algorithm.

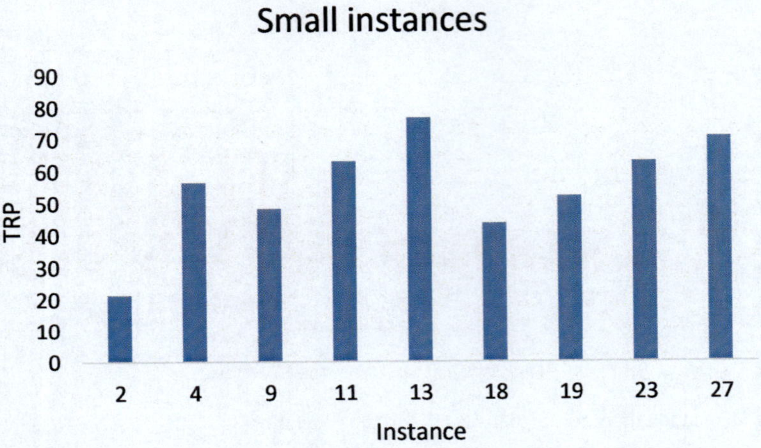

Fig. 12 Time reduction percentage under the condition (Fitness EDA-GMMp) ≤ (Fitness EDA-GMMc) for small instances

Table 4 Execution times under the condition (Fitness EDA-GMMp) ≤ (Fitness EDA-GMMc) for small instances

Small instances					
Instance	EDA-GMMp		EDA-GMMc		TRP
	Execution time (et)	Fitness	Execution time (et)	Fitness	
2	0.8022	808	1.02213	823	21.5
4	1.061	820	2.4423	820	56.6
9	2.417	1966	4.68746	2022	48.4
11	2.43736	1731	6.6098	1754	63.1
13	1.7536	1748	7.5528	1748	**76.8**
18	6.30643	2979	11.20636	3022	43.7
19	5.52713	2161	11.57343	2161	52.2
23	5.84856	2827	15.79303	2909	63
27	5.90056	3823	20.156	3896	70.7

Table 5 Execution times under the condition (Fitness EDA-GMMp) ≤ (fitness EDA-GMMc) for large instances

Large instances					
Instance	EDA-GMMp		EDA-GMMc		TRP
	Execution time (et)	Fitness	Execution time (et)	Fitness	
1	3.67983	2913	14.63943	2938	74.86
5	58.22253	6487	92.3275	6569	36.94
9	27.53273	10,916	222.35526	11,084	**87.62**
10	51.1718	10,451	338.33286	10,504	84.88
14	240.0879	14,648	585.7262	14,742	59.01
18	243.522	19,102	922.8543	19,288	73.61
19	187.67973	18,119	886.8696	18,219	78.84
23	485.81006	22,426	1533.0966	22,647	68.31
27	1973.45043	27,353	2363.10283	27,530	16.49
28	1920.64336	25,612	2443.967	25,712	21.41

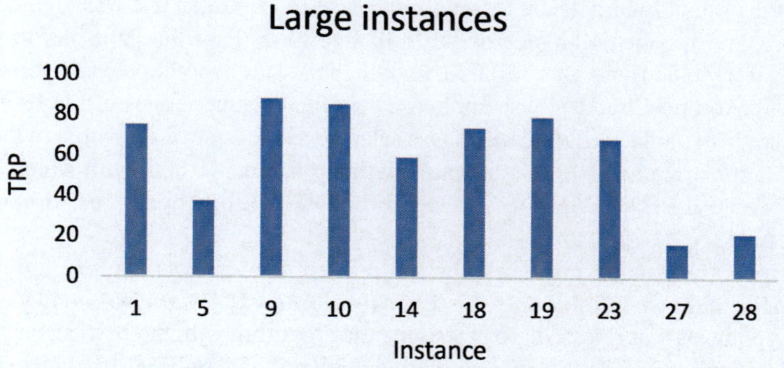

Fig. 13 Time reduction percentage under the condition (Fitness EDA-GMMp) ≤ (Fitness EDA-GMMc) for large instances

Taking into account the results in all instances, the Average TRP was 57.6%.

In the Case 3 algorithm, the Average TRP obtained was 44.4 and 44.6% for small and large instances, respectively.

For the Case 4 algorithm, the Average TPR results obtained were 26.6 and 19.14% for small and large instances, respectively. Table 6 shows the Average TRPs of each of the proposals about to the EDA-GMMc algorithm.

Table 6 Average TRP for all proposals using the EDA-GMMc algorithm as a reference

	Average TRP	
Algorithm	Small instances (%)	Large instances (%)
EDA-GMMp	**55.1**	**60.2**
Case 3	44.4	44.6
Case 4	26.6	19.14

7.1 *Statistic Analysis*

To establish whether the results obtained are statistically significant, non-parametric tests were applied for related samples, since some of the descriptive results in the normality tests obtained values below $\alpha = 0.05$, concluding that no they behave like a normal distribution.

The results of the statistical tests oriented to the analysis of the quality and execution time of the algorithms are presented. The quality tests (Fitness) are presented first and then the execution time tests.

In the case of quality tests, for small instances, a Friedman test was applied with $\alpha = 0.05$, posing the null hypothesis that all algorithms have the same performance.

The test yielded a p-value of 0.000, for which the null hypothesis was rejected and the Wilcoxon post hoc test was applied. From the average ranges of the Friedman test, shown in Table 7, EDA-GMMp was selected as the control algorithm in the post hoc test performed on pairs, in all cases with a p-value of 0.000, with which it was concluded that EDA-GMMp was the algorithm with the best quality performance in small instances.

In the case of large instances, the same Friedman test was applied, the ranges obtained are shown in Table 8. In this test, a p-value of 0.000 was also obtained, so the null hypothesis was rejected. To determine the algorithm with the best performance, the Wilcoxon post hoc test was applied, again taking EDA-GMMp as the control algorithm. In the post hoc test, a p-value of 0.000 was obtained in all cases, with which it was concluded that EDA-GMMp was the algorithm with the best quality performance in large instances.

Statistical analyzes of runtime tests are presented in two parts. The first is the results considering 100 generations for each experiment for all algorithms and the

Table 7 Friedman tests average ranges, fitness, and small instances. Experiments with 100 generations

Algorithm	Average range
EDA-GMMc	3.59
EDA-GMMp	1.07
Case 3	2.26
Case 4	3.07

Table 8 Friedman test average ranges, fitness, large instances. Experiments with 100 generations

Algorithm	Average range
EDA-GMMc	3.83
EDA-GMMp	1.07
Case 3	2.00
Case 4	3.10

second considers the results of applying the criterion (Fitness EDA-GMMp) $\leq$ (Fitness EDA-GMMc). The second part seeks to analyze the execution time that it takes for the EDA-GMMp algorithm to obtain the same fitness or less, using as a reference the results of the EDA-GMMc algorithm with 100 generations, i.e., it is desired to confirm with the test that EDA-GMMp improves significantly the time of obtaining solutions of the EDA-GMMc algorithm.

In the case of the experiments with 100 generations, it can be observed in the descriptive information of Fig. 10, that in the small instances Case 3 has a longer execution time and the rest a similar time.

To determine if exist a significant difference in the execution time of the algorithms for small instances, the Friedman test was applied, posing the null hypothesis that all the algorithms have the same execution time.

The test showed a p-value of 0.000, which indicates that the algorithms do not have similar performance. To identify the algorithm with the best performance, the Wilcoxon post hoc test was applied, which takes into account the average ranges of the Friedman test (Table 9), where it can be seen that the one with the lowest range is Case 4, therefore which was selected as the control algorithm.

The Wilcoxon test showed different p-values in the pairwise comparison (Table 10), these show that Case 4 and the EDA-GMMc algorithm do not have a significant difference in execution time and that Case 4 has better performance than the rest of the algorithms.

Table 9 Average range for execution time in the Friedman test for small instances. Experiments with 100 generations

Algorithm	Average range
EDA-GMMc	2.04
EDA-GMMp	2.26
Case 3	4.00
Case 4	1.70

Table 10 Wilcoxon Test p-values for small instances. Experiments with 100 generations

Case 4 versus EDA-GMMc	Case 4 versus EDA-GMMp	Case 4 versus Case 3
0.124	0.002	0.000

Table 11 Average range for execution time in the Friedman test for large instances. Experiments with 100 generations

Algorithm	Average range
EDA-GMMc	2.10
EDA-GMMp	3.27
Case 3	3.10
Case 4	1.53

Table 12 Wilcoxon Test p-values for large instances. Experiments with 100 generations

Case 4 versus EDA-GMMc	Case 4 versus EDA-GMMp	Case 4 versus Case 3
0.688	0.000	0.000

In the case of large instances, the Friedman test yielded a p-value of 0.000, which indicates that the algorithms do not have similar performance, based on the average ranges (Table 11), Case 4 corresponds to the applied position parameter with the Borda algorithm and the spread parameter with the proposal of the inverse of the mean Kendall distance, was established as a control algorithm to perform the Wilcoxon test.

As in the small instances, the Wilcoxon test yielded different p-values (Table 12), the results show that only in the Case 4 test versus EDA-GMMc there is no significant difference.

To determine if the difference in time to obtain the same results was significant, the Friedman test was applied, establishing the stoppage criterion (Fitness EDA-GMMp) $\leq$ (Fitness EDA-GMMc). In this case, the null hypothesis was rejected because the test yielded a p-value of 0.003 for small instances and 0.002 for large instances. To identify the algorithm with the best runtime performance, EDA-GMMp was established as the control algorithm (Tables 13 and 14) and the Wilcoxon test was applied.

The Wilcoxon test yielded a p-value of 0.008 for small instances and 0.005 for large instances, rejecting the null hypothesis and concluding that the EDA-GMMp algorithm obtains the same or better results than the EDA-GMMc algorithm in a significantly shorter time, confirming what was obtained with the TRP.

Table 13 Friedman Test results, under the condition (Fitness EDA-GMMp) $\leq$ (Fitness EDA-GMMc) for small instances

Small instances algorithm	Average range
EDA-GMMc	2.0
EDA-GMMp	1.0

Table 14 Friedman Test results, under the condition (Fitness EDA-GMMp) $\leq$ (Fitness EDA-GMMc) for large instances

Large instances algorithm	Average range
EDA-GMMc	2.0
EDA-GMMp	1.0

8 Conclusion

A learning model for an EDA Algorithm with the Mallows model was proposed to solve the MNPFSP problem. This proposal focuses on the modification of the strategies to estimate the central permutation and the spread parameter in the Mallows model. In the case of the central permutation, instead of using the Borda algorithm, which is commonly used for this problem, it was proposed to select the individual with the best fitness from each sample to guide the EDA learning process, because it is the same process that an EDA algorithm uses to evaluate individuals, allowing mathematical calculations to be reduced. In the case of the estimation of the spread parameter, instead of applying the maximum likelihood method (MLE), an inverse function of the mean Kendall distance was proposed. This practical proposal was raised for two main reasons, the first was the relationship that was observed between the mean Kendall distance and the spread parameter when applying the MLE and the second was to eliminate the calculations that the application of the MLE implies. The relationship that was proposed estimates the spread parameter with a magnitude greater than that obtained with the MLE, which allows greater exploration of the search space.

In the case of the search for better solutions, in the experiments where the algorithms ran 100 generations, the EDA-GMMp algorithm managed to improve 96% of the 27 small instances and in the 30 large instances, it managed to improve 100% of them. Subsequent statistical tests showed that the EDA-GMMp algorithm performed significantly better than the rest of the algorithms by obtaining better solutions.

Regarding the time to execute 100 generations, it was observed that the EDA-GMMp algorithm did not use, in general, less time than the rest of the algorithms. This was attributed to the longer exploration time associated with the higher values of the spread parameter that are obtained with the research proposal.

To obtain an objective measurement of the execution time of the proposed algorithm EDA-GMMp concerning algorithm EDA-GMMc, a test scheme with stopping criteria (Fitness EDA-GMMp) $\leq$ (Fitness EDA-GMMc) was proposed. The objective was to observe the time it takes for the proposed algorithm to obtain better or equal results than the reference algorithm. The general results show that the EDA-GMMp algorithm took 57.6% less average time than the EDA-GMMc algorithm to find the same or better solution, unlike the Case 3 and Case 4 algorithms, where average percentages of decrease were obtained. of 44.5 and 22.8% respectively. Statistical tests confirm that the EDA-GMMp algorithm performed significantly better than the EDA-GMMc algorithm by obtaining a shorter execution time.

It is concluded that, although the EDA-GMMp algorithm generated better solutions in less execution time, compared to the EDA-GMMc algorithm, is necessary to create a stopping criterion that does not require comparison with a reference algorithm to identify the best solution generated by the proposal.

To take advantage of the good performance of the EDA-GMMp algorithm, and to propose an adequate stopping criterion, it is proposed to analyze the convergence and the pattern of function calls concerning the number of machines and jobs of the test instances when applying the algorithm.

The results show the feasibility to be applied in a decomposition model for the No-Idle Permutation Mixed FlowShop Scheduling Problem [21].

Acknowledgements The authors wish to thank the Consejo Nacional de Ciencia y Tecnología (CONACYT) of Mexico, through scholarships for postgraduate studies: 634738 (E. Sánchez) and the Tecnológico Nacional de México/Instituto Tecnolóico de León, for the support provided to this research.

References

1. Pan, Q.-K. (2014). An effective iterated greedy algorithm for the mixed no-idle permutation flowshop scheduling problem. *Omega, 44*, 41–50.
2. Shih-Hsin Chen, M.-C. C. (2013). Addressing the advantages of using ensemble probabilistic models in Estimation of Distribution Algorithms for scheduling problems. *International Journal of Production Economics, 141*, 24–33.
3. Ceberio, J. (2014). A distance based ranking model estimation of distribution algorithm for the flow shop scheduling problem. *IEEE Transactions on Evolutionary Computation, 18*(2), 286–299.
4. Murilo Zangari, A. M. (2017). Multiobjective decomposition-based Mallows Models estimation of distribution algorithm. A case of study for permutation flowshop scheduling problem. In *Information Sciences*, 137–154.
5. Ricardo Pérez-Rodríguez, A. H.-A. (2019). A hybrid estimation of distribution algorithm for the vehicle routing problem with time windows. *Computers and Industrial Engineering, 130*, 75–96.
6. Josu Ceberio, A. M. (2011). In *Introducing the Mallows Model on Estimation of Distribution Algorithms, de Neuronal Information Processing* (pp. 461–470). Heidelberg: Springer.
7. Josu Ceberio, E. I. (2012). A review on estimation of distribution algorithms in permutation-based combinatorial optimization problems. *Progress in Artificial Intelligence,* 103–117.
8. Josu Ceberio, E. I. (2015). A review of distances for the Mallows and Generalized Mallows estimation of distribution algorithms. *Computational Optimization and Applications, 62*, 545–564.
9. Ceberio, J. (2014). Extending Distance-based ranking models in estimation of distribution Algorithms. *IEEE congress on evolutionary computation* (pp. 6–11).
10. Marina Meila, K. P. (2012). Consensus ranking under the exponential model, de. In *Proceedings of the 23rd Conference on Uncertainty in Artificial Intelligence* (pp. 285–294).
11. Aguirre. (2018). A hybrid estimation of distribution algorithm for flexible job shop scheduling problems with process plan flexibility. *Applied Intelligence.*
12. Alnur Ali, M. (2012). Experiments with Kemeny ranking; What works when? *Mathematical Social Sciences, 64*, 28–40.

13. Ricardo Péz Rodríguez, A. H. A. (2017). Un algoritmo de estimacion de distribuciones copulado con la distribucion generalizada de Mallows para el problema de ruteo de autobuses escolares con seleccion de Paradas. *Revista iberoamericana de automática e informática industrial* (vol. 14, pp. 288–298).
14. Ekhine Irurozki, B. C. (2016). An R package for permutations, Mallows and Generalized Mallows models. *Journal of Statistical Software*, 1–30.
15. Mark Hauschild, M. P. (2011). An introduction and survey of estimation of distribution algorithms. In *Medal Missouri Estimation of Distribution Algorithms Laboratory*, Missouri-St Louis (pp. 111–128).
16. Bajer, L. (2010). Estimation of distribution algorithms: a recent approach to evolutionary computation. In *WDS'10 Proceedings of Contributed Papers* (no 1, pp. 48–53).
17. Bartholdi, J. I. C. (1989). Voting schemes for which it can be difficult to tell who won the election. *Social Choice and Welfare, 6* (2), 157–165.
18. Ekhine, I. (2014). Sampling and learning distance based probability models for permutation spaces. Universidad del Pais Vasco, Donastia, San Sebastian.
19. Ekhine Irurozki, J. C. (2016). A Matlab toolbox of estimation of distribution algorithms for permutation-based combinatorial optimization problems. *ACM Transactions on Mathematical Software*, 1–13.
20. Quan-Ke Pan, R. R. (2014). An effective iterated greedy algorithm for the mixed no-idle permutation flowshop scheduling problem. *Omega, 44*, 41–50.
21. Tolga Bektas, A. H. (2020). Benders decomposition for the mixed no-idle permutation flowshop scheduling problem. *Journal of Scheduling, 23*, 513–523.

Interval Type-3 Fuzzy Decision Making in Material Surface Quality Control

Oscar Castillo and Patricia Melin

1 Introduction

Originally, fuzzy sets were outlined by Lotfi Zadeh in 1965. Later fuzzy logic and fuzzy systems were also proposed by Zadeh, and many applications follow, mainly in control [1]. Type-1 fuzzy systems evolved to type-2 fuzzy systems with the works by Mendel in 2001 [2]. Initially, interval type-2 fuzzy systems were studied and applied to several problems [3]. These systems were applied to many problems in areas such as: robotics, intelligent control and others [4, 5]. Simulation and experimental results show that interval type-2 outperform type-1 fuzzy systems in situations with higher levels of noise, dynamic environments or highly nonlinear problems [6–8]. Later, general type-2 fuzzy systems were considered to manage higher levels of uncertainty, and good results have been achieved in several areas of application [9–11]. Recently, type-3 fuzzy systems have been flourishing to become a useful tool for even more complex problems. For this reason, in this paper we are putting forward the basic constructs of type-3 fuzzy systems by extending the ideas of type-2 fuzzy systems [12–15]. Recently, there are been some applications of type-3 in diverse application areas, such as the one mentioned in [16–22]. This paper, as a difference to previous works is presenting an application in surface quality control viewed as a decision-making problem solved by experts.

The contribution is the realization of interval type-3 fuzzy in surface quality control in manufacturing has not been presented before in the literature. We consider that this is an important contribution to the frontier knowledge in the intelligent systems area.

O. Castillo (✉) · P. Melin
Tijuana Institute of Technology, TecNM, Tijuana, Mexico
e-mail: ocastillo@tectijuana.mx

O. Castillo and P. Melin (eds.), *Hybrid Intelligent Systems Based on Extensions of Fuzzy Logic, Neural Networks and Metaheuristics*, Studies in Computational Intelligence 1096,
https://doi.org/10.1007/978-3-031-28999-6_29

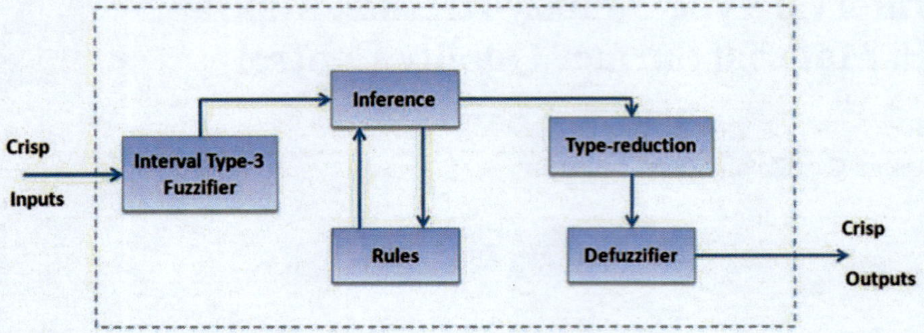

Fig. 1 Structure of an interval type-3 system

The rest of the article is structured as: In Sect. 2 we are outlining the type-3 fuzzy system design, in Sect. 3 an application to material quality estimation based on a fuzzy system is presented, and in Sect. 4 we are outlining the conclusions.

2 Interval Type-3 Fuzzy Systems

The structure of a type-3 fuzzy system, analogous to type-2, is formed by the fuzzifier, fuzzy rules, inference, type reduction and defuzzifier. In Fig. 1 we can find an illustration for this kind of system.

In more complicated situations or problems, it is possible that a hierarchical approach is needed, like in Fig. 3 where several simpler fuzzy systems are used and then an aggregator (higher type) fuzzy system is needed to combine the outputs of the simpler systems. We have used this architecture with type-1 individual controllers and then a type-2 fuzzy aggregator [23], but it is also possible that type-2 fuzzy controllers can be aggregated by a type-3 fuzzy system. In Fig. 2, the individual fuzzy fuzzy systems are shown with type *n-1* and the aggregator with type *n*.

3 Decision Making Application: Surface Quality Control

Fuzzy logic can be utilized to automate quality control in many manufacturing areas. In this situation, a set of fuzzy rules is established to relate relevant characteristics of the material (manufactured) with the product quality. The fuzzy rules represent the expert knowledge for this domain. If we assume that the product quality is a linguistic variable with values: bad, regular, and excellent, we can then put forward fuzzy rules for surface quality, as illustrated in Table 1. The input variables are roughness and porosity, both with linguistic values: bad, medium and good.

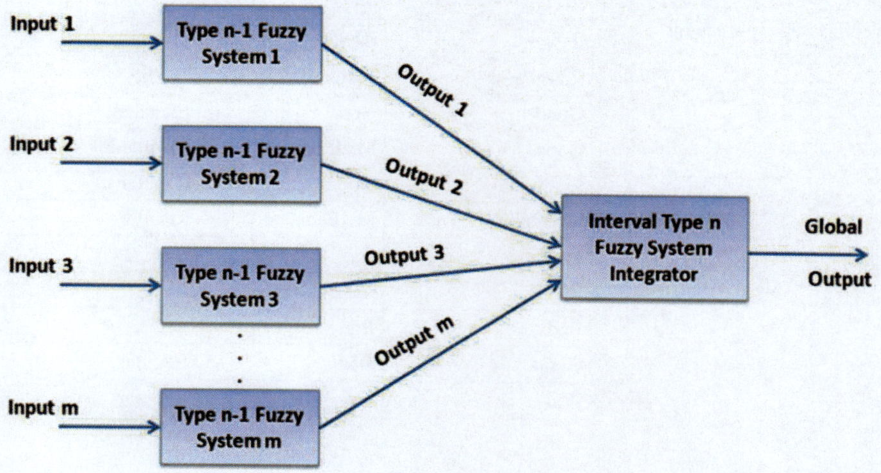

Fig. 2 Hierarchical architecture of multiple fuzzy systems

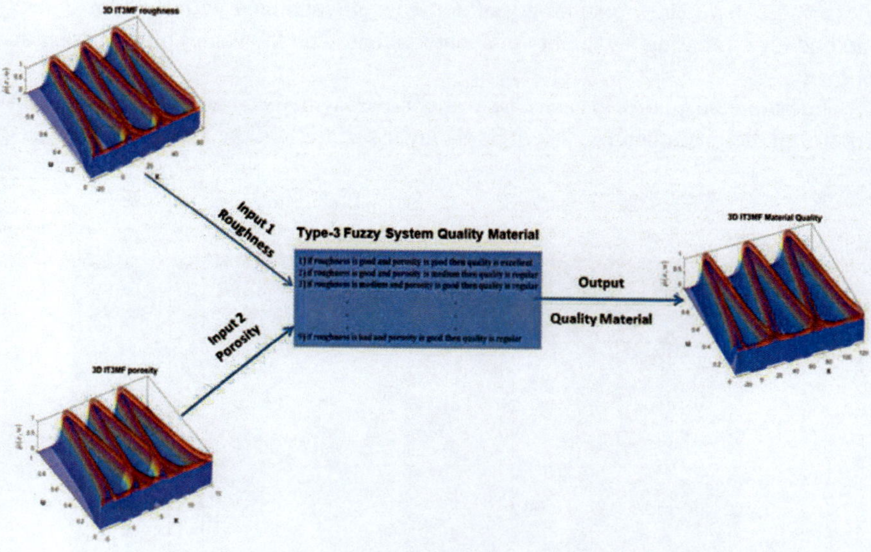

Fig. 3 Structure of type-3 system for material quality evaluation

The interpretation of this fuzzy system is that if the surface roughness is good and porosity is good then the product quality is assumed as excellent. Other fuzzy rules can be interpreted in an analogous way. We depict in Fig. 3 the particular structure of the system. We show in Fig. 4 the Gaussian interval type-3 MFs of the roughness input linguistic variable. In Fig. 5 we depict in a similar way the MFs for the porosity input linguistic variable. In Fig. 6 we show the MFs of quality output variable in the plane,

Table 1 Fuzzy rules for surface quality

IF	AND	THEN
Roughness	Porosity	Quality
Good	Good	Excellent
Good	Medium	Regular
Medium	Good	Regular
Medium	Medium	Regular
Medium	Bad	Bad
Bad	Medium	Bad
Bad	Bad	Bad
Good	Bad	Regular
Bad	Good	Regular

and in Fig. 7 with a 3D view. In Fig. 8 the surface summarizing the complete model is illustrated. In Fig. 9 we exhibit the 3 D nature of type-3 membership functions with the FOU highlighted in green color.

In Table 2 we show a summary of some results obtained with the type-3 fuzzy method for estimating the quality of a material based on the values of roughness and porosity.

The estimated quality in each case was compare with the human experts in quality control in the manufacturing of an oxide layer surface and the average error was of

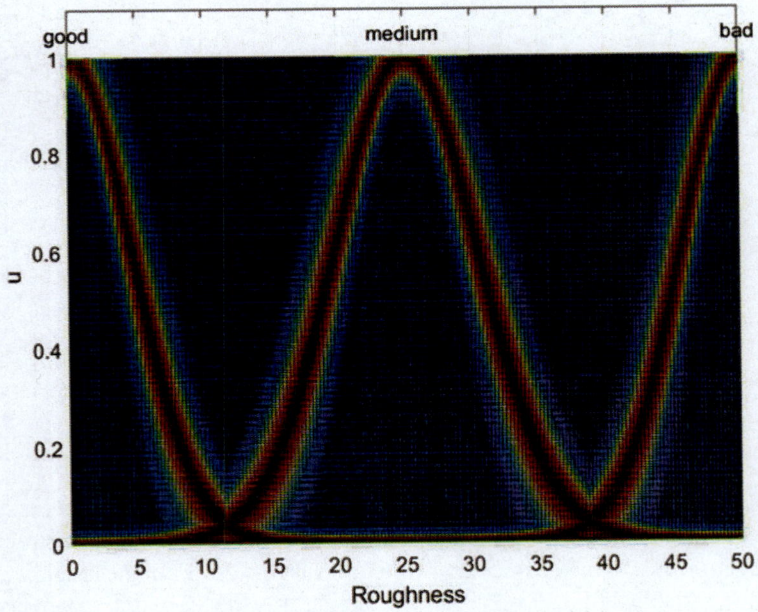

Fig. 4 Membership functions for roughness input variable

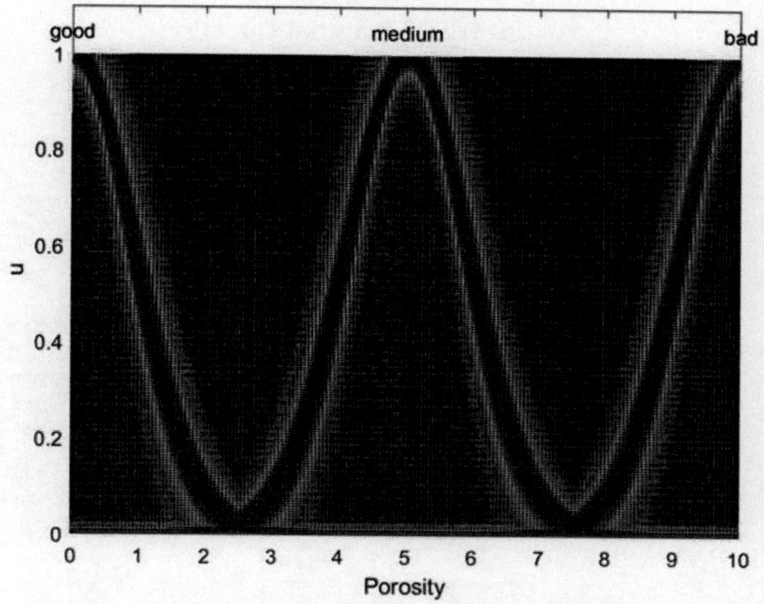

Fig. 5 Membership functions for porosity input variable

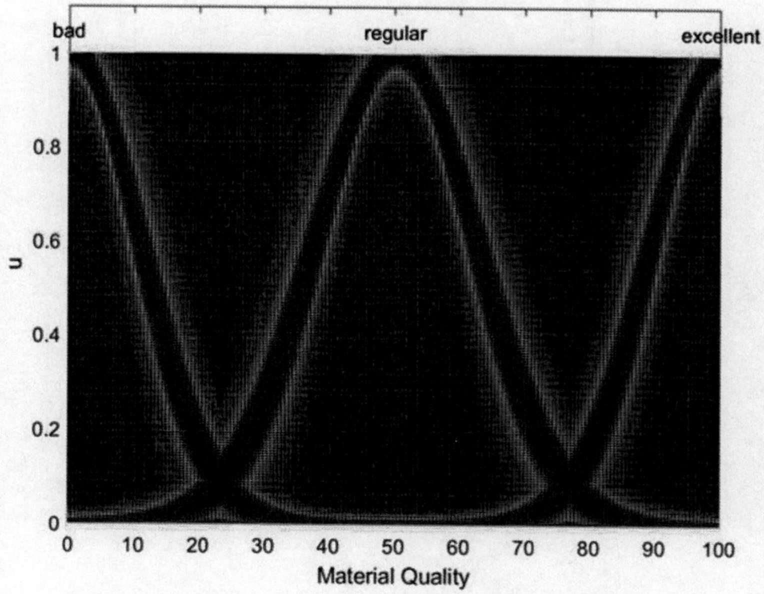

Fig. 6 Membership functions for quality variable

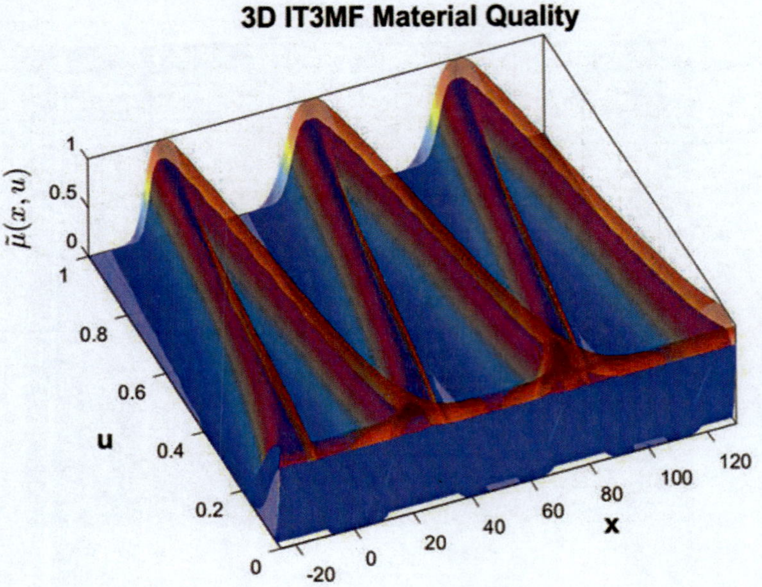

Fig. 7 Three-dimensional (3D) view of output MFs

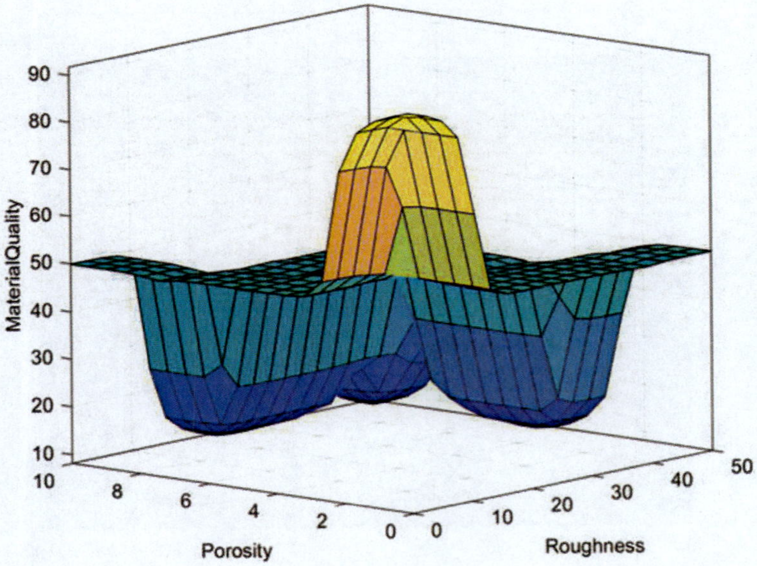

Fig. 8 Surface of the surface quality control fuzzy model

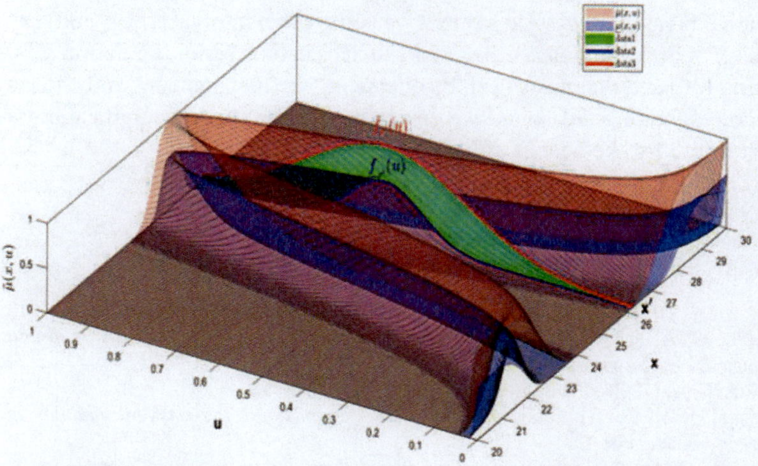

Fig. 9 Membership function as a 3 D plot for illustration purposes

Table 2 Summary of results with the interval type-3 fuzzy model

Roughness	Porosity	Quality
0.2000	0.5000	91.9048
1.0000	0.1000	92.1451
3.6700	0.0000	91.0108
5.0000	0.2500	89.7966
10.2000	0.3000	72.6401
16.9700	0.3600	49.1708
18.4000	7.5000	39.8726
20.0000	6.0000	48.9439
25.0000	2.1000	48.8231
30.0000	8.1500	15.2604
40.000	9.2000	23.3671
45.000	6.3000	10.4162

less than 1%, which is lower than we have previously achieved with type-1 and type-2 fuzzy approaches [24].

4 Conclusions

In this article we validate the impact of type-3 fuzzy in decision-making for surface quality control, which is the first time to be used in this area. Material manufacturing needs good quality control and a type-3 system is able to surpass type-2 and type-1 in this area. Finally, we have to say that we believe that interval type-3 could enhance the quality of the solutions in other areas, such as in time series prediction [25], fuzzy clustering [26] and diagnosis [27], by enhancing the management of the uncertainty in the corresponding areas, and we envision working on these application areas in the near future for the benefit of society and economy.

References

1. Zadeh, L. A. (1989). Knowledge representation in Fuzzy Logic. *IEEE Transactions on knowledge data engineering, 1*, 89.
2. Zadeh, L. A. (1998). Fuzzy Logic. *Computer, 1*(4), 83–93.
3. Mendel, J. M. (2001). *Uncertain Rule-Based Fuzzy Logic Systems: Introduction and New Directions*. Prentice-Hall.
4. Mendel, J. M. (2017). *Uncertain Rule-Based Fuzzy Logic Systems: Introduction and New Directions* (2nd ed.), Springer.
5. Karnik, N. N., & Mendel, J. M. (2001). Operations on Type-2 fuzzy sets. *Fuzzy Sets and Systems, 122*, 327–348.
6. Moreno, J. E., et al. (2020). Design of an interval Type-2 fuzzy model with justifiable uncertainty. *Information Sciences, 513*, 206–221.
7. Mendel, J. M., Hagras, H., Tan, W.-W., Melek, W. W., & Ying, H. (2014). *Introduction to Type-2 Fuzzy Logic Control*. NJ. Wiley and IEEE Press.
8. Olivas, F., Valdez, F., Castillo, O., & Melin, P. (2016). Dynamic parameter adaptation in particle swarm optimization using interval Type-2 fuzzy logic. *Soft Computing, 20*(3), 1057–1070.
9. Sakalli, A., Kumbasar, T., & Mendel, J. M. (2021). Towards systematic design of general Type-2 Fuzzy Logic Controllers: Analysis, interpretation, and tuning. *IEEE Transactions on Fuzzy Systems, 29*(2), 226–239.
10. Ontiveros, E., Melin, P., & Castillo, O. (2018). High order α-planes integration: A new approach to computational cost reduction of General Type-2 Fuzzy Systems. *Engineering Applications of Artificial Intelligence, 74*, 186–197.
11. Castillo, O., & Amador-Angulo, L. (2018). A generalized type-2 fuzzy logic approach for dynamic parameter adaptation in bee colony optimization applied to fuzzy controller design. *Information Sciences, 460–461*, 476–496.
12. Cao, Y., Raise, A., Mohammadzadeh, A. et al. (2021). Deep learned recurrent type-3 fuzzy system: Application for renewable energy modeling / prediction. *Energy Reports*.
13. Mohammadzadeh, A., Castillo, O., Band, S. S., et al. (2021). A novel fractional-order multiple-model Type-3 Fuzzy Control for nonlinear systems with unmodeled dynamics. *International Journal of Fuzzy Systems*. https://doi.org/10.1007/s40815-021-01058-1
14. Qasem, S. N., Ahmadian, A., Mohammadzadeh, A., Rathinasamy, S., & Pahlevanzadeh, B. (2021). A Type-3 logic fuzzy system: Optimized by a correntropy based Kalman filter with adaptive fuzzy kernel size Inform. *Science, 572*, 424–443.
15. Rickard, J. T., Aisbett, J., & Gibbon, G. (2009). Fuzzy subsethood for fuzzy sets of Type-2 and generalized Ttype-n. *IEEE Transactions on Fuzzy Systems, 17*(1), 50–60.
16. Mohammadzadeh, A., Sabzalian, M. H., & Zhang, W. (2020). An interval Type-3 fuzzy system and a new online fractional-order learning algorithm: Theory and practice. *IEEE Transactions on Fuzzy Systems, 28*(9), 1940–1950.
17. Liu, Z., Mohammadzadeh, A., Turabieh, H., Mafarja, M., Band, S. S., & Mosavi, A. (2021). A new online learned Interval Type-3 Fuzzy Control system for solar energy management systems. *IEEE Access, 9*, 10498–10508.
18. Amador-Angulo, L., Castillo, O., Melin, P., & Castro, J. R. (2022). Interval Type-3 Fuzzy adaptation of the bee colony optimization algorithm for optimal Fuzzy Control of an autonomous mobile robot. *Micromachines, 13*(9), 1490. https://doi.org/10.3390/mi13091490

19. Castillo, O., Castro, J. R., & Melin, P. (2022). Interval Type-3 Fuzzy Control for automated tuning of image quality in televisions. *Axioms, 11*, 276. https://doi.org/10.3390/axioms11060276
20. Castillo, O., Castro, J. R., & Melin, P. (2022). Interval Type-3 Fuzzy systems: Theory and design. *Studies in Fuzziness and Soft Computing, 418*, 1–100.
21. Castillo, O., Castro, J. R., & Melin, P. (2022). A methodology for building interval type-3 fuzzy systems based on the principle of justifiable granularity. *International Journal of Intelligent Systems*.
22. Castillo, O., Castro, J. R., & Melin, P. (2022). Interval Type-3 Fuzzy aggregation of neural networks for multiple time series prediction: The case of financial forecasting. *Axioms, 11*(6), 251.
23. Cervantes, L., & Castillo, O. (2015). Type-2 fuzzy logic aggregation of multiple Fuzzy Controllers for Airplane Flight Control. *Information Sciences, 324*, 247–256.
24. Castillo, O., & Melin, P. (2003). *Soft Computing and Fractal Theory for Intelligent Manufacturing*. Springer.
25. Castillo, O., Castro, J. R., Melin, P., & Rodriguez-Diaz, A. (2014). Application of interval Type-2 fuzzy neural networks in non-linear identification and time series prediction. *Soft Computing, 18*(6), 1213–1224.
26. Rubio, E., Castillo, O., Valdez, F., Melin, P., Gonzalez, C. I., & Martinez, G. (2017). An extension of the fuzzy possibilistic clustering algorithm using Type-2 fuzzy logic techniques. *Advances in Fuzzy Systems*. https://doi.org/10.1155/2017/7094046
27. Melin, P., Miramontes, I., & Prado-Arechiga, G. (2018). A hybrid model based on modular neural networks and fuzzy systems for classification of blood pressure and hypertension risk diagnosis. *Expert Systems with Applications, 107*, 146–164.

Interval Type-3 Fuzzy Decision Making in Quality Evaluation for Speaker Manufacturing

Patricia Melin and Oscar Castillo

1 Introduction

Originally, Lotfi Zadeh postulated fuzzy sets in 1965. Later fuzzy logic and fuzzy systems were also proposed by Zadeh, and many applications follow, mainly in control [1]. Type-1 fuzzy systems evolved to type-2 fuzzy systems with the works by Mendel in 2001 [2]. Initially, interval type-2 fuzzy systems were studied and applied to several problems [3]. These systems were applied to many problems in areas such as: robotics, intelligent control and others [4, 5]. Simulation and experimental results show that interval type-2 outperform type-1 fuzzy systems in situations with higher levels of noise, dynamic environments or highly nonlinear problems [6–8]. Later, general type-2 fuzzy systems were considered to manage higher levels of uncertainty, and good results have been achieved in several areas of application [9–11]. Recently, type-3 fuzzy systems have been able to solve more difficult situations. For this reason, in this paper we are putting forward the basic constructs of type-3 fuzzy systems by extending the ideas of type-2 fuzzy systems [12–14].

The main contribution is the utilization of type-3 fuzzy in speaker quality control, which cannot be find to the moment in the literature, and it is illustrated that type-3 potentially can be better.

In this work the idea is to utilize a fuzzy system to model expert knowledge on quality control. Usually, type-1 is utilized, which means that type-1 membership functions are used. In this situation, a type-1 description of an imprecise ("fuzzy") property like "small" means that to each possible value x, we assign a number $\mu(x)$ from [0,1] that represents, in the expert's opinion, to what extent the expert believes that the value x is small. Experts usually cannot describe their degree of belief by a precise number, so a better way is to allow the experts to describe their degree of

P. Melin · O. Castillo (✉)
Tijuana Institute of Technology, Tijuana, Mexico
e-mail: ocastillo@tectijuana.mx

O. Castillo and P. Melin (eds.), *Hybrid Intelligent Systems Based on Extensions of Fuzzy Logic, Neural Networks and Metaheuristics*, Studies in Computational Intelligence 1096,
https://doi.org/10.1007/978-3-031-28999-6_30

belief by an interval. This is what we know as interval type-2 fuzzy set. In addition to providing an interval, the expert can also say that some values on this interval are more probable and some are less probable. This is a general type-2 fuzzy set. Analogous to the fact that an expert cannot represent a degree of belief that x has a membership given by a number $\mu(x)$, the expert cannot represent the belief that u is a possible value of $\mu(x)$ by a unique value $\mu(x,u)$. A more adequate approach is allowing the expert a possibility to describe an interval of values of $\mu(x,u)$. This is what we mean by interval type-3 fuzzy set.

The remaining of the chapter is organized as: Sect. 2 outlines several ideas for fuzzy systems, in Sect. 3 the ideas are highlighted utilizing a real case, and in Sect. 4 conclusions are offered.

2 Interval Type-3 Fuzzy System Design

The architecture of an interval type-3 (IT3) fuzzy system, analogous to type-2, is comprised of a fuzzifier, rules, inference, type reduction and defuzzifier. In Fig. 1 the architecture of an IT3 system can be found.

In more complicated situations or problems, it is possible that a hierarchical approach is needed, like in Fig. 7 where several simpler systems are used and then an aggregator (higher type) fuzzy system is needed to mix the results of the simpler systems. We have used this architecture with type-1 individual controllers and then a type-2 fuzzy aggregator [18], but it is also possible that type-2 fuzzy controllers can be aggregated by a type-3 fuzzy system. In Fig. 2, the individual fuzzy systems are shown with type *n-1* and the aggregator with type *n*.

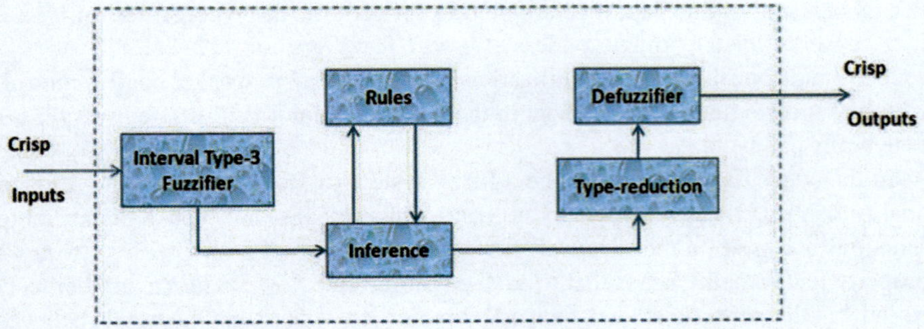

Fig. 1 Architecture of an IT3 system

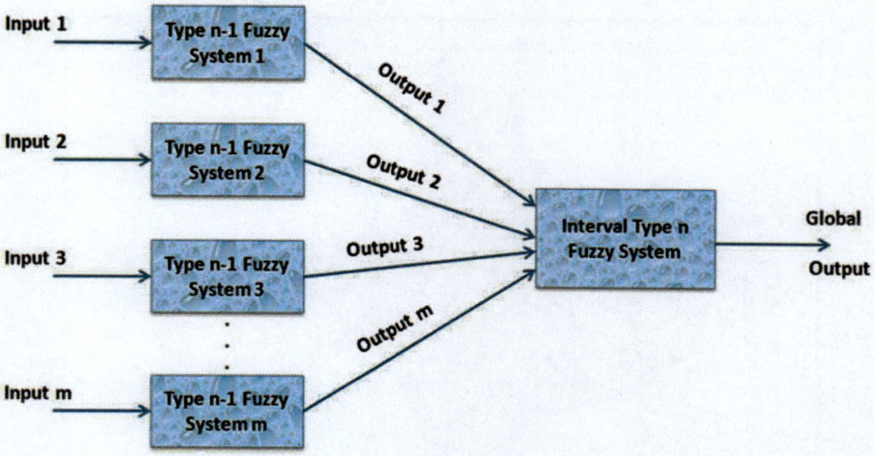

Fig. 2 Hierarchical architecture of multiple fuzzy systems

3 Application in Quality Control

The evaluation of quality requires a comparison between the real measured sound signal and the ideal good sound signal [19]. Speakers are labeled as of good quality when their sound signal does not differ too much from the ones of ideal speakers (see Fig. 3). The measured signal is composed by 108,000 points in 3 s. We highlight in Fig. 4 the signal for the case of a bad quality speaker. We notice the difference of this signal and the corresponding signal shown in Fig. 3.

The fuzzy rules are in Mamdani form and were elucidated from experts. The fuzzy rules were obtained in previous works and are as follows [19]:

- IF Difference is small AND Fractal Dimension is small THEN Quality Excellent
- IF Difference is regular AND Fractal Dimension is small THEN Quality is Good
- IF Difference is regular AND Fractal Dimension is high THEN Quality is Medium
- IF Difference is medium AND Fractal Dimension is small THEN Quality is Medium
- IF Difference is medium AND Fractal Dimension is high THEN Quality is Bad
- IF Difference is large AND Fractal Dimension is small THEN Quality is Medium
- IF Difference is large AND Fractal Dimension is high THEN Quality is Bad
- IF Difference is small AND Fractal Dimension is high THEN Quality is Medium

The structure of the fuzzy system for quality control is illustrated in Figure 5, where we can notice the two inputs and one output of the system.

In Figs. 6 and 7 the type-3 membership functions (MFs) for the input variables are illustrated.

In Fig. 8 we depict the IT3 MFs of the output variable. In Fig. 9 we highlight the 3-D nature of the IT3 MFs.

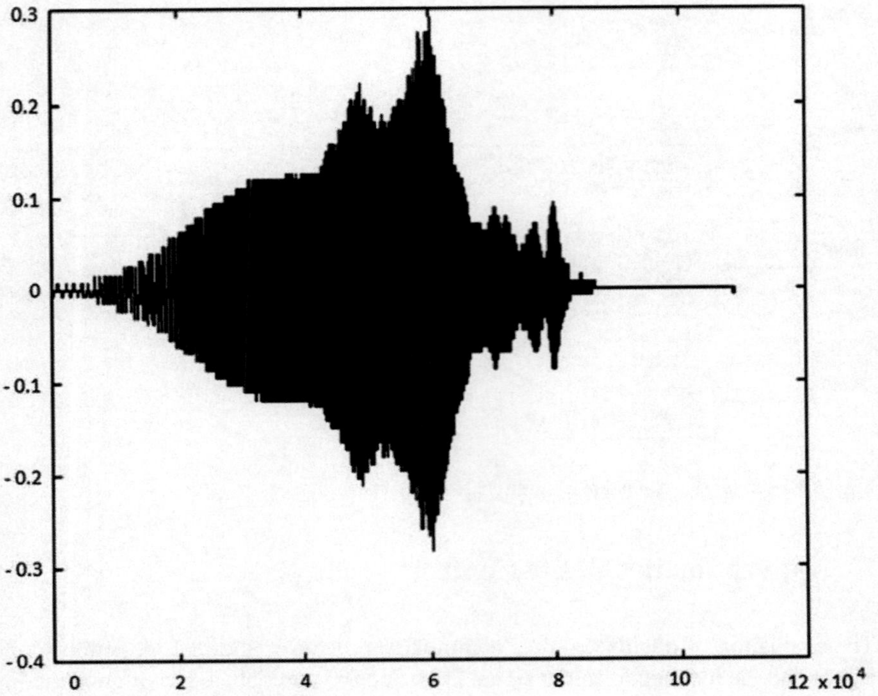

Fig. 3 Sound signal of speaker with excellent quality

In Fig. 10 we illustrate one of the IT3 MFs of the output variable, which is "medium" quality, which is a scaled Gaussian IT3 MF. In Fig. 11 we illustrate the surface exhibiting the model.

Finally, in Table 1 a comparative of results among type-3, type-2 and type-1 is presented, indicating that type-3 realizes better results in some cases. Summarizing, we obtain a higher quality with type-3 fuzzy when the speakers are good, like in rows 1 to 3, but in cases of bad quality (rows 4 to 7) we evaluate quality as lower.

We have to mention that the results presented in Table 1 are the initial experiments done with type-3 fuzzy logic for this application, meaning that we have not yet performed tuning of the MFs in the fuzzy system.

4 Conclusions

This chapter outlined an interval type-3 fuzzy in quality control. Speakers requires good quality control and a type-3 system is able to surpass type-2. As future work, we will continue constructing all the basis for type-3 fuzzy systems, and later their applications in different areas, like in [20–24]. Finally, we have to say that we believe that interval type-3 could enhance the quality of the solutions in other areas, such as in

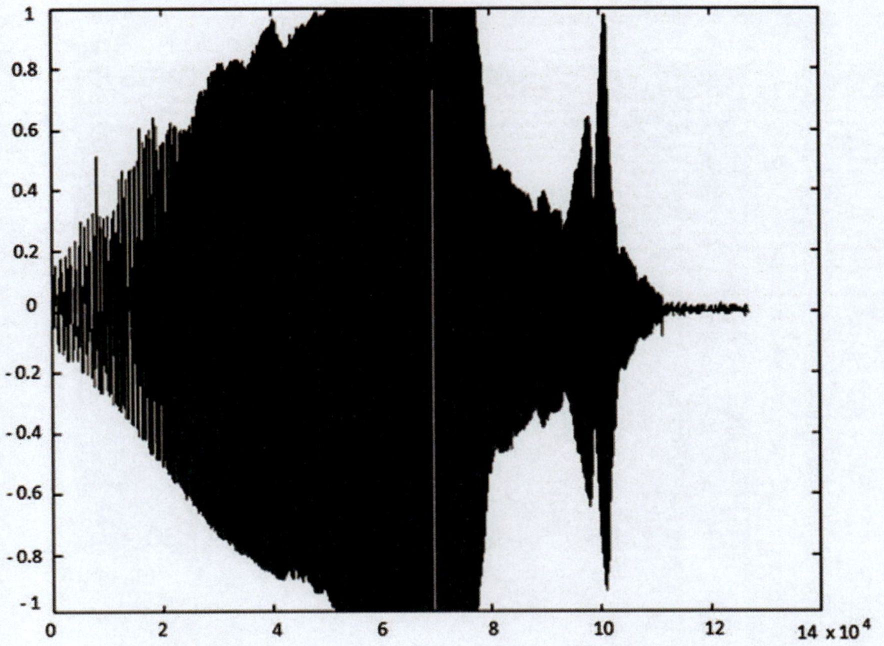

Fig. 4 Signal of a speaker with bad quality

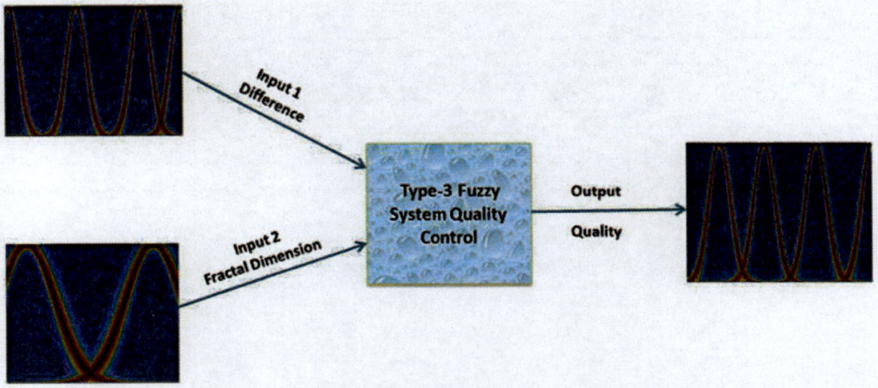

Fig. 5 Structure of the system for speaker quality

time series prediction [24], fuzzy clustering [25] and diagnosis [26], by enhancing the management of the uncertainty in the corresponding areas, and we envision working on these application areas in the near future for the benefit of society and economy. Also, the combination of type-3 and intuitionistic fuzzy shows some promise for handling other kinds of applications [27]. In addition, other kinds of applications, like in recognition [28] or robotics [29, 30], could be considered.

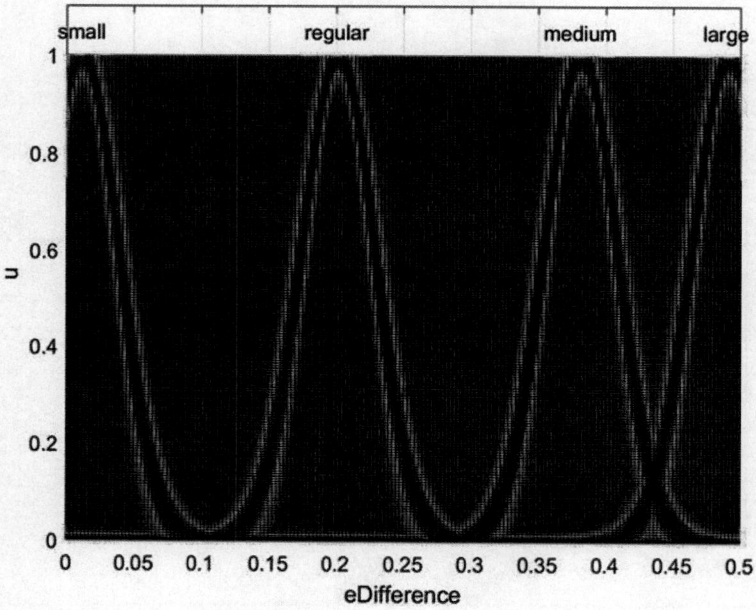

Fig. 6 MFs of the error difference variable

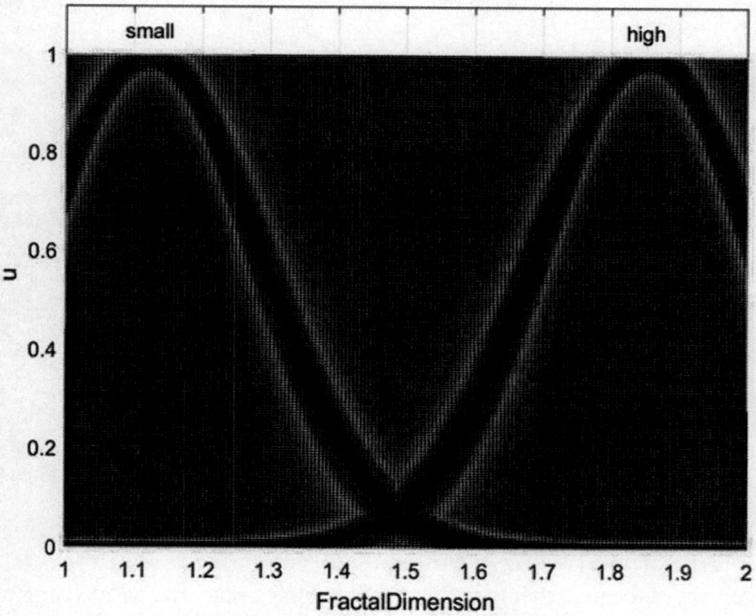

Fig. 7 MFs for the fractal dimension variable

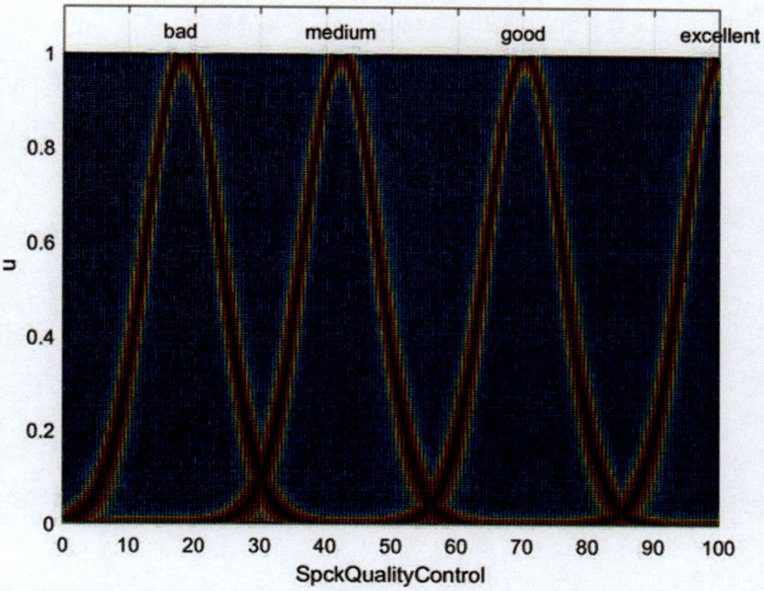

Fig. 8 MFs of speaker quality

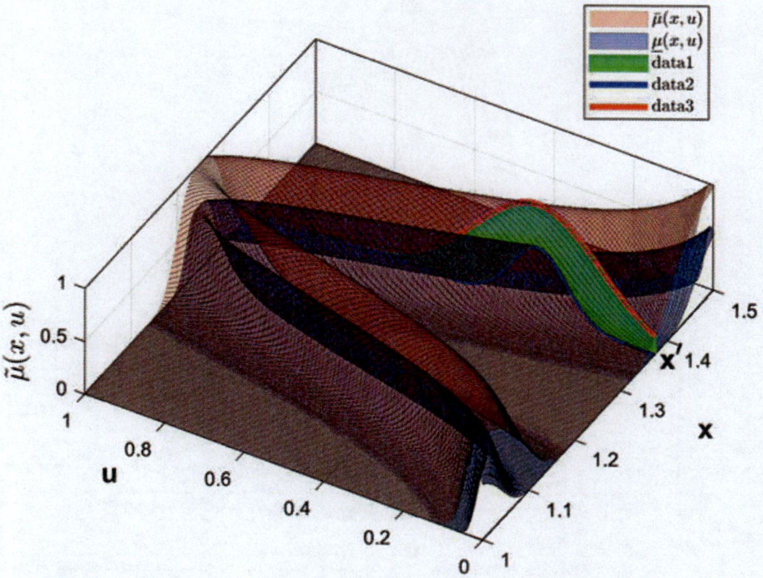

Fig. 9 A cut (in green) applied on the "small" MF of the fractal dimension

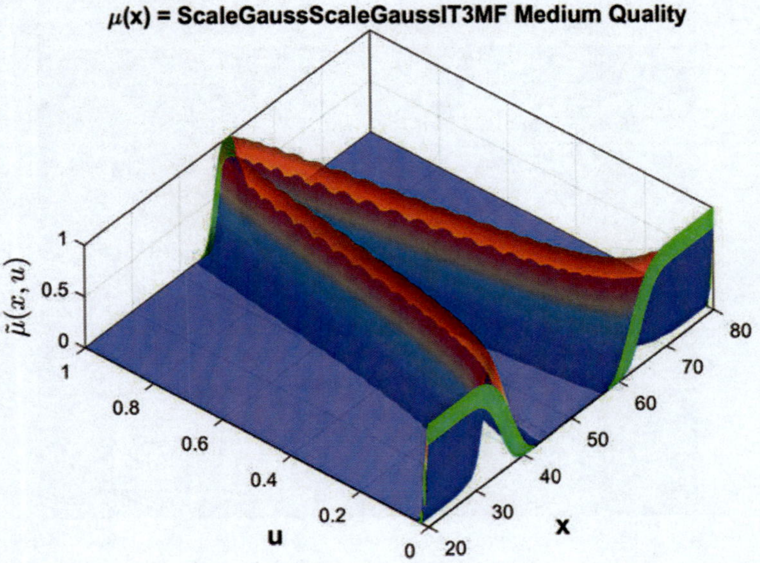

Fig. 10 MF of the "medium" value in the output

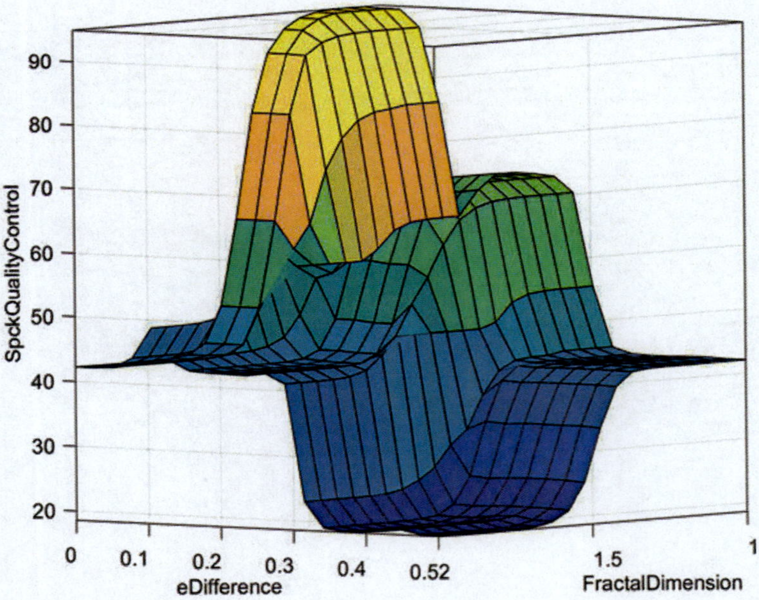

Fig. 11 Surface of the model

Table 1 Comparison among types of systems

Fractal dimension	Difference	Type-3 (this work)	Type-1 [19]	Type-2 [19]
1.1	0.01	94.2554	91.6159	91.05334
1.2	0.05	91.5032	88.4461	87.8591
1.3	0.03	92.2637	90.5619	90.1224
1.5	0.08	57.8789	72.0445	73.7142
1.6	0.02	43.3316	88.0722	87.8811
1.7	0.01	42.3178	87.0820	86.9462
1.8	0.10	47.8413	57.7984	59.3768
1.3	0.20	69.2439	70.2707	67.0798
1.7	0.40	18.9262	84.9467	83.5299
1.5	0.50	29.0963	86.4039	84.7432

References

1. Zadeh, L. A. (1989). Knowledge representation in Fuzzy Logic. *IEEE Transactions on knowledge data engineering, 1*, 89.
2. Zadeh, L. A. (1998). Fuzzy Logic. *Computer, 1*(4), 83–93.
3. Mendel, J. M. (2001). *Uncertain Rule-Based Fuzzy Logic Systems: Introduction and New Directions*. Prentice-Hall.
4. Mendel, J. M. (2017). *Uncertain Rule-Based Fuzzy Logic Systems: Introduction and New Directions* (2nd ed.). Springer.
5. Karnik, N. N., & Mendel, J. M. (2001). Operations on Type-2 Fuzzy sets. *Fuzzy Sets and Systems, 122*, 327–348.
6. Moreno, J. E., et al. (2020). Design of an interval Type-2 fuzzy model with justifiable uncertainty. *Information Sciences, 513*, 206–221.
7. Mendel, J. M., Hagras, H., Tan, W.-W., Melek, W. W., & Ying, H. (2014). *Introduction to Type-2 Fuzzy Logic Control*. NJ. Wiley and IEEE Press.
8. Olivas, F., Valdez, F., Castillo, O., & Melin, P. (2016). Dynamic parameter adaptation in particle swarm optimization using interval Type-2 fuzzy logic. *Soft Computing, 20*(3), 1057–1070.
9. Sakalli, A., Kumbasar, T., & Mendel, J. M. (2021). Towards systematic design of general Type-2 Fuzzy Logic controllers: Analysis, interpretation, and tuning. *IEEE Transactions on Fuzzy Systems, 29*(2), 226–239.
10. Ontiveros, E., Melin, P., & Castillo, O. (2018). High order α-planes integration: A new approach to computational cost reduction of General Type-2 Fuzzy Systems. *Engineering Applications of Artificial Intelligence, 74*, 186–197.
11. Castillo, O., & Amador-Angulo, L. (2018). A generalized type-2 fuzzy logic approach for dynamic parameter adaptation in bee colony optimization applied to fuzzy controller design. *Information Sciences, 460–461*, 476–496.
12. Cao, Y., Raise, A., Mohammadzadeh, A. et al. (2021). Deep learned recurrent type-3 fuzzy system: Application for renewable energy modeling / prediction. *Energy Reports*.
13. Mohammadzadeh, A., Castillo, O., Band, S. S., et al. (2021). A novel fractional-order multiple-model Type-3 Fuzzy control for nonlinear systems with unmodeled dynamics. *International Journal of Fuzzy Systems*. https://doi.org/10.1007/s40815-021-01058-1
14. Qasem, S. N., Ahmadian, A., Mohammadzadeh, A., Rathinasamy, S., & Pahlevanzadeh, B. (2021). A type-3 logic fuzzy system: Optimized by a correntropy based Kalman filter with adaptive fuzzy kernel size Inform. *Sci., 572*, 424–443.

15. Rickard, J. T., Aisbett, J., & Gibbon, G. (2009). Fuzzy subsethood for fuzzy sets of Type-2 and generalized type-n. *IEEE Transactions on Fuzzy Systems, 17*(1), 50–60.
16. Mohammadzadeh, A., Sabzalian, M. H., & Zhang, W. (2020). An interval type-3 fuzzy system and a new online fractional-order learning algorithm: Theory and practice. *IEEE Transactions on Fuzzy Systems, 28*(9), 1940–1950.
17. Liu, Z., Mohammadzadeh, A., Turabieh, H., Mafarja, M., Band, S. S., & Mosavi, A. (2021). A new online learned interval Type-3 Fuzzy Control system for solar energy management systems. *IEEE Access, 9*, 10498–10508.
18. Cervantes, L., & Castillo, O. (2015). Type-2 fuzzy logic aggregation of multiple fuzzy controllers for Airplane Flight Control. *Information Sciences, 324*, 247–256.
19. Melin, P., & Castillo, O. (2007). An intelligent hybrid approach for industrial quality control combining neural networks, fuzzy logic and fractal theory. *Information Sciences, 177*, 1543–1557.
20. Amador-Angulo, L., Castillo, O., Melin, P., & Castro, J. R. (2022). Interval Type-3 Fuzzy adaptation of the bee colony optimization algorithm for optimal fuzzy control of an autonomous mobile robot. *Micromachines, 13*(9), 1490. https://doi.org/10.3390/mi13091490
21. Castillo, O., Castro, J. R., & Melin, P. (2022). Interval Type-3 Fuzzy control for automated tuning of image quality in televisions. *Axioms, 11*, 276. https://doi.org/10.3390/axioms11060276
22. Castillo, O., Castro, J. R., & Melin, P. (2022). Interval Type-3 fuzzy systems: Theory and design. *Studies in Fuzziness and Soft Computing, 418*, 1–100.
23. Castillo, O., Castro, J. R., & Melin, P. (2022). A methodology for building interval type-3 fuzzy systems based on the principle of justifiable granularity. *International Journal of Intelligent Systems.*
24. Castillo, O., Castro, J. R., & Melin, P. (2022). Interval Type-3 fuzzy aggregation of neural networks for multiple time series prediction: The case of financial forecasting. *Axioms, 11*(6), 251.
25. Castillo, O., Castro, J. R., Melin, P., & Rodriguez-Diaz, A. (2014). Application of interval Type-2 fuzzy neural networks in non-linear identification and time series prediction. *Soft Computing, 18*(6), 1213–1224.
26. Rubio, E., Castillo, O., Valdez, F., Melin, P., Gonzalez, C. I., & Martinez, G. (2017). An extension of the fuzzy possibilistic clustering algorithm using Type-2 fuzzy logic techniques. *Advances in Fuzzy Systems*. https://doi.org/10.1155/2017/7094046
27. Melin, P., Miramontes, I., & Prado-Arechiga, G. (2018). A hybrid model based on modular neural networks and fuzzy systems for classification of blood pressure and hypertension risk diagnosis. *Expert Systems with Applications, 107*, 146–164.
28. Castillo, O., & Melin, P. (2022). Towards interval Type-3 intuitionistic fuzzy sets and systems. *Mathematics, MDPI, 10*(21), 4091. https://doi.org/10.3390/math10214091
29. Melin, P., Urias, J., Solano, D., Soto, M., Lopez, M., & Oscar Castillo, O. (2006). Voice recognition with neural networks, Type-2 Fuzzy Logic and genetic algorithms. *Engineering Letters, 13*, 108–116.
30. Castillo, O., Melin, P. (1998). A new fuzzy-fractal-genetic method for automated mathematical modelling and simulation of robotic dynamic systems. In *Proceedings of the 1998 IEEE International Conference on Fuzzy Systems* (FUZZ-IEEE 1998) (Vol. 2, pp. 1182–1187).
31. Castillo, O., & Melin, P. (2003). Intelligent adaptive model-based control of robotic dynamic systems with a hybrid fuzzy-neural approach. *Applied Soft Computing, 3*(4), 363–378.